Geological Processes on Continental Margins: Sedimentation, Mass-Wasting and Stability

Geological Society Special Publications
Series Editor A. J. FLEET

It is recommended that reference to all or part of this book should be made in one of the following ways.

STOKER, M. S., EVANS, D. & CRAMP, A. (eds) 1998. *Geological Processes on Continental Margins: Sedimentation, Mass-Wasting and Stability*. Geological Society, London, Special Publications, **129**.

BALTZER, A., HOLMES, R. W. & EVANS, D. 1998. Debris flows on the Sula Sgeir Fan, NW of Scotland. *In*: STOKER, M. S., EVANS, D. & CRAMP, A. (eds) *Geological Processes on Continental Margins: Sedimentation, Mass-Wasting and Stability*. Geological Society, London, Special Publications, **129**, 105–115.

GEOLOGICAL SOCIETY SPECIAL PUBLICATION NO. 129

Geological Processes on Continental Margins: Sedimentation, Mass-Wasting and Stability

EDITED BY

M. S. STOKER, D. EVANS
(British Geological Survey, Edinburgh, UK)

AND

A. CRAMP
(University of Wales, Cardiff, UK)

1998
Published by
The Geological Society
London

THE GEOLOGICAL SOCIETY

The Society was founded in 1807 as The Geological Society of London and is the oldest geological society in the world. It received its Royal Charter in 1825 for the purpose of 'investigating the mineral structure of the Earth'. The Society is Britain's national society for geology with a membership of around 8000. It has countrywide coverage and approximately 1000 members reside overseas. The Society is responsible for all aspects of the geological sciences including professional matters. The Society has its own publishing house, which produces the Society's international journals, books and maps, and which acts as the European distributor for publications of the American Association of Petroleum Geologists, SEPM and the Geological Society of America.

Fellowship is open to those holding a recognized honours degree in geology or cognate subject and who have at least two years' relevant postgraduate experience, or who have not less than six years' relevant experience in geology or a cognate subject. A Fellow who has not less than five years' relevant postgraduate experience in the practice of geology may apply for validation and, subject to approval, may be able to use the designatory letters C Geol (Chartered Geologist).

Further information about the Society is available from the Membership Manager, The Geological Society, Burlington House, Piccadilly, London W1V 0JU, UK. The Society is a Registered Charity, No. 210161.

Published by the Geological Society from:
The Geological Society Publishing House
Unit 7, Brassmill Enterprise Centre
Brassmill Lane
Bath BA1 3JN
UK
(*Orders*: Tel. 01225 445046
    Fax 01225 442836)

First published 1998

British Library Cataloguing in Publication Data
A catalogue record for this book is available from the British Library.

ISBN 1–897799–97–7

Typeset by Type Study, Scarborough, UK

Printed in Great Britain by
The Alden Press, Osney Mead, Oxford, UK.

Distributors

USA
  AAPG Bookstore
  PO Box 979
  Tulsa
  OK 74101–0979
  USA
  (*Orders*: Tel. (918) 584–2555
      Fax (918) 560–2652)

Australia
  Australian Mineral Foundation
  63 Conyngham Street
  Glenside
  South Australia 5065
  Australia
  (*Orders*: Tel. (08) 379–0444
      Fax (08) 379–4634)

India
  Affiliated East–West Press PVT Ltd
  G–1/16 Ansari Road
  New Delhi 110 002
  India
  (*Orders*: Tel. (11) 327–9113
      Fax (11) 326–0538)

Japan
  Kanda Book Trading Co.
  Tanikawa Building
  3–2 Kanda Surugadai
  Chiyoda-Ku
  Tokyo 101
  Japan
  (*Orders*: Tel. (03) 3255–3497
      Fax (03) 3255–3495)

Contents

Evans, D., Stoker, M. S. & Cramp, A. Geological processes on continental margins: sedimentation, mass-wasting and stability: an introduction mass-wasting and stability: an introduction   1

van Weering, Tj. C. E., Nielsen, T., Kenyon, N. H., Akentieva, K. & Kuijpers, A. H. Large submarine slides at the NE Faeroe continental margin   5

Reeder, M., Rothwell, R. G., Stow, D. A. V., Kahler, G. & Kenton, N. H. Turbidite flux, architecture and chemostratigraphy of the Herodotus Basin, Levantine Sea, SE Mediterranean   19

Dobson, M. R., O'Leary, L. R. & Veart, M. Sediment delivery to the Gulf of Alaska: source mechanisms along a glaciated transform margin   43

Holmes, R. W., Long, D. & Dodd, L. R. Large scale debrites and submarine landslides on the Barra Fan, W of Britain   67

Armishaw, J. E., Holmes, R. W. & Stow, D. A. V. Morphology and sedimentation on the Hebrides Slope and Barra Fan, NW UK continental margin   81

Baltzer, A., Holmes, R. W. & Evans, D. Debris flows on the Sula Sgeir Fan, NW of Scotland   105

Paul, M. A., Talbot, L. A. & Stoker, M. S. Shallow geotechnical profiles, acoustic character and depositional history in glacially influenced sediments from the Hebrides and West Shetland slopes   117

Audet, D. M. Mechanical properties of terrigenous muds from levee systems on the Amazon Fan   133

Mulder, T., Savoye, B., Piper, D. J. W. & Syvitski, J. P. M. The Var Submarine Sedimentary System: understanding Holocene sediment delivery processes and their importance to the geological record   145

Nielson, T., van Weering, Tj. C. E. & Andersen, M. S. Cenozoic changes in the sedimentary regime on the northeastern Faeroes margin   167

Clausen, L. The Southeast Greenland glaciated margin: 3D stratal architecture of shelf and deep sea   173

Ercilla, G., Baraza, J., Alonso, B. & Canals, M. Recent geological processes in the central Bransfield Basin (Western Antarctic Peninsula)   205

Egerton, P. Seismic stratigraphy of Palaeogene depositional sequences: northeast Rockall Trough   217

Stoker, M. S. Sediment-drift development on the continental margin off NW Britain   229

Rasmussen, T., Thomsen, E. & van Weering, Tj. C. E. Cyclic sedimentation on the Faeroe Drift 53–10 ka BP related to climatic variations   255

Howe, J. A., Harland, R., Hine, N. M. & Austin, W. E. N. Late Quaternary stratigraphy and palaeoceanographic change in the northern Rockall Trough, North Atlantic Ocean   269

VIANA, A. R. & FAUGÈRES, J. C. Upper slope sand deposits: the example of Campos Basin,   287
a latest Pleistocene/Holocene record of the interaction between alongslope and
downslope currents

STOW, D. A. V. & TABREZ, A. R. Hemipelagites: processes, facies and model   317

HALL, I. R. & McCAVE, I. N. Late Glacial to Recent accumulation fluxes of sediments at   339
the shelf edge and slope of NW Europe, 48–50°N

Geological processes on continental margins: sedimentation, mass-wasting and stability: an introduction

D. EVANS[1], M. S. STOKER[1] & A. CRAMP[2]

[1]*British Geological Survey, Murchison House, West Mains Road, Edinburgh EH9 3LA, UK*

[2]*Department of Earth Sciences, University College of Wales, PO Box 914, Cardiff CF1 3YE, UK*

Continental margins form the relatively narrow transition zones between the markedly different domains of land masses and deep-ocean basins. They are the main regions of input and transfer of sediments to the oceans and, as such, they represent important zones of sediment flux. Continental shelves are generally the immediate storage area for material derived from the continents, although this commonly only represents a temporary resting place for sediments. The interaction of factors such as sediment supply, tectonic activity and relative sea-level fluctuations largely determine whether the sediment remains on the shelf or is moved, either by currents or gravity, beyond the shelf onto the continental slope. Once on the slope, material is commonly transferred farther into deeper-water environments, such as the continental rise, ocean trench or abyssal plain, through resedimentation by gravity-transport or bottom-current processes. The sedimentary sequence on continental margins thus is rarely simple, for the sedimentation processes and post-depositional changes create complex results.

There has lately been a growth in the level of study of this critical zone. Apart from increased scientific interest, primarily from university and governmental organizations, studies of the margin have now taken on a much higher commercial profile, largely because of the oil industry's move off the shelf into increasingly deeper waters world-wide. The variation and disturbance of sediments on slopes make them ideal areas for the generation, migration and entrapment of hydrocarbons.

This volume addresses three topics of particular significance to continental margin development: sedimentation, mass-wasting and stability. The 19 papers published in this volume provide a sample of the work that has been carried out recently, and is in many cases continuing, with an emphasis on processes active beyond the shelf-break. Individual researchers have unsurprisingly tended to concentrate on specific types of process in their research, having chosen study areas well suited to the investigation of specific phenomena. Perhaps the key distinction is that between downslope, alongslope and hemipelagic processes, the subdivisions that are used in this introduction.

However, there is a need to appreciate that continental margin systems are likely to include all three components, albeit in differing proportions in space and time. A number of papers make the point that these processes are interrelated; for instance, the seismic stratigraphy for the upper Neogene to Pleistocene deposits off southeast Greenland presented by **Clausen** demonstrates the influence of contour-current and turbidity-current processes on this margin, which has undergone a number of periods of glaciation. In another high-latitude setting, the Central Bransfield Basin in Antarctica, **Ercilla *et al*.** recognize the alongslope reworking of structures built by mass-transport, turbiditic, and volcanic processes on this active margin.

It is important not only to recognize the interaction of these processes, but also to appreciate their variation in magnitude, and their frequency. Continuous background processes, unlike larger catastrophic events, may be more difficult to recognize in the geological record, but they remain important to our understanding of the complete system. Figure 1 presents a model that aims to illustrate the processes addressed by each of the papers in this volume and places them in the context of an overall continental margin system. It does not provide a comprehensive picture of the range of slope processes. The model depicts a passive-margin setting, as this best represents the focus of most, but not all, of the papers in this volume. In particular, it attempts to emphasize the combination of processes active on a continental margin.

Predominantly downslope processes

Downslope processes are synonymous with resedimentation by gravity mass-transport, and are those driven by gravitational forces that

EVANS, D., STOKER, M. S. & CRAMP, A. 1998. Geological processes on continental margins: sedimentation mass-wasting and stability: an introduction *In*: STOKER, M. S., EVANS, D. & CRAMP, A. (eds) *Geological Processes on Continental Margins: Sedimentation, Mass-Wasting and Stability.* Geological Society, London, Special Publications, **129**, 1–4.

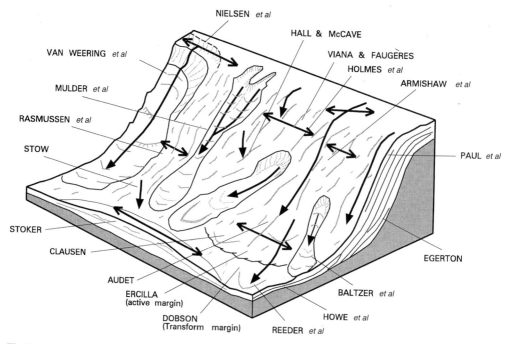

Fig. 1. A schematic margin model depicting the main processes described in this volume, together with an indication of the main processes described in each paper. The passive margin setting is used as this best represents most of the work reported. The arrows indicate the directions of sediment movement.

move sediment downslope over the sea floor into deeper water. They range in scale from minor creep to major translational slides, and may take place on a range of timescales. Sediment creep is a semi-continuous process caused by load-induced stress, whereas discrete debris flows, slumps and slides may occur with a duration of minutes to hours, and turbidity currents may flow for up to a few hours (high-density flows) or a few days (low-density flows) (Nardin *et al.* 1979). During any single downslope event these various processes may operate together or in temporal sequence. Continental margins subjected to a high frequency of downslope activity, comprising low- to high-magnitude events, are commonly characterized by slope aprons and submarine fans. Slides, slumps, debris flows and turbidites are common constituents of both these environments (Stow 1986).

The largest downslope feature described in this volume is a slide on the north Faeroes margin described by **Van Weering *et al*.** This paper exemplifies the range of sedimentological and morphological products resulting from a major failure, with large intact blocks on the middle to lower slope grading to debris flows towards the ocean basin of the Norwegian Sea. In the markedly different environment of the

Eastern Mediterranean off the Nile Cone, **Reeder *et al*.** show the distribution of turbidites and debris flows on the ocean plain that are derived from three sources, and demonstrate sedimentological and chemical signatures that can be used to differentiate flow units and their origins. **Dobson *et al*.** describe the distinctive morphology of deep-water fans sourced from grounded tidewater glaciers in the Gulf of Alaska. The transport pathways are small channels that coalesce into single leveed channels whose migration is related to plate movement on this transform margin.

Holmes *et al*. focus on the downslope aspects, both constructive and destructive, of sedimentation on the Barra Fan, west of Scotland. The fan has largely been built up since Neogene times by successive debris flows, which have been of glacigenic origin during the mid- to late Pleistocene. The transfer of sediment to the central parts of the Rockall Trough has been achieved through substantial sliding as a result of instability on the fan. Although recognizing the overall importance of downslope processes in the Neogene and Pleistocene build-up, **Armishaw *et al*.** present data that highlight the interplay of downslope and alongslope Holocene movement on the Barra Fan. They also demonstrate the reworking of the products of downslope events by spatially and secularly variable alongslope currents on what they term a composite slope-front fan.

A recognition of both constructive and erosional processes on glaciated margins has emerged from work around the North Atlantic (e.g. Vorren *et al.* 1989; King *et al.* 1996). In the region of the Sula Sgeir Fan off western Scotland, **Baltzer *et al*.** differentiate between debrites derived from the essentially constructive, latterly glacigenic, processes that have built up the fan, and those originating from erosional instability events. These authors also show that the glacial sediments in this region are stable at present and require a substantial external event to create instability.

The question of stability of slope sediments is addressed quantitatively by two papers describing the geotechnical properties of slope sediments. **Paul *et al*.** present data from the glaciated margin west of Scotland to show that the geotechnical properties on this glaciated margin are genetic and reflect the processes operative during their deposition. **Audet** used core from the recent Ocean Drilling Program drilling of the Amazon Fan to determine the mechanical strength of levee sediments, on which levee stability and channel propagation depend.

Canyon processes in a region with a narrow shelf are considered by **Mulder *et al*.**, who provide new data from the Var System, off the South of France, that comprises a river delta, a submarine valley, and a deep-sea fan. They describe high-frequency turbid plumes and shallow failures resulting from river flooding, as well as less frequent earthquake-triggered major slides. They also recognize background processes, but find they are not interpretable in the geological record as imaged by seismic systems.

There is increasing evidence that major slides are a particularly significant mechanism for the transfer of sediment from the shelf to the ocean basin (e.g. Kenyon 1987; Bugge *et al.* 1988; Weaver *et al.* 1992), and it is important to recognize that this must have been the case throughout the geological record. The paper by **Nielsen *et al*.** describes a large slide on the north Faeroes margin which has an entirely infilled slide scar, and pre-dates the surface slide from the same region described in greater detail by Van Weering *et al*. It is suggested that a change in the palaeoceanographic regime may have triggered the former event; if this was the case, then there was a direct relationship between changes in the alongslope regime and this large downslope event.

Predominantly alongslope events

Bottom-current activity greatly influences both sediment accumulation and erosion on continental margins. The effects of these contour-following, deep, geostrophic currents are particularly marked on slopes and basin plains within and adjacent to continental margins, where a variable sea-floor bathymetry may locally intensify and focus current activity. Sediment drifts and associated bedforms are a major depositional product of bottom currents, and commonly accumulate where there is a change in the gradient of the sea bed, such as at the base of continental slopes. Where particularly strong, bottom currents may cause erosion of the sea bed and the formation of channels, moats and furrows. The variable record of deposition and erosion preserved in bottom-current deposits (contourites) can be used as an indicator of the palaeoceanographic history of a region.

A long-term record of drift sedimentation on the continental margin west of Britain is presented by **Stoker**, who shows evidence that the onset of bottom currents was an early late Eocene event in this region. This resulted in the formation of deep-marine contourites preserved as mounded, elongate, and broad, sheeted sediment drifts with associated sediment waves. Temporal changes in the spatial pattern of sedimentation and erosion have been influenced in part by tectonism. Contemporaneous tectonism is also invoked by **Egerton** to explain the Palaeogene development of the NE flank of the Rockall Trough. Using multichannel seismic records with well control, he recognizes middle to upper Oligocene deep-water contourite drifts above lowstand mass-flow fans of late Eocene to early Oligocene age.

A higher-resolution study of bottom-current activity is presented by **Rasmussen *et al*.** from the north Faeroes margin, where the 53–10 ka record of palaeoceanographic change is of comparable resolution to the GRIP (Greenland Ice-core Project) ice core from the Greenland Icesheet. Cyclic variations in temperature, the Dansgaard–Oeschger events, are closely reflected in foraminiferal and physical changes in the core as a result of variations in the ocean current regime. Oceanographic changes during the late Quaternary in the northern Rockall Trough are documented by **Howe *et al*.**, who detail spatial and temporal variations in sedimentation on the basis of identification in cores of turbidites, contourites and hemipelagites. They consider that the last deglaciation comprised a series of irregular, non-linear events with associated fluctuating bottom-current activity.

Viana & Faugères, working on the upper slope in the Campos Basin of Brazil, describe the remobilization by the southward-flowing Brazil Current of elongated sand lobes derived from the shelf by gravity-induced mechanisms. Variations in the bedforms with water depth are associated with the change to the northward-flowing Brazil Counter-Current. This model, dealing as it does with the distribution of sand bodies of reservoir potential, is of particular interest to petroleum geologists.

Hemipelagic processes

Hemipelagic flux is the topic considered by **Stow & Tabrez**, who have developed a facies model from their studies of sediments on the Makran and Oman margins in which hemipelagites form 50% of the succession. Hemipelagites are fine-grained sediments formed by a combination of vertical settling (synonymous with pelagic settling) and slow lateral advection. They are commonly devoid of primary sedimentary structures other than lamination, are thoroughly bioturbated and poorly sorted, and have a composition dependent upon their transport pathway. This study is important because hemipelagic sedimentation is the primary 'background' process on margins, and is commonly referred to in several papers in this volume.

The late Glacial to Recent sediment budget on the Irish slope at the Goban Spur is largely the result of hemipelagic processes. **Hall & McCave** provide detailed analyses that give accurate rates of deposition and carbon flux, showing a high flux rate during stage 2 indicating supply by either ice rafting or downslope movement.

Summary

Although there is a tendency to consider margin processes in terms of downslope, alongslope and vertical flux, there is increasing recognition that all these need to be integrated to produce a model of margin development. Processes will differ in importance in space and time, but their interrelationship needs to be understood and background processes must be appreciated. A notable contribution from this volume is an increased awareness of the importance to margin development of alongslope processes within this model.

The oil industry's attitude to deep water is changing rapidly, and an understanding of

margin processes is becoming crucial to the safe development of new oilfields in deep water. In 10 years time as much as 25% of oil may be extracted from deep water (C. Hunt of BP, pers. comm. 1996). The scientific presentations made in this volume therefore have considerable actual or potential commercial value, and will become increasingly important to the world's economies.

The editors would like to thank the many referees whose comments have greatly enhanced the quality of this volume, as well as A. Kuijpers, who commented on this introduction. The volume is based on a meeting organized by the Marine Studies Group (MSG) and held at the Geological Society's Applied Geoscience conference at Warwick University in April 1996; sponsorship of the MSG meeting by Geoteam-Wimpol is gratefully acknowledged. This paper is published with the permission of the Director of the British Geological Survey (NERC).

References

BUGGE, T., BELDERSON, R. H. & KENYON, N. H. 1988. The Storegga Slide. *Philosophical Transactions of the Royal Society of London, Series A*, **325**, 357–388.

KENYON, N. H. 1987. Mass-wasting features on the continental slope of northwest Europe. *Marine Geology*, **74**, 57–77.

KING, E. L., SEJRUP, H. P., HAFLIDASON, H., ELVERHOI, A. & AARSETH, I. 1996. Quaternary seismic stratigraphy of the North Sea Fan: glacially-fed gravity flow aprons, hemipelagic sediments, and large submarine slides. *Marine Geology*, **130**, 293–315.

NARDIN, T. R., HEIN, F. J., GORSLINE, D. S. & EDWARDS, B. D. 1979. A review of mass movement processes, sediment and acoustic characteristics, and contrasts in slope and base-of-slope systems versus canyon–fan–basin floor systems. *In*: DOYLE, L. J. & PILKEY, O. H. (eds) *Geology of Continental Slopes*. Special Publication of the Society of Economic Palaeontologists and Mineralogists, **27**, 61–73.

STOW, D. A. V. 1986. Deep clastic seas. *In*: READING, H. G. (ed.) *Sedimentary Environments and Facies*. Blackwell Scientific Publications, Oxford, 399–444.

VORREN, T. O., LEBESBYE, E., ANDREASSEN, K. & LARSEN, K. B. 1989. Glacigenic sediments on a passive continental margin as exemplified by the Barents Sea. *Marine Geology*, **85**, 251–272.

WEAVER, P. P. E., ROTHWELL, R. G., EBBING, J. & HUNTER, P. M. 1992. Correlation, frequency of emplacement and source directions of megaturbidites on the Madeira Abyssal Plain. *Marine Geology*, **109**, 1–20.

Large submarine slides on the NE Faeroe continental margin

TJEERD C.E. VAN WEERING[1], TØVE NIELSEN[2], NEIL H. KENYON[3], KATJA
AKENTIEVA[4] & ANTOON. H. KUIJPERS[2]

[1]*Netherlands Institute for Sea Research (NIOZ), P.O. Box 59, 1790 AB Den Burg, Texel,
The Netherlands*

[2]*GEUS (formerly DGU), Thoravej 8, Copenhagen 2400, Denmark*

[3]*Southampton Oceanography Centre, Southampton SO14 3ZH, UK*

[4]*Moscow State University, Moscow, Russia*

Abstract: High-resolution seismic profiles of the NE Faeroe margin show large-scale slumping and sliding of the middle and lower continental slope, affecting sediments of presumed Miocene, Pliocene and Quaternary age. Mass-flow deposits on the upper slope are partially covered by more recent deposits, presumed to be contourites. A TOBI deep-tow side-scan sonar mosaic, in combination with deep-towed penetrating echosounder results, shows that the slump complex on the upper slope consists of a buried slide scar, bottleneck slides, debris-flow lobes and a number of shallow slides. Sliding at the middle and lower continental slope seems more recent and shows a steep, irregular main slump scar and very large, intact, angular blocks. The base of the slide at the lower continental slope shows numerous narrow and diverging tracks, 10–15 km length, that end at individual blocks. Several types of debris flows have been mapped, some with longitudinal flow fabrics.

The study area is located between approximately 2° and 6°W, and 64.20° and 63.40°N, at the northeastern continental margin of the Faeroe Islands (Fig. 1). It extends from *c.* 600 m down to 2500 m water depth. The Faeroe Islands are located on the northern end of the volcanic Faeroe–Rockall Plateau, and form part of the Faeroe Platform, which is surrounded by the Norwegian Basin, the Faeroe–Shetland Channel and the Faeroe Bank Channel (Boldreel & Andersen 1995). To the south, the study area is separated from the Faeroe–Shetland Channel (FSC) by the Fugloy Ridge, a WSW–ENE elongated continuation of the Faeroe Platform, which is considered a compressional ridge by Boldreel & Andersen (1993). The Faeroe continental margin has developed since the opening of the northeast Atlantic at *c.* 55 Ma (Talwani & Eldholm 1972). Its seismoacoustic basement is formed of basalts of Palaeocene or early Eocene age, thought to have been subaerially exposed, that extend over much of the Faeroe–Rockall Plateau (Andersen 1988; Boldreel & Andersen 1995).

Seismic facies analysis of the post-Palaeocene sedimentary succession of the Faeroe margin on the basis of high-resolution single- and multichannel data collected by NIOZ–GEUS (The Netherlands Institute for Sea Research–Geological Survey of Denmark and Greenland) during the ENAM (European North Atlantic Margin) project in 1993–1996, indicates that post-Palaeocene sedimentation has been controlled by rapid subsidence of the Norwegian

Basin floor, and slower subsidence of the Faeroe Platform (Nielsen *et al.* this volume).

Miller & Tucholke (1983) inferred that bottom-water flow in the Norwegian Sea was initiated in the late Eocene to Oligocene, and has been affecting continental margin build-up since then. Recent work by Stoker (1997, this volume) and Stoker *et al.* (1993) on the initiation and development of contourites along the Rockall and Hebrides margins indicates that the onset of strong bottom currents affecting the NW British margin was indeed in the early part of the late Eocene. However, overflow from the Norwegian Basin into the Atlantic is thought to have started only in the mid- or late Miocene (Wold 1995). A late Oligocene or early to mid-Miocene unconformity recorded on the FSC margin was, however, considered the result of (renewed) intense bottom-water flow from the Arctic via the FSC into the North Atlantic Ocean (Damuth & Olson 1993).

Piston cores from the Faeroese margin show locally enhanced sedimentation rates caused by this flow of Norwegian Sea Deep Water (NSDW) along the margin and through the FSC, and reflect abrupt temperature shifts similar to the Dansgaard–Oescher cycles reported from the Greenland ice cores (Dansgaard *et al.* 1993; Rasmussen *et al.* 1996a, b, c, this volume).

This contribution presents evidence for large-scale sliding and slumping of the NE Faeroe continental margin on the basis of the

VAN WEERING, TJ. C. E., NIELSEN, T., KENYON, N. H., AKENTIEVA, K. & KUIJPERS, A. H. 1998. Large submarine slides on the NE Faeroe continental margin. *In*: STOKER, M. S., EVANS, D. & CRAMP, A. (eds) *Geological Processes on Continental Margins: Sedimentation, Mass-Wasting and Stability*. Geological Society, London, Special Publications, **129**, 5–17.

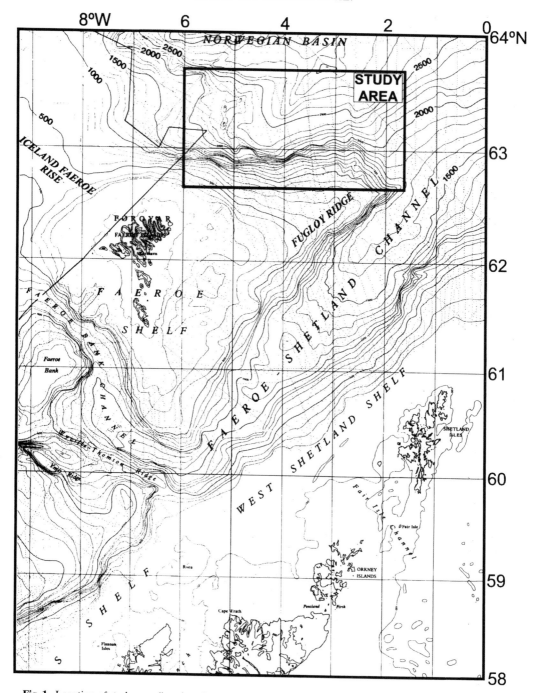

Fig. 1. Location of study area (inset) at the northeast continental margin of the Faeroe Islands. Bathymetry in hundreds of metres after Roberts *et al.* (1977).

interpretation of high-resolution seismic profiles, in combination with a first interpretation of deep-towed TOBI side-scan sonar profiles. It specifically aims at understanding the importance and timing of mass-wasting processes in the development of the NE Faeroe margin.

Material and methods

Seismic recording

During 1991, 1993 and 1994, multi-channel (24) and single-channel sleeve-gun profiles of the NE Faeroe margin were collected during co-operation between the NIOZ, GEUS (then the Danish Geological Survey) and the Department of Geophysics, University of Aarhus, during cruises with the Dutch vessel R.V. *Pelagia* (Fig. 2). An array of 2 × 40 and 2 × 25 cu. inch Halliburton sleeve guns formed the sound source. Single-channel high-resolution seismic records were collected using a high-resolution six-channel Teledyne streamer. The received signals were summed, amplified and filtered, and recorded (analogue) on a Dowty 195 dry-paper recorder. Multi-channel seismic profiling was carried out simultaneously using a 24-channel Teledyne streamer. Near-trace signals were directly printed on board via the Aarhus University MAF 01 system, to allow quality control and check of the operational performance of the outgoing signal. Registration was by a Geometrics ES 2420 acquisition control unit with a 19-bit analogue-to-digital converter. Post-cruise processing of the multi-channel profiles was carried out at Aarhus University and GEUS.

TOBI

TOBI is the Towed Ocean Bottom Instrument that was designed, constructed and operated by the Southampton Oceanography Centre (SOC) (Flewellen *et al.* 1993). It contains a 30 kHz side-scan sonar and a 7.5 kHz sub-bottom profiler as its main instruments. The side-scan sonar system can be towed in water depths up to 6000 m, and is usually operated at *c.* 400–500 m above the sea floor. The swath width of the side-scan imagery at that tow depth is *c.* 6 km, representing 3 km on either side. The vehicle is ballasted for neutral buoyancy in water and is connected to a depressor fish. As a general rule, the amount of cable deployed is 1.5 times the water depth, which limits the towing speed to between 1.5 and 2.5 knots. The signals from the vehicle are recorded onto magneto-optical disk on board ship, and a direct (uncorrected) printout is made of both side-scan channels and the penetrating echosounder. In 1995, ten lines parallel to the bathymetry were collected (Fig. 2) under ideal weather conditions, applying the above method at an average tow speed of 1.9 knots. During all cruises, navigation was by means of Differential Global Positioning System.

Post-cruise processing of TOBI data included correction for the distortion of the image along the path of the vehicle caused by the varying

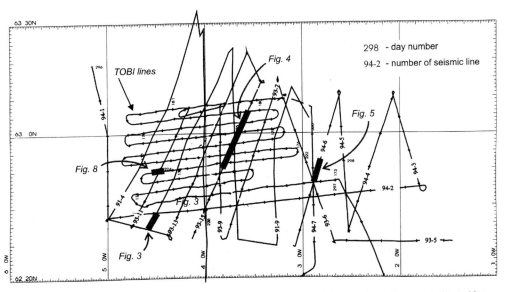

Fig. 2. Tracks of multi- and single-channel seismic lines and of the TOBI lines. Seismic lines are indicated by year of recording and by number (e.g. 93-15). TOBI lines are parallel to the bathymetry (approximately E–W) and extend to the lower slope at *c.* 2000 m. Heavy lines indicate locations of profiles shown in Figs.3–5 and 8. Marks along tracks are 12 h position signs.

speed over the ground. Missing swath lines were repeated to account for the varying distances covered. Subsequently, time variable gain was applied using shading correction files. Because the survey area lies in water depths below 2000 m, the reflection of the sea surface was removed before slant-range correction. Post-cruise processing was carried out by K.A. at SOC, following the methods of Le Bas *et al.* (1995).

Interpretation

Seismic profiles

The seismic data show the presence of a wedge-shaped sedimentary succession of Tertiary to Recent age locally up to 2 km thick, overlying the lower Tertiary basaltic basement of the NE Faeroe margin. The basaltic surface is characterized by a strong, high-amplitude reflector which changes from a zone with parallel reflectors on the shallow part of the Faeroe Platform, via seaward dipping reflectors in the middle and lower slope, to an irregularly reflecting basin-floor section. All profiles show a sedimentary sequence above the basaltic basement; the thickness of the post-basalt sediments is between about 0.2 km in the SE near Fugloy Ridge, and 1.5–2 km in the north. Following Nielsen *et al.* (this volume) the sedimentary section on top of the basalts can be divided into five major seismic sequences separated by unconformities. These major seismic sequences allow recognition of nine seismic facies units. A summary overview is provided to give a context for the interpretation of the mass-flow events that are the subject of this paper.

The earliest indications of deep-seated mass-movements are found within the first seismic sequence (0) overlying the basalts. This sequence locally contains minor internal erosional truncations and/or smaller mass-movements in the deeper parts of the study area. Sequence 1, which overlies sequence 0, has strong, parallel, onlapping reflectors at the upper slope, with local lenses of disrupted, fractured seismic reflections lower down the slope and in the basin-floor area. Local erosional truncation is seen in the midslope areas. Sequence 2 is defined by a continuous medium-strength reflector at the upper and middle slope, and is cut by an erosional unconformity at the middle slope, so that a major part is absent in the lower slope and basin-floor area. Sequence 2 contains stacked slides in the upper section of the lower slope. Sequence 3 is preserved only on the upper slope and its top is defined by an erosional unconformity. Sequence 4 is thought to reflect the onset of bottom-water and contourite formation in the Norwegian Sea and across the

Iceland–Faeroe–Scotland Ridge (IFR), and is considered to be of Plio-Pleistocene age (Nielsen *et al.* this volume).

The seismic profiles therefore indicate that the NE Faeroe continental margin is covered by relatively undeformed, well-stratified sediments characterized by strong, continuous, parallel reflectors at the upper slope and the upper part of the middle slope. However, at *c*.1500 m depth, a major escarpment marks the headwall of a large submarine slide; the steep escarpment is 200–300 m high. Below it, hummocky chaotic reflectors at and below the surface and a strongly irregular sea bed with large detached blocks and with numerous, stacked, slide deposits and debrites characterize the middle and lower slope and the adjacent basin floor.

The upper-slope sediments are considered to be contourites, on the basis of their internal seismic character, their distribution and their morphology. At about 1000 m water depth, the sedimentary sequence is cut at an erosional channel flanking a contourite drift deposit (Fig. 3). The channel cuts some 60–70 m into the sequence, and may have originally formed as a (now inactive) slide scar. Subsequently, the scar may have acted as a conduit for erosive bottom currents, which reshaped the scarp into the present channel morphology.

The contourite deposits become thinner downslope, but extend to the main scarp at the lower–middle slope. The greatest thickness, of *c*. 0.2 s TWTT (two-way travel time), is attained directly adjacent to the erosional channel. Indeed, piston core ENAM 93-21, taken at the top of the contourite deposit immediately adjacent to the erosional channel, has yielded elevated sedimentation rates (Rasmussen *et al.* 1996*a*, this volume). In the east and west of the study area the erosional channel is less well developed and fades away with a simultaneous decrease in the thickness of the contourite sediments.

The slump at the base of sequence 4 (Fig. 3) is characterized by a chaotic or transparent, disorganized internal-reflection configuration. The irregular surface of these deposits, and stacked lobes of debris flows, can be recognized beneath the contourite. Elongated slump or debris-flow deposits underlying the contourites directly upslope from the erosional channel can also be distinguished. On the 3.5 kHz echosounder record the upper slope down from 1000 m is characterized by relatively weakly backscattering sediments with well-stratified, parallel-bedded reflectors. The sea bed is generally smooth, reflecting the contourite sediments at the surface. Channels 50–60 m deep, with an irregular topography and characterized by hyperbolic echoes, can be

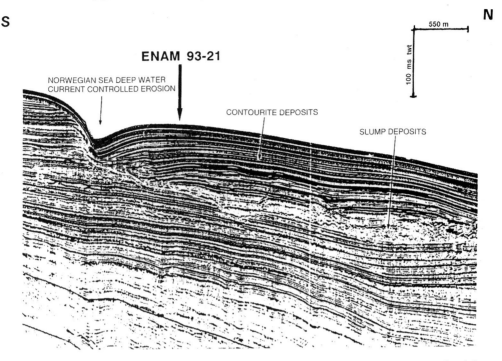

Fig. 3. Blow-up of part of seismic profile ENAM 91-06, with contourite deposits forming sequence 4 and the main erosional channel. Slump deposits form the lower part of sequence 4. The present erosional channel may have been formed initially as a slump scar. (For location see Fig. 2.) Vertical scale in ms TWTT.

discriminated locally beneath the reduced contourite cover; these probably represent longitudinal debris flows.

Figure 4 (panel A) illustrates an example of a feature interpreted to show initial extension of the contourite cover related to the initiation of sliding; it is associated with loss of well-bedded seismic characteristics and the development of hyperbolic reflectors directly at and below the sea bed. Alternatively, the seismic character reflects longitudinal channels, and the hyperbolic reflectors indicate a debris flow at or near the surface, as is also shown on the TOBI mosaic (see Fig. 6). Locally, pressure ridges and detached slabs (Fig. 5) can be discerned on the middle slope, indicating relatively recent activity of the mass-wasting processes.

The headwall of the main slump, at about 1500 m water depth, forms a very steep (Fig. 4, panel B), irregular escarpment, locally characterized by multiple, retrogressive faults. Towards the west (profiles 94-01 and 91-04) the headwall locally shows vertical displacements of about 100–200 m, whereas in the central part of the study area displacements of up to 250 m are found. Rotated and tilted blocks occur directly at, below and near the main slump (Profile 94-01). Debris-flow lobes, characterized by hummocky seismic reflection

character, pinch out at the lower slope and extend as tongues with a transparent signature.

Directly below the main slump scar, there are detached blocks up to 225 m high and 1 km across, some with a coherent seismic structure, (Fig. 4, panel C). The blocks locally appear covered by more recent sediments and may have a sub-rounded shape, but have strongly irregular shapes in other cases. Their steep slopes are locally exposed.

All seismic profiles show stacked debris flows in the middle and lower slope sequence, separated by virtually undisturbed sediments indicating that they have undergone repeated extensive, large-scale slumping and sliding. The mass-wasting deposits extend far into the Norwegian Sea, creating an irregular basin-floor relief for the entire lower continental margin. The sea floor below the main slump scar is locally covered by an irregular, thin veneer of sediments on top of the most recent debris-flow and slumped deposits.

TOBI deep-tow side-scan sonar profiling

Figure 6 presents a TOBI mosaic that extends from the upper slope to below the main scarp, of 225–300 m height at 1500 m water depth, and

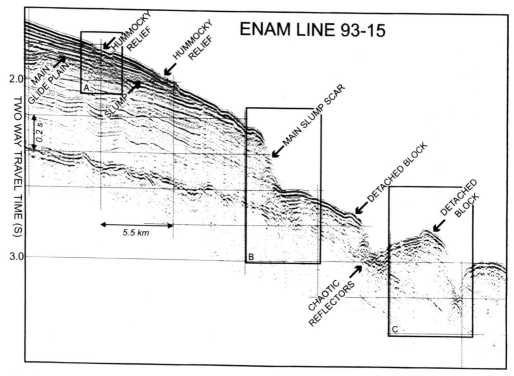

Fig. 4. Middle-slope section of profile ENAM 93-15. Blocks appear covered by some recent sediment, and have partially retained their original seismic structure. Also noteworthy is the occurrence of slide deposit directly below the area of hummocky relief (4A) above the main slide scar. Panel A shows disturbance of seismic character, hummocky reflectors and a depression in the sea bed, possibly the result of initial sliding and extension of the upper part of the sediment cover. The scale of the detached block in Panel C and its position near the main slide scar (Panel B) should be noted. The seismic character and coherence of structure may indicate that downslope block movement has been limited. (For location see Fig. 2.)

more detailed images from this area and farther to the NW are displayed in Figs 7–10. Figure 11 presents an initial interpretation for the entire area of TOBI coverage, illustrating the extent of the large submarine slide complex on this margin.

The upper slope down from 1000 m water depth is characterized by relatively weakly backscattering sediments. The sea bed is generally smooth, reflecting the contourite sediments, but collapse depressions occur locally on the upper slope. These features range in width from 100 to 500 m and have width-to-length ratios of 1–1.5. The side-scan sonar records clearly show that these bowl-shaped areas, bounded by scarps, have been displaced vertically and form distinct depressions on the sea floor.

At the middle slope there are a number of cross-slope-directed, contourite-draped debris-flow lobes (Fig. 6) characterized by increased backscatter levels; in the middle of the study area seven large slump scars with

horseshoe-shaped depressions occur. Their widths vary from 2 to 3.5 km in the upper part, and they show transverse cracks and arcuate scarps approximately parallel to the main slump scar (Fig. 7).

The lobe margins here have a smoothed character. The thickness of the lobes is about 30–40 m, and they form a characteristic, hummocky relief (Fig. 8). The upper parts of the lobes contain depressions of 10–15 m depth. The subdued relief and weak backscattering of both the slump scars and the sea bed between, and of the transverse cracks and the rim parallel blocks, indicate local contourite sedimentation on top of part of the lobes. The elongate plan of the upper part of the debris-flow lobes suggests a type of creep-like deformation, and shows that downslope movement has been limited. The associated slumps at the toe of the debris flows, above the main scarp, again show a very weak backscatter and do not indicate the presence of blocks near or at the surface.

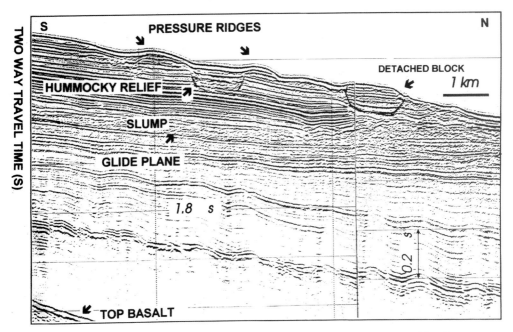

Fig. 5. Middle-slope section of profile ENAM 94-06, showing detached block and pressure ridges by downslope movement of blocks, covered partially by recent sediments. Noteworthy features are the hummocky relief at detachment scars, similar to Fig. 4 (panel A), and slight rotation of the toe of the sedimentary section below a detached block (at right side of figure). Deeper-seated slides below appear to have moved along the glide plane indicated. (For location see Fig. 2.)

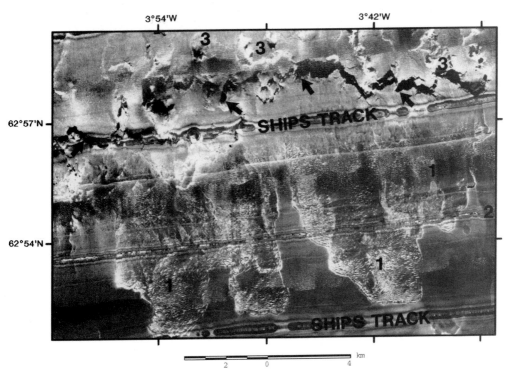

Fig. 6. Part of a TOBI mosaic showing: (1) partially contourite-covered debris-flow lobes upslope from the main slump scar, which is indicated by arrows; (2) elongated bottleneck slide; (3) slumped blocks north of the main slump scar. The irregular character of the slump scar and its angular, indented character should be noted. (For location of image see Fig. 11.)

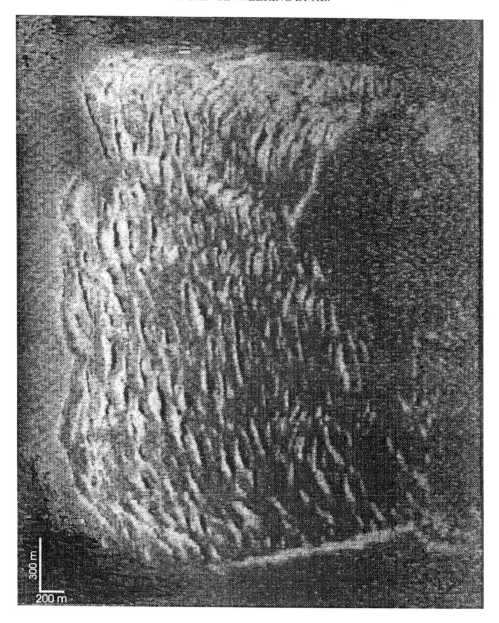

TOBI image (port) day 175
ENAM cruise June-July 1995 (leg 1)

Fig. 7. Blow-up of TOBI mosaic showing upper-slope debris flow, with smoothed character of debris flow and transverse cracks caused by contourite cover after initial formation of the debris flow. (For location see Fig. 11.) North is to right hand side.

A series of debris-flow tongues of up to 1.5 km wide with increased backscatter extend downslope from the lower part of these slumps over distances of up to 10 km. Features that are generally referred to as bottleneck slides (Kenyon 1987) occur to the sides of debris-flow lobes. These features are morphologically similar to the collapse depressions, but the boundary scarps do not form

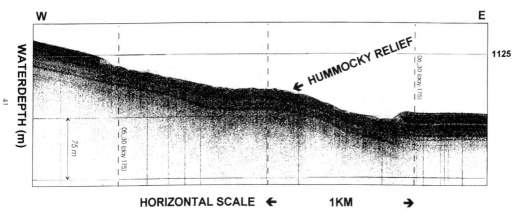

Fig. 8. Seismic profile showing hummocky relief at upper part of debris-flow lobe (indicated by (1) in Fig. 6). The upper parts of the lobes are expressed as depressions of 10–15 m-depth. Continuity of reflectors beneath the lobe indicates that only the surficial part of the seismic sequence is involved in the slide. (For location see Fig. 2.)

a totally closed perimeter around the zone of instability and they generally have a more elongate shape. Bottleneck slides at the margin of the main debris-flow lobes vary in length from 1 to 9 km, and they have width-to-length ratios of about 2.0–5.0.

The main slump scar at a water depth of about 1500 m extends over a distance of about 60 km and has an irregular, locally sharply indented, or crescent shape (Figs 6 and 9). The TOBI mosaic and the seismic profiles show that the main scar cannot be discriminated in the western part of the study area, partially because of an increased sediment cover. However, coverage is restricted here and the slumps may well extend along the Faeroe continental margin to the NW and W.

The upper part of the major slide mass directly below the scar contains a number of elongated, smooth and angular blocks that are aligned approximately parallel to the scar. These blocks are in some cases less than 200 m away from the scar. Slumped margin blocks up to 800 m long can be recognized in the upper part of the main slump. On TOBI profiles these irregular-shaped blocks appear to 'fit' the shape of the main slump-scar headwall upslope. This might indicate that these blocks have been displaced relatively recently, that downslope movement did not result in major internal disruption of the blocks (as is also shown by their seismic character), and that the more recent, larger slide blocks did not slide down over great distances.

The structure of the upper part of the main slumps shows that blockfaulting at the rims of the slump occurred in successive steps, as was also recognized on the seismic profiles. The slumps develop downslope into a series of

stacked debris-flow lobes with high backscatter and irregular shapes; these extend transversely over and across the lower slope. These debris flows have a radiating pattern, with individual outrunner blocks having trails up to 3 km long (Fig. 10). Block sizes vary from 70 to 120 m and the width of the trails varies from 60 to 80 m. Swarms of trails have a dominant northeasterly to easterly direction in the central study area, whereas in the eastern part they are directed locally to the NW. Trails having a common starting point appear to diverge strongly downslope. The slumped sections and debris flows at the lower continental slope have lower reflectivity than at the upper and middle slope, indicating that they are locally covered by more recent sediments. It is likely that turbidites associated with the NE Faeroe margin slide will be found farther downslope, and they may extend well into the Norwegian Sea.

Discussion

A reconnaissance of the European Atlantic margin with side-scan sonars (Kenyon 1987) showed that slope failure has affected some 95% of margin area of the Bay of Biscay, and some 20% of the slopes west of Scotland and Norway. Since that study, smaller-scale slides have been noted on the west Shetland margin (Stevenson 1991) and more recently, at the northeastern Faeroe–Shetland Channel margin south of Fugloy Ridge (Haflidason *et al.* 1996). Also, the Storegga Slide region on the western Norwegian margin (Evans *et al.* 1996) appears to have been affected by at least eight major events, the dating of which is as yet uncertain, although some slides

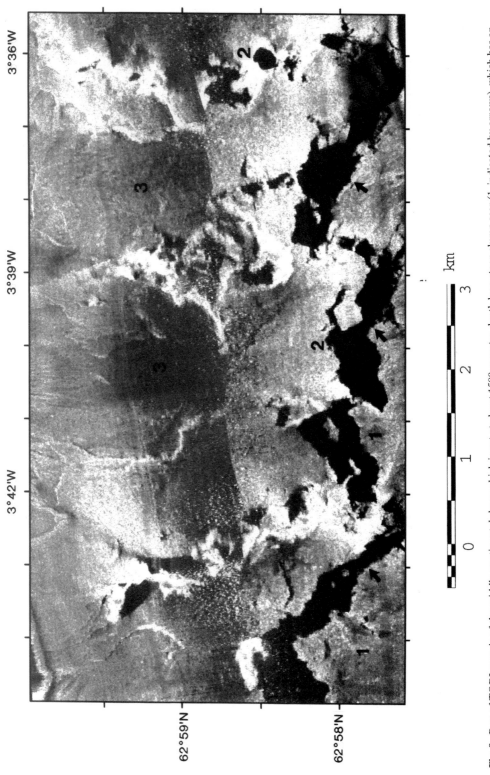

Fig. 9. Part of TOBI mosaic of the middle continental slope which is cut at about 1500 m water depth by a steep slump scar (1, indicated by arrows), which has an irregular indented shape (sharp edges in lower part of figure). Downslope from the main slump scar, large blocks occur near the main scar (2), and smaller blocks are present farther away from the slump margin. (3) Lower-slope debris flows. (For location see Fig. 11.)

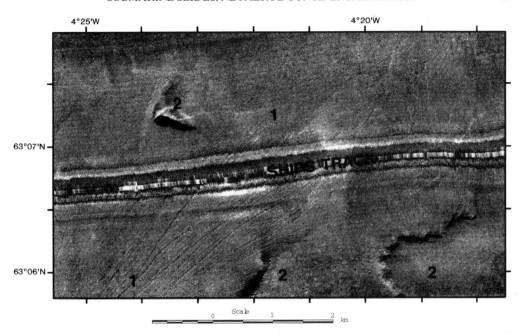

Fig. 10. Part of TOBI mosaic showing diverging trails (1) of blocks moving downslope in a northwesterly direction. Swarms of these trails with outrunner blocks at their ends can be recognized. Crosscutting of trails locally indicates the occurrence of multiple slide events. (2) Sediment-covered older slides or blocks. (For location see Fig. 2.)

LARGE SUBMARINE SLIDE COMPLEX MAPPED BY TOBI SIDESCAN

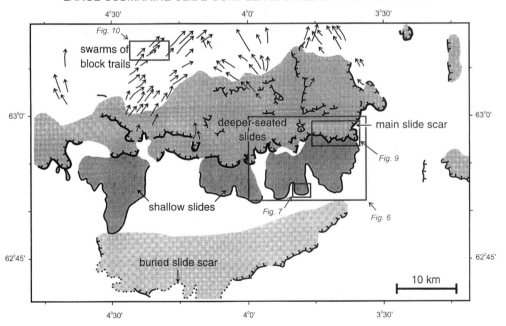

Fig. 11. Preliminary sketch of the main structures of the north Faeroe margin. The buried slide scar now locally eroded by contour currents (see Fig. 3) is shown in the south, and the area where slumps are buried beneath contourites occur is indicated by pale shading. Shallow slides and debris-flow lobes (shown in Figs 6, 7, and 9) occur above the main scar. The main scar is indicated by the irregular line with tick-marks separating the shallow slide area from the deeper-seated slides below. Diverging swarms of block trails (see Fig. 10) occur on the lower slope. This sketch covers the area of TOBI tracks in the middle of the NE Faeroe continental margin study area (see Fig. 2).

are considered to pre-date the onset of mid-Pleistocene glaciation (King *et al.* 1996).

Triggering of the slides by earthquakes has been considered a major mechanism, although the local presence of gaseous sediments and/or gas hydrates on the Norwegian margin (Bugge *et al.* 1987; Mienert *et al.* 1996) may have facilitated the repeated mass failures (Evans *et al.* 1996). In the Faeroe slide area no indication for the presence of gas in sediments has so far been detected on the seismic or acoustic profiles; this mechanism therefore is considered less likely to have triggered the slides and slumps at the Faeroe margin. The cause and timing of the failures of the NE Faeroe continental margin, similarly, are uncertain. However, the (continuing) subsidence of the cooling Norwegian Sea basin floor, as reflected in the increased dip and structure of the seismic reflectors characterizing the basaltic basement underneath the margin, may have caused steepening of the overlying middle and lower slope sedimentary sequences (Nielsen *et al.* this volume). This may have caused the main failure, in combination with an increased sediment supply by contourite currents directed along the Norwegian Sea Basin margin after the initiation of deep-water current transport. Glacigenic input from Faeroese sources may have contributed to locally high sedimentation rates; the seismic profiles, however, show that the glacigenic sediments are confined to the upper slope as far down as about 1000 m depth.

Dating of the first event is not possible at present, but all observations from the seismic profiles and TOBI side-scan sonar images indicate multiple sliding events. The TOBI images confirm that part of the recent contour-current flow at the upper slope is controlled by a now partially buried slide scar. It seems most likely that the sliding on the margin migrated upslope in the course of time, as is also expressed in the now partially contourite-covered debris-flow lobes above the main slump scar.

The age of the most recent sediment deformation remains unclear, but a single ^{14}C accelerator mass spectroscopy age dating of sediment (AAR-2017, boxcore ENAM 94-07, at 37 cm below the sea bed) at the foot of the main slump scar directly on top of deformed compacted sediments gave a reservoir corrected age of 9850 ± 140 a BP. This indicates a relatively young, post-Weichselian age for the most recent mass flow, leading to the possibility that the latest slumping may be related to sea-bed instability as a result of late Weichselian deglaciation. Isostatic adjustment of the basement of the Faeroe Islands and margin following the deglaciation could therefore be the cause for the most recent

mass flow. Post-late Weichselian slope failure is also observed at the Storegga Slide (Evans *et al.* 1996). Subsidence of the Norwegian Basin by cooling is, however, considered to be the main cause of the large-scale sliding and slumping observed in the study area; processes related to glaciation or deglaciation may have a superimposed effect on the development of the NE Faeroe margin. It is likely that the Faeroe slide and related deposits extend far enough into the Norwegian Basin to join, or interfinger with, the Storegga Slide and related deposits on the basin floor.

Conclusions

The northeast Faeroese continental margin is strongly influenced by large-scale mass-wasting processes. These are considered to be primarily the result of margin steepening as a result of the continuing subsidence of the floor of the Norwegian Sea. A combination of this process with the Pliocene and Quaternary isostatic rebound of the Faeroe Islands following deglaciation seems likely for the younger slides. The most recent slide, locally dated at 9850 ± 140 a BP, may have formed as a consequence of the most recent deglaciation. A slump scar at about 1000 m water depth has evolved into an erosional channel, and now marks the scouring effect of a contourite flow of presumably NSDW water. Contourite sediments now cover the upper and middle slope debris flows and slide deposits. A highly irregular and indented slide scar of 200–300 m height, marks the boundary between deeper-seated slides occurring on the lower slope, and shallower slides on the middle slope. Multiple, stacked debris flows characterize the middle slope adjacent to the main scar.

This research was carried out within the framework of the EU-supported ENAM programme (Contract MAS CT93-0064). The TOBI deep-tow facility was provided through a grant supplied under the large facilities programme of the EU. Shiptime on board R.V. *Pelagia* was made available by the Netherlands Institute for Sea Research. Seismic profiling was carried out in co-operation with the Department of Geophysics of the University of Aarhus; we express our thanks to H. L. Andersen for making these facilities available and to P. Trinhammer for his continuing interest and excellent operational skills, which greatly helped us. The participation of K. Akentieva was funded through a co-operation agreement between NIOZ and Moscow State University. Comments by D. Evans, T. Bugge and an anonymous reviewer helped improve the paper. We are grateful to R. Groenewegen, N. Millard, G. Phillips, W. Polman, J. Schilling, C. Willems, S. Whittle and many others for their highly qualified technical support during the cruises. This is NIOZ Publication 3189.

References

ANDERSEN, M. S. 1988. Late Cretaceous and early Tertiary extension and subsidence around the Faeroe Islands. *In*: PARSON, L. M. & MORTON, A. C. (eds) *Early Tertiary Volcanism and the Opening of the NE Atlantic*. Geological Society, London, Special Publications, **39**, 115–122.

BOLDREEL, L. O. & ANDERSEN, M. S. 1993. Late Paleocene to Miocene compression in the Faeroe–Rockall Area. *In*: PARKER, J. R. (ed.) *Proceedings of the 4th Conference on the Petroleum Geology of Northwest Europe*. Geological Society, London, 1025–1034.

—— & —— 1995. The relationship between the distribution of Tertiary sediments, tectonic processes and deep water circulation around the Faeroe Islands. *In*: SCRUTTON, R. A., STOKER, M. S., SHIMMIELD, G. B. & TUDHOPE, A. W. (eds) *The Tectonics, Sedimentation and Palaeoceanography of the North Atlantic Region*. Geological Society, London, Special Publications, **90**, 145–158.

BUGGE, T., BEFRING, S., *et al.* 1987. A giant three-stage submarine slide off Norway. *GeoMarine Letters*, **7**, 191–198.

DAMUTH, J. E. & OLSON, H. C. 1993. Preliminary observations of Neogene-Quaternary depositional processes in the Faeroe Shetland Channel revealed by high resolution seismic facies analysis. *In*: PARKER, J. R. (ed.) *Proceedings of the 4th Conference on the Petroleum Geology of Northwest Europe*. Geological Society, London, 1035–1045.

DANSGAARD, W., JOHNSEN, S. J., *et al.* 1993. Evidence for general instability of past climate from a 250-kyr ice core record. *Nature*, **364**, 218–220.

EVANS, D., KING, E. L., KENYON, N. H., BRETT, C. & WALLIS, D. 1996. Evidence for long term instability in the Storegga Slide region off western Norway. *Marine Geology*, **130**, 281–292.

FLEWELLEN, C., MILLARD, N. & ROUSE, I. 1993. TOBI, a vehicle for deep ocean survey. *Electronics and Communications Engineering Journal*, April 1993.

HAFLIDASON, H., KING, E. L., BRETT, C., STEVENSON, A. G., WALLIS, D. G., CAMPBELL, N. C., SEJRUP, H. P. & WAAGE, B. 1996. *Marine geological/geophysical cruise report on the North Sea Margin: Upper North Sea Fan, Miller Slide and Faeroe–Shetland Channel*. Dept. Geology, Univ. Bergen, Norway.

KENYON, N. 1987. Mass wasting features on the continental slope of Northwest Europe. *Marine Geology*, **74**, 57–77.

KING, E. L., SEJRUP, H. P., HAFLIDASON, H., ELVERHOI, A. & AARSETH, I. 1996. Quaternary seismic stratigraphy of the North Sea Fan: glacially-fed gravity flow aprons, hemipelagic sediments and large submarine slides. *Marine Geology*, **130**, 293–315.

LE BAS, T. P., MASSON, D. C. & MILLARD, N. C. 1995. TOBI image processing – the state of the art. *IEE Journal of Oceanic Engineering*, **20**, 85–93.

MIENERT, J., AARSETH, I., *et al.* 1996. *European North Atlantic Margin, sediment pathways, processes and fluxes*. Final Report ENAM project (EU Contract MAS CT93-0064).

MILLER, K. G. & TUCHOLKE, B. E. 1983. Development of Cenozoic abyssal circulation south of the Greenland Scotland Ridge. *In*: BOTT, M. H. P., SAXOV, S., TALWANI, M. & THIEDE, J. (eds) *Structure and Development of the Greenland Scotland Ridge*. Plenum, New York, 549–589.

NIELSEN, T., VAN WEERING, TJ. C. E. & ANDERSEN, M. S. 1997. Cenozoic changes in the sedimentary regime on the northeastern Faeroes margin. *This volume*.

RASMUSSEN, T. L., THOMSEN, E. & VAN WEERING, TJ.C.E. 1998. Cyclic sedimentation on the Faeroe Drift 53–10 ka BP related to climatic variations. *This volume*.

——, ——, & LABEYRIE, L. 1996*a*. Rapid changes in surface and deep water conditions at the Faeroe margin during the last 58,000 years. *Paleoceanography*, **11**, 757–771.

——, VAN WEERING, TJ. C. E. & LABEYRIE, L. 1996*b*. High resolution stratigraphy of the Faeroe Shetland Channel and its relation to North Atlantic paleoceanography. *Marine Geology*, **131**, 75–88.

——, & —— 1996*c*. Climatic instability, ice sheets and ocean dynamics at high northern latitudes during the last Glacial period (58–10 ka BP). *Quaternary Science Reviews*, **15**, 1–10.

ROBERTS, D. G., HUNTER, P. M. & LAUGHTON, A. S. 1977. *Bathymetry of the northeast Atlantic, Sheet 2: Continental Margin around the British Isles*. Admiralty Chart C6567, Hydrographic Department, Taunton.

STEVENSON, A. G. 1991. *Miller, Sheet 61°N, 02°W, Quaternary Geology, 1:250 000 Map Series*. British Geological Survey.

STOKER, M. S. 1997. Mid- to late Cenozoic sedimentation on the continental margin off NW Britain. *Journal of the Geological Society, London*, **154**, 509–515.

—— 1998. Sediment-drift development on the continental margin off NW Britain. *This volume*.

——, HITCHEN, K. & GRAHAM, C. C. 1993. *The geology of the Hebrides and West Shetland shelves and adjacent deep water areas*. United Kingdom offshore regional report. HMSO, London, for British Geological Survey.

TALWANI, M. & ELDHOLM, O. 1972. Continental margin off Norway, a geophysical study. *Geological Society of America Bulletin*, **83**, 3575–3606.

WOLD, C. N. 1995. Paleobathymetric reconstruction on a gridded database: the northern North Atlantic and southern Greenland–Iceland–Norwegian Sea. *In*: SCRUTTON, R. A., STOKER, M. S., SHIMMIELD, G. B. & TUDHOPE, A. W. (eds) *The Tectonics, Sedimentation and Palaeoceanography of the North Atlantic Region*. Geological Society, London, Special Publications, **90**, 271–302.

Turbidite flux, architecture and chemostratigraphy of the Herodotus Basin, Levantine Sea, SE Mediterranean

MIKE REEDER[1], R. GUY ROTHWELL[2], DORRIK A. V. STOW[1], GISELA KAHLER[2] & NEIL H. KENYON[2]

[1]*Department of Geology, University of Southampton*
[2]*Challenger Division for Seafloor Processes, Southampton Oceanography Centre, Empress Dock, European Way, Southampton, SO14 3ZH, UK*

Abstract: The Herodotus Basin is the deepest part of the SE Mediterranean and receives allochthonous sediments as turbidity currents and debris flows from around its margin. During the late Quaternary, characteristic supply has been from at least four sources. These are (1) dark coloured, calcium carbonate-poor fine-grained turbidites derived from the Nile Cone to the south and south-east, (2) lighter coloured, calcium carbonate-rich, slightly coarser-grained turbidites derived from the Libyan–Egyptian shelf to the south, (3) small, light brown foraminifer-rich, muddy-silty turbidites derived from the Cyprus–Eratosthenes seamount (Anatolian Rise) carbonate shelf to the east, and (4) small localized debris flow deposits derived from the Mediterranean Ridge to the north. During the late Quaternary (0–60 ka), and specifically the period of 0–27 ka, the basin has filled predominantly with allochthonous material derived from the Nile Cone, although one megaturbidite of basin-wide extent was derived from the Libyan–Egyptian shelf. Turbidites have been correlated across the Herodotus Basin using the technique of chemostratigraphy. Matching the results of geochemical analysis may show whether or not the beds in different cores were deposited by the same mass-wasting event, for individual turbidites commonly have diagnostic and unique geochemical 'fingerprints' in terms of major, minor and trace element composition. Sediment budgets for the three main turbidite sources are calculated. The cumulative volume of the sedimentary input for the Nile Cone- derived turbidites over the last 27 ka is c. 500 km^3, giving an average sedimentation rate of c. 45 cm ka^{-1}, and a volume per unit time of 18 km^3 ka^{-1}. A megaturbidite, derived from the Libyan–Egyptian shelf, is of basin-wide extent and has a volume of c. 400 km^3.

As part of the European Union-sponsored MAST II PALAEOFLUX programme, set up to investigate biogeochemical fluxes in the Mediterranean Sea, the RV *Marion Dufresne* (Cruise 81, January–February 1995) collected a total of 30 long piston cores and nine box cores from the Western and Eastern Mediterranean. This paper presents the results of a study based on five of these long piston cores (ranging from 13.7 to 25.8 m in length) recovered from the Levantine Sea in the SE Mediterranean. These cores were taken along a c. 325 km long SW–NE transect across the Herodotus Basin (Fig. 1, Table 1). Preliminary results are also presented from interpretation of 3.5 kHz high-resolution seismic profiles collected during the same cruise.

The main purpose of this research is to identify and correlate allochthonous units on the plain, so as to estimate the amount of material being eroded from sediment source areas per unit time, and to quantify downslope transport of terrigenous material onto the plain during the late Quaternary.

Methods

All cores were split and described on the ship. P-Wave velocity (compressional wave) and magnetic susceptibility measurements were then made at 5 cm intervals using the shipboard multi-sensor core logger operated and designed by the geotechnical consultants Geotek Ltd (Haslemere, UK). Detailed re-logging of the cores took place post-cruise together with standard sedimentological analyses for grain size, calcium carbonate content and geochemistry using inductively coupled plasma atomic emission spectroscopy (ICP-AES) and micropalaeontological analysis of foraminifera and nannofossils. Grain-size analyses of 88 sediment samples were made by wet sieving for any sand fraction present and then using a Micromeritics 5100 sedigraph size analyser for measuring the fine fraction (2–63 μm). Calcium carbonate content ($CaCO_3$) was determined by acid treatment of hand-ground material and coulometric detection of CO_2. Replicate analyses and in-house standards were used to control for error.

REEDER, M., ROTHWELL, R. G., STOW, D. A. V., KAHLER, G. & KENYON, N. H. 1998. Turbidite flux, architecture and chemostratigraphy of the Herodotus Basin, Levantine Sea, SE Mediterranean *In*: STOKER, M. S., EVANS, D. & CRAMP, A. (eds) *Geological Processes on Continental Margins: Sedimentation, Mass-Wasting and Stability*. Geological Society, London, Special Publications, **129**, 19–41.

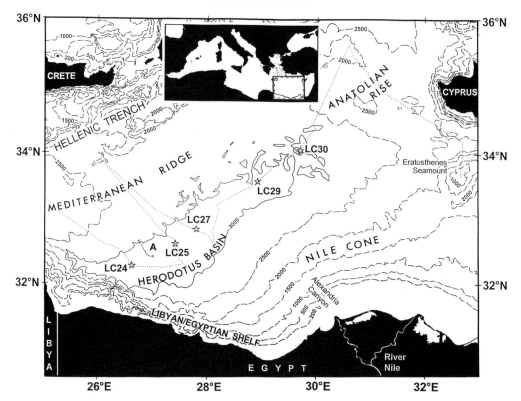

Fig. 1. Location map for the Herodotus Basin showing the bathymetry at 200 m, 500 m, and every 500 m thereafter. The core positions (LC24–LC30) and ships track for RV *Marion Dufresne* Cruise 81 are shown, together with the BAN-82 core (transect A) studied by Cita *et al.* (1984*a*) and Lucchi & Camerlenghi (1993).

Table 1. *Location, depth and length of the five cores from this study (MD-81) and the nine cores from two previous studies (BAN-82) by Cita* et al. *(1984a) and Lucchi & Camerlenghi (1993)*

Core no.	Core type	Water depth (corr. m)	Latitude	Longitude	Core length (m)
MD-81					
LC24	Long piston	3191	32°17.71′N	26°37.95′E	18.31
LC25	Long piston	3129	32°36.01′N	27°23.25′E	13.71
LC27	Long piston	3131	32°48.91′N	27°40.45′E	14.95
LC29	Long piston	3138	33°35.63′N	28°56.32′E	24.66
LC30	Long piston	3144	34°04.70′N	29°42.72′E	25.82
BAN-82					
PC-10	Piston	3198	32°20.55′N	26°58.98′E	11.09
PC-11	Piston	3161	32°26.19′N	26°54.55′E	7.88
PC-12	Piston	3088	32°29.95′N	26°52.62′E	10.92
GC-13	Gravity	3044	32°34.55′N	26°51.00′E	2.58
GC-14	Gravity	3088	32°37.70′N	26°48.14′E	4.31
PC-15	Piston	2915	32°42.60′N	26°44.61′E	10.50
GC-16	Gravity	3008	32°34.42′N	26°50.86′E	4.47
PC-17	Piston	3096	32°32.45′N	26°51.30′E	9.07
GC-18	Gravity	3066	32°32.83′N	26°50.49′E	4.62

Table 2. *The 22 elements measured for geochemical analysis of the turbidite horizons using inductively coupled plasma atomic emission spectroscopy (ICP-AES)*

Element	Units	Measured high count	Measured low count	Al_2O_3 corrected high count	Al_2O_3 corrected low count	Scaling Factor (X)
TiO_2	%	1.65	0.02	0.14	0.06	100.00
Al_2O_3	%	15.77	0.28	–	–	–
Fe_2O_3	%	10.66	0.16	0.73	0.42	10.00
MnO	%	0.73	0.01	0.07	0.004	1000.00
MgO	%	3.73	0.19	0.75	0.20	20.00
CaO	%	39.75	2.12	12.17	0.15	2.00
Na_2O	%	3.13	0.11	0.48	0.15	50.00
K_2O	%	1.97	0.06	0.22	0.07	100.00
P_2O_5	%	0.25	0.02	0.09	0.013	100.00
Ba	ppm	378.48	13.65	56.10	12.93	0.25
Co	ppm	54.80	1.13	8.30	2.00	2.00
Cr	ppm	107.02	1.85	8.94	3.24	2.00
Cu	ppm	105.64	2.03	11.07	2.47	2.00
Li	ppm	57.17	–8.39	–	–	–
Mo	ppm	18.04	–2.72	–	–	–
Ni	ppm	98.56	1.00	9.22	3.54	2.00
Pb	ppm	141.44	–8.96	–	–	–
Sc	ppm	20.02	0.28	1.33	0.95	10.00
Sr	ppm	4254.32	110.99	1302.70	9.87	0.02
Y	ppm	30.20	0.82	3.27	1.60	5.00
Zn	ppm	101.35	2.47	12.00	4.78	1.00
V	ppm	193.60	4.06	15.60	8.59	1.00

The measured high and low count numbers for each element have been corrected by dividing each element by the Al_2O_3 measured value for the separate samples and then scaled by a factor X to fall between 0.00 and 25.00 (Grant 1986). Li, Pb and Mo were disregarded because of their negative low count number.

Just under 200 samples were taken from turbidite beds for analysis of 22 major, minor and trace elements. The analyses were performed using ICP-AES after digestion of 0.5 g of sample with a combination of hydrofluoric, perchloric and nitric acids. Accuracy was checked using standard reference materials and monitored with in-house standards. Precision was better than 5% for all elements. The results were first normalized to their Al_2O_3 value, and then scaled so that a number for the whole range of elements was obtained that fell between zero and 25, following the method of Grant (1986) (see Table 2).

Geological setting and previous work

The Herodotus Basin or Abyssal Plain is an elongate depression in the Eastern Mediterranean defined by the 3000 m isobath. It is bounded to the NW by the accretionary prism complex forming the Mediterranean Ridge and to the SE by the Nile Cone. The shorter SW and NE ends of the basin are bounded by the Libyan–Egyptian continental slope and Anatolian Rise,

respectively (Fig. 1). Convergence of the European and African plates has created and deformed the Mediterranean Ridge accretionary prism resulting in complex topography along the NW margin of the basin, whereas growth and progradation of the Nile Cone has occurred across the SE part of the basin. Bathymetric study shows that much of the floor of the basin is no longer a plain in the true sense (as a result of deformation) and therefore the term 'basin' is used rather than 'abyssal plain'.

Two previous papers discussed sedimentary processes and patterns in the Herodotus Basin (Cita *et al.* 1984*a*; Lucchi & Camerlenghi 1993). Both concerned a study based on the RV *Bannock* BAN-82 cruise, during which five piston cores and four gravity cores were collected on a SE–NW transect across the width of the Herodotus Basin and onto the Mediterranean Ridge (Fig. 1). Cita *et al.* (1984*a*) detailed the petrology and sources for two distinct types of turbidites (Type-A and Type-B). The Type-A turbidites have a black plastic mud fraction rich in smectite, a terrigenous silt fraction, and are sourced from the Nile Cone. The light olive grey

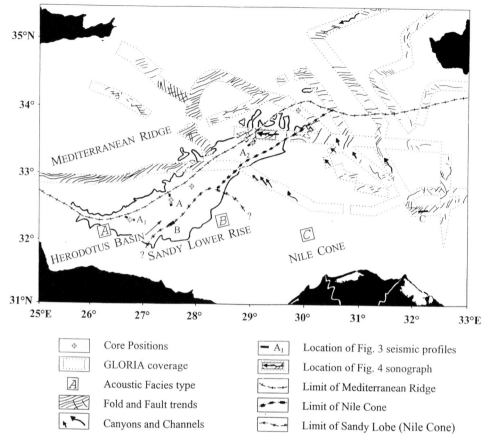

Fig. 2. Acoustic facies–physiographic map of the Herodotus Basin based on analysis of 3.5 kHz data from the RV *Marion Dufresne* Cruise 81 and 12 kHz and GLORIA side-scan sonar data from the RRS *Discovery* Cruises 40, 55 and 104 (Kenyon *et al.* 1975; Stride *et al.* 1977). The 3.5 kHz analysis is after criteria of Jacobi & Hayes (1993). Acoustic facies A–C are explained in the text; profiles shown in Fig. 3 are located. The GLORIA data show the predominant SW–NE trends of folds and faults on the Mediterranean Ridge and the canyon and channel orientations of the Nile Cone. The Herodotus Basin is delineated by the 3000 m isobath (bold line).

Type-B turbidites are coarser grained, carbonate rich with a shallow-water, bioclastic, basal sand poor in smectite, and sourced from the Libyan–Egyptian shelf. A conspicuous megabed termed 'β' by Cita *et al.* (1984*a*), also sourced from the Libyan–Egyptian shelf, was noted with an estimated volume of 10 km³. Lucchi & Camerlenghi (1993) concluded that the turbidites from the Libyan–Egyptian shelf travelled up the slope of the Mediterranean Ridge for as much as 57 km to a height of 283 m above the basin floor. In addition, the present study has revealed a third turbidite type, 'Type-C', a pale greyish yellow brown, carbonate-rich, foraminifer-rich silty mud sourced from the east towards the Anatolian Rise.

Material derived from the more tectonically active European margin is trapped in the

Fig. 3. Selected 3.5 kHz high-resolution seismic profiles across the Herodotus Basin, showing the characteristics of common echo-types and morpho-acoustic provinces. Profiles are located in Fig. 2. (**a**) Facies A: the deformation front associated with the Mediterranean Ridge accretionary complex showing ponded turbidites between uplifted blocks and ridges. Alongside, 3.5 kHz profiles across sites LC24 and LC 29 (A₁ and A₂, respectively) stationed within the ponded areas, showing thick acoustically transparent layer. (**b**) Facies B: impeded acoustic penetration interpreted as reflecting the presence of a surficial sandy lobe of the Nile Fan. (**c**) Facies C: thinly bedded sub-parallel reflectors from the Nile Cone, with small channels.

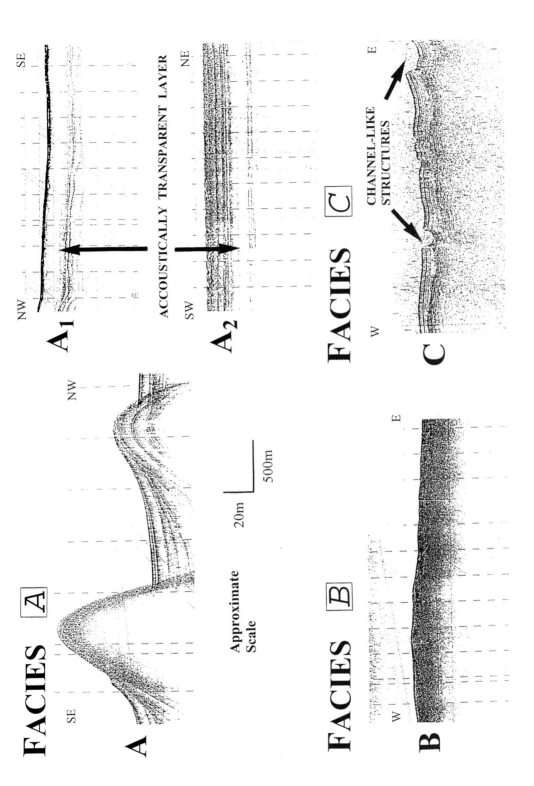

Hellenic Trench north of the Mediterranean Ridge and consequently cannot reach the Herodotus Basin (Cita *et al.* 1984*a*; Lucchi & Camerlenghi 1993). However, some local downslope reworking from the Ridge itself may occur together with a minor input to the NE end of the basin from the Anatolian Rise.

Morphology and acoustic facies

A preliminary acoustic facies map has been constructed for the Herodotus Basin (Fig. 2) using the 3.5 kHz profiles collected during RV *Marion Dufresne* Cruise MD-81 and GLORIA side-scan sonographs from other cruises in this region (RRS *Discovery* Cruises 40 and 55, Kenyon *et al.* (1975) and Stride *et al.* (1977), respectively). This shows that the truly flat portion of the basin, the abyssal plain, is smaller by approximately half in extent than previously mapped.

Echo types on the 3.5 kHz records have been classified on the basis of clarity and continuity of sub-bottom echoes, the depth of sub-sea-floor penetration by the acoustic signal, and microtopography (cf. Kidd *et al.* 1985; Jacobi & Hayes 1993). Three acoustic facies or province types have been identified, named A–C.

(A) A morpho-acoustic province consisting of a series of parallel ridges, separated by elongate basins. The Herodotus Abyssal Plain was a true plain in the past, but because of the collision of the African and European plates the northern part of the plain has been deformed into a system of parallel ridges and asymmetric troughs (Fig. 3a). These ridges increase in relief towards the Mediterranean Ridge up to about 100 m above the adjacent plain. Where they have greatest relief they display a hyperbolic echo type with a single, thick, sandy-type reflector, whereas the lower-relief ridges still show deformed but parallel sub-bottom reflectors similar to the flat portion of the plain. Deformation appears to continue to be active, and the developing microtopography clearly interacts with incoming turbidity currents, such that the parallel continuous sub-bottom reflectors of the troughs pinch out and onlap the rising ridges (Fig. 3a).

(B) The second acoustic facies type is a strong prolonged surface reflector with impeded acoustic penetration (Fig. 3b). This is interpreted as part of a sandy, lower-fan lobe system extending from the Nile Cone, based on analogous, but cored echo-types in other areas reported by Jacobi & Hayes (1993) and others.

(C) The third distinct acoustic facies type is the thinly bedded, continuous, sub-parallel reflectors that characterize the Nile Cone

(Fig. 3c). The 3.5 kHz seismic profiles show the presence of abundant low-relief channels which incise the thinly bedded sediments (Kenyon *et al.* 1975; Stride *et al.* 1977).

The GLORIA side-scan images show relatively small-scale relief providing widespread evidence of distinctive structural trends on the floor of the Herodotus Basin region (Kenyon *et al.* 1975; Stride *et al.* 1977). The data show that much of the Mediterranean Ridge comprises one or more sets of linear or curvilinear ridge features, some of which lie parallel to the ridge axis (SW–NE) whereas others lie perpendicular to it (NW–SE). The trend of these ridges is also approximately parallel to that of the Herodotus Basin, demonstrating a principal stress orientation, and hence the NW–SE direction of the collision between Africa and Eurasia in this region (Fig. 2).

The side-scan sonar images show a number of channel and canyon trends on the Nile Cone. These are predominantly seen around 33°00′N, 29°30′E trending in a northwesterly direction towards the Herodotus Basin between core sites LC27 and LC29 (Fig. 2). Other meandering channel systems and canyons are seen to the north and north east of this region, radiating from the distal parts of the cone. One such channel, adjacent to core site LC29 (Fig. 4), appears to have encroached into the basin.

The core stations that form the basis for this study were sited on areas of continuous parallel sub-bottom reflectors situated in the troughs between ridges or uplifted blocks (Fig. 3, A_1 and A_2), and represent abyssal plain sediments.

Sediment types

Six main sediment types are recognized in the cores and can be distinguished on the basis of colour, composition, grain size and geophysical properties. The two dominant sediment types are the two contrasting types of turbidites (Type-A and Type-B) first described by Cita *et al.* (1984*a*), whose terminology is retained. The third, Type-C turbidite, is recorded in cores LC29 and LC30, and has similar composition and geophysical properties to the Type-B turbidites. In our study, prominent individual turbidites are identified by letter notation, from turbidite 'p' at the base of the core to turbidite 'a' at the top. Correlation of these turbidites between cores using the above properties has been confirmed by chemostratigraphical techniques (see discussion below). Between individual turbidites there are thin pelagic–hemipelagic intervals and, towards the top of each core, a distinctive dark-coloured sapropel is present. A chaotic, clast-rich mud unit

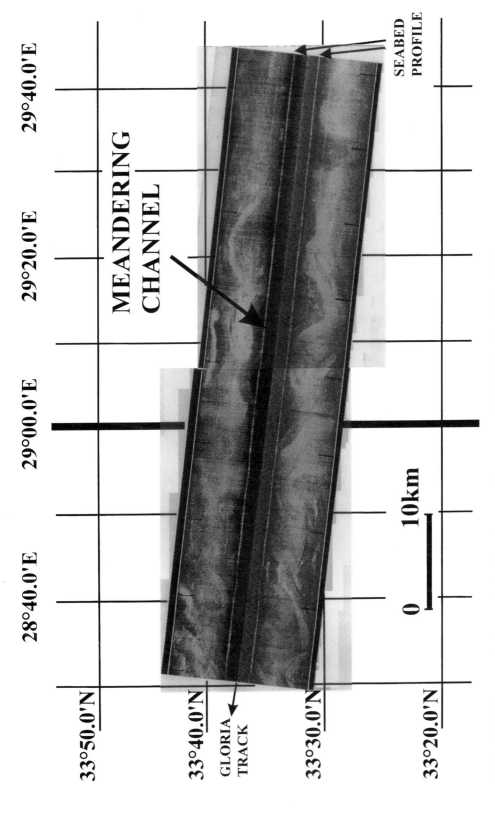

Fig. 4. GLORIA sonograph of a meandering channel–canyon adjacent to core site LC29. (See Fig. 2 for location.)

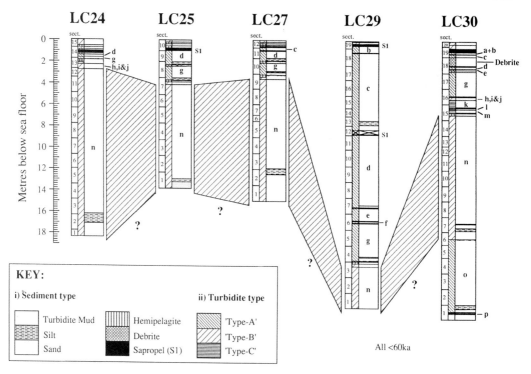

Fig. 5. Lithological logs for the five long piston cores (LC24–LC30) from the Herodotus Basin, showing correlation (oblique hatching) of a thick mud megabed n on the basis of major, minor and trace element geochemistry. The transect length is 325 km.

present in one core (LC30) is interpreted as a muddy debris-flow deposit.

Type-A turbidites

These are mud-rich, dark brown–grey in colour (typical Munsell colour value 10YR3/2, US Geological Survey Rock Color Chart Committee 1991) and range in thickness from <15 cm to >7 m (Fig. 5). Very few sedimentary structures are apparent except weak parallel lamination and discontinuous wavy lamination towards the base of beds. Some of the turbidites contain small iron sulphide nodules (1–5 mm) throughout (Fig. 6a). Type-A turbidites show a broadly bipartite grain-size structure with a thicker upper ungraded or homogeneous mud division (mean size 4–7 μm) overlying a thinner, normally graded basal division, which ranges up to medium–coarse sand (<1 mm) in grain size at the base of thicker beds. These thicker beds also show a subtle oscillation in mean grain size through both divisions (Fig. 7). Type-A beds are best interpreted as fine-grained, distal turbidites, mostly displaying Stow T_5–T_8 turbidite divisions (Stow & Shanmugam 1980; Stow 1985) or Piper E_2–E_3 divisions (Piper & Stow 1991). The coarser basal intervals of some beds belong to Stow T_2–T_4 or Piper E_1 fine-grained turbidite divisions.

Smear-slide examination of the silt and sand fractions of Type-A turbidites shows them to comprise mainly terrigenous quartz, feldspar

Fig. 6. Split-core photographs of sediments and sedimentary structures in selected cores. (a) Fine-grained, randomly spaced millimetric iron sulphide nodules in a dark brown–grey (10YR3/2) Type-A turbidite mud. (b) Base of a light olive–grey (5Y5/2) Type-B turbidite showing laminations. (c) Bioturbated pelagic and hemipelagic sediments with thin graded turbidite layers. (d) Oxidized, laminated and convoluted Type-A turbidite b within S1 Sapropel. (e) The Debrite in LC30, deposited within S1 Sapropel, comprising clasts of Type-A and Type-B turbidites and sapropelic material.

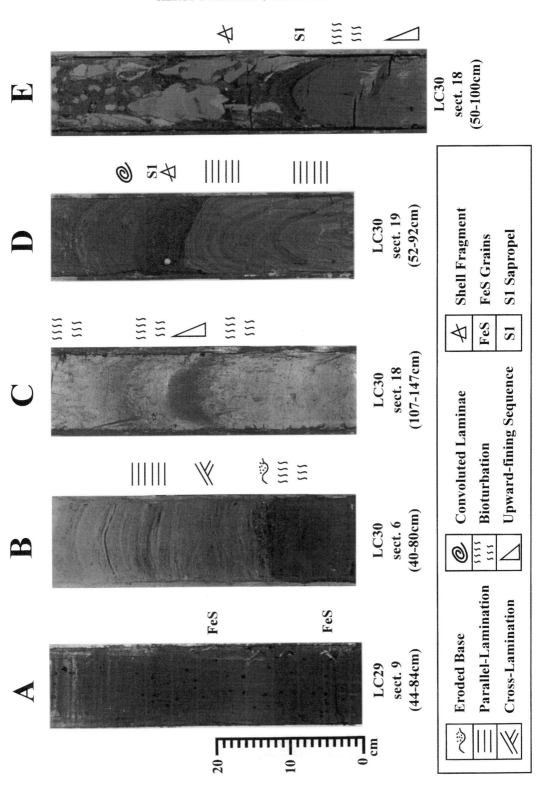

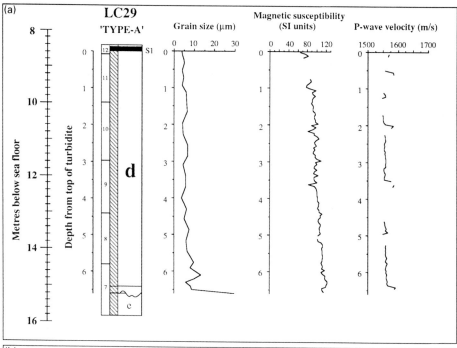

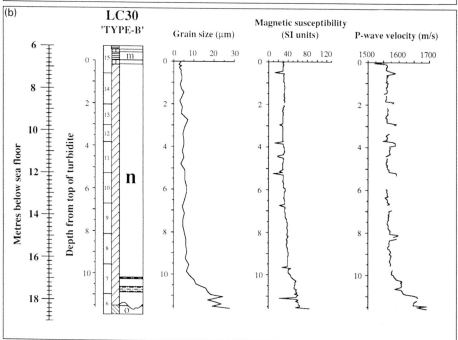

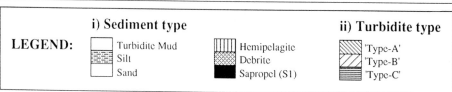

and mica (see Cita *et al.* 1984*a* for detailed analysis). The coarse fraction contains abundant plant debris and other dark organic material. Foraminifera and pteropods are also present in the coarser-grained size fractions. Calcium carbonate content of Type-A turbidites ranges from 4 to 16% (mean 10%), magnetic susceptibility is relatively high (80–150 SI units), and P-wave velocity is around 1550 m s^{-1} (Fig. 8).

Type-B turbidites

The less common Type-B turbidites are a light olive–grey in colour (typical Munsell colour value 10YR6/2) and range in thickness from a few centimetres to >16 m in LC24 (Fig. 5). Like the Type-A turbidites, Type-B are predominantly thick, structureless, homogeneous muds with thinner silty and sandy bases. However, the sand–silt unit at the bottom of cores LC24 and LC27 is >1 m thick and the base of this turbidite was not penetrated. Indistinct parallel lamination occurs in the lower part of the mud division, whereas the sand–silt division may show both parallel and cross-lamination (Fig. 6b). Beds show either a bipartite or tripartite grain-size structure with ungraded mud (4–6 μm) overlying slightly graded mud (4–10 μm) and normally graded silt–sand (up to coarse sand, <1 mm). A slightly irregular oscillation of mean grain size is apparent throughout the large Type-B turbidite n (Fig. 7).

Bed n of the Type-B turbidites is a very thick turbidite or 'megaturbidite', comprising Stow and Piper divisions T_5–T_8 or E_2–E_3 respectively, through the thicker upper mud part, and a coarse-grained sandy base (T_{CD} Bouma divisions, Bouma 1962). Megaturbidites of this sort have been recognized from a number of other Mediterranean basins. For example, the Ionian Sea tsunami-induced 3.5 ka 'Homogenite' of Kastens & Cita (1981) and Cita *et al.* (1984*b*) is a body of homogeneous hemipelagic mud characterized by a thin, normally graded, foraminiferal sand at the base. Similar deposits, such as the Unifites described by Stanley (1980, 1981), have been stripped of the sand fraction through deposition in small slope basins during emplacement, leaving a homogeneous or uniform mud body. The Libyan–Egyptian megaturbidite n shows similar characteristics to those mentioned above, except the sand fraction of the turbidite is thicker than that of the Homogenite and Unifites, and megaturbidite n is a factor of ten times greater in volume.

Analysis of smear slides of the coarse-grained basal sand of megaturbidite n shows quartz, calcite and a high proportion of shelf-derived bioclastic material, such as gastropods, sponge spicules, bivalves, bryzoans, pteropods, and shallow-water benthic and planktonic foraminifera. Some Type-B turbidites also have significant amounts of white and grey pumice grains, dark igneous minerals (e.g. pyroxenes and amphiboles), and cuspate and lunate volcanic glass shards. Calcium carbonate is typically around 50%, magnetic susceptibility values are low (25–45 SI units), and P-wave velocities around 1560–1580 m s^{-1} (Fig. 8).

Type-C turbidites

The rare Type-C turbidites are a moderate yellow–brown in colour (typical Munsell colour value 10YR6/4) and only reach 72 cm (turbidite **k**) and 35 cm (turbidite **m**) in thickness in core LC30 (Fig. 5). These small Type-C turbidites are predominantly structureless, homogeneous muds with centimetre thick-silty and sand-sized foraminifer-rich bases.

Like the Type-B turbidites, the Type-C turbidites comprise Stow and Piper fine-grained turbidite divisions T_5–T_8 or E_2–E_3, respectively, through the thicker upper mud part, and a centimetre-thick T_{CD} Bouma-division medium-grained sandy base.

Analysis of smear slides of the coarse-grained basal foraminiferal sand fractions shows a high proportion of both benthic and planktonic foraminifera (>70%), calcite (10%), muscovite mica (5%) and quartz (5%). There is a small fraction of shelf-derived bioclastic material consisting particularly of gastropods (5%). The Type-C turbidites also have significant amounts of dark igneous minerals (e.g. pyroxenes and amphiboles) and green volcanic glass (<5%). The composition of the Type-C turbidites is very similar to that of the Type-B megaturbidite n, although the colour is much paler. Other differences occur in the calcium carbonate percentages, which are generally lower than the Type-A turbidites at around 40%. The magnetic susceptibility values fall between those of the Type-A and Type-B turbidites (40–45 SI units), and

---

Fig. 7. Selected downcore parameter profiles for turbidite d (Type-A) and turbidite n (Type-B) in cores LC29 and LC30. Average grain size was measured and is shown against the P-wave velocity and magnetic susceptibility (multi-sensor core logger data).

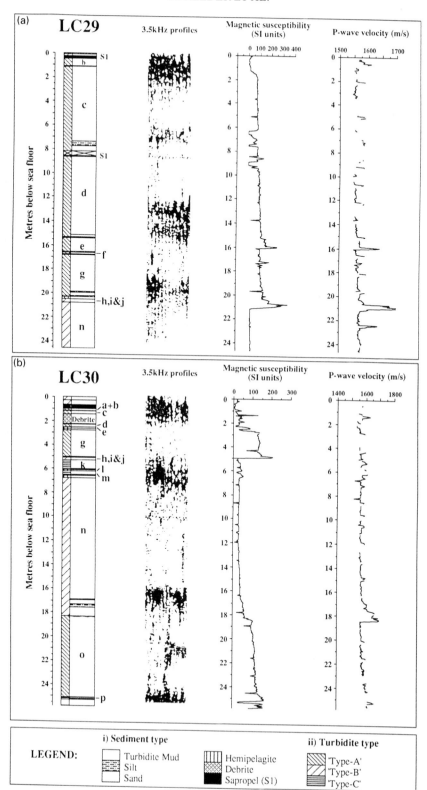

P-wave velocities are similar at around 1550–1575 m s^{-1}.

Pelagic and hemipelagic deposits

Thin pelagic–hemipelagic intervals occur between many, but not all, turbidite beds. They are typically mixtures of a pale yellowish (10YR6/6) biogenic mud and either brownish grey (10YR3/2) or light olive grey (5Y5/2) muds depending on the nature and colour of the underlying turbidite. They range in thickness from <1 cm to >20 cm, and are characterized by bioturbation and the occurrence of scattered foraminifera throughout (Fig. 6c). Trace fossils include *Chondrites*, *Planolites* and *Phycosiphon*. From their mixed calcareous–terrigenous (quartz and muscovite mica) composition and silty grain size, these sediments may be classified as hemipelagites (Stow *et al.* 1996), although the distinct colour gradation from the underlying turbidites suggests hemiturbidite deposition may have occurred (Stow & Wetzel 1990).

Sapropels

Dark, greyish black (typical Munsell value N0–N1), organic-carbon-rich sediments (C$_{org}$ value >2%, Higgs *et al.* 1994) known as sapropels are well documented from the eastern Mediterranean (e.g. Aksu *et al.* 1995; Thomson *et al.* 1995). A number of thin sapropelic intervals occur in the upper parts of all cores studied from the Herodotus Basin at a cored depth of between 0.2 and 8.7 m. Individual layers range up to 21 cm in thickness and are generally well laminated (Fig. 6d). Some of these layers contain abundant well-preserved pteropods (although in others these are absent), together with warm-water pelagic and benthic foraminifera (Kahler & Dossi 1996, see below). These layers occur interbedded with a variable number and thickness of turbidites in different cores. However, on the basis of faunal content and near-surface occurrence (Kahler & Dossi 1996), they are all considered as part of the same S1 Sapropel interval. The S1 Sapropel has been dated as between 5–6.5 and 9 ka (Bethoux 1993; Higgs *et al.* 1994; Rohling 1994; Aksu *et al.* 1995), and is interpreted as indicating a period of anoxic bottom-water conditions in the Eastern Mediterranean

Debrites

This sediment type occurs only in core LC30 from the northeastern part of the basin. It has been deposited within the S1 Sapropel interval of sections 18 and 19 (Figs 5 and 6e). It is a disorganized mud breccia 82 cm thick, containing irregular to rounded clasts of dark-coloured sapropelic mud and clasts of hemipelagic mud and Type-A, Type-B and Type-C turbidite muds. The sub-rounded–sub-angular clasts are poorly sorted and range in size from <1 cm to 12 cm. Both clasts and matrix generally show plastic deformation structures. Core-parallel elongation of many clasts is apparent, and is interpreted as the result of disturbance during coring. This sediment type is interpreted as a debrite (e.g. Stow & Piper 1984; Stow *et al.* 1996) that seems to be restricted in occurrence to the northern flank of the Herodotus Basin, as were similar debrites noted by Cita *et al.* (1984*a*) and Lucchi & Camerlenghi (1993) during the BAN-82 cruise.

Correlation and dating

To determine the extent and geometry of individual turbidite beds in the Herodotus Basin, and to more accurately estimate turbidite volumes emplaced, three techniques have been used to make a precise correlation between the cores sampled along the 325 km transect.

Geophysical correlation

Generally there is a close correspondence between the distinct, parallel reflectors on the 3.5 kHz high-resolution seismic profiles across the core sites and the contacts between thicker individual turbidite beds, especially where beds are underlain by basal sands (Fig. 8). High-amplitude reflectors mark the bed contacts and/or sandy bases, and acoustically transparent zones typically correlate with the thick structureless turbidite muds. Other geophysical properties, such as P-wave velocity and magnetic susceptibility, further help to characterize individual beds, but resolution of the thinnest beds is not possible on the geophysical records and hence correlation cannot be precise in all cases.

The distinctive megaturbidite n (a Type-B turbidite) correlates with a particularly thick acoustically transparent layer seen on the 3.5 kHz high-resolution seismic profiles that can be

Fig. 8. Geophysical downcore profiles for cores LC29 and LC30 (P-wave velocity and magnetic susceptibility) with corresponding 3.5 kHz high-resolution seismic profiles across the core sites.

traced with confidence over the full length of the basin (Figs 3 and 8). The top of this layer increases in depth from about 2 m below the top of the core in the SW (LC24) to over 21 m below the top of the core towards the NE (LC29), and then shallows to a depth of about 5 m in the extreme NE (LC30). The 3.5 kHz profiles show a very similar pattern to the cores recovered, suggesting that very little of the sediment above the megaturbidite has been lost as a result of the coring process (centimetres rather than metres). The thickness of megaturbidite n, estimated from the 3.5 kHz records using a sediment interval velocity of 1500 m s^{-1}, appears to decrease slightly from c. 20 m in the SW (core LC24) to c. 10 m in the NE (core LC30). This has proved to be the case when measuring the thickness of sediments above megaturbidite n in the cores (18 m and 11.5 m for cores LC24 and LC30, respectively). Megaturbidite n and other acoustic beds thin markedly against local deformation ridges that characterize the NW part of the basin (Fig. 3b).

There is some evidence from 3.5 kHz profiles that fewer, thicker megaturbidites characterized the late Quaternary. The topmost megaturbidite sampled in our cores was derived from the Libyan–Egyptian shelf, although the underlying megabeds may have been derived from any of the sources. Another three or four megabeds are recognized on the 3.5 kHz seismic profiles, but these were not penetrated by the corer.

Lithological correlation

The broad similarity of sediment types and their distribution in all five cores examined allows a preliminary lithological correlation. The apparent recognition of the Type B megaturbidite n on the 3.5 kHz seismic profiles show this bed's considerable extent. The sapropel layers that occur towards the tops of all five cores are identified as the S1 Sapropel of previous workers, and therefore represent a single dateable horizon (5–6.5 to 9 ka). There is little variation in sediment thickness above this horizon, but in the central and NE parts of the basin (LC27 to LC30) one or more thick turbidites are intercalated with the sapropels.

Geochemical correlation

High-resolution geochemical analysis by ICP-AES of all five Herodotus Basin cores provides a precise correlation tool that is effective over the entire 325 km transect across the Herodotus Basin. Individual turbidite events have their own geochemical signatures (Fig. 9), even though they may share similar colour, compositional, and geophysical properties (cf. Pearce & Jarvis 1992, 1995; Wray & Gale 1993). To assess the degree of geochemical similarity shown by individual turbidites in different cores, cross-plots of the same elements in different cores were made. The regression coefficient of the best fit line so obtained can be used as a measure of similarity.

These multiple-element cross-correlation plots are illustrated for the upper ungraded mud interval (E$_3$) of two Type A turbidites (Fig. 10a) and two Type B turbidites from cores LC27 and LC25 (Fig. 10b). In each case element concentrations are virtually identical and the beds can be identified as the same in both turbidite d and megaturbidite n (Figs 10a and 10b, respectively). By contrast, poor correlations are seen if the two turbidites compared are not from the same event; this is evident in Fig. 10c, where two unrelated Type-A turbidites are plotted against each other to give a poor coefficient of correlation. The beds were subsequently identified as turbidites g and d, using a combination of the other correlation techniques mentioned above.

However, the grain size of the samples compared must be taken into consideration. Figure 10d is a cross-plot of elements from the mud and sand fractions of the same turbidite, and the correlation is poor. Fine fraction must be compared against fine fraction. Using the same grain-size fraction for comparison allows a possible correlation for each turbidite on a basin-wide basis (Fig. 11).

Dating

Three different methods have been used to estimate the age of beds recovered from the Herodotus Basin: relationship to the S1 Sapropel, micropalaeontological study, and determination of the thickness of pelagic intervals. It is hoped that future stable-isotope work will confirm these preliminary findings.

The S1 Sapropel is well documented throughout the Eastern Mediterranean and dated to between 5–6.5 and 9 ka (Bethoux 1993; Higgs *et al.* 1994; Rohling 1994; Aksu *et al.* 1995). This provides the most recent date in our cores, although some controversy remains concerning the exact termination of this period of anoxia (Higgs *et al.* 1994; Aksu *et al.* 1995). The S3 and older sapropels (S3 is dated at 79 ka, Hilgen, 1991) were not recovered in any of the basin cores used in this study.

Examination of the planktonic foraminifera (Kahler & Dossi, 1996) shows warm-water assemblages in the upper parts of all cores, with

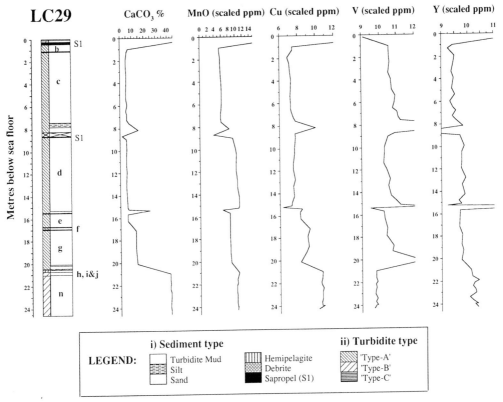

Fig. 9. Selected downcore geochemical profiles for core LC29. Individual turbidites show subtle differences in the concentration of the selected elements. Individual mud beds commonly have diagnostic geochemical 'fingerprints', particularly with regard to minor and trace element composition. The data have been scaled to fall between the values of zero and 25, using the method of Grant (1986).

a very warm-water assemblage in the S1 Sapropel horizon including *Globigerinoides ruber*, *Globigerinoides sacculifer*, *Hastigerina siphoniphera*, *Globigerinoides trilobus trilobus*, *Globigerinoides trilobus immaturus*, *Globigerinoides elongatus*, *Globigerina calida*, *Orbulina universa*, *Globorotalia truncatulinoides* and *Globigerina rubescens*. Turbidites g and e in core LC29 yield both warm- and cold-water species, the latter including left-coiling *G. truncatulinoides* and a high proportion of juveniles. Megaturbidite n also shows a mixed assemblage, whereas turbidite o (core LC30) has a distinctly cold-water fauna including *Globigerina bulloides*, *Globigerina glutinata*, *G. elongatus* and *O. universa*. These data would suggest a Pleistocene–Holocene boundary near the top of turbidite e, assuming a few warmer-water species beginning to return to the Mediterranean as the climate became warmer from *c.* 12–10 ka.

A total of 11 samples were taken for nannofossil analysis from supposed hemipelagic intervals near the base of all five cores. All samples examined contain abundant coccoliths, but some of the material is highly contaminated with older nannofossils as a result of either bioturbation or hemiturbidite processes. However, the assemblage is consistently dominated by *Emiliania huxleyi*, indicating that the pelagic fraction of all samples originates from the late Pleistocene *E. huxleyi* acme interval (Weaver 1983). This places all the recovered section above the earliest part of oxygen isotope stage 3, that is approximately within the last 60 ka.

A first-order estimate of the age at the base of the deepest penetrating core (LC30) can be obtained by adding up the cumulative thickness of pelagic–hemipelagic deposits, and by assuming a constant rate of accumulation for this material, assuming that no erosion by turbidity currents has occurred (cf. Weaver & Kuijpers 1983). Biostratigraphical study of a large number of Eastern Mediterranean cores suggests a rate of *c.* 2.5–3.0 cm ka^{-1} for pelagic sedimentation (E. Rohling, pers. comm. 1996). This places the base of core LC30 at about +27 ka (Table 3).

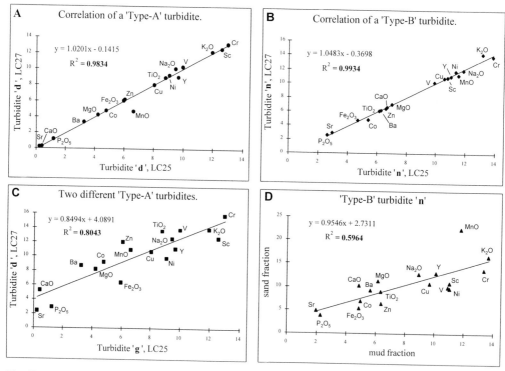

Fig. 10. Isocon diagrams (Grant 1986) for selected turbidite beds, showing the best-fit lines, equations and regression coefficients. Each symbol represents one element as stated in Table 2.

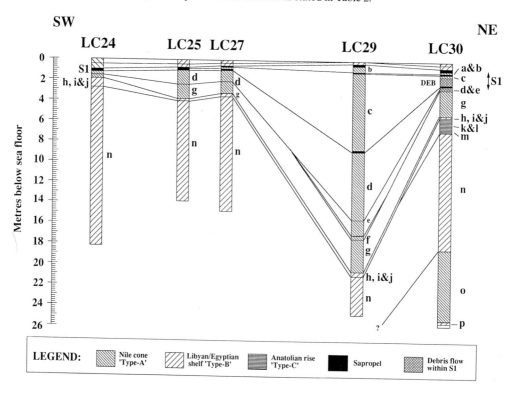

Table 3. *Minimum age calculations for each of the turbidites recorded in the Herodotus Basin*

Turbidite	Type	Present in cores	Max. Thickness (m)	Max. pelagic material between this and overlying turbidite (cm)	Cumulative pelagics down core (cm)	Minimum age (ka)
Turbidite a	A	LC30	0.15	within S1*	–	6.5
Turbidite b	A	LC29 and LC30	0.61	within S1*	–	7.0
Turbidite c	A	All	7.49	within S1*	–	7.5
Debrite		LC30	0.82	within S1*	–	8.0
Turbidite d	A	All	6.69	0.0	0.0	9.0
Turbidite e	A	LC29 and LC30	1.21	8.0	8.0	12.0
Turbidite f	A	LC29	0.20	3.0	11.0	13.5
Turbidite g	A	All	3.43	3.0	14.0	14.5
Turbidite h	B	All	0.14	6.0	20.0	17.0
Turbidite i	B	All	0.07	11.0	31.0	21.5
Turbidite j	B	All	0.06	5.5	36.5	23.5
Turbidite k	C	LC30	0.75	1.0	37.5	24.0
Turbidite l	C	LC30	0.05	6.5	44.0	26.5
Turbidite m	C	LC30	0.35	2.0	46.0	27.5
Turbidite n	B	All	>15.71	0.0?	>46.0	>27.5
Turbidite o	A	LC30	1.79	0.0?	>46.0	>27.5
Turbidite p	B	LC30	0.05	0.0?	>46.0	>27.5

Ages were obtained by using an accumulation rate of 2.5–3.0 cm ka^{-1} (assuming no erosion) and by taking the base of the S1 Sapropel to be 9 ka.
* Dated as being deposited between 5–6.5 and 9 ka (Bethoux 1993; Higgs *et al.* 1994; Rohling 1994; Aksu *et al.* 1995; Thomson *et al.* 1995).

Discussion

The sedimentology, mineralogy and palaeontology of the three types are so different that separate sources have to be considered. The dark brown–grey, fine-grained, quartz, mica and feldspar-rich Type-A turbidites are most probably sourced from the Nile River drainage basin via the Nile Cone (Cita *et al.* 1984*a*; Lucchi & Camerlenghi 1993). The light olive–grey, coarser-grained CaCO$_3$-rich Type-B turbidites, which have negligible terrigenous content, are probably derived from the carbonate-rich Libyan–Egyptian shelf. The moderate yellow–brown, foraminifer- and CaCO$_3$-rich Type-C turbidites are probably derived from the Anatolian Rise to the east of the basin. They cannot be a reworked product of the Type-B megaturbidite n as the Type-C turbidites have small coarse-grained basal fractions which are absent in the thick mud top of the megaturbidite. These proposals for likely provenances are supported by the known dispersal patterns of clay minerals in the sediments of the Eastern Mediterranean Sea (Venkatarathnam & Ryan 1971), and by previous work of Cita *et al.* (1984*a*).

Architecture of the Herodotus Basin fill

The three-dimensional geometry of the three types of turbidite is significantly different (Fig. 11). Megaturbidite n, derived from the Libyan–Egyptian margin, extends across the entire basin, decreasing in thickness from *c.* 18 m in the SW (measured in core LC24 and noted on the 3.5 kHz profiles) to 11.5 m at LC30 in the NE. This apparently uniform decrease in thickness, however, is modified by the ridge–trough system aligned parallel to the length of the basin adjacent to the Mediterranean Ridge. Within each trough segment between adjacent ridges, megaturbidite n is thickest on the NW side, gradually thins across the trough and then thins dramatically to zero as it rises up towards the crest of the lower ridge to the SE (Fig. 3a). This thickness variation is probably due to flow ponding in the deeper part of the asymmetric basins. The flow thins over the gentler flank of the lower ridge but does not rise up the steeper margin of the higher bounding ridge. This further suggests that the flows identified were less thick than the height of the highest ridge (i.e. <100 m).

Fig. 11. Ribbon diagram showing turbidite correlation across the Herodotus Basin. (Note the greater thickness of recent Nilotic turbidites at the site of LC29, suggesting this formed the main depocentre during the late Quaternary.)

By contrast, the post-n turbidites, predominantly derived from the Nile Cone, show a main depocentre centred on 33°36'N, 28°57'E (core LC29) downslope from a fan distributary channel (Fig. 4). These Type-A turbidites thin to the SW and NE, indicating that they were derived from rather smaller flows than megaturbidite n and that individual turbidity currents spread in two directions on entering the plain (or on reaching the NW margin) so that part is seen to the SW and part to the NE. The Coriolis force was clearly insufficient to constrain flow direction only to the right. Many of these thinner flows then travelled the full length of the basin in either direction, although small-volume turbidites, (a and b) are rather more restricted. The post-n sediments are some 21 m thick in LC29, decreasing to about 4 m in thickness at LC27 some 100 km SW, and to about 6 m in thickness at LC30 which is about 80 km to the NE.

During the latter part of the Pleistocene (since at least 27 ka), and the entire Holocene, Nile Cone turbidites represent the main allochthonous input into the basin. Turbidites derived from the Anatolian Rise region to the east (Type-C) are only seen in LC30 in the NE of the basin. The medium-sized turbidites, k and m, have silty, erosive bases but are not seen to extend along the basin in a SW direction. Turbidites of similar sizes or smaller, derived from other sources, are present in other cores in the basin. This may suggest that the mass-wasting events producing the Type-C turbidites k and m were either oblique on entering the basin or were ponded between the deformation ridges, and hence did not proceed down the length of the basin.

Cita *et al.* (1984a) and Lucchi & Camerlenghi (1993) showed the extent of the turbidites across the width of the Herodotus Basin towards its SW end. Their sample sites formed a transect from the middle of the basin and onto the Mediterranean Ridge accretionary prism (Fig. 1). Although the cores from the BAN-82 cruise only penetrated a maximum of 12 m, they penetrated older sediments than the cores used in the present study (Fig. 12). These cores show that other Type-A (Nile Cone) and Type-B (Libyan–Egyptian shelf) turbidites from below megaturbidite n were present. Older sapropels (S3–S8) were also recovered. Turbidite n is seen to be the major event in these cores, with PC-10 showing similar stratigraphical characteristics to those seen in LC24. The Libyan–Egyptian megaturbidite n correlates with the β turbidite described by Cita *et al.* (1984a) and Lucchi & Camerlenghi (1993). Although it is not possible to determine an exact correlation between the Nile Cone derived turbidites of Cita

et al. (1984a) in Fig. 12 and those from this study, it is evident that the turbidites become much thinner on the transect from the centre of the Herodotus Basin onto the Mediterranean Ridge.

Cita *et al.* (1984a) noted that in two cores (GC-13 and GC-14) the debrites occurred within the S1 Sapropel (Fig. 12). However, in cores taken from farther up the Mediterranean Ridge and farther onto the Herodotus Basin floor, these were absent. These debrites are not thought to originate from the same event as that found in core LC30 at the NE end of the basin. Debrites are generally not as laterally extensive as turbidite deposits and this irregular pattern of deposition could be due to local collapse on the NW flank of the Herodotus Basin from the Mediterranean Accretionary Prism (Fig. 13), or possibly from the smaller tectonically active ridges.

Using the core data from the two studies, together with the 3.5 kHz profiles from this study, it is possible to construct a preliminary isopach map of sediments above the n megabed (Fig. 13). This does not take into account local variations in sediment thickness related to the ridge–trough topography. The main recent depocentre for the Herodotus Basin is seen to be centred around 33°36'N, 28°57'E (LC29). This corresponds to the region of major sediment input from the Nile Cone. This is probably related to the presence of meandering feeder channels providing sediment preferentially to this region as shown by the GLORIA side-scan sonar image in Fig. 4.

Sediment budgets

Summing of pelagic intervals and using an average rate of sedimentation of 2.5–3.0 cm ka^{-1} (E. Rohling, pers. comm. 1996) allows approximate dates for individual turbidite emplacement (Table 3). As the allochthonous beds have been correlated layer-by-layer across the basin using chemostratigraphy, the approximate volumes for each of the larger, laterally extensive turbidites can be calculated (Table 4). Cita *et al.* (1984a) placed a rather conservative figure of just 10 km^3 for the volume of the Libyan–Egyptian megabed n. This study revises this figure upward considerably to *c.* 400 km^3, with an average thickness for the turbidite of 10 m and the dimensions covered by the turbidite as a *c.* 100 km in width and 400 km in length. The average frequency of turbiditic emplacement of all types is calculated as one event every 1.6 ka.

The volumes and frequencies of the Type-A Nile Cone turbidites have also been calculated (Table 4), so that sediment budgets can now be determined for the two main sources of

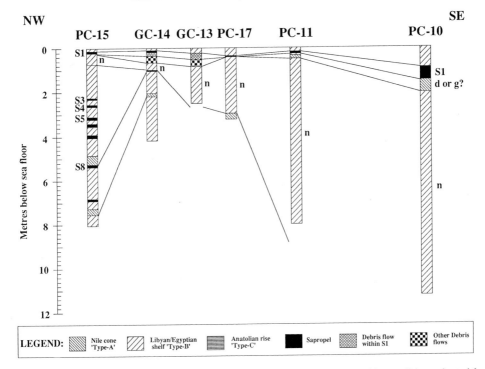

Fig. 12. Schematic ribbon diagram of six of the nine cores recovered by the BAN-82 expedition, adapted from Cita *et al.* (1984*a*) and Lucchi & Camerlenghi (1993). (See Fig. 1 for location of cores.) Turbidite n (this study) correlates with turbidite β of Cita *et al.* (1984*a*).

turbidites. The cumulative volume of the sedimentary input for the Nile Cone source is *c.* 500 km³, giving an average sedimentation rate of *c.* 45 cm ka⁻¹. The volume per unit time is therefore *c.* 18 km³ ka⁻¹. The megaturbidite n, of Libyan–Egyptian shelf derivation, has a volume of *c.* 400 km³. Type-C turbidites, derived from the Anatolian Rise, have a smaller volume of 12 km³, with a sedimentation rate and volume per unit time of 1 cm ka⁻¹ and 0.4 km³ ka⁻¹ respectively.

Sedimentary classification of the late Quaternary turbidite basin fill

Pilkey (1987) discussed the factors that control basin-plain geometry: (a) the arrangement of entry points around the edge of the plain and (b) the ratio of the drainage basin area to the area of basin-plain floor. The Herodotus Basin is dominated by turbidites, including large-scale events, sourced from the Nile Cone, the Libyan/Egyptian carbonate shelf, and the Anatolian Rise (Fig. 13). In Pilkey's classification of basin plains, the entry point configuration for the Herodotus Basin would be Type A, a radial configuration showing entry points from four sources. The

drainage basin area of the Nile River is *c.* 1.9 × 106 km², and the area of Herodotus Basin plain (below 3000 m isobath) is calculated at *c.* 4 × 10⁴ km². This gives a drainage/basin-plain ratio of *c.* 48.

The overall basin fill characteristics for the Herodotus Basin, such as basin area, drainage area, drainage/basin-plain ratio, volume of turbidites, volume per unit time and sedimentation rates, are very similar to those for a number of other basin plains (Pickering *et al.* 1989). For example, Rothwell *et al.* (1992) noted that the Madeira Abyssal Plain was mainly sourced from two compositionally different areas. These are organic-rich turbidites from the African shelf and volcanic-rich from the Canary Islands, with the volcanic source becoming dominant in the basin's later evolution. The Madeira Abyssal Plain is the most well studied of all abyssal plains (e.g. Jones *et al.* 1992; Rothwell *et al.* 1992; Weaver *et al.* 1992, 1995; Masson 1994; Schminke *et al.* 1995) and therefore forms a reference with which others can be compared. The dominant source of basin fill switched in the Madeira Basin, as has that in the Herodotus Basin from a Libyan–Egyptian shelf source to a predominantly Nile Cone source since 27 ka.

Table 4. *Approximate dimensions of the individual turbidites from the Herodotus Basin*

Turbidite	Type	Approximate average thickness (m)	Approximate area* (km²)		Approximate volume (km³)	Minimum age (ka)
Turbidite a	A	0.05	2000	(5)	0.1	6.5
Turbidite b	A	0.2	30,000	(75)	6.0	7.0
Turbidite c	A	2.5	32,000	(80)	80.0	7.5
Debrite		0.6	4000	(10)	2.4	8.0
Turbidite d	A	3.5	36,000	(90)	126.0	9.0
Turbidite e	A	0.7	36,000	(90)	25.2	12.0
Turbidite f	A	0.05	2000	(5)	0.1	13.5
Turbidite g	A	2.0	36,000	(90)	72.0	14.5
Turbidite h	B	0.07	32,000	(80)	2.2	17.0
Turbidite i	B	0.03	28,000	(70)	0.8	21.5
Turbidite j	B	0.03	28,000	(70)	0.8	23.5
Turbidite k	C	0.4	20,000	(50)	8.0	24.0
Turbidite l	C	0.02	2000	(5)	0.04	26.5
Turbidite m	C	0.2	20,000	(50)	4.0	27.5
Turbidite n	B	10.0	40,000	(100)	400	>27.5
Turbidite o	A	5.0?	38,000	(95)	190?	>27.5
Turbidite p	B	0.02?	4000	(10)	0.1?	>27.5

* The area of the basin was calculated as an appropriate proportion of 40,000 km² (percentage shown in parentheses).

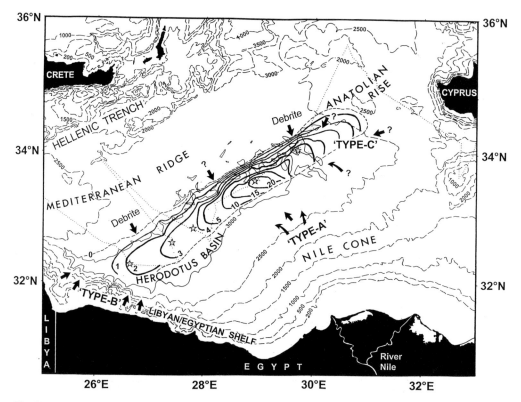

Fig. 13. Isopach map showing the thickness in metres of sediments above the megaturbidite n, the inferred limit of that turbidite, and the location of the late Quaternary depocentre centred on core site LC29. The contour has been drawn using the 3.5 kHz profiles or otherwise estimated from the known bathymetry of the basin.

Table 5. *Comparison of basin characteristics of the Herodotus Basin and the Madeira Abyssal Plain*

Basin characteristics	Madeira Abyssal Plain	Herodotus Basin
Basin area	68,000 km²	40,000 km²
Drainage area	3.36 × 10⁶ km²	1.9 × 10⁶ km²
Drainage/basin area	49*	48
Volume of turbidites – within depositional period studied	600 km³	840 km³
Depositional period studied	300 ka	27.5 ka
Volume/unit time	2 km³ ka⁻¹*	30.5 km³ ka⁻¹
Width of bounding continental shelf or slope	1000 km	150 km
Sedimentation rate throughout basin	9 cm ka⁻¹*	83 cm ka⁻¹

* Cf. Rothwell *et al.* (1992).

The sedimentation and volume rates are higher for the Herodotus Basin because of the high sediment supply from the River Nile and the smaller width of bounding shelf–slope compared with that of the Madeira Abyssal Plain (Table 5).

Conclusions

The Herodotus Basin represents the deepest part of the SE Mediterranean and receives allochthonous sediments from turbidity currents primarily from four sediment sources: (1) dark-coloured, fine-grained turbidites from the Nile Cone to the south and southeast, (2) lighter-coloured, $CaCO_3$-rich, slightly coarser-grained turbidites from the Libyan–Egyptian shelf to the southwest, (3) pale, foraminifer- and $CaCO_3$-rich turbidites from the Anatolian Rise region to the east and northeast, and (4) small localized debris flows from the Mediterranean Ridge to the north.

Approximately 50% of the Herodotus Abyssal Plain is no longer bathymetrically a true plain in the conventional sense as a result of neotectonic deformation against the Mediterranean Ridge accretionary prism. Echo-sounding and 3.5 kHz high-resolution seismic profiles collected during RV *Marion Dufresne* Cruise 81 show that part of the SW proximal plain is probably part of the sandy lower rise of the Nile Cone and that the northern part of the plain has been deformed into a belt of small ridges and troughs, up to 100 m in height above the surrounding sea floor. This region, which is at least 40 km in width, and laterally extensive, is interpreted as the deformation front associated with the Mediterranean Ridge accretionary complex.

A megaturbidite derived from the Libyan–Egyptian shelf has been dated at *c.* 27 ka and has an approximate volume of 400 km³. This does not take into account local variations in sediment thickness related to the ridge–trough topography. The turbidites have been correlated across the Herodotus Basin using major, minor and trace element geochemistry. Each turbidite event has a characteristic geochemical signature allowing excellent chemostratigraphical correlation. Sediment colour, mineralogical and geophysical properties can also be used to correlate the turbidites but the chemostratigraphy provides the most reliable method. This study shows the value of chemostratigraphy as a correlation tool, provided the same size grain fraction is used for analysis.

The main depocentre of the Herodotus Basin for the last 27 ka is centred around 33°36′N, 28°57′E, near the location of core LC29. GLORIA side-scan sonar records show this is related to the presence of meandering channels which provide sediment input to this region from the Nile Cone.

The cumulative volume of the sedimentary input for the Nile Cone is calculated at *c.* 500 km³, giving an average sedimentation rate of *c.* 45 cm ka⁻¹ and a volume per unit time of 18 km³ ka⁻¹. The small Type-C turbidites derived from the Anatolian Rise contribute little to the Herodotus Basin fill, with a volume of *c.* 12 km³ and a sedimentation rate of 1.1 cm ka⁻¹.

This work was supported by the European Union Marine Science and Technology Programme (contract MAS2-CT-93-0051) and M.S.R.'s PhD study is NERC funded (award GT4/95/288/E). S. Nixon, N. Higgs and D. Green are thanked for their help in sample preparation and for running geochemical analyses. We also thank A. Dunkley for drafting some of the diagrams and D. Gunn for the geophysical analysis of the core sections using the SOC multi-sensor core logger.

References

Aksu, A. E., Yasar, D. & Mudie, P. J. 1995. Paleoclimatic and paleoceanographic conditions leading to development of sapropel layer S1 in the Aegean Sea. *Palaeogeography, Palaeoclimatology, Palaeoecology*, **116**, 71–101.

BETHOUX, J.-P. 1993. Mediterranean sapropel formation, dynamic and climatic viewpoints. *Oceanologica Acta*, **16**, 127–133.

BOUMA, A. H. 1962. *Sedimentology of some Flysch Deposits: a Graphic Approach to Facies Interpretation*. Elsevier, Amsterdam, 168 pp.

CITA, M. B., BEGHI, C., CAMERLENGHI, A., KASTENS, K. A., McCOY, F. W., NOSETTO, A., PARISI, E., SCOLARI, F. & TOMADIN, L. 1984a. Turbidites and megaturbidites from the Herodotus Abyssal Plain (Eastern Mediterranean) unrelated to seismic events. *Marine Geology*, **55**, 79–101.

——, CAMERLENGHI, A., KASTENS, K. A. & McCOY, F. W. 1984b. New findings of the Bronze age Homogenites in the Ionian Sea: geodynamic implications for the Mediterranean. *Marine Geology*, **55**, 47–62.

GRANT, J. A. 1986. The Isocon diagram – a simple solution to Gresens' equation of metasomatic alteration. *Economic Geology*, **81**, 1976–1982.

HIGGS, N. C., THOMSON, J., WILSON, T. R. S. & CROUDACE, I. W. 1994. Modification and complete removal of Eastern Mediterranean sapropels by postdepositional oxidation. *Geology*, **22**, 423–426.

HILGEN, F. J. 1991. Astronomical calibration of Gauss to Matuyama sapropels in the Mediterranean and implication for the Geomagnetic Polarity Time Scale. *Earth and Planetary Science Letters*, **104**, 226–244.

JACOBI, R. D. & HAYES, D. E. 1993. Northwest African continental rise: effects of near-bottom processes inferred from high-resolution seismic data. *In*: POAG, C. W. & DE CRACIANSKY, P. C. (eds) *Geologic Evolution of the Atlantic Continental Rises*. Van Nostrand Reinhold, New York, 293–326.

JONES, K. P. N., McCAVE, I. N. & WEAVER, P. P. E. 1992. Textural and dispersal patterns of thick mud turbidites from the Madeira Abyssal Plain. *Marine Geology*, **107**, 149–173.

KAHLER, G. & DOSSI, M. 1996. Micropalaeontology. *In*: ROTHWELL, R. G. (ed.) *R/V Marion Dufresne Cruise 81 – Mediterranean giant piston coring transect*. Cruise Report **40–63**.

KASTENS, K. A. & CITA, M. B. 1981. Tsunami induced sediment transport in the abyssal Mediterranean Sea. *Geological Society of America Bulletin*, **92**, 845–857.

KENYON, N. H., BELDERSON, R. H. & STRIDE, A. H. 1975. Plan views of active faults and other features on the Lower Nile Cone. *Geological Society of America Bulletin*, **86**, 1733–1739.

KIDD, R. B., SIMM, R. W. & SEARLE, R. C. 1985. Sonar acoustic facies and sediment distribution on an area of the deep ocean floor. *Marine and Petroleum Geology*, **2**, 210–221.

LUCCHI, R. & CAMERLENGHI, A. 1993. Upslope turbiditic sedimentation on the southeastern flank of the Mediterranean ridge. *Bollettino di Oceanologia Teorica ed Applicata*, **11**, 3–25.

MASSON, D. G. 1994. Late Quaternary turbidity current pathways to the Madeira Abyssal Plain and some constraints on turbidity current mechanisms. *Basin Research*, **6**, 17–33.

PEARCE, T. J. & JARVIS, I. 1992. Applications of geochemical data to modelling sediment dispersal patterns in distal turbidites: Late Quaternary of the Madeira Abyssal Plain. *Journal of Sedimentary Petrology*, **62**, 1112–1129.

—— & —— 1995. High-resolution chemostratigraphy of Quaternary distal turbidites: a case study of new methods for the analysis and correlation of barren sequences. *In*: DUNAY, R. E. & HAILWOOD, E. A. (eds) *Non-biostratigraphical Methods of Dating and Correlation*. Geological Society, London, Special Publications, **89**, 107–143.

PICKERING, K. T., HISCOTT, R. N. & HEIN, F. J. 1989. *Deep Marine Environments: Clastic Sedimentation and Tectonics*. Unwin Hyman, London, 416 pp.

PILKEY, O. H. 1987. Sedimentology of basin plains. *In*: WEAVER, P. P. E. & THOMSON, J. (eds) *Geology and Geochemistry of Abyssal Plains*. Geological Society, London, Special Publications, **31**, 1–12.

PIPER, D. J. W. & STOW, D. A. V. 1991. Fine-grained turbidites. *In*: EINSELE, G. *et al.* (eds) *Cycles and Events in Stratigraphy*. Springer, Berlin, 360–376.

ROHLING, E. J. 1994. Review and new aspects concerning the formation of Eastern Mediterranean Sapropels. *Marine Geology*, **122**, 1–28.

ROTHWELL, R. G., PEARCE, T. J. & WEAVER, P. P. E. 1992. Late Quaternary evolution of the Madeira Abyssal Plain, Canary Basin, NE Atlantic. *Basin Research*, **4**, 103–131.

SCHMINKE, H.-U., WEAVER, P. P. E., FIRTH, J. V., *et al.* 1995. *Proceedings of the Ocean Drilling Program, Initial Reports*, **157**. Ocean Drilling Program, College Station, TX.

STANLEY, J. D. 1980. Mediterranean sedimentation models: carbonate shelves, slope bypassing, ponding, transformation products and "unifites". *In*: WEZEL, F. C. (ed.) *Sedimentary Basin of Mediterranean Margins*. CNR–Urbino University, Urbino, 47–48.

—— 1981. Unifites: structureless muds of gravity-flow origin in the Mediterranean basins. *Geo-Marine Letters*, **1**, 77–83.

STOW, D. A. V. 1985. Fine-grained sediments in deep water: an overview of processes and facies models. *Geo-Marine Letters*, **5**, 17–23.

—— & PIPER, D. J. W. 1984. Deep-water fine-grained sediments: facies models. *In*: STOW, D. A. V. & PIPER, D. J. W. (eds) *Fine-grained Sediments: Deep-water Processes and Facies*. Geological Society, London, Special Publications, **15**, 611–646.

—— & SHANMUGAM, G. 1980. Sequence of structures in fine-grained turbidites: comparison of recent deep-sea and ancient flysch sediments. *Sedimentary Geology*, **25**, 23–42.

—— & WETZEL, A. 1990. Hemiturbidite: a new type of deep-water sediment. *Proceedings of the Ocean Drilling Program, Scientific Results*, **116** Ocean Drilling Program, College Station, TX, 25–34.

——, READING, H. G. & COLLINSON, J. D. 1996. Deep seas. *In*: READING, H. G. (ed.) *Sedimentary Environments: Processes, Facies and Stratigraphy*, Blackwell, Oxford, 395–453.

STRIDE, A. H., BELDERSON, R. H. & KENYON, N. H. 1977. Evolving miogeoanticlines of the East Mediterranean (Hellenic, Calabrian and Cyprus Outer Ridges). *Philosophical Transactions of the Royal Society of London, Series A*, **284**, 255–285.

THOMSON, J., HIGGS, N. C., WILSON, T. R. S., CROUDACE, I. W., DeLANGE, G. J. & VAN SANTVOORT, P. J. M. 1995. Redistribution and geochemical behaviour of redox-sensitive elements around S1, the most recent Eastern Mediterranean sapropel. *Geochimica et Cosmochimica Acta*, **59**, 3487–3501.

US GEOLOGICAL SURVEY ROCK COLOR CHART COMMITTEE 1991. *Rock Color Chart*. Geological Society of America, Boulder, CO.

VENKATARATHNAM, K. & RYAN, W. B. F. 1971. Dispersal patterns of clay minerals in the sediments of the Eastern Mediterranean Sea. *Marine Geology*, **11**, 261–282.

WEAVER, P. P. E. 1983. An integrated stratigraphy of the Upper Quaternary of the King's Trough flank area, NE Atlantic. *Oceanologica Acta*, **6**, 451–456.

—— & KUIJPERS, A. 1983. Climatic control of turbidite deposition on the Madeira Abyssal Plain. *Nature*, **306**, 360–363.

——, MASSON, D. G., GUNN, D. E., KIDD, R. B., ROTHWELL, R. G. & MADISON, D. A. 1995. Sediment mass wasting in the Canary Basin. *In*: PICKERING *et al.* (eds) *Atlas of Deep Water Environments: Architectural Style in Turbidite Systems*. Chapman & Hall, London, 287–296.

——, ROTHWELL, R. G., EBBING, J., GUNN, D., & HUNTER, P. M. 1992. Correlation, frequency of emplacement and source directions of megaturbidites on the Madeira Abyssal Plain. *Marine Geology*, **109**, 1–20.

WRAY, D. S. & GALE, A. S. 1993. Geochemical correlation of marl bands in Turonian chalks of the Anglo-Paris Basin. *In*: HAILWOOD, E. A. & KIDD, R. B. (eds) *High Resolution Stratigraphy*. Geological Society, London, Special Publications, **70**, 211–226.

Sediment delivery to the Gulf of Alaska: source mechanisms along a glaciated transform margin

M. R. DOBSON[1], D. O'LEARY[2] & M. VEART[1]

[1]*Institute of Earth Studies, University of Wales, Aberystwyth SY23 3DB, UK*
[2]*United States Geological Survey, Denver, CO, USA*

Abstract: Sediment delivery to the Gulf of Alaska occurs via four areally extensive deep-water fans, sourced from grounded tidewater glaciers. During periods of climatic cooling, glaciers cross a narrow shelf and discharge sediment down the continental slope. Because the coastal terrain is dominated by fjords and a narrow, high-relief Pacific watershed, deposition is dominated by channellized point-source fan accumulations, the volumes of which are primarily a function of climate. The sediment distribution is modified by a long-term tectonic translation of the Pacific plate to the north along the transform margin. As a result, the deep-water fans are gradually moved away from the climatically controlled point sources. Sets of abandoned channels record the effect of translation during the Plio-Pleistocene.

The GLORIA surveys carried out by the US Geological Survey in co-operation with the British Institute of Oceanographic Sciences during the 1980s revealed many unexpected features of deep seafloor sediment deposition and sediment delivery paths across continental margins. Among the surprising discoveries was the abundance of rill-like canyons that cross the continental slope. Networks of canyons and gullies coalesce to form great channels that cross the rise and enter the abyssal plain. The extent and intricacy of the canyon networks, their associations with submarine fan genesis, and their relations to terrigenous sediment sources and delivery mechanisms have been a focus of speculation in the last decade. However, most analyses have been confined to the mid-latitude passive margin (trailing edge) setting which includes some of the world's largest river mouths. This paper examines the function of submarine canyons and nearshore glaciers as sediment delivery mechanisms, and their relation to sedimentation along the active tectonic margin of the Gulf of Alaska (Fig. 1), a setting dominated by strike-slip faulting as well as abrupt piedmont glacier advances and retreats.

Glacially sourced sedimentation on the deep ocean floor is well documented for the Polar North Atlantic where the provenance consists of large persistent ice fields, such as the Greenland icecap, and where the plate tectonic environment is dominantly of a passive margin type. Commonly, massive slide complexes characterize these environments (Vorren *et al.* 1989; Vanneste *et al.* 1995). This contrasts with the tectonic and depositional environment of SE Alaska, which is small (160,000 km^2), where glaciers tend to be temperate and wet-based compared with the Polar North Atlantic, and where strike-slip faulting deforms the continental margin.

Two mechanisms strongly influence the mode and volume of sediment delivery to the Gulf of Alaska floor: glaciation, which episodically brings large volumes of sediment very rapidly to narrowly confined shelf-slope depocentres, and an active plate boundary transform (Miall 1984), which tends to distribute point sources laterally. Thus, the hinterland source can be considered an abrupt runoff mechanism that delivers highly localized high-volume charges of unweathered, dominantly fine- to medium-grained clastics directly to the outer shelf and upper slope. Tectonic displacement tends to spread these depocentres laterally with time, and changes in climate vary the volumes of sediment available from place to place.

Numerous studies of glaciomarine sedimentation along the Alaskan coast, particularly in the vicinity of tidewater glaciers (Molnia 1977; Molnia & Carlson 1978; Carlson 1989; and Zellers 1995) provide a firm basis for assessing the distribution of sediment in the Gulf. Bruns & Carlson (1987) determined that sediment is carried downslope in confluent channels that coalesce into major riverine channels which meander for hundreds of kilometres across a nearly flat Gulf floor, ultimately debouching the sediment on the abyssal plain. Our data indicate that the channels may well be single-event features, abandoned after each glaciation and filled in after having been displaced by plate motion from their initial sources. Few studies have dealt

DOBSON, M. R., O'LEARY, D. & VEART, M. 1998. Sediment delivery of the Gulf of Alaska: source mechanisms along a glaciated transform margin. *In*: STOKER, M. S., EVANS, D. & CRAMP, A. (eds) *Geological Processes on Continental Margins: Sedimentation, Mass-Wasting and Stability.* Geological Society, London, Special Publications, **129**, 43–66.

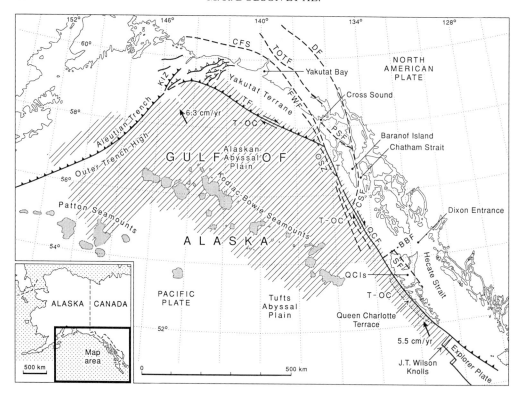

Fig. 1. Generalized location map for the Gulf of Alaska, showing the sourcing zones for the four principal fans and the three exits along the fjordic coast. Also includes the principal tectonic plates and their boundaries. The boundaries between the three plates are shown as single lines where a transform fault occurs, single lines with filled triangles if convergent and double lines for ridges. Open triangles refer to thrust belts. Dashed lines are fault traces and structural zones. Large seamounts are shown as hatched areas; diagonal lines indicate the area of GLORIA coverage. BBF, Beresford Bay Fault; CFS, Contact Fault System; CSF, Chatham Strait Fault; DF, Denali Fault; FWF, Fairweather Fault; KIZ, Kayak Island Zone; OSZ, Outer Structural Zone (this persists to the south as two structural zones); PSF, Peril Strait Fault; QCF, Queen Charlotte Fault; SF, Sandspit Fault; TF, Transition Fault; TOTF, Totschunda Fault; T, transform; T-OC, transform with oblique convergence. The plate boundary transform is affected by differing degrees of oblique convergence (Yorath 1987). Relative motion between the Pacific and American plates bordering the Queen Charlotte Islands (QCIs) (Riddihough & Hyndman 1992) is 5.5 cm/a at 20° west of north, in contrast to the direction of movement on the Queen Charlotte Fault which is 40° west of north. Curvature of the transform boundary fault results in the component of convergence being at a maximum immediately north of the J. Tuzo Wilson Knolls; convergence declines northwards as far as the Yakutat Terrane.

with this phenomenon or with the inferred mechanism of sediment delivery; the Gulf of Alaska therefore may offer insights into depositional architecture for mid-latitude Pleistocene glacial settings and possibly for those preserved in the geological record.

The study is based on new GLORIA side-scan sonar images and supporting seismic reflection data, and addresses three issues relevant to sedimentation along transform and mixed transform–oblique convergent margins located in high latitudes: sediment supply mechanisms; sediment transport mechanisms; the interaction between plate kinematics and sedimentation.

Survey data

Data for the Gulf of Alaska were collected on two cruises of the RV *Farnella* (F6- and F7-89) during summer 1989. About 300,000 km^2 of the continental slope, rise and deep ocean floor was mapped (Fig. 1) along *c.* 18,000 km of track using GLORIA (Geological Long Range Inclined Asdic) side-scan sonar.

Although details of the GLORIA system have been described elsewhere (Somers *et al.* 1978; and Laughton 1981), it is important to point out that the system has a spatial resolution of about 125 m along track and about 50 m across track, yet the brightness of individual pixels is at least partly a function of sea floor roughness on the order of several centimetres (Kidd *et al.* 1987). Furthermore, the GLORIA beam is capable of penetrating at least a few tens of centimetres below the sea floor, so that image brightness may also be related to inhomogeneities within the surficial sediment or the roughness of a shallowly buried surface (Gardner *et al.* 1991). Additionally, GLORIA images provide essentially no information on relief or slopes inclined less than the minimum acoustic incidence angle of a few degrees. This makes interpretation of slope forms (i.e. relief less than about 1:50) particularly difficult, or impossible, without the aid of echo-sounder profiles.

A 3.5 kHz seismic profiler with a calculated resolution of 0.8 m and a 160 in^3 airgun system with a vertical resolution of 25 m were routinely deployed with GLORIA. Compressed airgun records or 'squash' records were also made available to assist with the interpretation of large-scale features. The 3.5 kHz profiles were used to determine local relief, areas of outcrop, and depositional features directly linkable to sonar image brightness. For this study the echo types reported by Pratson & Laine (1989) and O'Leary & Dobson (1992), and the GLORIA sonar facies listed by Kidd *et al.* (1987) were used.

Physiography, tectonic setting and stratigraphy

The Alaskan margin

The Gulf of Alaska covers an area of 700,000 km^2 located in the northeast Pacific Ocean south of the Alaska peninsula and west of the Alaskan–Canadian coastal archipelago (Fig. 1). The Gulf is floored by oceanic crust of Cenozoic age that is covered, in the east, by a thick (>5 km) post-Miocene sedimentary succession (Bruns & Carlson 1987; Plafker 1987; Yorath 1987).

As a tectonic feature, the Gulf floor is part of the Pacific plate, bounded to the north by the Aleutian Trench and to the east by a plate boundary transform (Miall 1984); to the west and southwest the limit is arbitrarily placed at the line of the Patton Seamounts (Fig. 1). The plate boundary transform originates at the J. Tuzo Wilson Knolls spreading ridge (Fig. 1) and extends north, having increasing oblique convergence and an embryonic accretionary prism (the Queen Charlotte Terrace). Near Chatham Strait the transform reverts to a simple strike-slip

structure, the Fairweather–Queen Charlotte fault (QCF), which extends from Chatham Strait as far north as the Yakutat Terrane. The Yakutat Terrane splits strike-slip motion between the Fairweather fault (FWF) and the outboard terrane-bounding Transition fault (TF). The Transition fault is a mixed strike-slip convergent plate boundary which extends northwards to, and merges with, the Aleutian Trench (Bruns & Carlson 1987; Yorath 1987; Plafker 1987).

The landward side of the plate boundary transform is a continental assemblage of five fault-bounded allochthonous terranes that were accreted to the North American plate in the Mesozoic and Cenozoic (Monger & Berg 1984). A sixth terrane, the Yakutat, lies to the west of the Fairweather–Queen Charlotte fault (Fig. 1) and is at present accreting to southern Alaska. The tectonic setting is significant because it causes the juxtapositioning of high mountains with a deep ocean basin. The complex terranes are eroded to form the mountain ranges of southeastern Alaska. The high-relief (>3000 m), rugged sea-coast is cut by a maze of fjords. There are few rivers of any size, with the exception of the Alsek, the Copper and the Stikene, that discharge into the Gulf principally because the continental divide is no more than 200 km from the present coastline. High precipitation nourishes both rivers and large, persistent icefields, which in turn promote and sustain valley glaciers.

Quaternary sequences fill narrow linear basins developed between mixed thrust and strike-slip structural belts associated with plate convergence and which lie across the continental slope and upper rise (Bruns & Carlson 1987). The upper Miocene to Recent Yakataga Formation of the northeastern Gulf consists of interbedded glacimarine and marine clastics and contains a record of glacially influenced sedimentation over the last 2.5 Ma (Zellers 1995).

Interpretation of tectonic and related sedimentological features of the Alaskan transform margin in light of GLORIA data

An interpretation of tectonic margin features and the slope–upper rise depocentres (Figs 1 and 2) together with the airgun and 3.5 kHz echo sounder profiles, and multichannel seismic data collected earlier (Bruns & Carlson 1987), is presented. The margin is divided into four structurally discernible segments; each segment is associated with distinct channellized fan sources. The proximal distribution of sediment and the depositional patterns are also described.

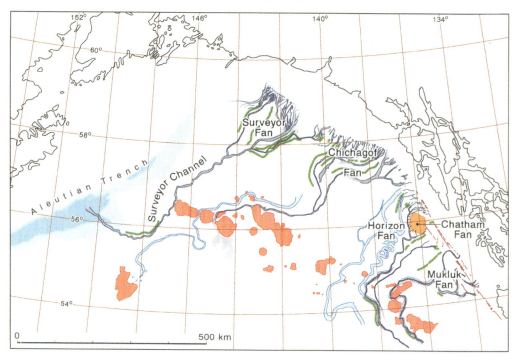

Fig. 2. GLORIA interpretation, showing the distribution of the channel systems in the Gulf of Alaska together with the main seamounts imaged by GLORIA. The named channels are shown in blue, and a light blue colour is used to show the abandoned channels; seamounts are shown in red stipple, green lines represent levee crests; the Queen Charlotte Fault is picked out in dark red, as are prominent structural lineaments. Submarine fan boundaries are not outlined because of depositional overlapping. Chatham Fan is coloured orange.

Queen Charlotte Terrace margin, a mixed transform–convergent margin with incipient accretionary prism

The principal structural element is the Queen Charlotte transform system, which defines the margin of the continental crust (Figs 3 and 4). Along this section, a true continental shelf is absent. Instead, a 30 km wide, slightly back-tilted terrace fronted by a sinuous or cuspate scarp orientated at a small angle to the margin (5–8°) is present at a depth of 1500 m. The terrace is considered to be a tectonic transpression feature similar to the western Aleutians (Dobson *et al.* 1996). It is bounded landward and seaward by faults. Beyond the seaward fault scarp lies a 15 km wide trough or incipient trench, its surface at water depths of between 2500 and 3000 m, and beyond that an outer flexural bulge (Figs 3 and 4; Yorath & Hyndman 1983). Opposite Dixon Entrance (Fig. 3) several transpressional ridges are present. Multichannel seismic surveys have clarified the internal structure of the ridges, showing they consist of bundles of strike-slip faults (Bruns & Carlson 1987).

GLORIA images (Fig. 5) of the J. Tuzo Wilson Knolls indicate that sediment on the landward side of the Queen Charlotte transform fault has been transferred to the sea-floor, beyond the Revere–Dellwood transform fault, along a channel–levee system that shows prominent flank-failure zones. GLORIA images (Fig. 6) west of Queen Charlotte Island show that sediment on the terrace appears to have slumped both landward into the terrace back depression, and seaward towards the Queen Charlotte trough. Locally, debris flows overwhelm the scarp and form small fans that spread into the trough.

Queen Charlotte-Fairweather transform margin with convergence

This section extends from Queen Charlotte Island across Dixon Entrance north for >200 km towards Chatham Strait (Fig. 1). Near Chatham Strait the shelf widens from about 4 km to 50 km, and the slope is fronted by a broad terrace which is margined by tectonic ridges seen as part of transpressive plate activity (Bruns & Carlson

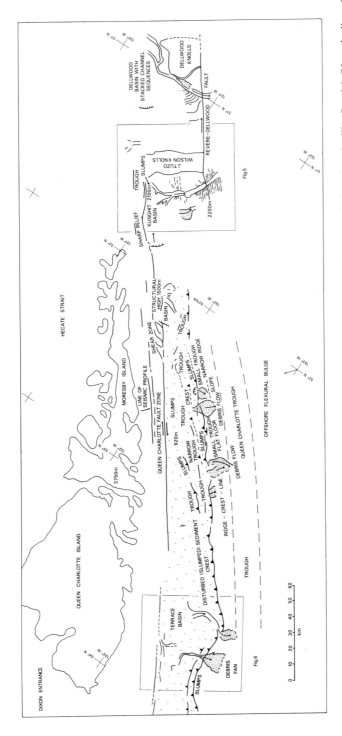

Fig. 3. Interpretation of the GLORIA mosaic for the margin between J. Tuzo Wilson Knolls and Dixon Entrance, with locations for Figs 5 and 6. (Note the line of the seismic profile, Fig. 4.) Continuous lines are transforms, lines with triangles are thrusts. Thrusts usually define the outer margin of the terrace or incipient accretionary prism, which is dotted. Prominent dashed lines refer to the margins of the Queen Charlotte trough. Finely dotted areas are debris fans.

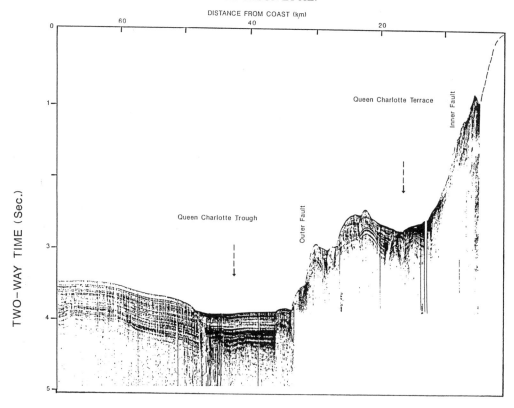

Fig. 4. Seismic profile located in Fig. 3 and showing the main structural elements. Inner fault refers to the Queen Charlotte Fault, outer fault is the thrust fault along the outer margin of the terrace. Modified from Yorath & Hyndman (1983).

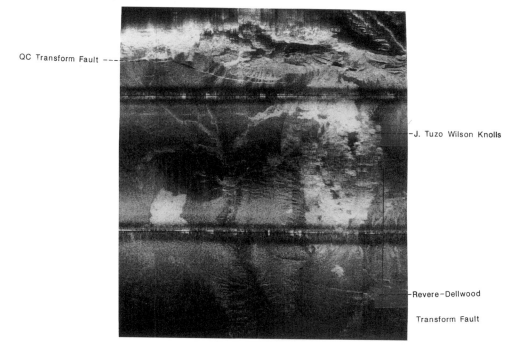

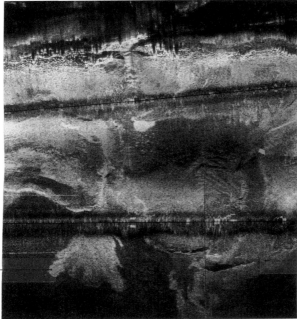

Debris Fan

Terrace Margin Thrust

Fig. 6. GLORIA mosaic showing part of the Queen Charlotte terrace, the cuspate thrust faulted margin and debris fans extending into the trough.

1987). Mukluk fan, which dominates the section (Figs 2 and 7), is the most southerly of the principal fan complexes. It largely fills the depression between the continental slope and the southern end of the Kodiak–Bowie seamounts. The fan is traversed by four major channels labelled from south to north as channels N, M, L, and Swan Neck channel (Fig. 7).

The most southerly Mukluk channel, channel N, is formed from the coalescence of as many as ten chutes, or small upper slope canyons, that range in length from 2 to 10 km. The chutes are located high on the terrace, but below the transform fault, which at this point is coincident with the shelf break. Channel M results from the coalescing of several smaller channels, the most prominent of which originates as an incised canyon on the continental shelf.

At 54°13′N, 134°56′W channels M and N join (Fig. 7). A prominent flexural bulge appears to influence the direction taken by the combined channel up to the confluence with Swan Neck and beyond (Fig. 7). Channel L is a minor feature that extends across the terrace between two ridges before debouching into Mukluk

Basin (Fig. 7). GLORIA images reveal that between channel L and Swan Neck channel 11 relatively recent debris flows extend from the terrace margin for an average of 7 km downslope before spreading out and creating a bajada-type debris-apron (Fig. 7). Beyond lies the sinuous Swan Neck channel.

Swan Neck channel receives its name from the distinctive 'swan neck' bend it makes as it passes the shelf break and encounters the trace of the QFC (Fig. 8). The channel follows the QFC northwestward for 30 km before breaking free of its constraining influence and arcing seaward and downslope in a broad counterclockwise course toward Mukluk Basin (Fig. 7). Channel erosion has created steep banks on the outsides of bends of the Swan Neck channel, which implies that any material that has slumped into the channel has been reworked.

Airgun profiles (Fig. 9) reveal details of the internal architecture of the channel–levee system. Channel deposits are recognized in the profiles as a series of stacked, closely spaced, sub-parallel high-amplitude reflections or HAR (Weimer 1993). By contrast, levee or overbank

Fig. 5. GLORIA mosaic showing the J. Tuzo Wilson Knolls as a zone of strong reflections bounded by narrow bright lines interpreted to be the traces of the two transforms. An extensive channel system, with broad failure zones, is imaged along the northern flank of the Knolls.

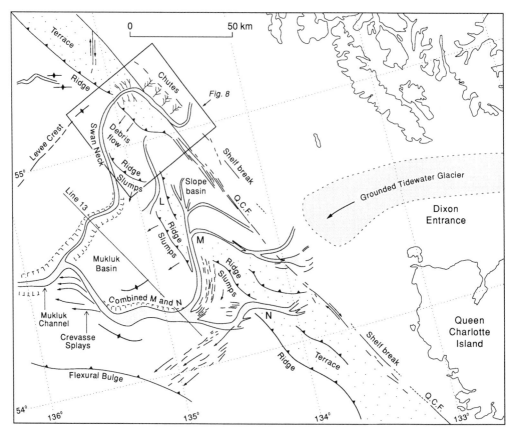

Fig. 7. Interpretation of seabed features imaged by GLORIA together with structures seen on the seismic profiles in the vicinity of Dixon Entrance, showing the proximal Mukluk channel system and a terrace with thrust faulted margin. The terrace is dotted, thrust faults are shown as lines with triangles (this symbol is also used to show the convergent Flexural Bulge). Multichannel seismic profiles indicate the thrusts have a marked strike-slip component (Bruns & Carlson 1987). Where a slope basin extends across the terrace the dot symbol has not been applied. The major channels, from south to north, are N, M, and combined M and N, L and Swan Neck. Channels are highlighted with a close-spaced dot symbol, arrows refer to crevassing, the L-shaped symbol along the channels indicates zones of failure. Levee crests are shown as lines with filled diamonds. QCF, Queen Charlotte Fault. Inferred possible westward extension of a tidewater glacier during periods of glacial maxima is shown as area of light stipple bounded by a dashed line.

deposits are shown predominantly as low-amplitude, laterally persistent, parallel to wedge-shaped reflections (LAR). Levee sequences commonly are indicated by vertical variations in signal amplitude that may be repeated several times through a sediment thickness of 100 m. In the absence of cores, the physical cause of these variations cannot be determined, but may be due to textural contrasts and/or bedding thickness. Levee sequences and their associated channels may be traced westwards through the seamount chain and beyond (Figs 2 and 10). The squash record for profile line 13 (Fig. 11) shows that Swan Neck channel is narrow and relatively deep, and has small levees, the southern one of which is

being overwhelmed by the levee of the combined M and N channels.

Profile line 13 (Fig. 9), reveals a broad shallow channel with 50 m of vertically stacked HAR that occupies the site of an earlier channel. Irregular hummocky reflections are probably related to slumping. A prominent levee, with a well-defined crestal zone, is developed on the northern side of the channel. The axis of the levee crest has migrated north as the channel has widened, and this, together with slumping, has created an angular unconformity with the older levee sediments.

Variations in backscatter levels, recorded by GLORIA along the floors of the channels, offer insight into possible channel-fill sedimentation.

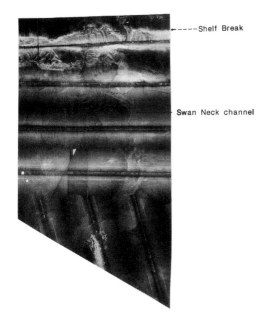

----Shelf Break

Swan Neck channel

Fig. 8. GLORIA mosaic showing the course of Swan Neck channel from the shelf to the lower rise. The channel exploits the zone of weakness along the transform fault. Here the ship's track coincides with the trace of the transform, making interpretation of the GLORIA mosaic difficult. (For position and relative scale see Fig. 7.)

Assuming that high backscatter is due to the presence of coarse, sand-size, material, then the Mukluk channels are floored with sand. Much of the floor of the combined M and N channels features 500 m long patches and 2–3 km long ribbons of high-backscatter material, presumably sand or gravel. These features are especially common immediately above and for 10 km below the confluence with the Swan Neck channel, where overbanking has occurred. Channel floor features of this type persist down channel although the signal strength declines.

TIME

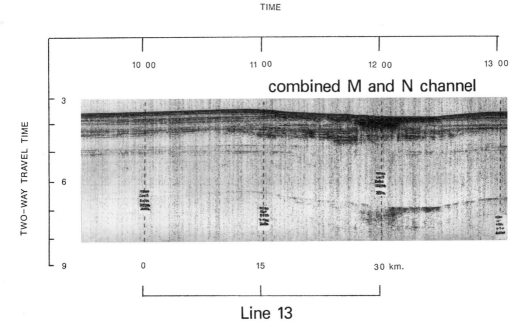

Line 13

Fig. 9. Airgun record of the combined M and N channel, Line 13 (see Fig. 7 for location). The internal reflection architecture includes HAR, centred on the channel axis, and bundles of strong and weak reflections. Based on evidence from DSDP Site 178 (Kulm *et al.* 1973), weak reflections are associated with diatom-rich pelagic and hemipelagic mud beds. The crestline of the more prominent northern levee occurs at Fix 172/1100. (Note the position of the crest has moved south as the combined channel aggrades and reduces the directional control exercised by the Flexural Bulge.)

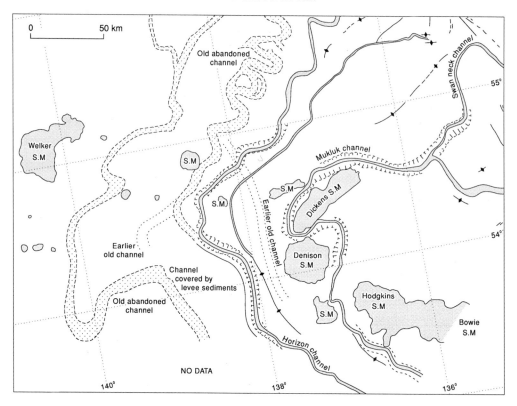

Fig. 10. Map of seabed features imaged by GLORIA together with structures seen on the seismic profiles to the west of Fig. 7 (see Fig. 2 for general location). The figure shows the distal channel systems of Mukluk and Horizon together with old abandoned earlier channels. Channels are highlighted with a close-spaced dot symbol, arrows refer to crevassing, the L-shaped symbol along the channels indicates zones of failure. Levee crests are shown as lines with filled diamonds. A close-spaced dot symbol is also used for seamounts; depending on size, they are named or indicated by SM (seamount).

Queen Charlotte–Fairweather transform margin with localized transpression

Between Dixon Entrance and Chatham Strait (Fig. 1) the margin is tectonically a transform; here minor amounts of convergence promote the development of upper slope terraces. To the north, between Chatham Strait and Cross Sound (Fig. 1), a distance of 120 km, the margin is a simple transform fault with no evidence for oblique convergence.

Chatham Strait represents an important break in the coastal archipelago where glaciers sourced

from the adjacent mainland mountains can extend out to the continental shelf and thence shed sediment on the slope. A major result of Chatham Strait glaciation is Chatham Sand Fan (Figs 12 and 13). This structure covers an area of 7500 km².

GLORIA images suggest the arcuate fan structure consists of three superimposed overlapping systems, each supplied from a separate feeder channel. The two older channels (Fig. 12) lie north of the youngest channel at distances of 7.5 km and 15 km. The youngest channel is 2 km wide. It extends 20 km beyond the shelf break,

Fig. 12. Chatham Sand Fan, the proximal part of the Horizon fan (see also Fig. 13, the GLORIA mosaic from which this diagram is partly derived); other evidence comes from seismic profiles. Channels are highlighted with a close-spaced dot symbol, levee crests are shown as lines with filled diamonds except for single ridge on the terrace. Continuous lines are transforms, lines with triangles are thrusts. Thrusts, with a pronounced strike-slip component, define the outer margin of the terrace or incipient accretionary prism, which is dotted. Lines 957 and 955 refer to multichannel seismic profiles used to assess the structural character of the terrace (Bruns & Carlson 1987). Earlier channel systems are indicated by 1 and 2.

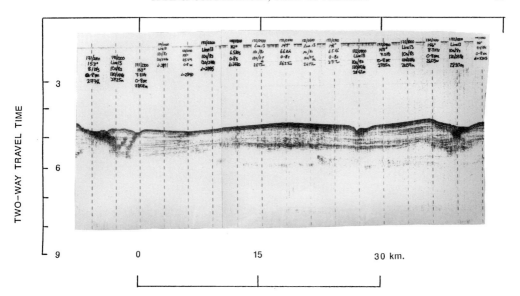

Fig. 11. Compressed or 'squash' record of Line 13 (0–8 s two-way travel time) (see Fig. 7 for location and Fig. 12 for south end of the profile). Here the N and M combined channel lies to the south and Swan Neck channel to the north. There are strong continuous high-frequency reflections close to the upper surface and high-amplitude reflections (HAR) associated with the combined channel. The northern end extends across the Chatham Sand Fan, imaging both Channel B and Channel A (see Fig. 12 for location).

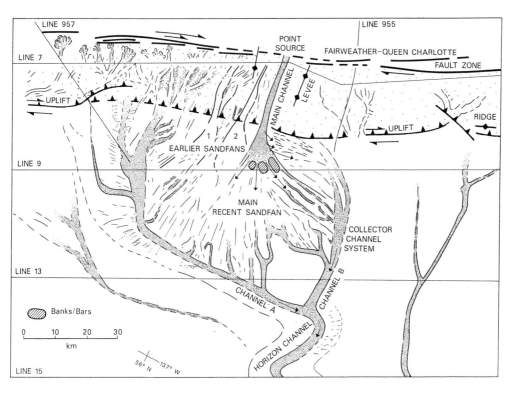

Line of transform fault

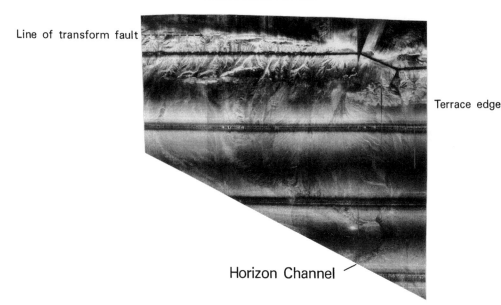

Terrace edge

Horizon Channel

Fig. 13. GLORIA mosaic of Chatham Sand Fan. The tendrils of sediment flowing off the fan are clearly imaged. (For relative scale, compare this mosaic with Fig. 12.)

defined by the Fairweather–Queen Charlotte Fault, and thereupon it abruptly broadens. Channel broadening coincides with the western limit of a terrace, recognized on multichannel seismic profiles by Bruns & Carlson (1987), who defined it as the Outer Structural Zone (Fig. 1). The terrace is clearly imaged by GLORIA (Fig. 12). The terrace represents a significant decrease

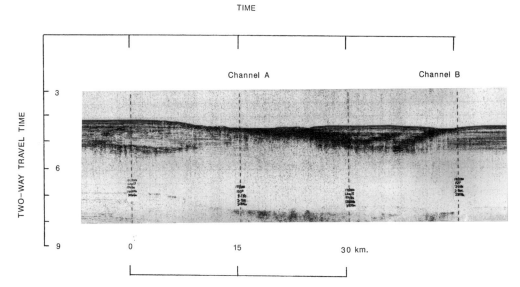

Fig. 14. Seismic profile of line 13, Channels A and B. (For location, see Fig. 12). There are strong continuous high-frequency reflections associated with the levees and high-amplitude reflections (HAR) linked to both Channel A and Channel B. Channel A appears to have migrated north. Channel B has migrated south, with evidence for an earlier channel that was abandoned and subsequently covered by the levee system located between the two channels. (See also the northern section of the squash record, Fig. 10).

TIME

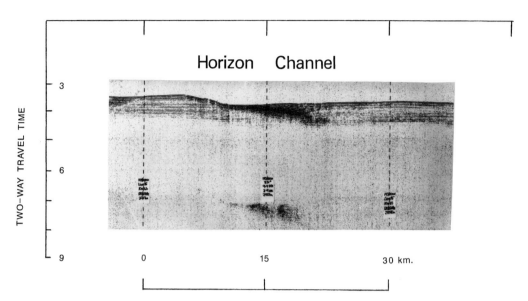

Fig. 15. Seismic profile of Line 15, Horizon channel (for location see Fig. 12). The channel is characterized by HAR; it has also migrated north and built up a large levee on the northern bank. Channel migration and associated flank failure is reflected in the internal architecture of the levee. Bundles of high-level reflections are separated by intervals of low refectivity.

in gradient for the youngest channel, which above the point of broadening has an overall thalweg gradient of c. 5°. The average fan surface slope appears to be c. 1°.

The surface of the most recently active sector of the fan structure is covered by a braid pattern that extends to two collector channels designated A and B (Fig. 12). A and B are directed toward each other around the distal margin of the Chatham Fan, then merge into a single channel termed the Horizon channel (Figs 10 and 12). GLORIA images reveal that on the basis of backscatter values both collector channels are floored with sand-size sediment; indeed, they are fed from an extensive series of high-backscatter tendrils emanating from the Chatham Fan (Fig. 13). Below the confluence, channel-wide elongate patches, 2–3 km long occur around bends. On the abyssal plain, where the slope gradient is very low, bright patches on the channel floor appear to be confined to the insides of bends.

Airgun profiles reveal that the two collector channels have migrated, one north and the other south, probably in response to incremental expansion of the arcuate Chatham Fan. Profile line 13 (Fig. 14) shows that Channel B has HAR and a history of persistent southward migration. The profile shows evidence of an earlier buried

channel sequence. Channel A on line 13 is wide and shallow, and the HAR indicate migration to the north. Flank failure is extensive, but sub-parallel reflectors extend beyond the crest and onlap an older channel–levee system.

Profile line 15 provides a section across Horizon channel (Fig. 15). The HAR are prominent and show that through time the channel has migrated northward. The large levee on the northern side of the channel has an extensive failure zone marked by listric faults; reflectors associated with this levee onlap the channel and levees of two older abandoned systems (Fig. 2). Both these old systems include HAR patterns and levees.

The Chichagof Fan system is developed opposite Cross Sound (Fig. 16). Two channels are involved, a more southerly one, Channel G, and a northern one, Channel H. Channel G is broad (6 km) and deep (80 m); it emanates on the upper slope across a 50 km wide sector consisting of steep-sided coalescent chutes that extend directly down a gradient of 1:40. The single channel arcs westward around a 5 km wide failure zone then swings northwest around the nose of a prominent bathymetric high (buried ridges, Fig. 17), then arcs to the southeast before flowing toward the seamount chain (Fig. 2). Channel H is about 3 km wide and 60 m deep; it

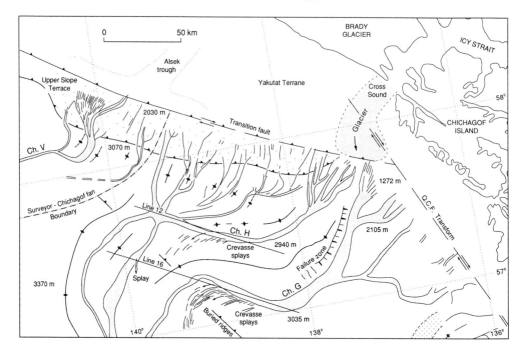

Fig. 16. Proximal section of the Chichagof Fan based on GLORIA and seismic profiles. Channels are highlighted with a close-spaced dot symbol, levee crests are shown as lines with filled diamonds. Continuous lines are transforms, lines with triangles are thrusts. Thrusts, with a pronounced strike-slip component, define the outer margin of the terrace that fronts the Yakutat terrane. QCF, Queen Charlotte Fault. Inferred possible westward extension of a tidewater glacier during periods of glacial maxima is shown as an area of light stipple bounded by a dashed line.

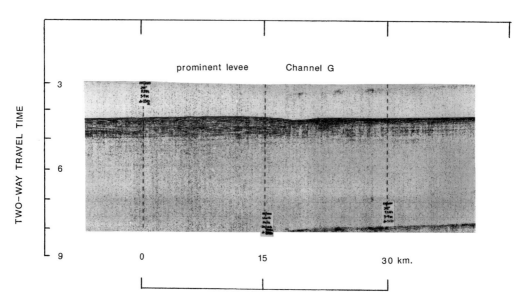

Fig. 17. Seismic profile Line 16 runs partly along the axis of Channel G (for location, see Fig. 16). The profile shows the relationship between the HAR of the channel floor and the levee on the northern flank. This levee displays strong reflections, which may be caused by a high sand-size component in the sediment.

follows a course broadly parallel to G as it flows across the lower slope and around the bathymetric high (Fig. 16). Channel H has a restricted source sector (25 km long) that comprises small confluent chutes or channels. These chutes are sourced from the southern end of the Yakutat Terrane, which represents a mixed type margin with incipient convergence, bounded by the Transition Fault (Fig. 16). The sea-floor expression of this convergence is an upper slope terrace that widens northward (see next section) and that is susceptible to mass movement and debris chute formation, features that combine to form the source zone of channel H.

Seismic profile line 12 crosses Channel H (Fig. 18); the HAR show marked channel migration to the south, probably as a result of aggradation that would reduce the base level control on flow direction. In deeper water beyond the bathymetric high, the larger Channel G joins Channel H to form Chichagof Channel, which flows westward toward the abyssal plain at a gradient of 1 m in 20 km.

The Yakutat Terrane margin: a mixed transform–convergent type

The Yakutat Terrane is fronted by a well-developed upper slope terrace formed by oblique (right-lateral) subduction; it extends from immediately north of Cross Sound to the Aleutian Trench (Fig. 1), a distance of c. 750 km between the Fairweather Fault and the Aleutian Trench. Surveyor Fan is the most northerly of all the fans imaged by GLORIA; it spans >600 km, from the continental slope to the Aleutian Trench (Fig. 2). The fan is headed by a 186 km long gully–ridge complex developed across the terrace and apron of the upper continental slope.

Most of the Yakutat slope is dominated by an apron from which extends a complex pattern of chutes and gullies that supply sediment to Surveyor Fan (Figs 2 and 19). These gully systems take two forms: fan-shaped groups of coalescent ravines within amphitheatres generated by slope failure, and a sub-parallel type oriented directly downslope (Fig. 20). High-resolution seismic profiles reveal that the principal sediment transport style in these gullies is by debris flow. The gullies are linear, uniformly straight sided, coalesce below 3000 m water depth and extend over a downslope distance of 20 km with an average gradient of 1:10. The gullies feed into three prominent channels identified, from south to north, as V, U, and X (Fig. 19). The three channels extend from about 3500 m to 3800 m water depth, at which depth they coalesce to form a single major channel, Surveyor Channel

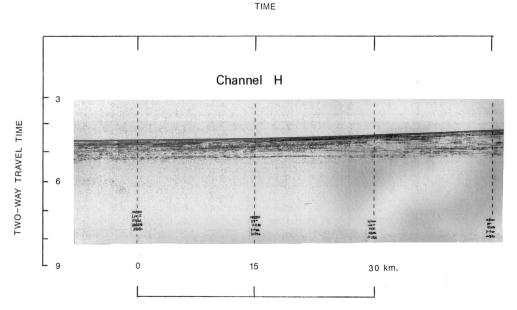

TIME

TWO–WAY TRAVEL TIME

Channel H

Fig. 18. Seismic profile Line 12 also extends along the course of Channel H (for location see Fig. 16). This channel–levee system is on a smaller scale compared with Channel G and is strongly influenced by the northern levee of G.

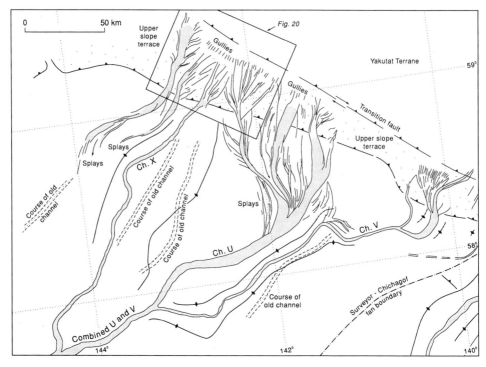

Fig. 19. Diagram of the proximal section of the Surveyor Fan. Channels are highlighted with a close-spaced dot symbol, levee crests are shown as lines with filled diamonds. Continuous lines are transforms, lines with triangles are thrusts. Thrusts, with a pronounced strike-slip component, define the outer margin of the terrace along the front of the Transition Fault. During periods of glacial maxima Yakutat Terrane and part of the terrace would be covered by broad-fronted piedmont glaciers.

(Fig. 2), that persists to the Aleutian Trench. Channel V is narrow (<2 km wide), steep sided and deep (<80 m) with small, but sharply defined levees. It is sourced from the Alsek Trough (Fig. 16), which extends from the Alsek river (Fig. 1). Channel U originates from a 70 km wide zone of chutes, gullies and intervening ridges on the upper continental shelf, sourced largely from Yakutat Bay (Fig. 1). Channel U is about 10 km wide and ranges from 60 to 90 m deep. Channel depth–width ratios for all the three conduits are broadly constant except that Channel U narrows and deepens before joining Channels V and X. Surveyor Channel is considered to extend from the confluence of Channel X with a combined U and V channel to the Aleutian Trench where the channel depth exceeds 600 m (Fig. 21).

High backscatter levels are a prominent feature of all the channels in the Surveyor system. Backscatter levels are very high above the confluence; below the confluence, signals from the channel floor remain strong, suggesting that sand-size material has been entrained as far as the trench. This view is supported by core data (Ness & Kulm 1973). Sonar images of Surveyor

Channel appear sharp and clear in contrast to the southern channels, suggesting that it has remained active very much longer, perhaps up to the present. This judgement is supported by the fact that several glaciers extend to the coast and regularly discharge material across the adjacent shelf opposite the heads of Channels X, U, and V. Moreover, there is no evidence of levee or bank failure along the course of this channel.

Direct evidence for sediment texture and depositional style associated with the channels also comes from DSDP Sites 178 and 179 (Kulm *et al.* 1973). Site 178 is located 28 km north of the Surveyor Channel beyond the low rounded levee crest where the sea floor slopes gently to the Aleutian Trench. Here the oceanic crust is covered with 777 m of sediment that consists of turbidites interbedded with hemipelagic and pelagic layers; the upper 96 m consists of silty clays and ice-rafted erractics. Site 179, located on a small saddle on a seamount, lies 50 km south of the Surveyor Channel. Despite the 200 m elevation above the abyssal plain, turbidites, composed of silty clay, graded silt layers and sands rich in microfossils, dominate.

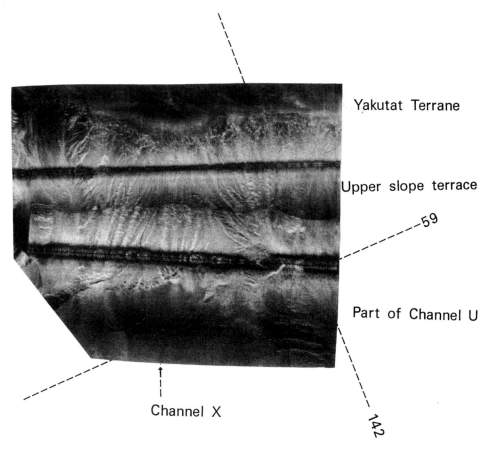

Yakutat Terrane

Upper slope terrace

59

Part of Channel U

Channel X

142

Fig. 20. GLORIA mosaic of the gully zone that extends across the terrace (for location and relative scale, see Fig. 19). The irregular margin of the upper slope terrace coincides with the termination of the gully zone. A broad, open, shallow channel extends down the NW side of the mosaic.

Discussion

Supply mechanisms: sourcing from glaciers

The supply of terrigenous sediment to the Gulf of Alaska is dependent on glaciers. The discharge of large volumes of sediment by Alaskan glaciers has been fuelled by rapid tectonic uplift over the last 5 Ma, thus this region supports the most active erosion system in the world. Yet, in a global context, the drainage area (160,000 km²), which includes the coastal ranges, is modest, as the watershed divide lies <200 km from the coast; it is small in comparison with sources such as the Greenland icecap (Dowdeswell *et al.* 1996). Assuming the overall time period involved (2.5 Ma) for the repeated growth and decay of Late Cenozoic icesheets in Alaska is comparable with that of Greenland, the calculated volume of sediment flooring the Gulf approaches under-nourished levels compared with the Polar North

Atlantic (Dowdeswell *et al.* 1996). The volume of sediment involved is also modest in comparison with river-sourced fans such as the Indus, Bengal, Amazon and Mississippi (Damuth *et al.* 1983; Damuth & Flood 1985; Twitchell *et al.* 1991).

Although the total sediment volume delivered to the Gulf from glacial sources is relatively small, the rate of sediment delivery is high and the kind of sediment delivered is strongly dependent on the glacial supply mechanism. A small drainage area affects the volume and velocity of ice flow and hence the scale of sediment delivery to the continental margin.

The rate of discharge of sediment from glaciers is related to the thermal regime of the parent ice mass (Dowdeswell & Murray 1990), which effectively controls the amount of debris entrained in the basal layers. Observations on glaciers in Alaska (Powell 1984, 1990; Powell & Molnia 1989; Syvitshi 1989) confirm that they are sediment rich and that vigorous melting

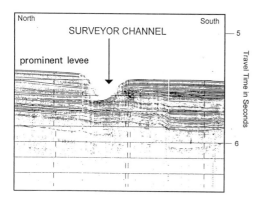

Fig. 21. Seismic profile across the Surveyor Channel located (see Fig. 2) close to the western limit of the levee crest, *c.* 150 km SE of the Aleutian Trench. Modified from Stevenson & Embley (1987).

produces high sediment discharge rates. It is also known that modern Alaskan glaciers are extremely dynamic, temperate types noted for their propensity to surge (Hambrey 1994). When temperate glaciers surge, they accelerate their flow velocity and become deeply crevassed; more importantly, they rapidly release large volumes of meltwater and sediment. Evidence from Alaskan fjord studies also supports the notion that temperate glaciers have high sediment discharge rates (Powell & Molnia 1989) and further indicates that released sediment plumes, invariably, are dense enough to flow along the fjord bottom.

Two independent factors appear to exert a disproportionate affect on the supply of sediment by glaciers to the shelf edge: the presence of structurally determined fjords and a narrow (<50 km) to absent continental shelf. In this coastal setting valley glaciers, effectively, will be channelled to the shelf and, therefore, be able to extend their grounding line to the shelf edge. Where the continental shelf is wide (>50 km) and a fjordic coastline has not developed, broad, unconfined, piedmont glaciers will occur.

South of the Alsek river (Fig. 1), faulted terrane boundaries provide the ideal conditions for a fjordic coastline, and because of the structural grain there are only three 30 km wide breaks or discharge points along the entire sector; from south to north these are Dixon Entrance (which lies on the USA–Canadian border), Chatham Strait and Cross Sound. As the routes taken by valley glaciers are along structurally controlled narrow fiords, it is expected that they would, during episodes of glacial maxima and lowered sea level, extend beyond the confines of the fjords to the shelf

edge (Carlson *et al.* 1982), thus allowing discharge of contained sediment directly to the upper continental slope (Figs 7, 16, 22a and 22b). It is through such a mechanism (i.e. point sourcing) that vast quantities of sediment may be delivered direct to the slope along very narrow fronts. Large prograding arcuate fans, known as trough-mouth fans, are commonly associated with this form of point sourcing (Hambrey 1994). They are located beyond the shelf break and the adjacent shelf shows evidence of overdeepening.

North of the Alsek river, the width of the continental shelf increases from *c.* 50 km to >150 km, and a fjordic coast is absent. Although tidewater glaciers extend almost to the coast today, during glacial maxima ice would have extended out across the shelf along a 400 km wide front. Supply of sediment to the shelf break would be principally a function of shelf width as rates decline exponentially with distance from the ice front (Boulton 1990). Indeed, although large volumes of discharged sediment have been trapped in the basins of the Yakutat Terrane (Zeller 1995), diamict aprons are present along the Transition Fault.

Supply mechanisms also play an important role in determining sediment texture. Rapid subglacial sediment delivery from a tectonically active source area characterized by high relief and mechanical weathering results in thick diamictites that are commonly supplemented by supra-glacial debris originally derived from rockfalls initiated by earthquakes. In the section of the coast from Dixon Entrance north as far as Cross Sound, glacial sediments will be reworked repeatedly during ice advances down the extensive fjord network, eventually to be transported to the shelf break as fine-grained rock flour beneath temperate ice streams (Alley *et al.* 1989). North of Cross Sound temperate tidewater glaciers would extend across the wide Yakutat piedmont and shelf, and carry sediment as a deforming basal till layer. Here, the sediment will be reworked by the repeated growth and retreat of ice streams across the shelf; it eventually reaches the shelf edge having been transported successively by glaciers, melt streams and marine currents.

Sediment transport mechanisms: beyond the shelf break

There appear to be two fundamentally distinct sediment transport mechanisms on the slope and rise: dispersion by gravity downslope from glacial point sources, and a confluent,

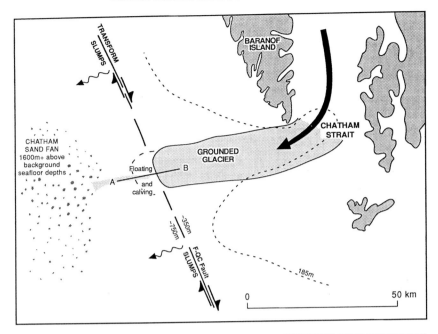

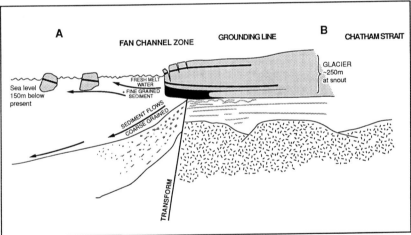

Fig. 22. (**a**) Hypothetical model of Chatham Strait and the probable westward extent of a tidewater glacier assuming a sea-level drop of 130 m. This pattern may have obtained during the last glacial maxima 20 ka ago when sea level was at least 100 m below present-day level. F-QC, Fairweather–Queen Charlotte Fault; dark stipple refers to the land, lighter stipple is the glacier. (**b**) Hypothetical cross-section through a grounded tidewater glacier at the entrance to Chatham Strait. This shows a possible depositional model for the generation of sediment flows, especially density flows emanating from the base of the grounded glacier.

channellizing flow mechanism capable of extremely far-reaching delivery across the rise. A third, tectonic mechanism, plate motion, has an important role to play with respect to point sourcing. Point sourcing promotes well-developed, single-channelled fans. Over time, plate motion offsets these fans from their fixed source on the shelf; therefore, the offset fans are abandoned but the result is a laterally extended composite fan having a number of channels of differing ages. Laterally offset point-sourced fans are different from slope aprons, which tend to maintain several small channels concurrently, and where abandonment is more a function of compensation cyclicity. Several abandoned feeder channels appear to confirm the influence exerted by plate motion where narrow point sourcing occurs.

Where tidewater glaciers discharge at the shelf break, several distinct forms of sediment transport and associated deposition will be formed. These range from channellized transport from the shelf to the abyssal plain as imaged at Dixon Entrance, channellized transport leading to an arcuate fan as recognized opposite Chatham Strait, and the formation of a mixed channel and slope apron pattern, as opposite Cross Sound.

Direct outflow of sediment-charged meltwater from a glacier would be at least homopycnal (equal density) to even hyperpycnal (higher density) compared with seawater (Bates 1953), despite the potentially large volumes of freshwater released by the glacier, which would have the effect of diluting the salinity of the adjacent seawater, because of a combination of low-temperature meltwater and the enhanced density provided by the contained sediment. Density currents probably emanate directly from glaciers, so this mechanism may operate with the point-sourced fans.

At the entrance to Chatham Strait, sediment appears to have been released directly from the snout of a tidewater glacier, at the shelf break, comparable with the mechanism proposed by Powell (1990), and then transported along a short feeder channel to the continental slope where deposition occurred in the form of an arcuate fan. GLORIA reflectivity levels suggest that the fan is composed of sand.

Where piedmont glaciers emanate from coastal icefields, the shelf break is dominated by laterally extensive but unstable slope aprons. Long-term strength degradation, seen as a function of creep, promotes failure of these deposits despite induration, dewatering and varying degrees of cementation. Failure may be triggered in a variety of ways including wave-induced depositional loading (Schwab & Lee 1988), but this is not thought to be a primary cause. Prime candidates for triggering include earthquake shocks and water piping along permeable layers susceptible to grain shearing and hydrostatic overpressuring (O'Leary & Dobson 1992; Orange & Breen 1992). The consequence of these processes is the downslope release of sediment along a broad front on a timescale unrelated to glacial events. Downslope movement ranges from detachment of a coherent mass of material to complete disaggregation and the generation of debris flows. Slides and slumps are included within this spectrum of mass movement but most mass movement seems to enhance gully erosion.

Although the literature refers to the translation of debris flows to turbidity currents as a common downslope development (Stow 1986), this is by no means always the case, as mass sediment failure on passive continental margins testify (O'Leary & Dobson, 1992). It must be assumed, therefore, that the transition from one style of transport to another is determined by factors that include grain-size range and water content of the original deposit. Slope is significant, but perhaps the single most important requirement is the early formation of a narrow chute or conduit that serves to ensure initial flow is contained and prevented from losing energy through being dissipated, with concomitant loss of carrying capacity. Discharge of sediment along a channel that originates on the inner shelf, extends cross the outer shelf, incises the shelf break and continues downslope towards the lower rise, is unique to the fjordic margin.

Interaction between plate kinematics and sedimentation

Plate motion is an important factor where sediment is discharged repeatedly to short sections of the shelf break. As the Pacific plate is moving towards the Aleutian Trench at about 6 cm/a or 60 km/Ma, point source supply sites, such as Chatham Strait, which is 30 km wide, will receive large volumes of sediment intermittently for 500 ka, to be followed by effective abandonment. This suggests there has been several channel avulsion events that may have been tectonically induced. Assuming a uniform motion for the Pacific plate, relative to the American plate, of 6 cm/a, this translates into sediment supply episodes 125 ka and 250 ka years ago. This pattern of sediment nourishment followed by sediment starvation contrasts with a similar length of shelf edge where a wide shelf obtains, such as along the western flank of the Yakutat Terrane. At these sites, poorly nourished slope aprons are likely to form; they would receive sediment intermittently over a longer period, but with a reduced prospect of abandonment.

A simple staged evolution for sediment delivery to the Gulf of Alaska is presented below. The stages are not only successive but are related to distance from source, i.e. the earliest stage 1 is confined to the slope, the latest stage 5 is confined to the lower rise.

Stage 1. Plate motion causes the receiving plate to approach a rich sediment source associated with either a river or a glacier-filled fiord. During a period of glacial flow to the margin, a suspended sediment plume forms over the slope from point source: it evolves by granular rain and segregation into a lower (traction) part and an upper dispersed part. Rapid buildout of sediment discharged from a narrow exit point is seen as a prograding slope fan or apron that becomes a fan with associated debris flows, chutes and canyon heads.

Stage 2. Traction part of the sediment plume generates linear gullies; steep slope (terrace front) gradients promote acceleration and fluidization such that channel confluence is enhanced. Mass movement and large-scale gravity flows and local turbidity currents initially fill the depressions between the compression ridges, then proceed to fill the sector from the base of the continental slope and any ocean floor irregularities. Gradients fall from 1 : 20 to 1 : 300.

Stage 3. Outboard confluent channels grow; dispersed plume settles as overbank or levee deposits; elongate fans form; channellized grain flows scour broad channels at distal margins of 'sand fans'. Further sediment supply that is sustained during glaciation promotes the formation of contained channels.

Stage 4. Grain flows (relatively low velocity) form sinuous channels on the rise; these debouch as grain-flow or debris-flow deposits; channels can undergo avulsion during this stage. Deposition is dominated by background sedimentation, principally pelagic sources and distal material from active channels.

Stage 5. Final abandonment as plate motion and/or climatic changes terminates the sediment supply. Resurgence of sediment supply and development of another channel system and associated elongate fan. As source delivery under climatic influences is likely to be far more abrupt than long-term tectonic slip rates, the likelihood of plate motion terminating point source supply is low.

Depositional environments and fan architecture

A range of factors controls both the siting and scale of sediment deposition in the Gulf of Alaska. Factors include convergence ridges or upper slope terraces, flexural bulges associated with convergence, an outer trench high linked to subduction, seamount chains and sea-floor depressions. These plate-tectonically induced features influence the overall planimetric architecture of the fans including flow patterns, meandering and lobeing, channel abandonment, the formation of shingled accumulations, and channel capture. Moreover, plate motion ensures abandonment of these channel systems and, importantly, the overwhelming of the

subdued and abandoned southern levee by the northern levee of the next south channel–levee system. Levee onlapping and overlapping may be amplified by channel migration. Where separate channels are associated with individual fans, this type of fan overlapping may be considered shingling, although the angular relationships involved may be too slight to be observable on an outcrop scale.

Fluvial (subaerial) levees are flood overbank deposits; they crest and are thickest near the channel and thin laterally outward. Submarine levees are more symmetric with respect to the depositional crest, which is commonly conspicuously distant from the channel. Therefore, the relationship of submarine channels and their inferred levees is unclear. They each may represent processes that are fundamentally different, or there may be genetically different kinds of submarine levees that are at present difficult to discriminate.

Recognition criteria for transform margin sedimentary sequences

Tectonically dominated transform margins are linear features with sharp boundaries between normal thickness continental crust and oceanic crust. Hence any continental shelf will be narrow to absent, and, if present, have a steep to vertical slope gradient. Varying degrees of oblique convergence will occur, promoting incipient subduction and a fractured, locally bulged oceanic crust. Oblique convergence compresses any sediment accumulation into an incipient accretionary prism composed of deformed sediment ridges that may form a margin-bounding terrace. As this type of plate boundary transform is also associated with terrane migration, so sediment provenance will be difficult to assess.

Sedimentation patterns reflect these topographic contrasts and linear wedge-shaped accumulations form (Fig. 23), seen as slope aprons with debris flows and slumps; channel systems also develop. Sediment depocentres may develop in response to tectonically induced depressions formed across the continental slope through compression. Steep gradients result in mass movement and large-scale gravity flows seen as slides and turbidity currents that initially fill the depressions between the compression ridges, then proceed to fill the sector from the base of the continental slope and any ocean floor irregularities. We would expect to see abrupt contrasts among turbidites, mass movement deposits ranging from debrites to glide blocks, and especially sand- or diamictite-filled

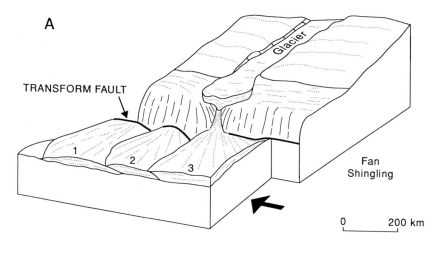

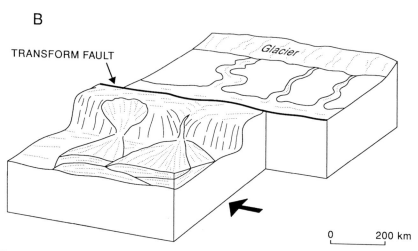

Fig. 23. Hypothetical model diagram showing shingling and compensation cycles modified from Howell *et al.* (1980). In (**a**), a point or single source episodically supplies sediment to the oceanic crust outboard of the transform fault. This supply pattern results in fan shingling. In (**b**), a broad piedmont glacier source, such as obtains across the Yakutat Terrane, creates a sediment apron outboard of the transform fault. Here, fan deposition tends to be dictated by the bathymetry resulting from earlier sedimentation such that compensation cycles may occur.

channels. A significant amount of faulting and folding, both syn- and postdepositional, may be present.

Conclusions

Each of the principal channel systems persisted for about 500 ka. Assuming glacial maxima approximately every 100 ka, it is to be expected that at least five episodes of high sediment influx will occur during the lifetime of the channel.

Moreover, within each period of glacial maxima major melting events will obtain. If the evidence of the separate flows can be applied to this scenario, at least eight melting or sediment supply events are related to each period of glacial maxima.

Channel capture appears to be a function of supply. In an environment where the principal supply points are directing material to new sections of the ocean floor as a result of plate motion, the well-supplied channels and their associated levees will eventually overwhelm the earlier, partially abandoned channel–levee system. This processes will be seen as capture. Plate shift eventually ensures final abandonment of the channel.

References.

ALLEY, R. B., BLANKENSHIP, D. D., ROONEY, S. T. & BENTLEY, C. R. 1989. Sedimentation beneath ice shelves – the view from ice stream B. *Marine Geology*, **85**, 101–120.

BATES, C. C. 1953. Rational theory of delta formation. *Bulletin, American Association of Petroleum Geologists*, **37**, 2119–2162.

BOULTON, G. S. 1990. Sedimentary and sea level changes during glacial cycles and their control on glacimarine facies architecture. *In*: DOWDESWELL, J. A. & SCOURSE, J. C. (eds) *Glacimarine Environments: Processes and Sediments*. Geological Society, London, Special Publications, **53**, 15–52.

BRUNS, T. R. & CARLSON, P. R. 1987. Geology and petroleum potential of the southeast Alaska continental margin. *In*: SCHOLL, D. W. GRANTZ, A. & VEDDER, J. G. (eds) *Geology and Resource Potential of the Continental Margin of Western North America and Adjacent Ocean Basins Beaufort Sea to Baja California*. Circum-Pacific Council for Energy & Mineral Resources, Houston, TX, 269–282.

CARLSON, P. R. 1989. Seismic reflection characteristics of glacial and glacimarine sediment in the Gulf of Alaska and adjacent fjords. *In*: POWELL, R. D. & ELVERHOI, A. (eds) *Modern Glacimarine Environments: Glacial and Marine Controls on Modern Lithofacies and Biofacies*. *Marine Geology*, **85**, 391–416.

——, BRUNS T. R., MOLNIA, B. F. & SCHWAB, W. C. 1982. Submarine valleys in the northeast Gulf of Alaska: characteristics and probable origin. *Marine Geology*, **47**, 217–242.

DAMUTH, J. E. & FLOOD, R. D. 1985. Amazon Fan, Atlantic Ocean. *In*: BOUMA, H. A., NORMARK, W. R. & BARNES, N. E. (eds) *Submarine Fans and Related Turbidite Systems*. Springer, New York, Chap. 15.

——, KOLLA, V., FLOOD, R. D., KOWSMANN, R. O., MONTEIRO, M. C., GORINI, M. A., PALMA, J. J. C. & BELDERSON, R. H. 1983. Distributary channel meandering and bifurcating patterns on the Amazon deep-sea fan as revealed by long-range side-scan sonar (GLORIA). *Geology*, **11**, 94–98.

DOBSON, M. R., VALLIER, T. & KARL, H. 1996. Sedimentation along the fore-arc region of the Aleutian Island Arc, Alaska. *In*: GARDNER, J. V. FIELD, M. E. & TWITCHELL, D. C. (eds) *Geology of the United States' Seafloor, the View from GLORIA*, **16**. Cambridge University Press, 279–304.

DOWDESWELL, J. A. & MURRAY, T. 1990. Modelling rates of sedimentation from icebergs. *In*: DOWDESWELL, J. A. & SCOURSE, J. C. (eds) *Glacimarine Environments: Processes and Sediments*. Geological Society, London, Special Publications, **53**, 121–137.

——, KENYON, N. H., ELVERHOI, A., LABERG, J. S., HOLLENDER, F.-J., MIENERT, J. & SIEGERT, M. J. 1996. Large-scale sedimentation on a glaciated passive continental margin: the Polar North Atlantic. *Geophysical Research Letters*, **23**, 3535–3538.

GARDNER, J. V., FIELD, M. E., LEE, H., EDWARDS, B. E., MASSON, D. G., KENYON, N. & KIDD, R. B. 1991. Ground-truthing 6.5 kHz sidescan sonographs: What are we really imaging? *Geophysical Abstracts* 7–8, (Supplement to *EOS*, January 1991).

HAMBREY, M. J. 1994. *Glacial Environments*. UCL Press, London, 296 pp.

HOWELL, D. G., ALPHA, T. R. & JOYCE, J. M. 1980. Plate tectonics, physiography and sedimentation of the northeast Pacific region. *In*: FIELD, M. E., DOUGLAS, R. G., BOUMA, A. H., INGLE, J. C. & COLBURN, I. P. (eds) *Quaternary Depositional Environments of the Pacific Coast. SEMP Pacific Section, Pacific Coast Paleogeography Symposium*, **4**, 43–53.

KIDD, R. B., HUNTER, P. M. & SIMON, R. W. 1987. Turbidity current and debris flow pathways to the Cape Verde Basin. *In*: Weaver, P. P. E. & Thomson, J. (eds) *Geology and Geochemistry of Abyssal Plains*. Geological Society, London, Special Publications, **31**, 33–48. Blackwell.

KULM, L. D., VON HEUNE, R. *et al.* 1973. *Initial Reports of the Deep Sea Drilling Project*, **18**. US Government Printing Office, Washington, DC.

LAUGHTON, A. S. 1981. The first decade of GLORIA. *Journal of Geophysical Research*, **86**(B), 11511–11534.

MIALL, A. D. 1984. *Principles of Sedimentary Basin Analysis*. Springer, New York, 490 pp.

MOLNIA, B. F. 1977. Rapid shoreline erosion at Icy Bay, Alaska – a staging area for offshore petroleum development. *Proceedings 1977 Offshore Technology Conference* **IV**, 115–126.

MOLNIA, B. F. & CARLSON, P. R. 1978. Surface sedimentary units of the northern Gulf of Alaska continental shelf. *Bulletin, American Association of Petroleum Geologists*, **62**, 633–643.

MONGER, J. W. H. & BERG, H. C. 1984. Lithotectonic terrane map of western Canada and southeastern Alaska. *In*: SILBERLING, N. J. & JONES, D. L. (eds) *Lithotectonic Terrane Maps of the North American Cordillera*. US Geological Survey Open-File Report **84-523**, scale 1 : 2,500,000.

NESS G. E. & KULM, I. D. (1973). Origin and development of Surveyor deep sea channel. *Geological Society of America Bulletin*, **84**, 3339–3354.

O'LEARY, D. W. & DOBSON, M. R. 1992. Southeastern New England Continental Rise: origin and history of slide complexes. *In*: POAG, C. W. & DE GRANCIANSKY, P. C. (eds) *Geologic Evolution of Atlantic Continental Rises.* Van Nostrand Reinhold, New York, 214–261.

ORANGE, D. & BREEN, N. A. 1992. The effects of fluid escape on accretionary wedges, 2, seepage force, slope failure, headless submarine canyons and vents. *Journal of Geophysical Research*, **97**, 9225–9277.

PLAFKER, G. 1987. Regional geology and petroleum potential of the northern Gulf of Alaska continental margin. *In*: SCHOLL, D. W., GRANTZ, A. &

VEDDER, J. G. (eds) *Geology and Resource Potential of the Continental Margin of Western North America and Adjacent Ocean Basins Beaufort Sea to Baja California*. Circum-Pacific Council for Energy & Mineral Resources, Houston, TX, 229–268.

POWELL, R. D. 1984. Glacimarine processes and inductive lithofacies modelling of ice shelf and tidewater glacial sediments based on Quaternary examples. *Marine Geology*, **57**, 1–52.

—— 1990. Glacimarine sedimentation processes at grounding-line fans and their growth to ice-contact deltas. *In*: DOWDESWELL, J. A. & SCOURSE, J. C. (eds) *Glacimarine Environments: Processes and Sediments*. Geological Society, London, Special Publications, **53**, 53–74.

—— & MOLNIA, B. F. 1989. Glacimarine sedimentation processes, facies and morphology of the south-southeast Alaska shelf and fjords. *Marine Geology*, **85**, 359–390.

PRATSON, L. F. & LAINE, E. P. 1989. The relative importance of gravity induced versus current controlled sedimentation during the Quaternary along the middle U.S. continental margin revealed by 3.5 kHz echo character. *Marine Geology*, **89**, 87–126.

RIDDIHOUGH, R. P. & HYNDMAN, R. D. 1992. Modern plate tectonic regime of the continental margin of western Canada. *In*: GABRIELSE, H. & YORATH, C. J. (eds) *Geology of the Cordilleran Orogen in Canada*, Geological Survey of Canada, **4**, 435–455.

SCHWAB, W. C. & LEE, H. J. 1988. Causes of two slope-failure types in continental shelf sediment, northeastern Gulf of Alaska. *Journal of Sedimentary Petrology*, **58**, 1–11.

SOMERS, M. L., CARSON, R. M., REVIE, J. A., EDGE, R. H., BARROW, B. J. & ANDREWS, A. G. 1978. GLORIA II an improved long range sidescan sonar. *Oceanology International*, **78**, 16–24.

STEVENSON, J. A. & EMBLEY, R. 1987. Deep-sea fan bodies, terrigenous turbidite sedimentation and petroleum geology, Gulf of Alaska. *In*: SCHOLL, D. W., GRANTZ, A. & VEDDER. J. G. (eds) *Geology and Resource Potential of the Continental Margin of Western North America and Adjacent Ocean Basins Beaufort Sea to Baja California*. Circum-Pacific Council for Energy & Mineral Resources, Houston, TX, 503–522.

STOW, D. V. A. 1986. Deep clastic seas. *In*: READING, H. G. (ed.) *Sedimentary Environments and Facies*. Blackwell, Palo Alto, CA.

SYVITSHI, J. P. M. 1989. On the deposition of sediment within glacier-influenced fjords. *Marine Geology*, **85**, 301–329.

TWITCHELL, D. C., KENYON, N. H., PARSON, L. M. & MCGREGOR, B. A. 1991. Depositional patterns of the Mississippi Fan surface. Evidence from GLORIA II and high resolution seismic profiles. *In*: WEIMER, P. & LINK, M. H. (eds) *Seismic Facies and Sedimentary Processes of Submarine Fans and Turbidite Systems. Frontiers in Sedimentary Geology, 18*. Springer, New York, 349–363.

VANNESTE, K., UENZELMANN-NEBEN, G. & MILLER, H. 1995. Seismic evidence of long term history of glaciation on central East Greenland shelf south of Scoresby Sund. *Geo-Marine Letters*, **15**, 63–70.

VORREN, T. O., LEBESBYE, E., ANDREASSEN, K. & LARSEN, K. B. 1989. Glacigenic sediments on a passive margin as exemplified by the Barents Sea. *Marine Geology*, **85**, 251–272.

WEIMER, P. 1993. Seismic facies, characteristics and variations in channel evolution, Mississippi fan (Plio-Pleistocene), Gulf of Mexico. *In*: WEIMER, P. & LINK, M. H. (eds) *Seismic Facies and Sedimentary Processes of Submarine Fans and Turbidite Systems. Frontiers in Sedimentary Geology, 18*. Springer, New York, 323–348.

YORATH, C. J. 1987. Petroleum geology of the Canadian Pacific continental margin. *In*: SCHOLL, D. W., GRANTZ, A. & VEDDER, J. G. (eds) *Geology and Resource Potential of the Continental Margin of Western North America and Adjacent Ocean Basins Beaufort Sea to Baja California*. Circum-Pacific Council for Energy & Mineral Resources, Houston, TX, 283–304.

—— & HYNDMAN, R. D. 1983. Subsidence and thermal history of Queen Charlotte Basin. *Canadian Journal of Earth Sciences*, **20**, 135–159.

ZELLERS, S. D. 1995. Foraminiferal sequence biostratigraphy and seismic stratigraphy of a tectonically active margin; the Yakatanga Formation, northeastern Gulf of Alaska. *Marine Micropalaeontology*, **26**, 255–271.

Large-scale debrites and submarine landslides on the Barra Fan, west of Britain

R. HOLMES[1], D. LONG[1] & L. R. DODD[2]

[1]British Geological Survey, Murchison House, West Mains Road, Edinburgh EH9 3LA, UK

[2]Geoteam-Wimpol Ltd, Regent House, Regent Quay, Aberdeen AB1 2BE, UK

Abstract: The Barra Fan is a large Neogene to Pleistocene composite fan that has built out into the deep-water basin of the Rockall Trough west of Britain. The bulk of the landward depocentre of the Barra Fan takes the form of a relatively undisturbed shelf-margin wedge, in places more than 600 m thick, that is perched above the deep-water basin and, together with the Donegal Fan, partially engulfs the Hebrides Terrace Seamount. A continuous series of slope-front channels and gullies incise into the shelf-margin wedge of the northernmost Barra Fan and are the likely transport paths for small- to medium-scale debrites found on the middle and lower slope-front. In contrast, at least four large-scale submarine landslide events making up the Peach Slide appear to have translated the bulk of the slope-front of the northern Barra Fan to the northwest, so that debrites and hemipelagic sediments exhibit maximum composite thicknesses of approximately 450 m on the deep-water basin floor of the eastern Rockall Trough. The extent and volume of successive large-scale debrite units decrease with age. From an interpretation of a limited historical dataset it is thought that the ground accelerations from earthquakes have been sufficient to trigger slope-front instability and sediment transfer.

This paper summarizes the overall geomorphological, structural and seismostratigraphical setting of the Barra Fan (Fig. 1), and relates the evidence derived from seabed and sub-seabed seismic-reflection facies to submarine landslide events at various scales on the fan. Although slide-related processes are often seen as important components of fan development, a lack of sufficient data usually prevents quantification. A major objective of this paper is to quantify the proportions and volumes of sediments transported as debris flows and originating from partial destruction of the slope-front on the northern Barra Fan by large-scale submarine landslides previously identified with the Peach Slide (Holmes 1994).

For this paper, the term submarine landslide follows the original definition by Prior & Coleman (1979), where the bulk of the sediments in translational slides disintegrate during mass-transport or movement, forming submarine sediment flows. The definition of mass-movement has been refined by division into non-disintegrative mass-movement and disintegrative mass-movement (Whitman 1985; Booth & O'Leary 1991), the latter including rubble slide, collapse or debris flow. Fast-moving disintegrative submarine landslides that generate debris flows typically occur where there is a rapid accumulation of thick sedimentary deposits, a sloping seafloor, and where loss of shear strength to less than gravitational force

originates from a transient loading event (e.g. Hampton et al. 1996). Following slide initiation and sediment remoulding, the debris flow is emplaced as a poorly sorted sediment, termed a debrite. However, evidence is not available in this study to differentiate the processes of debrite emplacement, for example, if originating from a Bingham viscoplastic fluid (e.g. Jiang & LeBlond 1993), or liquefaction that produces concentration of the larger particles at the debrite top (Elverhoi et al. 1997).

Although it is assumed that sediment translation in submarine landslides by mass-movement will involve more than one mechanism, Pickering et al. (1989) identified debris flow and turbidite flow as the only submarine sediment flows capable of transporting particulate sediments over long distances on low-angle slopes. However, the sizes of slides and their associated debrites vary considerably. For example, a statistical overview of the North American Atlantic continental margin shows that slide areas range from less than 0.1 km^2 to c. 19 000 km^2, and that nearly two-thirds of all slides have an area equal to, or less than, 10 km^2 (Booth & O'Leary 1991).

The diagnostic plan-components of submarine landslides originating from unconsolidated sediments are a detachment scar overlooking a source-bowl area, an elongate transport chute or gully commonly associated with sidewall instability, and downslope depositional lobes. The lobes terminate when gravity no longer exceeds

HOLMES, R., LONG, D. & DODD, L. R. 1998. Large-scale debrites and submarine landslides on the Barra Fan, west of Britain. *In*: STOKER, M. S., EVANS, D. & CRAMP, A. (eds) *Geological Processes on Continental Margins: Sedimentation, Mass-Wasting and Stability*. Geological Society, London, Special Publications, **129**, 67–79.

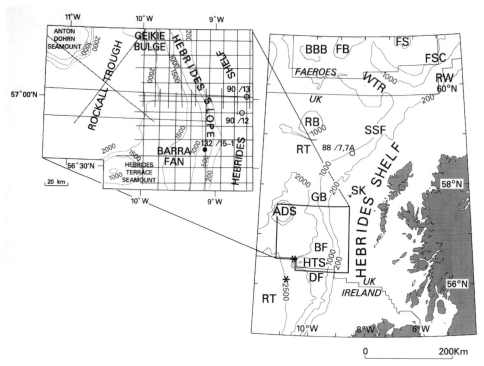

Fig. 1. Location of the study area, bathymetric setting (contours in metres), track plots of regional air-gun and sparker lines, and sample sites. South of *c.* 57°, swath bathymetric data are available between *c.* 160 and 1500 m water depth for most of the study area. BBB, Bill Bailey's Bank; FB, Faeroe Bank; FS, Faeroe Shelf; WTR, Wyville-Thomson Ridge; FSC, Faeroe–Shetland Channel; RW, Rona Wedge; RT, Rockall Trough; RB, Rosemary Bank; SSF, Sula Sgeir Fan; GB, Geikie Bulge; SK, St Kilda; ADS, Anton Dohrn Seamount; BF, Barra Fan; HTS, Hebrides Terrace Seamount; DF, Donegal Fan; ●, commercial well location; ○, BGS borehole location; ★, earthquake location (Table 2).

the shear strength of the debris flow, or when the pore pressures have dissipated. The upslope detachment scars are generally inward-looking over the slide and enclose areas with an overall negative cross-sectional seabed relief, whereas the toe-end escarpments are outward-looking and enclose an area of overall positive sea bed relief. The net effect is for slide debrites to lose section upslope and increase section downslope (e.g. Sangree & Widmier 1977; Hiscott & Aksu 1994; Laberg & Vorren 1995). The uppermost surfaces of slide debrites commonly have a highly irregular or hummocky relief at the seabed, whereas the lowermost surfaces on the slide sole are commonly flat or curved and may be sharply defined on seismic reflection profiles. For interpretation purposes we have assumed that homogeneously chaotic or reflection-free seismic facies together with the geometries described above are criteria for the recognition of debrites originating from submarine landslide. The internal homogeneity of these seismic facies is interpreted as being indicative of a single event.

Fan geometry and regional stratigraphy

A regional intra-Miocene unconformity defines the base of the Barra Fan (Fig. 2). The intra-Miocene unconformity can be traced westwards under the shelf and slope into the Rockall Trough (Fig. 2) and is correlated in this paper with the H2 reflector previously thought to separate an Oligocene – ?late Eocene unit from a late Cenozoic unit (Jones *et al.* 1986). However, on the basis of an extensive coring programme on the western margin of the Rockall Trough, the unconformity is now tentatively correlated with the 14 Ma Chron (middle Miocene; Stoker, 1997). Unconsolidated sandy and shelly sediments were sampled on the inner fan just above the intra-Miocene unconformity in BGS (British Geological Survey) borehole 90/13 (Fig. 2) and yield dates with an estimated average age of 10.1 Ma (late Miocene) from $^{87}Sr/^{86}Sr$ isotopic ratios on three bivalve fragments and three composite fractions of foraminifera. This age is compatible with a late

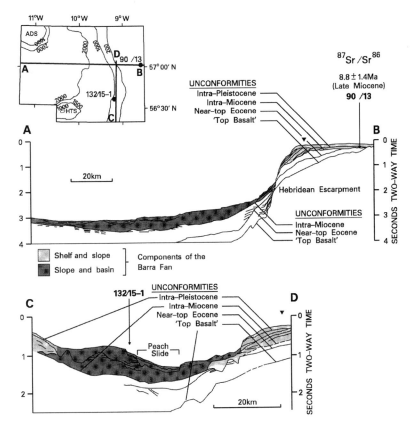

Fig. 2. Generalized geological cross-sections, Barra Fan. ▼, Line intersection.

Miocene to early Pliocene biostratigraphical age-range determined from the same interval (Barreiro 1995). Above the $^{87}Sr/^{86}Sr$-dated sediments BGS borehole 90/13 recovered thick glauconitic sands that are identified with an undifferentiated Pliocene to early Pleistocene biostratigraphy (BGS unpublished data). Clino-forms above the intra-Miocene unconformity on the outer shelf define stages in the seawards progression of the depositional slope-front and shelfbreak that contributed to construction of the shelf-margin wedge. The clinoforms form part of the Lower Macleod sequence, and had previously been attributed a predominantly Pliocene to early Pleistocene age (Stoker *et al.* 1993), an estimate that is consistent with the evidence from the newer $^{87}Sr/^{86}Sr$ isotopic and biostratigraphical age estimates summarized above. Therefore, the sedimentary record indicates that there was a delay, possibly of the order of 5 Ma or more, before the depositional slope-front of the Barra Fan prograded to allow Pliocene and early Pleistocene sediments to spill westwards into deep water (Fig. 2).

The top of the shelf-margin wedge is truncated on the middle and outer shelves by a sub-horizontal intra-Pleistocene unconformity. At 57°N, the prograding clinoforms are also truncated near the seabed on the slope above the Hebridean Escarpment, a prominent depositional feature formed from Palaeocene basalts (Musgrove & Mitchener 1996). Erosional truncation of the shelf-margin wedge on the slope is the cause of the structural separation between the shelf-margin wedge and the lower slope-front and basin sediments (Fig. 2). This style of structural separation typifies the overall geometrical style of sediments preserved in the Barra Fan north of *c.* 56°50′N. In contrast, estimated depths from mean sea level to the base of the Pliocene to Pleistocene between *c.* 56°20′N and 56°50′N indicate that the bulk of the earliest Barra Fan sediments are underpinned by a shelf-to-slope ramp margin (Holmes *et al.* 1993). The shelf-margin wedge sediments that are preserved above the ramp margin, and adjacent to the modern shelfbreak, form a depocentre with a thickness of at least 800 ms two-way-travel-time

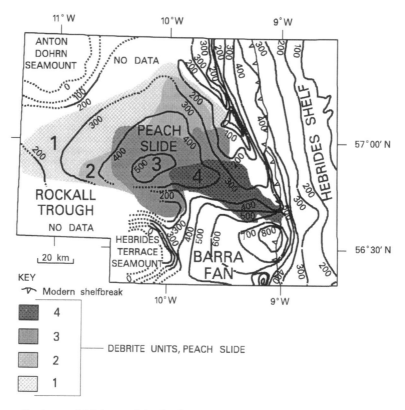

Fig. 3. Generalized map of debrites, and the distribution and sediment thicknesses from seabed to the intra-Miocene unconformity. Sediment thicknesses are shown in milliseconds two-way travel time; dotted contours are uncertain. The information extending from 08°03′W to 10°00′W is modified after Evans (1992), James (1992) and James & Hitchen (1992).

(TWTT) (Fig. 3), equivalent to a sediment thickness of at least 660 m if an average interval velocity of 1650 m s⁻¹ is assumed.

Seabed physiography of the Barra Fan and northern Donegal Fan

In contrast to the ramp-style physiographic setting at fan initiation south of *c.* 56°50′N, the morphological expression of the boundary between the modern shelf and slope-front on the Barra Fan is a sharply defined shelfbreak at the seabed (Fig. 4). A north–south section across the slope-front in the northern Barra Fan demonstrates that the slope and basin component has an overall concave-up section encompassing the Peach Slide, whereas the north–south section across the slope and basin component of the southern Barra Fan is marked by a convex-up section, its apex at the latitude of the Hebrides Terrace Seamount (Fig. 2, section C–D). The deep-water slope-front

encompassing the Barra Fan and Donegal Fan is marked by an overall bulge of the bathymetric contours extending westwards into the Rockall Trough (e.g. 2000 m in Fig. 1).

A three-dimensional (3-D) perspective view of the modern seabed morphology demonstrates the almost-continuously eroded appearance of the slope-fronts on the Barra Fan and the Donegal Fan (Fig. 4c). The seabed north of *c.* 56°55′N has a maximum gradient, outside the channels, of *c.* 10°. This area is eroded by a distinctive suite of channels that branch out upslope in 500–900 m water depth. Between *c.* 56°55′N and 56°45′N, channels and gullies also incise into the slope-front with average seabed gradients of *c.* 7–8° between 400 and 800 m water depth. They truncate reflectors in the shelf-margin wedge perched above the Hebridean Escarpment (Fig. 2, section A–B) and at the shelfbreak have contributed to shelf-break regression (Fig. 3).

The channels and gullies on the northern Barra Fan feed downslope across a gradient that

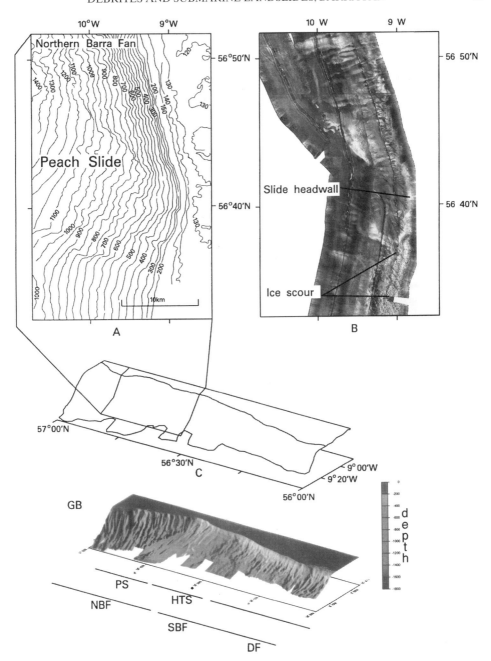

Fig. 4. Seabed morphology. (**a**) Isobaths in metres. Contour interval on the shelf is 10 m (after James 1992). Slope information derived from swath bathymetric survey. (**b**) TOBI side-scan image; high-backscatter seismic facies are shown as areas with paler shade; the north-trending parallel lines, some truncated, are artefacts of recording and sonar-mosaic boundaries. (**c**) Three-dimensional perspective of seabed topography derived from the swath bathymetric survey, with an upper-limit cut-off defined for the model by the 200 m isobath. GB, Geikie Bulge; NBF, Northern Barra Fan; PS, Peach Slide; HTS, latitude encompassing the Hebrides Terrace Seamount (the seamount itself is not illustrated); SBF, Southern Barra Fan; DF, Donegal Fan.

decreases from *c.* 7–8° to 3–4° on a slope-front apron that exhibits a mixed seabed physiography with fan, channel levee and gully bedforms (some of which undoubtedly occur with small- to medium-scale debrites; Armishaw *et al.* this volume). At *c.* 56°47′N in 1100 m water depth, the bedforms are truncated by the SE-trending northern detachment scar of the latest failure event of the Peach Slide as illustrated by the contoured isobaths, TOBI side-scan records, and the shadow from the artificial illumination of the 3-D model (Fig. 4). The southern sidewall of the Peach Slide is not highlighted on the 3-D morphological model because the direction of artificial sunlight is from the north. However, the offset in the contoured isobaths, together with the changes from high- to low-reflectivity seismic facies on the TOBI records, identify the fall in seabed occurring across the southern detachment scar (Fig. 4a). The overall impression from the TOBI records from the headwall area of the latest large-scale slide event on the Peach Slide is that it comprises a bowl-shaped

depression that has eroded into iceberg-scoured deposits at the shelfedge (Fig. 4b).

To the south of the Peach Slide, the topographical bulge associated with the overall build-out of the Barra Fan and Donegal Fan into the Rockall Trough is clearly demonstrated by the 3-D perspective model (Fig. 4c). Between the southern sidewalls of the Peach Slide and *c.* 56°30′N, the seabed from the shelfbreak to *c.* 1000 m is typified by a relatively smooth morphology compared with all other areas on the Barra Fan slope-front. Typical average gradients from shelfbreak at 200 m to the 300 m isobath are only *c.* 4°, and from the 300 m isobath to the 1000 m isobath *c.* 3°. This area overlies sections with minimal evidence for major breaks in the fan's depositional history (Fig. 5, section H–G). It coincides with the location of thickest landward accumulation of sediments along the shelf-margin wedge where the lower slope-apron appears to be partially buttressed against the Hebrides Terrace Seamount (Fig. 3).

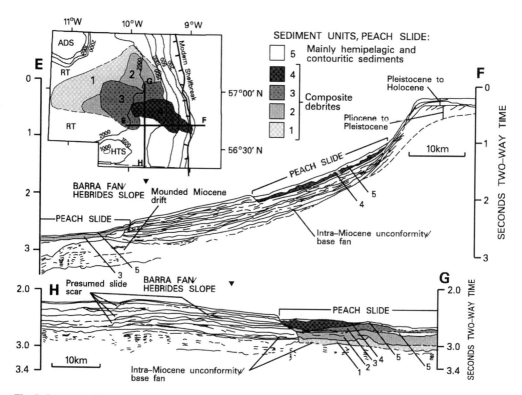

Fig. 5. Interpreted line sections across the Barra Fan illustrating the styles of mid-slope erosion scarps and the associated emplacement style of debrites in the Peach Slide. ADS, Anton Dohrn Seamount; HTS, Hebrides Terrace Seamount; RT, Rockall Trough. ▼, Line intersection.

Evidence for large-scale slide events in the Barra Fan

A west-to-east seismic profile from deep water on the northern Barra Fan at 57°00'N shows a unit *c*. 300 m thick and *c*. 40 km long that was interpreted as a large-scale gravitational slide by Jones *et al.* (1986). Subsequently, possible upslope slide detachment scars were mapped on the northern Barra Fan from the BGS dataset of airgun, sparker, deep-tow boomer and pinger profiles (Long & Bone 1990). On the basis of a larger dataset, some released from commerce, it is apparent that upslope detachment scars on the northern Barra Fan are continuous for a distance of some tens of kilometres and that extensive drape by hemipelagic and contouritic sediments, typically 1–5 m thick, extends to the seabed, indicating a lack of recent large-scale movement on the detachment scars. These features were identified with a composite slide unit named the Peach Slide by Holmes (1994).

An overall negative seabed topography in up-dip sections of the Peach Slide defines an embayment in the northern Barra Fan (Fig. 5) which is complemented downslope in the deep-water basin by a depocentre with a maximum sediment thickness of at least 500 ms TWTT (Fig. 3). The bulk of the sediments within the depocentre comprise four major sediment units (Fig. 5) whose acoustic homogeneity is typified by a predominantly chaotic or reflection-free seismic facies. Their bases are sharply defined by slightly curved reflectors, and, except where overlain by succeeding units, the downslope sections taper so that the toe margins commonly exhibit an overall positive relief compared with adjacent areas (Fig. 6, section K–L). The depositional setting, geometry and seismic facies described for units 1–4 fit the criteria previously summarized for the recognition of translated sediments originating from slide movement. They are therefore interpreted as debrites, hereafter termed Debrites 1–4. In contrast, Unit 5 consists of a well-layered seismic facies on airgun profiles. Locally, thin lenses of Unit 5 separate debrites on the main body of the slide, thicker sections being observed at the slide margins and between the seabed and underlying debrites in cases where the slide debrites are not stacked (Figs 5–7). The well-layered seismic facies is interpreted as predominantly hemipelagic and contouritic sediments, although thin debrites and turbidites also form components of this seismic facies locally (Armishaw *et al.* this volume).

An outstanding example of the positive topography at the down-dip margins of slide debrites occurs at *c*. 57°N, 10°W. Here the toe of Debrite 3 coincides with a hummocky seabed in 2000–2090 m water depth; the bulk of the chaotic seismic facies in Debrite 3 terminates across a sharply defined scarp, with a maximum gradient of *c*. 5° occurring within an overall drop in seabed of 20–30 m (Fig. 7). Reflectors in Unit 5 onlap the scarp but are truncated under the base of Debrite 3. Sub-debrite erosion is also typical for Debrites 1, 2 and 4, so that for the most part the seismic reflection profiles show little evidence for physical separation between Debrites 1 and 4 in upslope sections (Figs 5–7). Conversely, because of the overall tendency for offlap in Debrites 1–4, the regional pattern is for cumulative thicknesses of the well-layered seismic facies to increase significantly towards the northern and western margins of the depocentre (Figs 6 and 7). The evidence for distinctly separate phases of sediment transfer over the history of the Peach Slide is in the form of debrite separation by components of Unit 5 at the slide toes.

Debrite 4 on the Peach Slide has a total area of *c*. 1100 km^2 and a run-out distance of *c*. 66 km as measured from its headwall at the modern shelfbreak. This is smaller than the preceding Debrites 1–3 (Fig. 3). On the Atlantic margin of North America less than 5% of large slides occur with areas of more than 500 km^2 (Booth & O'Leary 1991), so that even Debrite 4 is large by comparison.

Discussion

Slide transfer pathways, timing and volumes

Confinement of the repeated failure events on the Peach Slide to areas north of the Hebrides Terrace Seamount may be most simply explained by slope-front stabilization from sediment buttressed against the Hebrides Terrace Seamount, with possibly some topographic control also originating from the mounded Miocene drift deposits occurring to the north and east (Fig. 5, section E–F). Debrite 1 pinches out above a slight topographic high on mounded Miocene drift at the eastern margin of the Anton Dohrn Seamount (Fig. 6, section I–J) and the earliest slide pathway appears to be diverted to the south around the Anton Dohrn Seamount, there being insufficient data to determine its maximum northwards extent (Fig. 3).

Data to precisely date the failure events associated with the Peach Slide are not available. The interval between the intra-Miocene unconformity and the point at which Debrite 1 merges into well-layered hemipelagic sediments represents *c*. 35% of the total thickness of the Miocene to Holocene succession above the intra-Miocene

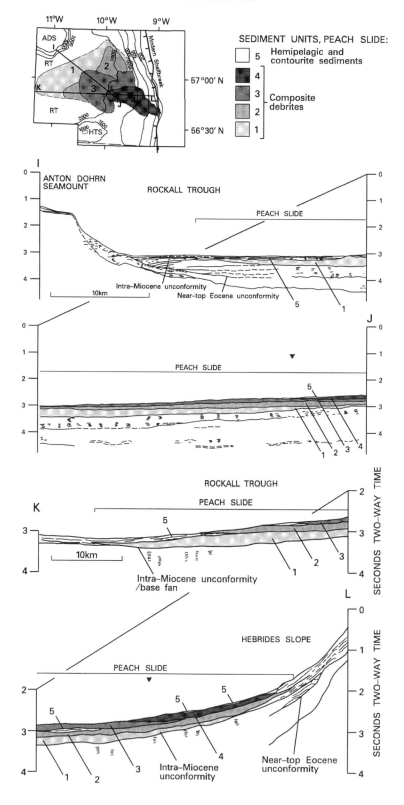

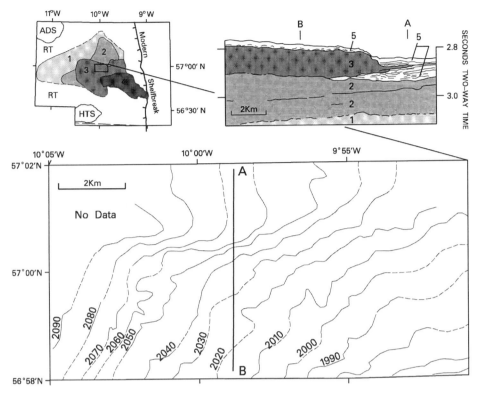

Fig. 7. Emplacement style at the toe of Debrite 3 on the Peach Slide. Composite Debrite 1 and Debrite 2 are also shown, as are the hemipelagic and contouritic deposits at the distal end of event 3. Bathymetry compiled in metres below mean sea level from BGS close survey in 1992 using 3.5 kHz pinger profile data.

unconformity (Fig. 6, section K–L), but this observation alone is an unsuitable basis for ascribing an age to the first major event in the Peach Slide. However, as described previously, a delay probably occurred before sediment at the Pliocene to Early Pleistocene depositional slope-front of the northern Barra Fan spilled over into the deep water of the Hebridean Slope. For the most part, Debrite 1 rests in contact with the intra-Miocene unconformity (Figs 5 and 6). It is therefore suggested that Debrite 1 has entrained Neogene and early Pleistocene sediments that were remoulded from the deep-water slope-front of the early Barra Fan.

Where Debrite 1 has extended over positive topography on a Miocene sediment drift occurring to the southeast of the Anton Dohrn Seamount, the debrite toe appears to be in contact with the intra-Miocene unconformity (Fig. 6, section I–J). Because Debrite 1 is interpreted as one event, differences of spatial

separation of the toe of Debrite 1 above the intra-Miocene unconformity are attributed to local differences of reworking under the debrite and not to separate ages. The occurrence of well-layered hemipelagic and contouritic sediments beneath all other debrites from the Peach Slide indicates that subsequent development of the northern Barra Fan was punctuated by catastrophic downslope failure. In this regard, the latest failure event associated with Debrite 4 eroded into the iceberg scour observed in seabed sediments at the shelfbreak, there being an absence of iceberg scour within the amphitheatre of the remoulded sediment in the head area of the slide (Fig. 4b). A post-glacial age is therefore likely for Debrite 4; this accords with similar evidence for a post-glacial age proposed for the bottleneck slides observed at the northern margin of the Geikie Bulge (Kenyon 1987).

Although sediment slides are perceived as vehicles for large-scale mass transfer on the NW

Fig. 6. Interpreted line sections showing the sequence and style of debrites deposited in the mid-Rockall Trough. ▼, Line intersection.

European margin, published volume calculations are commonly based on seabed morphology in their proximal headwall and sidewall environments (e.g. Bugge *et al.* 1987) or on minimum estimates of volumes based on seabed morphology and subsurface morphology (e.g. King *et al.* 1996). The proximal- to mid-slide environments are, however, likely to be unrepresentative of the total volume of sediments incorporated into the slide debrites, and although there is a lack of detailed subsurface survey data in the distal environments of the Peach Slide (Fig. 1), the available data provide an opportunity to make controlled estimates of preserved slide-component volumes from the major part of the Peach Slide.

Debrite 1 runs out oceanwards into the Rockall Trough some 160 km if measured from the modern shelfbreak, although the precise run-out distance is uncertain as the headwall scars for this slide event are not preserved beneath younger slide debrites. This fact, and erosion of pre-existing debrites under the slide sole, must lead to an underestimation of original debrite volumes. Nevertheless, the numerical estimates indicate that the older preserved debrites have the largest volumes (Table 1). The map of the distribution of Debrites 1–4 indicates their overall decreasing extent into the Rockall Trough with time (Fig. 3).

Slide processes

The evidence from BGS boreholes 90/13 and 88/7,7A (Fig. 1) is for sand-prone Pliocene to early–mid Pleistocene sediments on the shelf and upper slope. Mud-prone sediments correlated with Anglian shelf glaciation that took place between *c.* 0.44 and 0.38 Ma are the first recorded evidence for the glaciation of the Hebridean Shelf during the mid-Pleistocene (Stoker *et al.* 1994). As suggested above, Debrite 1 probably incorporates Neogene and early Pleistocene sediments that pre-date the severe climate changes associated with cycles of regional shelf glaciation during the mid-Pleistocene. Sediment cohesion is a factor in debrite mobility; sediment cohesion is higher in mud-prone sediments than in sand-prone sediments, and cohesion is generally higher in mud-prone glaciogenic sediments than in non-glaciogenic mud-prone sediments (Baltzer *et al.* 1995). It is speculated that the evidence for overall decreasing debrite volume and debrite run-out with decreasing age could originate from bulk changes of sediment texture and cohesion originating from an overall deteriorating climate from Neogene to middle Pleistocene times. As the extent of the textural and cohesion variation is unknown, this hypothesis remains to be tested.

Assessing the stability of Pleistocene slope sediments requires information on their geotechnical properties to the depth of sediment affected by slide failure. In the case of the Barra Fan this is to tens of metres below the seabed. Geotechnical data in the area are, however, restricted to 6 m maximum depth below the seabed and are determined from samples obtained during the BGS regional mapping programme, particularly vibrocores and gravity cores. On the shelf, and just beyond the shelfbreak to about 250 m water depth, the sediments at 1 m depth usually have shear strengths of more than 50 kPa. On the slope and deep-water basinal areas to *c.* 2000 m water depth, shear strengths commonly decrease down-core from around 20 kPa near the seabed to less than 10 kPa at 2 m depth, below which a slight increase in strength occurs (Dodd 1995; BGS unpublished data). Undrained cohesion (shear strength) to effective overburden pressure ratios below 2 m sub-seabed depth are usually in the range 0.3–0.4. Based on infinite slope stability analysis (Morgenstern 1967) these sediments are stable on slopes of less than 15° whereas the

Table 1. *Sediment components of the Peach Slide*

Dominant lithology	Event	Assumed interval velocity (m s^{-1})	Volume (km^3)	% of total
Hemipelagic or contouritic muds and sands	5	1600	835	31
Debrite	4	1600	135	5
Debrite	3	1600	199	8
Debrite (composite)	2	1700	673	25
Debrite	1	1700	823	31
All sediments in slide area	1–5	as above	2665	100
All debrites in slide area	1–4	as above	1830	69

modern upper slope rarely exceeds 12°. Therefore, for the sediments to become unstable requires that additional stresses be applied to the sediments.

Alternatively, the sediment may behave as a Bingham fluid, in which case pore pressure can increase yield stress above the Bingham yield stress (Jiang & LeBlond 1993). Pore pressure may increase as a result of rapid sedimentation, but as Holocene sediment thicknesses are typically less than 30 cm, the effects of modern sedimentation are considered negligible. In borehole 88/7,7A on the slope north of the Barra Fan (Fig. 1), the estimated sedimentation rates range from 5 to 25 mm ka^{-1} during the early Pleistocene to the earliest–middle Pleistocene to an average accumulation rate of c. 560 mm ka^{-1} during the Anglian shelf glaciation (Stoker et al. 1994). During this and similar shelf glaciations a mechanism for a pore pressure increase may therefore involve the rapid deposition of low-permeability muds. The evidence is, however, that the latest slide on the Peach Slide post-dated glaciation, indicating that although processes involving rapid sedimentation may contribute to slope loading and failure, the timings and distributions of slope failure are not uniquely related to glaciogenic events.

Excess pore pressure may also develop because of increases originating from buoyant free gas in sediment pore spaces. NW-trending Mesozoic growth faults underlying the Barra Fan (Musgrove & Mitchener 1996) are thought to have been reactivated as Cenozoic transfer faults, and dilational stress fields within them possibly allowed the vertical migration of gas (Doré & Lundin 1996). However, the presence of free gas, for example methane, in shallow sediments may also be theoretically dependent on physico-chemical controls on the sub-seabed boundary of the methane-hydrate stability zone which has been modelled for sections west of the Hebrides (Holmes et al. in press). Incorporation of detailed bottom-water temperatures into the numerical model indicates that the methane-hydrate layer theoretically crops out at the seabed at c. 600 m water depth in the eastern Rockall Trough, which is shallower than the bulk of the large-scale debrites. However, without any evidence from the seismic reflection profiles for modern hydrates or free gas it is inappropriate to imply that hydrates were the trigger for emplacement of the Peach Slide debrites. This is particularly so when the temperature and pressure (water depth) were likely to have varied considerably during the Pliocene and Pleistocene.

Forces generated by earthquake-induced periodical displacements leading to reduction of sediment strength are one of the significant testable causes that are commonly cited in the scientific literature for the initiation of submarine landslides (Hampton et al. 1996). Horizontal accelerations as low as 0.08 g could destabilize a slope of 5° where the sediments have ratios of undrained cohesion (shear strength) to effective overburden pressure of 0.3 or less (Almagor & Wiseman 1982). Higher slope angles occur predominantly on the upper slope, but ratios of undrained cohesion to effective overburden pressure in this zone are usually greater than 0.8, because of the higher shear strengths. Below about 1000 m water depth, where seabed slopes are typically only 1°, horizontal accelerations in excess of 0.1 g would be required to destabilize the slope (Morgenstern 1967). There are limited ground-motion data from this area, but predicted peak ground accelerations less than 0.1 g are given at the 10 000 a return period level, and below 0.05 g at the 1000 a return period (Musson et al. 1997). These values would be at rockhead and may be exaggerated or attenuated by the unconsolidated sediments above. Two earthquakes have been recorded in the vicinity of the Hebrides Terrace Seamount in the last 20 years (Table 2). Although no data exist on the ground accelerations associated with these events, on the basis of other submarine events, accelerations of 0.06 g may be expected within 10 km of the hypocentre (focus) and 0.1 g would probably be a maximum value. Therefore, the potential area for modern earthquakes such as these to destabilize sediments would not extend to the main slope-front of the Barra Fan (Fig. 1). However, if ground accelerations similar to those of the modern earthquakes have occurred under the slope-front, or if earthquakes were historically of higher magnitudes, then it is suggested that these may have triggered slide movement leading to the emplacement of debrites on the Barra Fan.

Conclusions

(1) Except where fan sediments comprising a shelf-margin depocentre appear to be buttressed against the Hebrides Terrace Seamount, channels, gullies and slides are ubiquitous and impart an overall rugged topography to the Barra Fan and the Donegal Fan. Variations in buried topography and internal seismic facies also attest to a process of fan development punctuated by slide failures of varying proportions following fan initiation during the mid-Miocene.

Table 2. *Barra Fan earthquakes*

Reference	Latitude (°N)	Longitude (°W)	Year/Month Day	h/min/s	Depth (km)	Magnitude (M_L)
1	56.44	10.58	1980/04/13	10/47/56.8	2.1	3.3
2	56.09	10.93	1986/12/19	02/15/43.9	5.0	3.1

References: 1, Jacob *et al.* (1983); 2, Turbitt (1988). The tabulated values are recalculated from the references quoted (G. Ford pers. comm. 1996).

(2) The shelf-margin wedge in the northern Barra Fan has been partially destroyed by the Peach Slide, which comprises at least four large-scale submarine slide events decreasing in their extent and volume with time. These have transferred debrites into the deep-water basin of the eastern Rockall Trough and constructed a depocentre 450 m or more thick that is detached from the slope-front of the Barra Fan.

(3) There is a topographic control to the overall NW-trending distribution of debrites originating from the Peach Slide, as they appear to be confined to the slope-front between the Hebridean Escarpment to the north and the Hebrides Terrace Seamount to the south. There may therefore be a structural factor in the large debrite run-out distances. Topographic restriction to the run-out of the largest and oldest debrite into the Rockall Trough appears to have been exerted by the Anton Dohrn Seamount.

(4) NW-trending Cenozoic transfer zones beneath the Barra Fan have been speculatively related to structural controls on the ascent of shallow gas. However, there is no modern evidence derived from the systematic interpretation of the shallow seismic reflection profiles, or from consideration of the theoretical position of the methane-hydrate boundary, to support a hypothesis that shallow gas has triggered translation in the Peach Slide.

(5) Although the observation that the last large-scale failure event on the Peach Slide has destroyed iceberg-scoured seabed at the shelf edge has been tentatively interpreted to mean that it is possibly post-glacial event, there are insufficient data to secure correlation of the timing and extent of slide failures with climate change.

(6) A limited dataset from records of modern earthquakes indicates that post-glacial ground acceleration may trigger local submarine landslides and the deposition of debrites on the Barra Fan.

We thank colleagues D. Evans and K. Hitchen, who made many useful suggestions to improve this paper, and E. Gillespie, who drafted the figures. The manuscript benefited greatly from reviews by R. Whittington and an anonymous referee. Some of the work for this paper was funded by the European Commission MAST II programme as part of the BGS contract for ENAM (European North Atlantic Margin: sediment pathways, processes and fluxes) Contract MAS2-CT93-0064. L. Dodd was in receipt of a research grant from NERC; the seabed topographic data in Fig. 7 originate from her PhD thesis. The swath bathymetric data were made available from the LOIS/SES (Land Ocean Interaction Study/Shelf Edge Study) Community Research Programme of the Natural Environment Research Council. Special thanks are due to P. Weatheral, British Oceanographic Data Centre, who supplied the data for the 3-D topographic model. The authors would like to thank the following oil companies, who together with the BGS, make up the Rockall Margin Consortium, without whose permission the interpretation of airgun profiles west of 10°W could not have been undertaken: Agip, Amerada Hess, Arco, British Gas, BP, Conoco, Elf, Enterprise, Esso, Mobil, Phillips and Statoil. This paper is published with the permission of the Director of the British Geological Survey (NERC).

References

ALMAGOR, G. & WISEMAN, G. 1982. Submarine slumping and mass movements on the continental slope of Israel. *In*: SAXOV, S. & NIEUWENHUIS, J. K. (eds) *Marine Slides and other Mass Movements*. Plenum, New York, 95–128.

ARMISHAW, J. E., HOLMES, R. W. & STOW, D. A. V. 1998. Morphology and sedimentation on the Hebrides Slope and Barra Fan, NW UK continental margin. *This volume*.

BALTZER, A., COCHONAT, P. & AUFFRET, G. A. 1995. Comparaison des facteurs necessaires a l'initiation de glissements sous-marins sur les marges de Nouvelle Ecosse et de Nord Gascogne au Quaternaire. *Comptes Rendus Hebdomadaires des Séances de l'Académie des Sciences, Serie IIa*, **321**, 1001–1008.

BARREIRO, B. 1995. *Sr isotopic compositions and Sr isotope stratigraphy of bioclastic material from Cenozoic sediments from boreholes near the Orkney Islands*. NERC Isotope Geosciences Laboratory Report Series, No. 60.

BOOTH, J. S. & O'LEARY, D. W. 1991. A statistical

overview of mass movement characteristics on the North American Atlantic Outer Continental Margin. *Marine Geotechnology*, **10**, 1–18.

BUGGE, T., BEFRING, S., BELDERSON, R. H., *et al.* 1987. A giant 3 stage submarine slide off Norway. *Geo-Marine Letters*, **7**, 191–198.

DODD, L. 1995. An investigation into the conditions and controls of submarine slope instability – the Var Canyon, Nice, France and the Barra Fan, Hebrides, Scotland. PhD Thesis, University College of North Wales at Bangor.

DORÉ, A. G. & LUNDIN, E. R. 1996. Cenozoic compressional structures on the NE Atlantic margin: nature, origin and potential significance for hydrocarbon exploration. *Petroleum Geoscience*, **2**, 299–311.

ELVERHOI, A., NOREM, H., ANDERSEN, E. S. *et al.* 1997. On the origin and flow behaviour of submarine slides on deep-sea fans along the Norwegian–Barents Sea continental margin. *Geo-Marine Letters*, **17**, 119–125.

EVANS, C. D. R. 1992. *St Kilda (57°N, 10°W) Quaternary geology 1:250 000 Map Series*, British Geological Survey.

HAMPTON, M. A., LEE, H. J. & LOCAT, J. 1996. Submarine landslides. *Reviews of Geophysics*, **34**, 33–59.

HISCOTT, R. N. & AKSU, A. E. 1994. Submarine debris flows and continental slope evolution of Quaternary ice sheets, Baffin Bay, Canadian Arctic. *Bulletin, American Association of Petroleum Geologists*, **78**, 445–460.

HOLMES, R. 1994. *Seabed topography and other geotechnical information for the Shelf Edge Study 55°N–60°N NW of Britain*. British Geological Survey Technical Report, No. WB/94/15.

——, ALEXANDER, S., BALL, K. *et al. The issues surrounding a shallow gas database.* Health and Safety Executive OTH Report No. 96 054, in press.

——, JEFFREY, D. H., RUCKLEY, N. A. & WINGFIELD, R. T. R. 1993. *Quaternary geology around the United Kingdom (North Sheet). 1:1 000 000 Map Series.* British Geological Survey: Edinburgh.

JACOB, A. W. B., NIELSON, G. & WARD, V. 1983. A seismic event near the Hebrides Terrace Seamount. *Scottish Journal of Geology*, **19**, 287–296.

JAMES, J. W. C. 1992. *Peach (56°N, 10°W) Quaternary geology 1:250 000 Map Series* British Geological Survey.

—— & HITCHEN, K. 1992. *Peach (56°N, 10°W) Solid geology 1:250 000 Map Series.* British Geological Survey.

JIANG, L. & LeBLOND, H. P. 1993. Numerical modelling of an underwater Bingham plastic mudslide and the waves which it generates. *Journal of Geophysical Research*, **98**, 10303–10317.

JONES, E. J. W., PERRY, R. G. & WILD, J. L. 1986. Geology of the Hebridean margin of the Rockall Trough. *Proceedings of the Royal Society of Edinburgh, Section B*, **88**, 27–51.

KENYON, N. H. 1987. Mass wasting features on the continental slope of Northwest Europe. *Marine Geology*, **174**, 57–77.

KING, E. L., SEJRUP, H. P., HAFLIDASON, H., ELVERHOI, A. & AARSETH, I. 1996. Quaternary seismic stratigraphy of the North Sea Fan: glacially-fed gravity flow aprons, hemipelagic sediments, and large submarine slides. *Marine Geology*, **130**, 293–315.

KVENVOLDEN, K. 1995. Natural gas hydrate occurrence and issues. *Sea Technology*, **36**(9), 69–74.

LABERG, J. S. & VORREN, T. O. 1995. Late Weichselian submarine debris flow deposits on the Bear Island Trough Mouth Fan. *Marine Geology*, **127**, 45–72.

LONG, D. & BONE, B. D. 1990. Sediment instability on the continental slope of northwestern Britain. *Proceedings of Oceanology International '90*, **2**, Spearhead Exhibitions Ltd, Kingston upon Thames.

MORGENSTERN, N. R. 1967. Submarine slumping and the initiation of turbidity currents. *In*: RICHARDS, A. (ed.) *Marine Geotechniques*. University of Illinois Press, 189–210.

MUSGROVE, F. W. & MITCHENER, B. 1996. Analysis of the pre-Tertiary rifting history of the Rockall Trough. *Petroleum Geoscience*, **l2**, 353–360.

MUSSON, R. M. W., LONG, D., PAPPIN, J. W., LUBKOWSKI, Z. A. & BOOTH, E. 1997. *UK Continental Shelf seismic hazard*. Health and Safety Executive, Offshore Technology Report, No. OTH93 416.

PICKERING, K. T., HISCOTT, R. N. & HEIN, F. J. 1989. *Deep Marine Environments: Clastic Sedimentation and Tectonics.* Unwin Hyman, London, 416 pp.

PRIOR, D. B. & COLEMAN, J. M. 1979. Submarine landslides – geometry and nomenclature. *Zeitschrift für Geomorphologie, NF*, **23**(4), 415–426.

SANGREE, J. B. & WIDMIER, J. M. 1977. Seismic interpretation of clastic depositional facies. *In*: PAYTON, C. E. (ed.) *Seismic Stratigraphy – Applications to Hydrocarbons Exploration*. Memoirs, American Association of Petroleum Geologists, **26**, 165–184.

STOKER, M. S. 1997. Mid- to late Cenozoic sedimentation of the continental margin off NW Britain. *Journal of the Geological Society, London*, **154**, 509–516.

——, HITCHEN, K. & GRAHAM, C. C. 1993. *United Kingdom offshore regional report: the geology of the Hebrides and West Shetland Shelves, and adjacent deep-water areas*. HMSO, London, for the British Geological Survey.

——, LESLIE, A. B., SCOTT, W. D. *et al.* 1994. A record of late Cenozoic stratigraphy, sedimentation and climate change from the Hebrides Slope, NE Atlantic Ocean. *Journal of the Geological Society, London*, **151**, 235–249.

TURBITT, T. 1988. *Bulletin of British earthquakes, 1986.* British Geological Survey Technical Report, No. WL/88/11.

WHITMAN, R. V. 1985. On liquefaction. *Proceedings of the Eleventh International Conference on Soil Mechanics and Foundation Engineering*, San Francisco, CA, **4**, 1923–1926.

Morphology and sedimentation on the Hebrides Slope and Barra Fan, NW UK continental margin

JULIE E. ARMISHAW [1,2], RICHARD W. HOLMES[2] & DORRIK A. V. STOW[1]

[1]*Department of Geology, University of Southampton, Southampton Oceanography Centre, European Way, Southampton SO14 3ZH, UK*

[2]*British Geological Survey, Marine Operations and Petroleum Geology Group, Murchison House, West Mains Road, Edinburgh EH9 3LA, UK*

Abstract: Rapidly deposited Neogene sands and Pleistocene glacigenic sediments originating from NW Britain were transported across the shelf and downslope to the Barra Fan depocentre. In contrast, the Holocene interglacial shelf environment is typified by sediment winnowing, sea-floor polishing and transport of relatively small volumes of sediment along the shelf to the northwest. Interpretation of swath bathymetry, 32 kHz side-scan sonar, 3.5 kHz pinger, sea-bed photography and near-bottom current survey data on the shelf edge and slope provides good morphological evidence for both downslope and alongslope sedimentary processes on this composite slope-front fan. On the northern fan a corrugated shelf edge gives rise to a network of channels on the upper slope, which incise into Pleistocene and older sediments and funnel sediment to middle and lower fan. An underpinning variable topography on the fan bulge originates from major and minor slide masses, sediment creep and debris flows. These are reworked by spatially and secularly variable strong-to-weak bottom currents and redistribute material across and down fan forming transverse and linear bedforms, draped sandy contourite sheets and moulded drifts.

The Barra Fan complex is a Neogene to Holocene sedimentary prism that extends oceanward adjacent to the Hebrides Terrace Seamount between the structural highs of the Geike Bulge in the north and the Donegal Platform to the south (Fig. 1). The Barra Fan north of the Hebrides Terrace Seamount and the Donegal Fan to the south are generally considered to be components of a single fan complex, which forms the largest sediment depocentre off NW Britain and covers an area of c. 6300 km[2]. The oceanward boundary of the Barra Fan in the NE Rockall Trough exceeds 2000 m water depths. Sediments equivalent in age to the Barra Fan post-date the mid- to late Miocene unconformity on the shelf and thicken seaward in a prograding sediment wedge of late Miocene to Pleistocene age between 400 and 700 m thick (Stoker 1995; Holmes *et al.* this volume). These sediments correlate with the oldest fan–basinal sediments on the margin that post-date the regional Rockall Trough unconformity, itself dated at c. 14 Ma (Stoker this volume). Lower Pleistocene shelf sediments are unconformably overlain by a sub-horizontal glacigenic erosional surface c. 250 m below modern sea level, which, in turn, is overlain by a sequence of mid-Pleistocene to Holocene sediments (Selby 1989; James & Hitchen 1992).

The immediate provenance of the sediments is evidenced by the modern shelf topography, which exhibits well-pronounced ridges adjacent to, and sub-parallel to, the shelfbreak. These sediments contain diamict lithofacies and are interpreted as submarine end-moraines deposited during the retreat of ice from the shelf-break during the last regional shelf glaciation (Selby 1989). Middle- to inner-shelf basins formed from glacial overdeepening are filled with post-glacial sediments. These basins occur landward of the outer shelf ridges and have eroded into either glaciomarine sediments or Cenozoic–Palaeozoic bedrock (Holmes *et al.* this volume). Shoulder-to-axis relief of the basins is between 20 and 60 m.

The shelfbreak is sharply defined at 220 m within the shelf embayment of the Barra Fan at the southern limits of the study area but changes northward to a poorly defined convex-upward ramp shelfbreak on the southern Geike Bulge. The Barra Fan complex itself progrades westwards in the centre of the study area and encroaches on the Hebrides Terrace Seamount, which is enclosed to the east by the 1800 m isobath and to the west by the 2300 m isobath.

Previous regional studies with GLORIA of the northern Barra Fan and adjacent slope apron indicated that the principal mode of sediment supply was via turbidity currents associated with the gullied shelf edge and that large-scale downslope erosional processes were absent (James *et al.* 1990). However, more

ARMISHAW, J. E., HOLMES, R. W. & STOW, D. A. V. 1998. Morphology and sedimentation on the Hebrides Slope and Barra Fan, NW UK continental margin. *In*: STOKER, M. S., EVANS, D. & CRAMP, A. (eds) *Geologial Processes on Continental Margins: Sedimentation, Mass-Wasting and Stability*. Geological Society, London, Special Publications, **129**, 81–104.

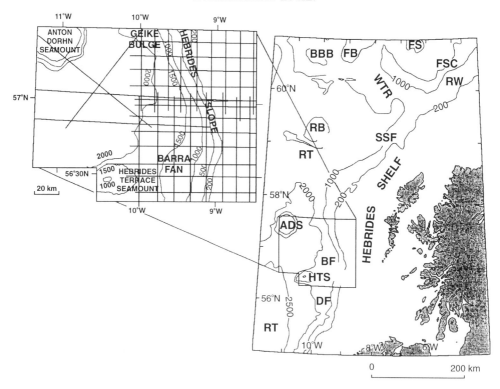

Fig. 1. Location of the study area, bathymetric setting (contours in metres) and track plots of regional airgun and sparker lines. South of *c.* 57°N, swath bathymetric data are available between 150 and 1900 m water depth for most of the study area. BBB, Bill Bailey's Bank; FB, Faeroe Bank; FS, Faeroe Shelf; WTR, Wyville-Thomson Ridge; RW, Rona Wedge; FSC, Faeroe–Shetland Channel; RT, Rockall Trough; RB, Rosemary Bank; SSF, Sula Sgeir Fan; ADS, Anton Dohrn Seamount; BF, Barra Fan; HTS, Hebrides Terrace Seamount; DF, Donegal Fan.

recent studies of the British Geological Survey (BGS) 3.5 kHz sonar dataset have shown mass-transport, where large segments on the North Barra Fan have been transported downslope into the mid-NE Rockall Trough in several slide events associated with a mega-slide known as the Peach Slide (Holmes 1994).

Physical oceanographers have known, since 1832, of a major surface current, the North Atlantic Drift, flowing northward along the coastal margins west of Ireland and Scotland (Deacon 1996). More than 130 years later, the impact of this current on sea-bed sediments began to be documented, when initial studies in the area of the Barra Fan complex began in the 1970s. Submersible operations described alongslope sediment transport in waters NW of Scotland in terms of E–W aligned sandy megaripples indicating a general northward current flow (Eden *et al.* 1971). Subsequently, a persistent current flowing along the edge of the continental shelf west of Britain, which represents a continuous mid-water filament of the North Atlantic Drift transport, and carrying water of mixed Atlantic origins was

described by Ellett *et al.* (1986). Kenyon (1986) established a link between the bedforms and this strong northward-flowing current on the upper slope. He recognized an internally well-layered acoustic facies of linear slope-parallel bedforms with vertical relief of up to 10 m and wavelengths of several kilometres, which can be related to the acceleration of contour-parallel currents in areas where slope gradients range from 4° to 20° (Leslie 1992). Comparison of bedforms on the Barra Fan and on the Geike Bulge, just north of the region, may provide evidence for a more complex origin.

The objective of this paper is to explore the interaction between downslope gravitational and parallel-to-slope geostrophic processes on the North Barra Fan between 150 m and 1800 m water depths and 56°05′N and 57°15′N (Fig. 2). We attempt to interpret the modern sea-floor morphology and the underlying buried morphology in terms of sedimentary processes based primarily on high-resolution 12 kHz swath bathymetry, 32 kHz TOBI sidescan surveys, and 7.5 kHz and 3.5 kHz sonar data. A time series of

(a)

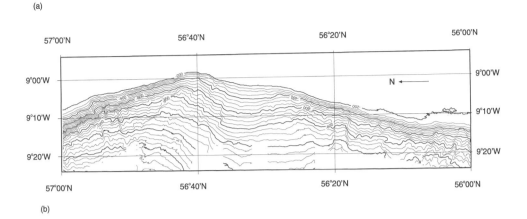

(b)

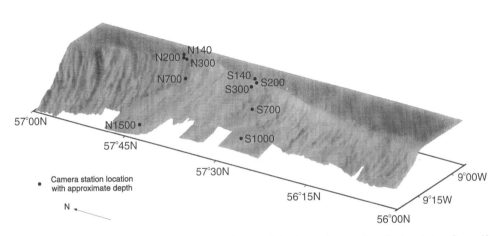

Fig. 2. (a) Bathymetry of the Hebrides Slope based on swath-bathymetric mapping. Contour intervals are 40 m spacing. **(b)** 3-D greyscale perspective image of the sea-bed topography derived from processed EM12-120 swath bathymetric survey data. The image is artificially truncated near the shelf edge at the 200 m isobath and at its westernmost downslope extent, corresponding to the 1500 m isobath. Artificial illumination of the sea-bed topography accentuates SE-facing scarps, which are in shadow, and conceals the lighter grey N-facing scarps. Areas of white within the frame are areas of no data. Data for the topographic model are courtesy of Pauline Weatherall of the British Oceanography Data Centre. Black dots mark the location of camera stations, with approximate depth.

bottom photographs taken during an 18 month period and combined with core samples allow calibration of the remotely sensed data, as well as observations on seasonal variation of sea-bed micromorphology.

Data and methods

The bathymetric chart (Fig. 2) of the region between 56°N and 57°N was obtained from the lines traversed during RRS *Charles Darwin* cruise 91, equipped with a Simrad EM12-120 (12 kHz) multibeam swath-bathymetry echosounder. This device consists of a hull-mounted echosounder, and provides an effective means to map in detail a swath of the ocean floor on either side of the ship's track. The EM12-120 system generates 81 stabilized beams, providing ±60° athwart coverage in water depths between 100 and 11,000 m, though depth coverage here ranges between 150 and 2000 m. Because of difficulties of following the desired survey lines in extreme weather conditions, the survey area was constructed by revisiting subareas several times during the cruise. This resulted in editing and repetition of the gridding process and produced charts on a 1:50,000 scale with a contour interval of 10 m.

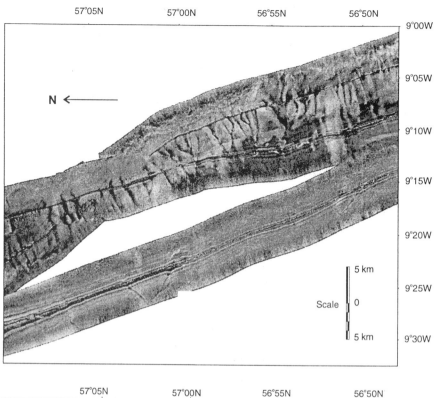

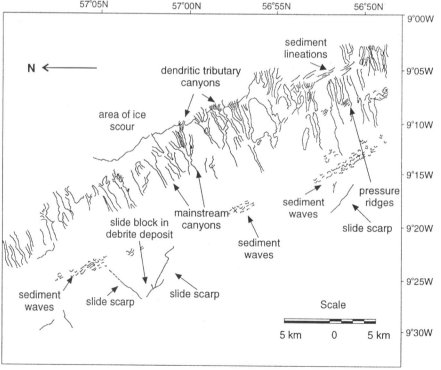

The geomorphic reconnaissance of the continental slope from 56°N to 57°20'N was conducted using the Institute of Oceanographic Sciences Deacon Lab TOBI side-scan sonar system, which has a neutrally buoyant fish towed 200 m horizontally behind a heavy depressor weight. The starboard and port side-scans operated at frequencies of 30 kHz and 32 kHz respectively, with a repetition rate of 4 s (Le Bas et al. 1995). Despite rough conditions restricting the direction of travel, a total of 85 hours of good, interpretable 6 km swath side-scan sonar and 7.5 kHz sub-bottom profiles was gathered. After reception and processing, signals were photographically anamorphosed to take into account the ship's speed over the sea floor and to correct for the horizontal scale, and the overlapping tracks were then mosaiced to produce the final copy maps (Figs 3–5).

Bottom photographs were obtained using the Proudman Oceanographic Laboratory UMEL deep-sea survey camera. A time series of shots were taken during a period of 18 months, between March 1994 and November 1996, to allow seasonally variable observations of sea-floor micromorphology to be monitored. The sea-bed area photographed is trapezoidal in shape and c. 3 m front to back. The width across the bottom of the frame is c. 120 cm and that across the top is c. 250 cm (J. Humphery, pers. comm.).

Classification and mapping of microtopography is based on the study of 3.5 kHz echograms taken during the BGS 85/06 survey in 1985. The principles of echo-character mapping and relationships of various echo types to sea-floor topography and near-bottom processes have been described in detail by J. Damuth and co-workers (Damuth 1975, 1978, 1980, Johnson & Damuth 1979; Damuth et al. 1983), and we use a modified version of his classification based on our own observations.

A total of 12 cores have now been collected from the study area (Fig. 2) and preliminary descriptions of these cores have been used to calibrate the remotely sensed data.

Bathymetry

The Hebrides Shelf east of the study area is broad and relatively smooth. It extends for about 150 km in width with shallow basins on the inner shelf and a broad smooth outer shelf (Fig. 1). The slope width in the study area ranges from

about 50 km in the north to more than 100 km in the central area (including the Hebrides Terrace Seamount). The Barra Fan complex extends downslope both north and south of this seamount.

Based on recent data from EM12-120 swath bathymetry surveys (cruises CD91A and 91B, spring 1995) and IOSDL TOBI surveys, we have been able to produce a much more refined bathymetric map of the area than was previously available (Fig. 2). The 10 m spacing between isobaths permits the resolution of small-scale morphological features in four distinct zones based on marked changes in the slope gradient.

The outer shelf and shelfbreak. The outer shelf and shelfbreak are of interest to the present study as the immediate source of material for downslope resedimentation. The shelfbreak occurs between 150 m and 250 m below sea level (b.s.l.), but mainly ranges between 180 and 220 m. The average slope of the shelf ranges from less than 0.2° to 0.5°, and is characteristically smooth. The gently convex shelfbreak in the study area is often poorly defined, and is generally deeper where slope angles are lower. The shelfbreak is shallowest (150 m) where sliding and slumping have cut back into the shelf edge. Bathymetric features observed include iceberg scour marks, longitudinal furrows, sand streaks, sand ribbons, sand patches and sand waves. In areas where the shelfbreak is well defined, canyons produce a crenulated topography.

Upper slope. The upper slope (200 to 1000 m) is characteristically steep with angles ranging from 2.5° to 3.5°; however, a maximum slope of 16° occurs near slide escarpments. The upper part of the slope has a convex-upward profile, whereas the lower part of this zone tends to be concave. The upper slope has a characteristically steep and rugged terrain, with canyons, debris-flow deposits, contourite deposits, slides and fault scarps all well defined. Of particular note are the most recent boundaries of the Peach Slide, clearly visible as steep-sided curvilinear traces on the bathymetric chart and on TOBI images (Figs 2, 4 and 5).

Middle slope. The middle slope occurs between depths of 1000 m and about 2000 m b.s.l. Slope angles are lower than those of the upper slope, and vary between <0.1° and 1.6°, but may locally reach 3°. Slope profiles indicate that the slope is

Fig. 3. TOBI side-scan mosaic and interpretive map of the northern region of the Barra Fan. Location is shown in Fig. 1.

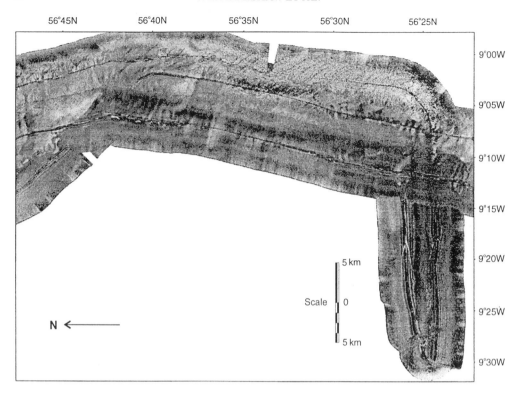

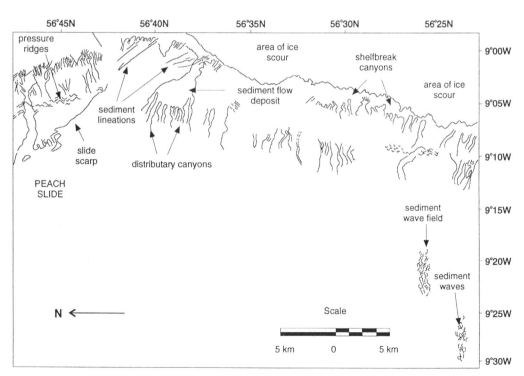

notably smooth in parts and irregular in others. The main bathymetric features include canyons, debris-flow lobes and slide masses. Canyons tend to be broader and shallower than those in the upper to middle slope. The Peach Slide cuts across the central region and shows a much broader and more complex morphology, having smaller channels, scarps and slides superimposed on it.

Lower slope. The lower slope at depths in excess of 2000 m is mostly beyond the boundary of the present study area. Gradients are low ($<0.5°$) and the bathymetry is mainly smooth, except for a more irregular, convex-upwards region just below the change in gradient from the middle slope. This region represents depositional debris-flow lobes, and the distal edges of debris flows or slide masses.

Regional variations. In addition to these depth zones there are distinct regional variations in bathymetry. Moving from north to south across the region there is a significant contrast in the appearance of both the slope itself and the features on the slope. In the northern sector of the region ($56°50'N$ to $57°15'N$), the slope tends to be steep and heavily dissected with canyons. These canyons are largely continuous from the upper slope, in depths of 300 m down to at least the mid-slope areas, and more commonly down to the basin plain at depths of 1800 m or greater. Many of the canyons disappear downslope without giving rise to any distinct constructional features, whereas others appear to fan out into very low-relief, elongated debris-flow lobes where there is a significant decrease in the slope gradient. Canyons in this northern area are most striking in the upper and middle slope reaches between water depths of 500 and 1100 m, where slope angles range between $2°$ and $4°$, but may locally exceed $6°$. Canyons are sharply defined, straight to slightly sinuous, broaden down the slope (coincident with decreasing slope angles) and are commonly found in association with ridges, channel–levees and terraces. Lengths frequently exceed 10 km, widths vary between 150 and 1000 m, and depths range from 20 to 100 m. Between 30 and 40 distinct canyons or part canyons can be identified in a 28 km wide band to about $56°50'N$. South of $56°50'N$ the abundance and continuity of canyons begins to change. The most apparent change is in their confinement to slope areas above 800 m, which is coincident with higher slope gradients and with the northern margin of the Peach Slide, which is one of the most significant features of the Fan.

The Peach Slide occurs in the central area of the study region between $56°50'N$ and $56°35'N$ and is definable by its distinct margins (Figs 2, 4 and 5). The northern margin exhibits several slide scarps that lie parallel and sub-parallel to one another for lengths of up to 7 km alongslope. In contrast, the southern margin is mappable as a mound-like feature >12 km wide alongslope and a relief of up to 50 m above the adjacent slope. The slide is a composite and complex feature covering *c.* 1600 km^2 and has a total of seven slide scarps with lengths varying from 10 to 20 km, displacements of between 10 and 50 m, and slopes averaging $38°$, with some in excess of $50°$. The majority of the larger scarps are parallel or sub-parallel to the strike of the slope (i.e. N–S to NW–SE), whereas smaller scarps are generally orientated at right angles to the slope (i.e. downslope). These smaller slide scarps commonly cross-cut recent topography and disrupt pre-existing canyons. Below the NW–SE orientated scarp, running from approximately $56°47'N$, $9°13'W$ to $56°43'N$, $9°02'W$ the slope becomes smoother and shallower, and gradients decrease from an average of $4°$ above the scarp to $2°$ below it. Canyons, although abundant in the upper-slope areas only on the north slide margin are extensive and larger in the southern slide region. The blocky nature of the slide, particularly on the slope area along $56°40'N$, may signify an area where slope creep is or has been active.

South of the Peach Slide, the main feature is the Hebrides Terrace Seamount, which extends up from the sea floor from 1660 m on the eastern flank and 2300 m on the western flank to a summit of less than 1000 m b.s.l. The presence of this seamount, a remnant of an early Tertiary volcanic centre, has had a major influence on the nature and development of the slope in this region. This seamount has served as a barrier against which sediment has banked up and so contributed to the shallower slope gradient, and has diverted the major influx of sediment to the Barra Fan complex to the north, forming the

Fig. 4. TOBI side-scan mosaic and interpretive map of the southern region of the Barra Fan, incorporating the Peach Slide. (Note the southern slide scarp of the Peach Slide, which is outlined by the sharp contrast between high reflectivity (white) returns of the slide and the low reflectivity (dark grey) of undisturbed sea bed.) South of the Peach Slide, ice scour appears as a mottled area on the mosaic; some lineations on the upper slope and canyons can also be distinguished. Location is shown in Fig. 1.

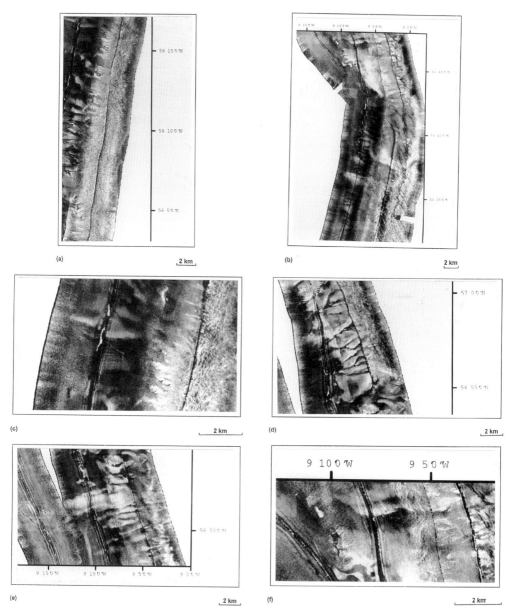

Fig. 5. TOBI side-scan images of the main morphological elements. (**a**) Ice scour marks. Randomly orientated features up to 500 m in length and 50 m wide, common on the outer shelf and upper slope regions in depths up to 460 m. Refer to Fig. 4 for location. (**b**) Peach Slide mega-slide. Note the body of the slide which is outlined by the contrast between areas of very bright (white) backscatter and darker (grey) backscatter. (**c**) Micro-slide occurring independently of both mega- and meso-slides in the upper slope between depths of approximately 700 and 800 m. Refer to (**a**) for location. (**d**) Canyons. Steep-sided erosive slope canyons which are fed by a number of shorter tributary canyons giving rise to a complex, dendritic network of canyons on the steep upper slope to the north of the region. Refer to Fig. 3 for location. (**e**) Debris-flow deposits and lobes. Typical irregular lobate deposit, identifiable by its very bright (white) backscatter, caused by surface irregularities, situated in the intercanyon region of the upper slope. Refer to Fig. 3 for location. (**f**) Pressure ridges and local irregularities. Marked by the thin, parallel-aligned curvilinear stips of high (white) backscatter superimposed on regions of darker (grey) backscatter. Refer to Fig. 4 for location.

Barra Fan (*sensu stricto*), and to the south, forming the somewhat smaller Donegal Fan (Fig. 1). The bathymetric constriction formed between the crest of the seamount and the upper slope–shelfbreak may, at times, have constrained and accelerated the northward flow of bottom water in the mid-slope current. The slope directly east of the seamount (approximately 56°34'N to 56°20'N) is particularly smooth and devoid of major geomorphological features.

In the most southerly region of the study area (south of 56°20'N) the slope is more irregular, with several distinct slide scarps, irregular slide masses and canyons. The canyons are more widely spaced than those in the northern area, and tend to head in the upper slope (around 350–450 m) rather than at the shelfbreak.

Morphological elements

Combined use of TOBI side-scan sonar, 7.5 kHz reflection profiles and EM12-120 swath bathymetry has allowed very precise documentation of the sea-floor morphological elements that characterize the Hebridean margin. Seven major elements are recognized, including ice scour features, canyons, slides, debris-flow masses, lobes, bottom-current bedforms and smoothed sediment surfaces (Figs 3–5). Until core and photo coverage is greater in the study area, our interpretations of sediment types must be based, in part, on inferences from previous studies with better core calibration and, in part, on the assumption that the strength of side-scan backscatter is controlled by a combination of sea-floor roughness and sediment grain-size.

Ice scour marks

Ice scour marks are narrow elongate, erosional features, typically 100–500 m length, 40–50 m wide and 1–3 m deep (Belderson *et al.* 1972; Kenyon 1987). Their random orientation and cross-cutting relationships provide a striking 'speckled' sonar facies with elongate patches of high and low backscatter, that characterizes the outer shelf and upper-slope regions to depths of around 400 m (maximum 460 m) (Figs 3 & 5). Scour marks are patchily distributed alongslope, with the most intense areas between 56°55'N and 57°05'N and between 56°35'N and 56°05'N (Figs 3–5). In this southern area, the speckled sonar facies is superimposed on a series of sub-parallel lineaments orientated SE–NW and extending some 9 km across the outer shelf and upper slope. These are provisionally interpreted

as a specific type of ice scour mark. In areas where ice scour density is less, it is possible that the marks have been removed by erosion or buried by subsequent sedimentation. The very bright backscatter, particularly on the scour depressions, suggests a coarse sand–gravel sediment type perhaps cut into a finer-grained (?) sandy sea floor.

Slides and slumps

Slides and slumps are recognized on the basis of distinctive curvilinear scarp edges, running sub-parallel to the strike of the slope in the head region and downslope along the lateral margins. There may be one or more sub-parallel closely spaced scarps marking the headwall, indicative of retrogressive sliding. The main body of the slide, and especially the toe region, locally shows an irregular or chaotic topography, although this is not the case everywhere.

Three scales of slides can be recognized. (1) Mega-slides have a surface area >500 km², headwall scarps of 10–25 km length and 10–100 m displacement. These are initiated on the upper slope and extend across the middle- to lower-slope area. They are generally composite in nature, as illustrated by the Peach Slide (discussed below) (Figs 2, 4 and 5). (2) Meso-slides have an area of 50–500 km², headwall scarps of 1–10 km length and typically 5–50 m vertical displacement. These occur throughout the region on the upper and middle slope (Figs 3 & 5). (3) Micro-slides are 10–20 km² in area with headwall scarps 1–10 km long showing vertical displacement of less than about 20 m (Fig. 5). These occur both independently and superimposed on the mega- and meso-slides. The sediments in all of these slide masses are the normal mud-rich slope sediments.

Canyons

The Hebrides Slope in the study area is highly channelled although many of the channels are only relatively short, shallow features. Three distinct types can be recognized as follows:

(1) Shelfbreak canyons are short (1–4 km), narrow incisions eroded into the shelfbreak and uppermost slope, ranging from 15 to 50 m in relief. They are typically closely spaced (every 0.5–1 km) and form part of a dendritic tributary network feeding the main slope canyons. Canyon gradients range from about 0.2° to 2° (Figs 3 and 4).

(2) Slope canyons are longer (10–20 km), broader (0.5–2 km), straight to slightly sinuous features that cut across the upper- and mid-slope

regions. They range up to 100 m or so in relief, are mainly erosive in nature but may show some constructional levees in their lower reaches. They are typically fed by a network of tributary canyons and, in some cases, give rise to a series of distributary canyons. In other cases they appear to broaden out into lobes-like bodies in the mid- to lower-slope region. Canyons' gradients range from about 2° to 4°, being steepest in the middle reaches and flattening out towards the lower slope (Figs 3–5).

(3) Slide-margin canyons are distinctive slope gullies that follow the lateral scarp margins of mega- and meso-slides. They may be continuous up to 15 km downslope with widths of 100–600 m and relief up to 100 m (Fig. 4).

We have little direct evidence of sediment types associated with the different canyons. Whereas the outer shelf sands and gravels are probably funnelled downslope via the channel network, sonar images reveal only low-reflectivity (dark) canyon axes, which would imply fine-grained sediment (Figs 3–5). However, the interchannel areas have a higher reflectivity (light), which suggests that topography, rather than the sediment grain-size, is dictating the sonar images. Thus the dark colour of the canyons axes may be a shadow effect.

Debris-flow deposits and lobes

The intercanyon areas on the upper slope show relatively high reflectivity that extends downslope in an irregular lobate form for various distances (Figs 3–5). The high backscatter is interpreted as reflections from small-scale surface irregularities (e.g. blockiness). The lobate terminations tend to coincide with an upper- to mid-slope gradient change. These observations suggest that much of the intercanyon slope area is made up of debris-flow deposits.

A few of the larger debris-flow deposits extend part way across the mid-slope region, causing local irregularities, possible pressure ridges and isolated blocks (Figs 3–5). Much of the mid-slope region, however, shows much more subtle downslope-orientated dark–light reflectivity patterns. These are tentatively interpreted as the downslope (probable turbidite) deposits or lobes derived from the debris flows. The canyons between the debris-flow masses mostly fade out into and across the mid- to lower-slope region. However, in a few cases, they appear to fan out into a lobate form with subtle convex-upward relief. These features are interpreted as terminal lobes at the ends of canyons

Elongate bedforms

Narrow, elongate features characterized by light-to-dark alternations in backscattering intensity (TOBI data) occur on parts of both the upper slope and mid-slope as broad zones extending 5–10 km alongslope, generally above 700 m water depth, and 5–10 km downslope at depths greater than about 800 m. Smaller-scale patches that extend for 1–2 km alongslope also occur in the upper-slope region (Figs 3 and 4).

The deeper-water across-slope features have a distinct undulatory sea-floor morphology and are therefore interpreted as sediment waves formed by parallel-to-slope bottom currents. They have sinuous crests, wavelengths of 0.1–1.2 km, and amplitudes of 10–100 m. Many of the shallower-water features have a more crenulated to cusp-like lineation and an irregular blocky microtopography superimposed upon the ridge–trough morphology. The spacing between adjacent ridges ranges from 0.1 to 1.5 km, and the height from trough to ridge crest ranges from *c.* 10–80 m. We interpret these as pressure ridges formed by buckling of the top few tens of metres of sediment as a result of downslope creep, slumping or sliding. In some cases they have then undergone subsequent modification by alongslope bottom currents resulting in differential sedimentation and local erosion. Clearly, they will also have affected any downslope flows.

Echo character

Echo character observed on continuous 3.5 kHz echogram profiles shows six distinct echo types on the Hebrides Slope. Our simple classification scheme and interpretation of echo character follows previous work by Damuth (1975, 1978, 1980), Johnson & Damuth (1979) and Damuth *et al.* (1983).

Type IA (Damuth et al. *1983, Types IA and IB).* Sharp, distinct to indistinct bottom echoes with numerous conformable parallel to sub-parallel reflectors (Fig. 6, profile A). This echo type is the most widespread across areas of flat to gently rolling sea floor where the slope is smooth, low angle, and topographically feature-less. The identical acoustic character is also observed in areas of hummocky and hilly sea-floor topography, where the hummocky returns are caused by the undulations in the underlying acoustic basement (e.g. Damuth *et al.* 1983). The well-stratified appearance and its conformable draping of this echo type over irregular subsurface topography indicate settling of

A

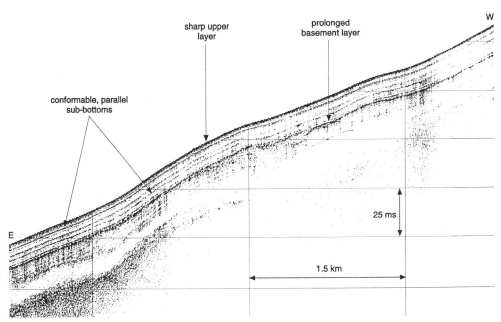

B

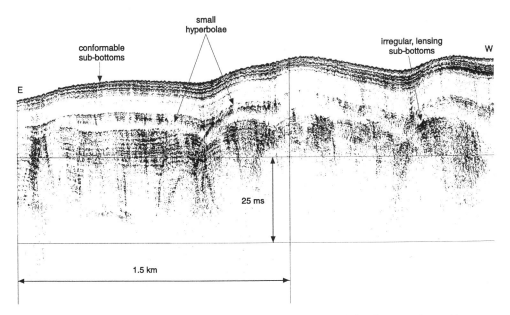

Fig. 6. Echo Types IA (profile A) and IB (profile B). Sharp, distinct to indistinct bottom echoes with numerous conformable to sub-parallel sub-bottoms characteristic of smooth to gently rolling sea floor, where the slope is low angle and topographically featureless.

C

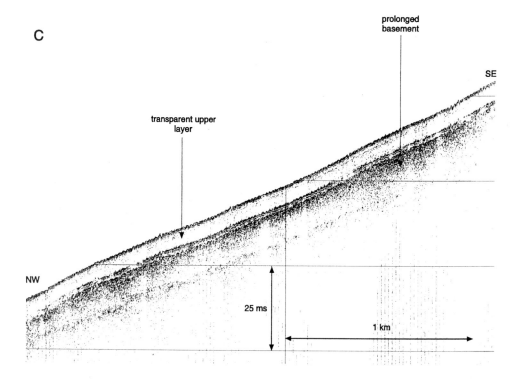

prolonged
basement

SE

transparent upper
layer

NW

25 ms

1 km

D

SE

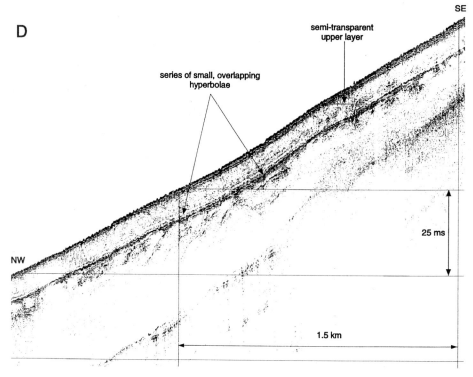

semi-transparent
upper layer

series of small, overlapping
hyperbolae

NW

25 ms

1.5 km

hemipelagic sediment through the water column where no significant bottom currents are active (Damuth 1975; Chough & Hesse 1985; Yoon *et al.* 1991). In some areas the uppermost surface is composed of very small, regular, overlapping hyperbolae which may indicate the presence of weak bottom-current activity. Where the reflections are stronger we interpret the section as comprising interbedded hemipelagites and turbidites.

Type IB. Thin or thick, distinct acoustically well-layered sequences with numerous conformable sub-bottom reflectors similar to Type IA echoes, but they are classified separately on the basis of their different acoustic basement. Acoustic basement beneath Type IB echoes comprises either conformable or unconformable sequences of chaotic, irregular lensing sub-bottom echoes, or small regular single or overlapping hyperbolae (Fig. 6, profile B). The distribution of Type IB echoes is extensive (for tens of kilometres) throughout smooth lower-slope areas, mainly below 1200 m, and across areas where slope irregularities are more prominent than for Type IA echoes. Type IB echoes are also returned from small scarp faces and in areas of where upper-slope channels and gullies have dispersed into broad lower-slope channels or lobes. The upper layers, like Type IA echoes, are interpreted as slope hemipelagites and turbidites, whereas the underlying zone probably represents a slope system dominated by canyon dissection, slumping and debris flows.

Type IIA (Damuth et al. *1983, Type II).* Sharp bottom echo or echoes, underlain by an acoustically transparent to semi-transparent zone with an intermittent, discontinuous prolonged basement (Fig. 7, profile C). This basement layer, itself, may have parallel to sub-parallel sub-bottom reflections, may be topographically smooth or gently rolling, and is also widespread and generally found in upper- and mid-slope areas below 500 and 600 m where the sea-floor morphology is relatively steep, rugged and dissected by straight and meandering canyons. These well-layered and strong reflections at the surface or subsurface are interpreted as turbidites, whereas the transparent zones more probably indicate debris-flow deposits.

Type IIB. These echoes have an upper zone similar to that of type IIA echoes, but are classified separately because of their different acoustic basement. Acoustic basement is generally unconformable and it overlies a sequence of chaotic, irregular, wedging or truncated sequence of sub-bottom reflections. These reflections may be interspersed with parallel sub-bottom reflectors or single or overlapping small hyperbolae (Fig. 7, profile D). Echo Type IIB is observed in close association with Type IIA echoes. Both types are more prominent in the northern part of the area, as well as on the steeper upper-slope region (Fig. 5). Type IIB echoes are interpreted as turbidites and debrites, with the underlying zone being more indicative of extensive slope disruption by slides, slumps and debris flows.

Type III (Damuth et al. *1983, Type III).* Distinct to indistinct bottom echoes with intermittent to continuous zones of conformable, migrating, truncated or wedging sub-bottom reflectors, in some cases with zones of intermittent regular overlapping hyperbolae (Fig. 8, profiles E and F).These echoes are those that occur in areas of undulating sea-floor topography. For the most part, Type III echoes occur along the steep upper slope in areas of rugged terrain, but there is also one large mid-slope occurrence in the north over smooth gently rolling sea floor. In some cases the wavy or irregular topography appears to have formed by slope creep and slide processes and the overlying reflectors indicate bottom-current control of hemipelagic input (e.g. Fig. 8, profile E). In other cases, the sea-floor waves may be due to entirely bottom-current control (e.g. Fig. 8, profile F).

Type IV echoes. Irregular to more regular, overlapping hyperbolae commonly with greatly varying vertex elevations (Johnson & Damuth, 1979, Types IIIA and IIIB) (Fig. 9). Echoes are patchily distributed across the region. In most cases these are best interpreted as an irregular

---

Fig. 7. Echo Types IIA (profile C) and IIB (profile D). Sharp bottom echoes underlain by an acoustically transparent to semi-transparent zone, characteristic of flat or gently rolling sea-floor topography. These echoes are divided into two subtypes to distinguish regions with prolonged parallel to sub-parallel acoustic basements (echo Type IIA) and regions with more unconformable, chaotic, wedging, and truncated sub-bottom reflections interspersed with single or overlapping hyperbolae (echo Type IIB).

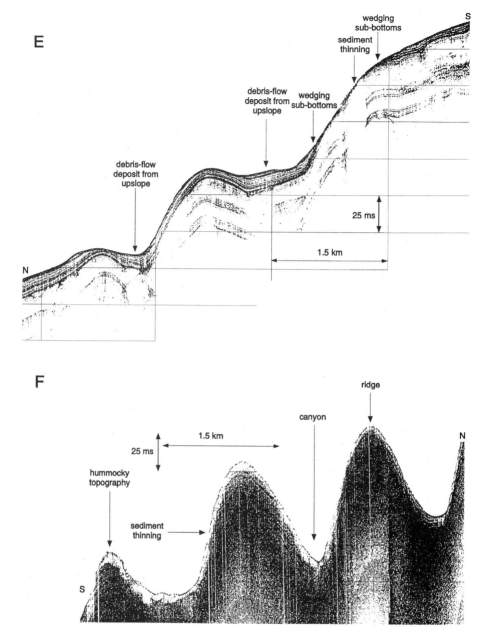

Fig. 8. Echo Type III (profiles E and F). Distinct to indistinct bottom echoes with intermittent to continuous zones of conformable, migrating or wedging sub-bottoms, or intermittent zones of regular single or overlapping hyperbolae.

sea-floor topography resulting from slump or slide masses. Type IV echoes occur on the steep upper slope, in association with the Peach Slide and near the foot of the Hebrides Terrace Seamount. The presence of smaller, more regular hyperbolae at some locations may indicate some bottom-current control.

Seabed photography

The Proudman Oceanography Laboratory deep-sea survey camera was deployed on several occasions on cruises between March 1995 and February 1996. Numerous bottom photographs were taken at each of five stations along a

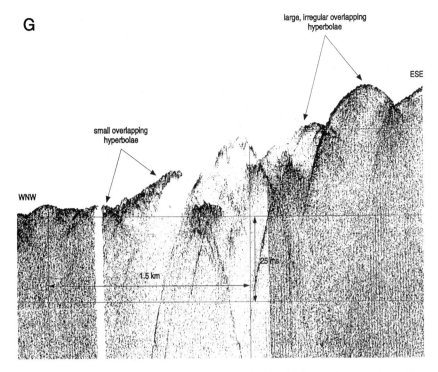

Fig. 9. Echo Type IV (profile G). Irregular, overlapping hyperbolae with large, often greatly varying vertex elevations.

southern transect running from 9°W to 9°25'W, between depths of 140 m and 1000 m, and at five stations along a northern array between 56°35'N and 56°45'N at depths between 140 m and 1500 m. All were taken on open slope sites. The photographs illustrate a variety of features attributable to current activity, normal marine hemipelagic sedimentation, and iceberg activity. Faunal (population type and density) and anthropogenic (trawl marks and relict core sites) indicators have been noted from several locations and are useful measures of current strength and the extent of sediment reworking.

Sediment texture

Both transects show a clear decrease in mean grain size of the surface sediment cover with depth downslope. The outer shelf and upper slope (140–300 m) have coarse cover with up to 25% cobbles and boulders (2–20 cm diameter, maximum 50 cm) (Figs 10–12). Sand cover persists, with less gravel, to a depth of *c.* 700 m (Fig. 13). From 700 m (Fig. 13) to 100 m (Fig. 14), the surface cover changes from mainly silty mud to mud, with both rare gravel patches and very rare cobbles and boulders (maximum 25 cm diameter). The deepest camera station, at 1500 m

(Fig. 15), shows a muddy surface which is distinctly granular in parts.

Current-induced features

Current-induced features are characteristic of sites between 140 and 1000 m, including sand and silt ripples and obstacle marks such as crescents and sediment shadows (e.g. Lonsdale & Hollister 1979). Other indicators of current activity observed in the study area are crag-and-tail structures, which are often infilled with a coarser-grained lag deposit, current lineations and areas of smoothed sediment surface (Fig. 10).

One of the most common features across the slope are scour crescents, which typically exhibit an upstream moat partly infilled with coarse gravel-lag and a downstream sandy mound in the lee of the obstacle (Fig. 11b), commonly a large stone or boulder. Scour is observed at sites from 140 m (Fig. 10), where it clearly occurs around the southern faces of the obstacles, down to 700 m, where it is indistinct and most likely to be relict from previous current action (Fig. 13b). The strongest scouring is observed in mixed grain-size populations on the shelf and upper-slope regions (140–300 m) (Figs 10–12), where scours up to 10 cm deep and 20 cm × 30 cm in

Fig. 10. Bottom photograph from station N140 in 140 m of water. Development of crag-and-tail structures around the bases of lithic clasts, current lineations and areas of smoothed sediment surface. The sea-bed area photographed is trapezoidal in shape and *c.* 120 cm across the bottom and *c.* 250 cm across the top of the frame (J. Humphery pers. comm.). Location is shown in Fig. 2b.

area occur on the southern sides of obstacles (Fig. 11b). Deep scour crescents are good indicators of strong and persistent current conditions at the sea bed, with velocities of at least 12–15 cm s^{-1} and maxima of greater than 20–25 cm s^{-1} (Lonsdale & Hollister 1979). Occurrence of scours around the southern faces suggests that the highest-velocity currents flow in a northerly direction (Fig. 12a). Where current velocities are markedly increased, elongate patches of coarse gravel-lag occur in place of lee drift deposits.

Crag-and-tail structures (Fig. 10), are commonly associated with current scour and are common between depths of 140 and 300 m. A tail of sediment up to 5 cm develops in the lee of a small obstacle and represents the reduction of current strengths from strong upstream to moderate in the lee of the obstacle (Lonsdale & Hollister 1979; Howe & Humphery 1995).

The most common and most extensive current features observed in the photographs are sand–silt ripples. These range from patterns of confused, asymmetric sand ripples on the upper shelf and slope (140–300 m) to extensive ripple fields of well-aligned, straight-crested silt ripples, as far down the slope as 1000 m (Fig. 14c). The most commonly observed ripples are asymmetric sand and silt ripples with wavelengths of 10–50 cm and amplitudes of 1–5 cm mostly found at the 200 and 300 m stations (Figs 11 and 12), although asymmetric ripples with very small amplitudes of 0.5 cm are present to depths of 1000 m (Fig. 14a).

Many of the ripple troughs have coarse sand or fine gravel lag deposit, indicating some form of current winnowing from higher-velocity currents. Ripple crest orientation is variable, depending upon the position and depth at which they are found. The majority show advection towards the N to NNE (e.g. Fig. 13), or more rarely, NE or NW (Fig. 14a), indicating a generally northerly-flowing current. At the 200 m station, linguoid ripples developed on a thick sand sheet indicate a NNE-advecting current (Fig. 11a). At the same station, in August, there is a more confused pattern with a dominant ripple set showing current flow towards the WNW and a second set towards the NE (Fig. 11b). N–S orientated sand ribbons are less common, but also occur at these shallow stations (Fig. 10).

Symmetric ripples are most common in the lower part of the mid-slope. Extensive ripple fields are observed at the 700 m station during April (Fig. 13a) and April (Fig. 14a), and at the 1000 m station during November (Fig. 14c). The dimensions are similar to those of the asymmetric ripples, with wavelengths of 20–30 cm and amplitudes of 1–5 cm.

Discussion

A variety of techniques have been used to study the slope margin. By combining all these strands of data, several conclusions regarding the classification of the slope and process evaluation

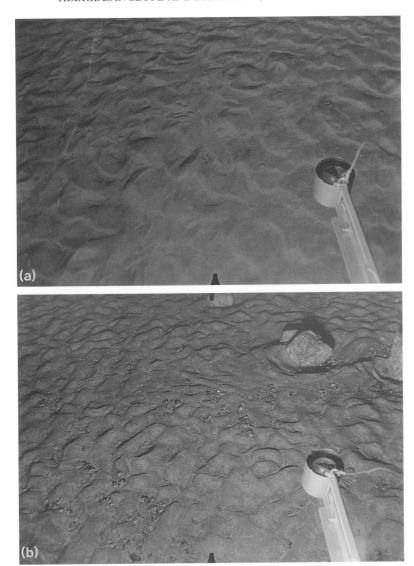

Fig. 11. Bottom photographs from camera stations N200 and S200 in 200 m of water. (**a**) March. Development of a thick sand sheet, subject to small-scale turbulence, with chaotic sand-ripples indicating a NNE-advecting current. (**b**) August. Range in grain-size from sand to large stones (up to 35 cm across) with the development of scour and crag-and-tail structures around larger stones. In parts, the sand is thick enough to bury stones, indicating sand-drifts. Where this occurs, the surface is reworked into current-generated ripples, which may be confused and show current motion in two directions, towards the WNW (major ripples) and NE (superimposed ripples). The sea-bed area photographed is trapezoidal in shape and *c.* 120 cm across the bottom and *c.* 250 cm across the top of the frame (J. Humphery, pers. comm.). Locations are shown in Fig. 2b.

have been drawn, a number of which are discussed below.

Slope apron or composite slope-front fan

The Barra Fan–Donegal Fan complex is the largest depocentre on the NW UK continental margin, and covers an area of *c.* 6300 km² (Stoker *et al.* 1993). The fan morphology forms a seaward bulge on the Hebridean slope that is about 150 km wide and extends downslope for 90–100 km to the Hebrides Terrace Seamount (Fig. 1). Stoker *et al.* (1993) described the Barra Fan, together with the Sula Sgeir, Rona and

Fig. 12. Bottom photographs from camera stations N300 and S300 in *c.* 300 m of water. (**a**) March. Fields of small (1–2 cm) and larger (5–8 cm) stones (maximum 12 cm) with interstitial sand. Sand ripples advecting NNE over a stony pavement with the occasional appearance of N–S aligned streaky sand ribbons. Benthic community is small and is limited to larger stones. (**b**) August. Small stones (3–5 cm) interspersed in sand, which is thick enough to form a mobile pavement exhibiting small ripples showing movement to the NE. Scour development around a few of the larger (up to 35 cm) stones. The sea-bed area photographed is trapezoidal in shape and *c.* 120 cm across the bottom and *c.* 250 cm across the top of the frame (J. Humphery pers. comm.). Locations are shown in Fig. 2b.

Fig. 13. Bottom photographs from camera stations N700 and S700 in *c.* 700 m of water. (**a**) April. Sea floor composed of fine sand and silt which has been reworked by bottom-currents into symmetrical and confused bedforms and marked increase in benthic population, notably brittle-stars and echinoids. (**b**) August. Obliteration of coherent bedforms and marked increase in benthic population, notably brittle-stars and echinoids. (**c**) November. Soft mud and silty sea floor with a granular texture exhibiting limited occurrence of bedforms and small benthic population. The sea-bed area photographed is trapezoidal in shape and *c.* 120 cm across the bottom and *c.* 250 cm across the top of the frame (J. Humphery pers. comm.). Locations are shown in Fig. 2b.

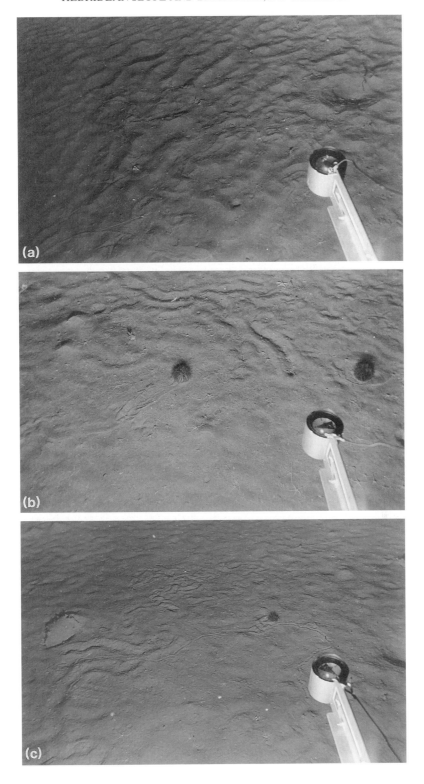

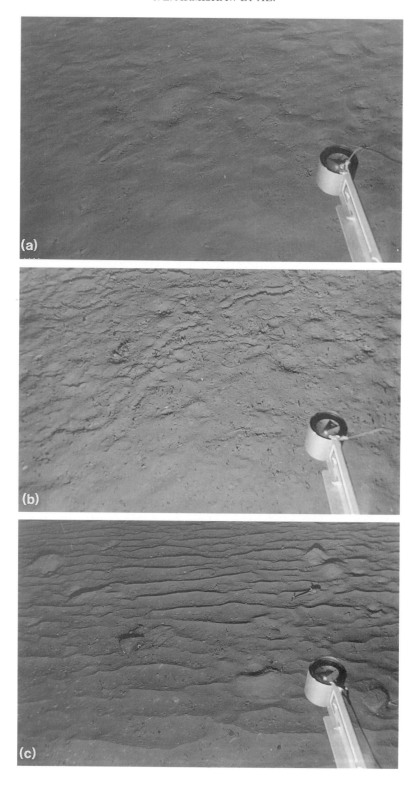

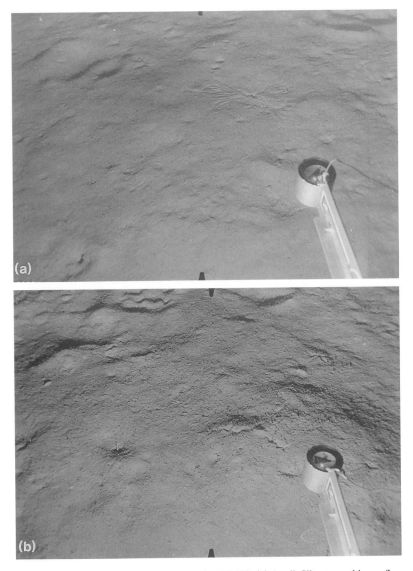

Fig. 15. Bottom photographs from camera station number N1500. (**a**) April. Silty to muddy sea floor with no current-induced bedforms. (**b**) August. Much modified sea floor, with the absence of any current-induced features. The sea-bed area photographed is trapezoidal in shape and *c.* 120 cm across the bottom and *c.* 250 cm across the top of the frame (J. Humphery pers. comm.). Location is shown in Fig. 2b.

Fig. 14. Bottom photographs from camera station S1000 in *c.* 1000 m of water. (a) April. Sea floor composed of fine sand and silt, with a scattering of small (1–5 cm) stones and few larger stones (up to 25 cm). Weak NW-flowing driving currents indicated by small-amplitude ripples. (**b**) August. Sea floor heavily reworked by a large benthic population, giving it a granular texture. Occasional presence of small (1–2 cm) stones. (**c**) November. Field of symmetrical, straight-crested and confused sand ripples, interspersed with small (1–3 cm) and large (up to 30 cm) stones, causing the development of crag-and-tail structures. The sea-bed area photographed is trapezoidal in shape and *c.* 120 cm across the bottom and *c.* 250 cm across the top of the frame (J. Humphery pers. comm.). Locations are shown in Fig. 2b.

Foula fans located further north along the margin, as slope-front fans.

The main growth of these slope-front fans appears to have been during the Pleistocene and the sediments were primarily glacially derived. On a broad scale the Barra Fan is very much a focused depocentre, either through fluvial or ice-stream activity, and in this respect has a point-source character. However, on a local scale there is no single point source for sediment supply but, rather a line source along the shelf edge. A variety of resedimentation processes then transferred this material downslope and resulted in the same combination of morphological elements that characterize a normal (or muddy) slope apron (Stow 1985) or ramp system (Stow *et al.* 1996, following the terminology of Reading & Richards 1994).

There is strong evidence also for the influence of alongslope bottom-current processes on the Barra Fan. These currents have been particularly effective during the Holocene period, but they also interacted with the downslope processes during glacial episodes, as shown by Howe (1996) for the margin north of the present-day study area. It is this interaction of downslope and alongslope processes, which led to the NW–NNW extension of the Barra Fan between the Hebrides Terrace Seamount and the Hebrides Shelf, that prompts us to term the Barra complex a composite slope-front fan. The different elements of this system are discussed below.

Downslope elements and processes

The primary source for downslope resedimentation is the outer Hebridean Shelf. This region is covered by a diamictite facies, in parts winnowed to give a present-day sand–gravel surficial layer. Ice scour marks are widespread, whereas the ridge and gully systems form a locally corrugated shelf edge and are interpreted as shelf channels developed in submarine end-moraines. Subsequent current action and slope steepening served to further deepen those channels to their present forms. Shelf spillover processes, perhaps mainly post-glacial, have transported the winnowed sands and gravels part way down the upper slope so that the mud line now lies at around 500–700 m.

The upper slope both below and above the mud line is heavily dissected by canyons. These are most numerous and have distinctive networks of tributary channels where the slope gradient is greatest (i.e. in the northern and southern sectors). In some cases, these canyon systems give rise to distributary channels across

the mid-slope region, but most appear to die out gradually without feeding any terminal lobe system. Although minor levees are present along some of the larger channels, most are mainly erosional features dissecting a mud-rich slope.

Slides, slumps and debris-flow masses are equally commonplace throughout the whole margin. These give rise to an irregular morphology, in part erosive with steep scarps and hollowed-out slide scars, and in part depositional with a blocky–hummocky convex-upward lobate form. Some of the parallel wrinkled or ridged appearance of parts on the upper slope appears to be pressure ridges formed in response to slide-slump and/or sediment creep processes.

The Peach Slide

The largest slide mass identified in the area is known as the Peach Slide (Holmes 1994; Holmes *et al.* this volume). This slide has translated a large portion of the northern Barra Fan downslope towards the NE Rockall Trough, probably during the late Pleistocene although precise dating of the movement has not yet been possible. This slide represents a complex body with a number of different features.

The headwall can be traced as a curvilinear scarp edge between 150 m and 400 m on the upper slope, which has cut back into the shelf. The steepest relief is at least 50 m. Smaller subparallel scarps are present and may have formed in response to either retrogressive or progressive slump events. Downslope movement has also carved out steep scarp margins, which cut across the slope and show an en echelon displacement in parts. Some of these have subsequently been pirated as slump margin channels.

Canyons across the upper part of the Peach Slide are mostly original slope channels now cut off from their shelf-edge source. Further down the slide, the morphology is distinctly irregular, with interspersed blocks, lobes and depressions making up the surface of an overall convex-upward bulge. The area covered by this composite slide is *c.* 1600 km[2].

Alongslope elements and processes

Photographic evidence clearly supports the widely accepted view that there is a generally northerly advecting current on the Hebrides Slope (Booth & Ellett 1983; Huthnance 1986; Kenyon 1986; Howe & Humphery 1995) and that current strength and orientation fluctuates considerably, depending on depth and location on the slope. In general, current-induced bed-forms disclose a shelf edge–upper-slope current that is advecting N to NNE and occasionally towards the NE (Fig. 11). Sites across the upper slope indicate a strong, continuous N- to NNE-flowing current throughout the year, although at some sites (during August), the occurrence of major WNW current ripples is superimposed upon ripples with a NE orientation, indicating current motion in two directions (Fig. 11b). Slightly further down the slope the current strength is significantly reduced.

A second area of current activity is observed on the lowermost part of the upper slope in the study area. This again supports the observations of Booth & Ellett (1983) for a northward current flow of 16 cm s^{-1} above the 1000 m depth contour over a 5 month period between May and September. From our photographs, taken at stations located at 700 m and 1000 m water depth (Fig. 13a–d), however, we would infer a seasonally affected current with a NNW–SSE oscillating flow during the winter and spring months and a much reduced flow during the summer. Ellett et al. (1986) suggested that the current to the west of Scotland broadens during the winter months, particularly under the influence of southeasterly and easterly winds, which may also cause the increase of the slope current transport in deeper waters (D. J. Ellett, pers. comm., 1995). The decrease of a slope current during the summer months may be explained by the notable lack of exchange between shelfedge waters and the slope current during summer and autumn (Booth & Ellett 1983). Water near the sea bed west of Scotland becomes isolated during the summer, and remains cool and dense until autumnal gales increase the mixing on the shelf (Edelsten et al. 1976), which may result in the confused and randomly orientated linguoid ripple bedforms seen at some stations (Fig. 11a).

Larger-scale morphological elements associated with alongslope bottom currents are present, but less readily observed on the Barra Fan. Sediment wave-like topography occurs along much of the upper slope, as well as in more restricted parts of the upper- to mid-slope region of the northern Barra Fan. However, some of these features, particularly those on steeper slopes, appear to have developed as pressure ridges from slump and creep processes, rather than as current bedforms. Furthermore, an area of true sediment waves on the northern edge of the Barra Fan has been interpreted as formed from turbidity-current flow (Howe 1996). Whereas unequivocal examples of sediment waves formed by bottom currents are elusive in the study area, the pattern of small-scale drift

sedimentation associated with some of the wave fields, as well as with other topographic features (e.g. slide scarps and blocks), does suggest bottom-current moulding in two main depth zones: (a) from 150 to 300 m and (b) from 700 to 1100 m.

Bottom-current smoothing of the sediment by plastering over minor topographic irregularities is apparent over a rather broader depth range on the upper- to mid-slope region (i.e. c. 600–1200 m), particularly in the southern and central parts of the study area.

The first author wishes to thank J. Humphery for information and preparation of bottom photographs, Ms P. Weatherall for preparation of the bathymetric data, and J. E. Damuth and C. James for their reviews of the manuscript. This work is based on a number of cruises (1995–1996) to the Hebridean continental slope area on the RRS *Charles Darwin* and RRS *Challenger* made possible through funding from the Natural Environmental Research Council Land–Ocean Interaction Study: Shelf Edge Study programme. This study is part of the first author's PhD thesis, which is supported by NERC and a CASE studentship with the British Geological Survey.

References

BELDERSON, R. H., KENYON, N. H., STRIDE, A. H. & STUBBS, A. R. 1972. *Sonographs of the Sea Floor*. Elsevier, Amsterdam.

BOOTH, D. A. & ELLETT, D. J. 1983. The Scottish slope current. *Continental Shelf Research*, **2**, 127–146.

CHOUGH, S. K. & HESSE, R. 1985. Contourites from Eirik Ridge, South of Greenland. *In*: HESSE, R. (ed.) *Sedimentology of Siltstone and Mudstone. Sedimentary Geology*, **41**, 185–199.

DAMUTH, J. E. 1975. Echo-character of the western equatorial Atlantic floor and its relationship to the dispersal and distribution of terrigenous sediments. *Marine Geology*, **18**, 17–45.

—— 1978. Echo character of the Norwegian–Greenland Sea: relationship to Quaternary sedimentation. *Marine Geology*, **28**, 1–36.

—— 1980. Use of high-frequency (3.5 kHz) echograms in the study of near-bottom sedimentation processes in the deep-sea: a review. *Marine Geology*, **38**, 51–75.

——, JACOBI, R. D. & HAYES, D. E. 1983. Sedimentation processes in the Northwest Pacific Basin revealed by echo-character mapping studies. *Geological Society of America Bulletin*, **94**, 381–395.

DEACON, M. B. 1996. How the science of oceanography developed. *In*: SUMMERHAYS, C. P. & THORPE, S. A. (eds) *Oceanography: an Illustrated Guide*. Manson, Publishing pp. 352.

EDELSTEN, D. J., ELLETT, D. J. & EDWARDS, A. 1976. Preliminary results from current measurements at the Scottish continental shelf-edges. *ICES* 1976 (**C:12**), 10 pp (mimeo).

EDEN, R. A., ARDUS, D. A., BINNS, P. E., MCQUILLIN, R. & WILSON, J. B. 1971. *Geological investigations with a manned submersible off the west coast of Scotland 1969–1970.* NERC Institute of Geological Sciences, Report **71/16**.

ELLETT, D. J., EDWARDS, A. & BOWERS, R. 1986. The hydrography of the Rockall Channel – an overview. *Proceedings of the Royal Society of Edinburgh, Section B*, **88**, 61–81.

HOLMES, R. W. 1994. *Seabed topography and other geotechnical information for the Shelf Edge Study 55°N–60°N NW of Britain.* British Geological Survey, Report **WB/94/15**.

——, LONG, D. & DODDS, L. R. 1998. Large-scale debrites and submarine landslides on the Barra Fan, west of Britain. *This volume.*

HOWE, J. A. 1996. Turbidite and contourite sediment waves in the northern Rockall Trough, North Atlantic Ocean. *Sedimentology*, **43**, 219–234.

—— & HUMPHERY, J. D. 1995. Photographic evidence for slope-current activity, Hebrides Slope, NE Atlantic Ocean. *Scottish Journal of Geology*, **30**(2), 107–15.

HUTHNANCE, J. M. 1986. Rockall Slope Current and shelf edge processes. *Proceedings of the Royal Society of Edinburgh, Section B*, **88**, 83–101.

JAMES, J. W. C. & HITCHEN, K. 1992. *Peach (56°N, 10°W): Solid Geology.* British Geological Survey 1 : 250,000 Offshore Map Series.

——, BOOTH, S. J. & WRIGHT, S. A. 1990. *Peach (56°N, 10°W): Sea Bed Sediments.* British Geological Survey 1 : 250,000 Offshore Map Series.

JOHNSON, D. A. & DAMUTH, J. E. 1979. Deep thermohaline flow and current-controlled sedimentation in the Amirante Passage: Western Indian Ocean. *Marine Geology*, **72**, 1–44.

KENYON, N. H. 1986. Evidence for a strong poleward current along the upper Continental Slope of NW Europe. *Marine Geology*, **72**, 187–198.

—— 1987. Mass-wasting features on the continental slope of Northwest Europe. *Marine Geology*, **74**, 57–77.

LE BAS, T. P., MASSON, D. C. & MILLARD, N. C. 1995. TOBI image processing – the state of the art. *IEEE Journal of Oceanic Engineering*, **20**(1), 85, 61–78.

LESLIE, A. 1992. *A sedimentological study of the Tertiary and Quaternary sediments in borehole 88/7, 7A, Hebrides Slope, northern Rockall Trough region.* British Geological Survey, Technical Report **WB/92/16**.

LONSDALE, P. & HOLLISTER, C. D. 1979. A near-bottom traverse of the Rockall Trough: hydrographic and geological inferences. *Oceanologica Acta*, **2**, 91–105.

READING, H. G. & RICHARDS, M. 1994. Turbidite systems in deep water basin margins classified by grain size and feeder systems. *Bulletin, American Association of Petroleum Geologists*, **78**, 792–822.

SELBY, I. C. 1989. The Quaternary geology of the Hebridean continental margin. PhD thesis, University of Nottingham.

STOKER, M. S. 1995. The influence of glacigenic sedimentation on slope-apron development on the continental margin off Northwest Britain. *In:* SCRUTTON, R. A., STOKER M. S., SHIMMIELD, G. B. & TUDHOPE, A. W. (eds) *The Tectonics, Sedimentation and Palaeoceanography of the North Atlantic Region.* Geological Society, London, Special Publication, **90**, 159–177.

—— 1998. Sediment-drift development on the continental margin off NW Britain. *This volume.*

——, HITCHEN, K. & GRAHAM, C. C. 1993. *United Kingdom Offshore Regional Report: The Geology of the Hebrides and West Shetland Shelves, and Ancient Deep-water Areas.* HMSO. London, for the British Geological Survey.

STOW, D. A. V. 1985. Deep-sea clastics: where are we going? *In:* BRENCHLY, P. J. & WILLIAMS, B. P. J. (eds) *Sedimentology: Recent Developments and Applied Aspects.* Geological Society, London, Special Publications, **28**, 67–93.

——, READING, H. G. & COLLINSON, J. D. 1996. Deep seas. *In:* READING, H. G. (ed.) *Sedimentary Environments.* Blackwell, Oxford, 395–453.

YOON, S. H., CHOUGH, S. K., THIEDE, J. & WERNER, F. 1991. Later Pleistocene sedimentation on the Norwegian continental slope between 67° and 71°N. *Marine Geology*, **99**, 187–207.

Debris flows on the Sula Sgeir Fan, NW of Scotland

A. BALTZER[1,2], R. HOLMES[1] & D. EVANS[1]

[1]*British Geological Survey, Murchison House, West Mains Road, Edinburgh EH9 3LA,
UK*

[2]*Present address: Université de Caen, Laboratoire de Géologie, BP 5186, 14032 Caen
Cédex, France*

Abstract: This paper describes the various debris-flow processes which have occurred on
the Sula Sgeir Fan, a well-documented feature situated on the Hebrides Slope, NW of Scotland. The Sula Sgeir Fan covers an area 65 km in length and 28 km in width, between the
400 m and 1200 m bathymetric contours. The fan appears as a build-up of stacked debris-
flow packages with transparent acoustic facies. These are termed glacigenic debris flows, as
they are derived from sediment deposited by ice at the shelfbreak. However, on the southwestern edge of the Sula Sgeir Fan, and on its northeastern flank, there are different
acoustic features which are described here as 'classic' debris flows. They are larger than the
glacigenic type, and a scarp or a sliding scour characterizes their heads; they are thought to
have been triggered by seismic events. Associated with these 'classic' debris flows are features on the seismic profiles interpreted as gas- or liquid-escape structures.

The Sula Sgeir Fan is a well-documented
accumulation of sediments on the Hebrides
Slope. This fan, previously described as an
accumulation of debris-flow deposits with intervening hemipelagic sediments (Stoker 1995), is
here recognized as comprising two kinds of
debris flows, termed 'classic' and glacigenic
debris flows. This study is based on sparker and
deep-tow-boomer seismic profiles collected as
part of the British Geological Survey (BGS) offshore regional survey of the United Kingdom,
and on geotechnical analysis of gravity cores of
maximum length 2 m and vibrocores of
maximum length 6 m collected between 200 m
and 1200 m water depth as part of the same
survey (Strachan & Evans 1991; Talbot *et al.*
1994).

The strike of the Hebrides Slope changes at
about 58°30'N from a northerly to a northeasterly trend (Fig. 1). This is marked by a distinct
bulge on the margin, the Geikie Bulge, which
owes its origin to a buried Tertiary volcanic
centre on the outer shelf (Evans *et al.* 1989). The
Barra Fan and the Sula Sgeir Fan are situated,
respectively, to the south and the north of this
bulge. The shelfbreak occurs at about 200 m,
although the maximum change in slope angle
occurs at about 600 m on the Geikie Bulge,
where the Geikie Escarpment has slopes up to
28° (Evans & McElvanney 1989). This escarpment was interpreted by Jones *et al.* (1986) and
Evans *et al.* (1989) as a probable erosional
feature cut by bottom currents during the mid-
Oligocene low sea-level stand. The upper slope
has remained relatively starved of sediment and

subjected to occasional erosion in the area of the
Geikie Escarpment.

This contrasts with the adjacent Barra and
Sula Sgeir fans, which have been major
depocentres throughout Plio-Pleistocene times
(Stoker 1995). In particular, mid- to late
Pleistocene ice sheets contributed large
amounts of glacially derived sediment from
land and shelf to these slope-apron fans during
periods of low eustatic sea level (Haq *et al.*
1987). The glacial activity on the Hebridean
margin was distal in character until Anglian
times (0.44 Ma) when ice reached the shelfbreak, as it did on at least two subsequent
occasions. At their maximum extent, these ice
sheets extended, at least locally, over the shelf
edge, and delivered ice-marginal sediments
directly to the slope (Stoker 1995). Even in the
'between-fan' area, the sediment accumulation
rates increased significantly during glacial
intervals (Stoker *et al.* 1994). This is reflected by
sediment accumulation rates, which increased
from 25 mm/ka during the early Pleistocene to
a maximum of 560 mm/ka during the mid- to
late Pleistocene (Talbot *et al.* 1994).

Sula Sgeir Fan

This fan covers an area of 65 km × 28 km,
between the 300 m and 1200 m bathymetric contours (Fig. 1). Its thickness may reach 300 m, and
a mid-Miocene age has been suggested for its
initiation, perhaps in response to eustatic and
tectonic events. A lowstand setting for the

BALTZER, A., HOLMES, R. & EVANS, D. 1998. Debris flows on the Sula Sgeir Fan, NW of Scotland. *In*:   105
STOKER, M. S., EVANS, D. & CRAMP, A. (eds) *Geological Processes on Continental Margins: Sedimentation,
Mass-Wasting and Stability.* Geological Society, London, Special Publications, **129**, 105–115.

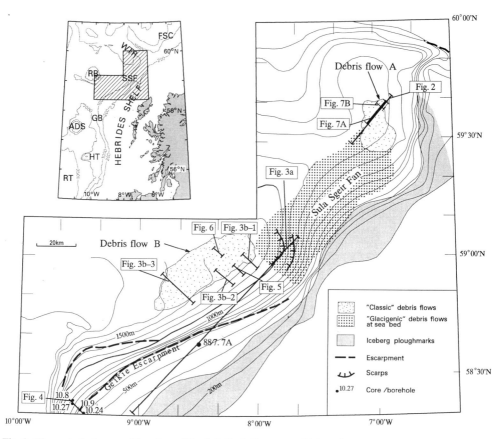

Fig. 1. The main features and location of the Sula Sgeir Fan area, with location of the text figures, and selected seismic tracks and cores. Bathymetry is in metres. FSC, Faeroe–Shetland Channel; WTR, Wyville-Thomson Ridge; SSF, Sula Sgeir Fan; RB, Rosemary Bank; GB, Geikie Bulge; ADS, Anton Dohrn Seamount; HT, Hebrides Terrace; RT, Rockall Trough.

middle to upper Miocene deposits is compatible with the eustatic sea-level curve of Haq *et al.* (1987). However, a phase of intra-Miocene tectonism and uplift proposed in this area (Boldreel & Andersen 1993) may have had an additional effect on relative sea level. During its early development, the fan was probably fed by fluvial systems draining from the Scottish Highlands and Islands (Hall 1991). The sediment supply may initially have been boosted by intense weathering and erosion under warm, humid climatic conditions which characterized the mid-Pliocene (Dowsett *et al.* 1992), but the bulk of the fan has been built during the mid- to late Quaternary by a large supply of glacially transported sediment. Sula Sgeir Fan build-up probably ended in the early Devensian, as the regional seismostratigraphy indicates that the extent of the last Scottish ice sheet (late Devensian) was restricted to the inner shelf (Stoker *et al.* 1994).

The fan appears as a build-up of debris-flow packages stacked up on each other, presenting a transparent acoustic facies (Figs 2 and 3a). These transparent packages are commonly separated by thin, continuous reflectors that probably represent intervals of reduced sediment supply to the slope. This may have been due to an interstadial or interglacial rise in sea level, or to more localized conditions within a glacial phase, such as less extensive shelf glaciation (Stoker *et al.* 1994). The size of the glacigenic debris flows varies from 0.5 to 3–5 km in length, and they have a thickness of 5–35 m. Accumulation of these mass-flow packages was episodic and related to specific rapid phases of downslope sedimentation concomitant with maximum glaciations and glacio-eustatic lowstands. Their progradational geometry (Stoker *et al.* 1994) demonstrates that such debris flows are constructional in character; they are similar to those observed west of Shetland by Holmes (1991), on

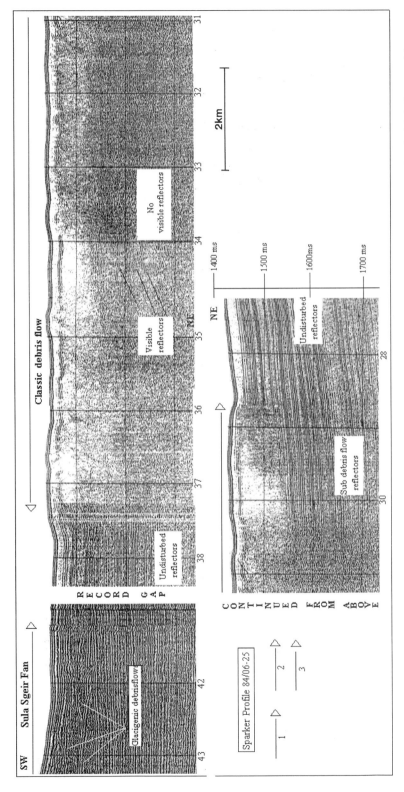

Fig. 2. Seismic profile 84/06-25 showing the transition between the Sula Sgeir Fan in the south west, built by the accumulation of glacigenic debris flows, and the 'classic' debris flow. (Note the undisturbed reflectors beneath the 'classic' debris flow. See Fig. 1 for location.)

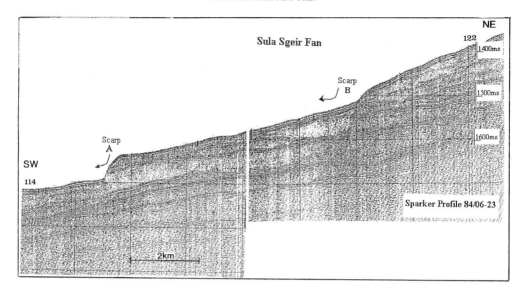

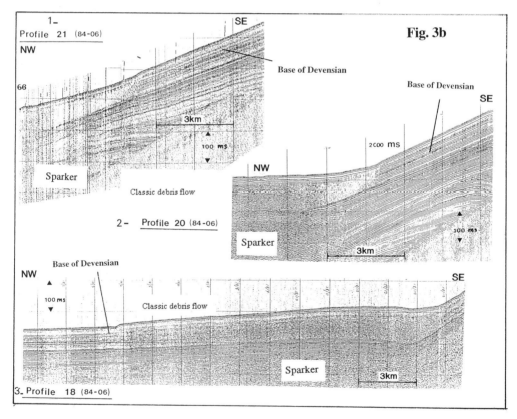

Fig. 3. An example of a 'classic' debris flow (debris flow B) from its initiation zone to its depositional area:
(**a**) shows the scarps related to the debris flow; (**b**) illustrates a series of cross-sections through the debris flow.

the Bear Island Trough Mouth Fan by Laberg & Vorren (1995), and on the North Sea Fan by King *et al.* (1996).

On the southwestern flank of the fan, and also on its northeastern edge, there are acoustic features which appear to be 'classic' debris flows with a different mode of origin (Fig. 2). The northern feature (debris flow A, Fig. 1) extends northward from 900 m to 1000 m water depths, although its shape is not well constrained because of the seismic line distribution. Nevertheless, sparker profile 84/06-25, perpendicular to the slope, shows a transition of acoustic facies between the Sula Sgeir Fan and this debris flow (Figs 1 and 2). The southern feature (debris flow B, Fig. 1) extends about 40 km to the southwest and has the same seismic transparent acoustic characteristics as that to the north. Figure 3b presents the thick transparent unit, showing strong remoulding of the sediments, as is typical of debris flows. The underlying reflectors are weakly recorded but remain observable (Figs 2 and 3b). This acoustic appearance is different from the piled-up packages observed for the glacigenic debris flows.

On transverse cross-sections, the southern debris flow shows a very sharp contact with undisturbed sediments and shows negative relief proximally but positive relief distally (Fig. 3b), as is typical of a gravity-flow event. The sharp contact on these sections is further indication of an origin that is predominantly not directly upslope but rather from the Sula Sgeir Fan edge to the northeast, although some sediment may have been derived from the slope above. Proximally, the gravity process caused net erosion of the sediments, for it had enough energy to remould and erode the pre-existing layers. As it moved distally, the flow incorporated sediment and lost energy and speed, especially on account of the flat character of the area, resulting in positive relief.

Two scarps around 50 m high have been identified on the southwestern flank of the Sula Sgeir Fan (Figs 1 and 3a). It is considered likely that the major classic debris flows originated from a translational slope failure event, or events, which formed these two scarps. The lack of sedimentary cover on the scarps suggests that this slope failure event, or events, occurred at the end of active fan deposition. On the profiles 85/06-21, -20 and -18 (Fig. 3b), we can distinguish undisturbed reflectors above the debris-flow unit. Also indicated in Fig. 4b is the probable base of the Devensian as derived from regional studies, indicating that the event may have taken place at some time during the early Devensian or just before.

Discussion

There are no geotechnical data available from the Sula Sgeir Fan, but geotechnical data have been collected (Fig. 1) from BGS borehole 88/7, 7A (Talbot *et al.* 1994) and on the Geikie Bulge (Strachan 1987). Although these are from well-layered sediments that are largely of hemi-pelagic and contouritic origin (Leslie 1993), in general, such deposits are likely to be less stable than glacigenic sediments (Whalley 1978; Baltzer *et al.* 1995) and may provide a reference point for ground conditions on the Sula Sgeir Fan. Shear strength, bulk density, water content and plasticity limits are used in two ways in order to analyse the slope stability.

(1) Booth diagrams (Booth 1979) give an idea of the consolidation state of sediments (Fig. 4). The horizontal scale represents the depth (m), and the vertical scale shows the 'missing thickness'. A negative missing thickness indicates an overconsolidation state corresponding to the erosion of x metres. Conversely, a positive thickness is related to underconsolidated sediments. Results between –2 m and +2 m reflect a 'normal' consolidation state (Baltzer *et al.* 1994). The analyses on cores 58-10-26 and 58-10-27 show normal consolidation states (Fig. 3b and 3c), even slightly underconsolidated for 58-10-26 although the reliability of the measurements carried out on this core is suspect. Sediments samples in cores along profile 85/07-05 (Fig. 4a) present a normal consolidation state of 5–7 kPa in the upper 2 m. Core 58-09-23, situated upper on the slope, shows an apparent overconsolidation (15 kPa at 2 m depth) which may be related to iceberg ploughing.

(2) Stability analysis diagrams (Fig. 4c) present cohesion (kPa) v. thickness (m) of the sediment. The upper curve represents the driving stress caused by the weight of the sediment, the lower is the resisting shear stress related to the intrinsic cohesion of the sediment. If the driving stress is equivalent or superior to the resisting shear stress, the sedimentary column may become unstable. Figure 4c shows that in each of the three examined cases, there is no evidence of intersection between the resisting stress and the driving stress, so the sediment appears stable to at least 6 m depth. An external event such as earthquake or gas overpressure is therefore needed to trigger a disequilibrium, and a similar situation is likely to pertain on the Sula Sgeir Fan itself.

Strachan & Evans (1991) and Long (1993) investigated seismic profiles in the area, but did not report any features attributable to gas, and none have been observed in this study other

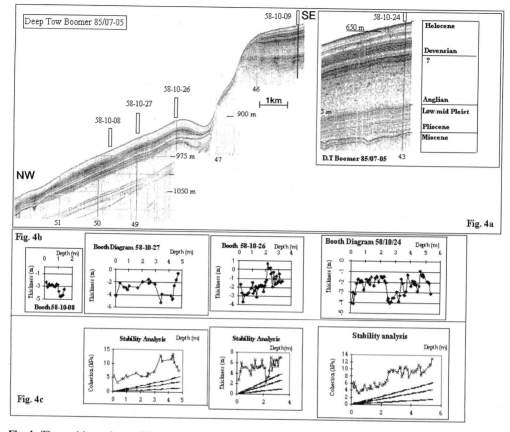

Fig. 4. The positions of cores (**a**) in relation to seismic profile 85/07-05 (see Fig. 1 for location). Stratigraphy after Stoker *et al.* (1994); (**b**) illustrates Booth diagrams used to establish the consolidation state of the cores, and (**c**) presents stability analysis diagrams.

than examples from deep water. It is therefore considered that gas may not have been a significant triggering mechanism in this case, although it cannot be excluded.

The horizontal acceleration needed to trigger instability may be calculated from the slope stability analysis software of Potter (1989), based on the equations of Morgenstern (1967). This has been used to consider three examples (Table 1) corresponding to 30, 55 and 75 m thickness of debris flow. The safety factor gives the slope's state of equilibrium. If it is greater than unity, the sediments on the slope will remain stable, and an external event would be needed to cause instability. If that external event were an earthquake, its magnitude may be deduced from the horizontal acceleration (A_h), following the expression log A_h + 0.25. IMM + 0.25 (Murphy & O'Brien, in Despeyroux 1989). In the first example, an earthquake of magnitude 6 on the Modified Mercalli Scale is needed to trigger instability on a slope angle

of 1.5°, and a magnitude 2 earthquake is required for a 6.5° slope. As the southern flank of the fan shows scarps around 50 m high and a slope less than 1°, an earthquake of magnitude 6 would have been sufficient to trigger disequilibrium in sediments of 55 m thickness on a 1° slope.

It is known that earthquakes have occurred in modern times along this margin. Jacob *et al.* (1983) recorded one of magnitude 4 in 1980 on the Hebrides Terrace Seamount (Fig. 1), which they related to isostatic adjustment. Another, of magnitude 3.1, was recorded in the same area in 1986 (Turbitt 1988). Paul & Jobson (1987) estimated that an earthquake of Richter magnitude 6 within a few tens of kilometres distance could have caused failure on the Hebrides Slope. A map of peak ground acceleration for the UK continental shelf (1000 a return period) produced by Ove Arup *et al.* (1993) shows values ranging from 0.10 *g* to 0.05 *g* for this area; this is similar to the theoretical acceleration required

Table 1. *Stability analysis: three theoretical cases are presented for different sediment thickness; undrained shear strengths and bulk densities are extrapolated from borehole 88/7, 7A and other BGS cores*

Geikie 1, debris flow[*]				Geikie 2, debris flow[†]				Geikie 3[‡]			
Slope angle	Safety factor	A_h (cm/s^2)	Seism. Magn.	Slope angle	Safety factor	A_h (cm/s^2)	Seism. Magn.	Slope angle	Safety factor	A_h (cm/s^2)	Seism. Magn.
1	6.99	57.52	6	1	5.92	48.16	6	1	5.49	44.57	6
1.5	4.66	52.74	6	1.5	3.95	43.29	6	1.5	3.66	39.62	5
2	3.5	47.97	6	2	2.96	38.42	5	2	2.75	34.68	5
2.5	2.8	43.21	6	2.5	2.37	33.56	5	2.5	2.2	29.75	5
3	2.33	38.46	5	3	1.98	28.71	5	3	1.83	24.82	5
3.5	2	33.71	5	3.5	1.7	23.86	5	3.5	1.57	19.9	4
4	1.75	28.97	5	4	1.48	19.01	4	4	1.38	14.98	4
4.5	1.56	24.23	5	4.5	1.32	14.17	4	4.5	1.22	10.06	3
5	1.4	19.49	4	5	1.19	9.32	3	5	1.1	5.14	2
5.5	1.28	14.76	4	5.5	1.08	4.48	2	5.5	1	0	
6	1.17	10.03	3	6	0.99	0		6	0.92	0	
6.5	1.08	5.3	2	6.5	0.92	0		6.5	0.85	0	
7	1.01	0		7	0.85	0		7	0.79	0	
7.5	0.94	0		7.5	0.8	0		7.5	0.74	0	
8	0.89	0		8	0.75	0		8	0.7	0	
8.5	0.83	0		8.5	0.71	0		8.5	0.66	0	
9	0.79	0		9	0.67	0		9	0.62	0	
9.5	0.75	0		9.5	0.63	0		9.5	0.59	0	
10	0.71	0		10	0.6	0		10	0.56	0	

[*] Thickness 30 m; length 1500 m; undrained shear strength 45 kPa; bulk density 2.29 g/cm^3.
[†] Thickness 55 m; length 2750 m; undrained shear strength 73 kPa; bulk density 2.34 g/cm^3.
[‡] Thickness 75 m; length 3750 m; undrained shear strength 95 kPa; bulk density 2.39 g/cm^3.

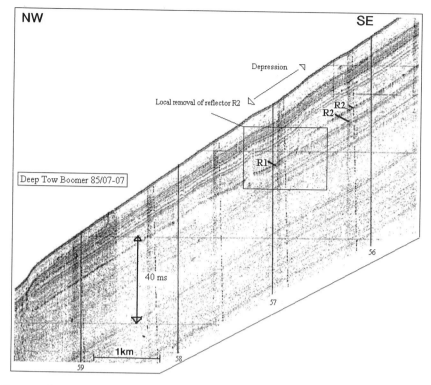

Fig. 5. Deep-tow-boomer profile 85/07-7 showing a seismic structure similar to the fluidization feature described by Strachan & Evans (1991). (See Fig. 1 for location.)

to induce instability in the Geikie Escarpment region (Table 1, 57.52 cm/s^2 = 57.52 gal = 0.0575 g). As the overall tectonic setting has not changed during Quaternary times, it is reasonable to consider that earthquakes have occurred in the area throughout this time, and a magnitude 6 event may well have taken place during times of major isostatic readjustment.

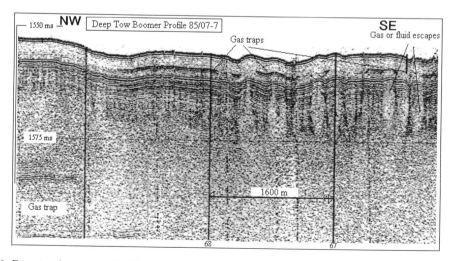

Fig. 6. Deep-tow-boomer profile 85/07-7 through debris flow B, with possible gas or liquid escapes features. (See Fig.1 for location.)

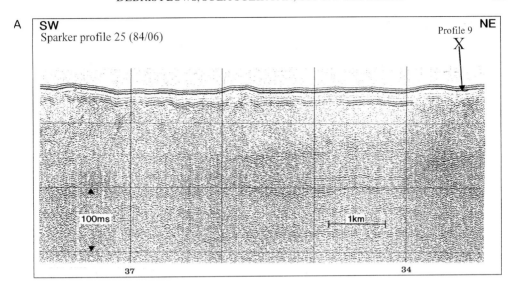

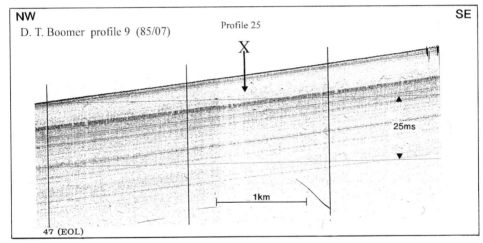

Fig. 7. Sparker profile 84/06-25 through debris flow A (**a**), and the perpendicular deep-tow-boomer profile 85/07-9 (**b**). The boomer shows features possibly associated with gas or fluid escape. (See Fig.1 for locations.) Intersection of profiles marked by X.

A seismic event was invoked by Strachan & Evans (1991) as the most likely triggering event for the sediment failure below the Geikie Escarpment. They described depressions most commonly 90 m wide but up to 300 m wide with relief of between 1 and 2 m increasing to a maximum of 10 m upslope. This was interpreted as being a fluidization feature, the result of earthquake-induced liquefaction acting retrogressively upslope, with the subsequent alleviation of excess pore pressure. Figure 5 shows a similar feature observed directly upslope from the southern debris flow (deep-tow-boomer profile 85/07-7, Fig. 1). It shows the disappearance of reflector R2 along a distance of 1100 m,

creating a lowering of 5 m in reflector R1, with a depression in the overlying sediments that extends to the sea bed. These observations may similarly be taken as indications of earthquake-induced liquefaction.

Downslope, where profile 85/07-7 crosses debris flow B (Figs 1 and 6), the same profile includes vertical columns, varying from 20 to 50 m in width. They are associated with enhanced and raised upper reflectors, and were observed only in the 'classic' debris flow B. These columns, showing no acoustic returns, have been called acoustic voids or wipeout zones (Bouma *et al.* 1983) and have been noted elsewhere on the Scottish margin by Long (1993).

They are believed to indicate areas where excess pore pressure has been dissipated by the explusion of the water and/or gas (Hovland & Judd 1988).

In the debris flow A (Fig. 1), 'gas chimneys' have widths of 15 m and show no acoustic returns (Fig. 7). A way of distinguishing them from data-recording problems (C. Brett, pers. comm., 1996) is that the latter generally affect all the reflectors (except sometimes the sea bed). These 'gas chimneys' do not appear at the same levels and have different lengths, which tends to confirm that they are not seismic artefacts. These liquid or gas features affect at least the upper Devensian and part of the Holocene, coming from 100–250 ms below sediment surface. It is tempting to link these liquid or gas escapes to the change in pore pressure and destruction of sedimentary structure triggered by a sliding event.

It is surmised that at least one major earthquake occurred, possibly just before the Devensian or during early Devensian times, triggering liquefaction and at least two sliding events. The resultant debris flows, in turn, facilitated or triggered liquid or gas escapes. Two mechanisms for fluid or gas release could be invoked here:

(1) The disturbance of the sediment by the debris flows facilitated the evacuation of the excess pore pressure and hence gas or liquid escaped. This is consistent with Long's (1993) observation that the occurrence of liquid or gas escapes is linked with areas of disturbed sediment.

(2) Rapid burial of the sea bed at the depositional margins of the debris flows may have destabilized a methane hydrate layer, thus releasing gas.

Conclusions

The Sula Sgeir Fan area shows evidence of a variety of sedimentary processes that are likely to occur in a glacial-margin environment. The development of the fan was largely through the build-up of constructional glacigenic debris-flow packages. At least one earthquake, probably occurring just before or during the early Devensian, triggered at least two mass movements and generated two 'classic' debris flows and two associated erosional scarps. The remoulding of sediment by these 'classic' debris flows appears to have facilitated, and to have triggered, liquid or gas escape which is recognized in these debris flows. The Sula Sgeir Fan represents a simple model of glacigenic fan construction and associated erosional processes on passive margins.

This work reported in this paper was carried out while A.B. was funded by an EC Human Capital and Mobility Fellowship. The paper is published with the permission of the Director of the British Geological Survey (NERC).

References

BALTZER, A., COCHONAT, P. & AUFFRET, G. A. 1995. Comparaison des facteurs necessaires a l'initiation de glissements sous-marins sur les marges de Nouvelle Ecosse et de Nord Gascogne au Quaternaire. *Comptes Rendus Hebdomadaires des Séances de l'Academie des Sciences, Serie IIa*, **321**, 1001–1008.

——, —— and PIPER, D. J. W. 1994. *In situ* geotechnical characterisation of sediments on the Nova Scotian Slope, eastern Canadian continental. *Marine Geology*, **120**, 291–308.

BOLDREEL, L. O. & ANDERSEN, M. S. 1993. Late Palaeocene to Miocene compression in the Faeroe–Rockall area. *In*: PARKER, J. R. (ed.) *Proceedings of the 4th Conference on the Petroleum Geology in Northwest Europe*. Geological Society, London, 1025–1034.

BOOTH, J. 1979. Recent history of mass wasting on the upper continental slope, northern Gulf of Mexico, as interpreted from the consolidation state of the sediment. *In*: DOYLE, L. J. & PICKLEY, O. L. (eds) *Geology of Continental Slopes*. Society of Economic Paleontologists and Mineralogists, 153–164.

BOUMA, A. H., STELTING, C. E. & FEELEY, M. H. 1983. High resolution seismic reflection profiles. *In*: BALLY, A. W. (ed.) *Seismic Expression of Structural Styles – a Picture and Work Atlas. Studies in Geology*. American Association of Petroleum Geologists, Series No. 15, **1**.

DESPEYROUX, J. 1989. Constructions en zone sismique. Les niveaux d'agression à considérer pour l'application des recommandations de l'AFPS relatives aux constructions à édifier dans les regions su jettes a séismes. *Cahier Spécial de l'AFPS, No. Spécial 'Accélérations Nominales'*, 48 pp.

DOWSETT, H. J., CRONIN, T. M., POORE, R. Z., THOMSON, R. S., WHATLEY, R. C. & WOOD, A. M. 1992. Micropalaeontological evidence for increased meridional heat transport in the North Atlantic Ocean during the Pliocene. *Science*, **258**, 1133–1135.

EVANS, D. & McELVANNEY, E. P. 1989. *Geikie, Sheet 58°N–10°W, 1:250,000 Map Series, Quaternary Geology*. Ordnance Survey, Southampton, for the British Geological Survey.

——, ABRAHAM, D. A. & KITCHEN, K. 1989. The Geikie igneous centre, west of Lewis: its structure and influence on Tertiary geology. *Scottish Journal of Geology*, **25**, 339–352.

HALL, A. M. 1991. Pre-Quaternary landscape evolution in the Scottish Highlands. *Transactions of the Royal Society of Edinburgh: Earth Sciences*, **82**, 1–26.

HAQ, B. U., HARDENBOL, J. & VAIL, P. R. 1987. Chronology of fluctuating sea levels since the Triassic. *Science*, **235**, 1156–1167.

HOLMES, R. 1991. *Foula, Sheet 60°N–10°W. 1:250,000 Map Series, Quaternary Geology.* Ordnance Survey, Southampton, for British Geological Survey.

HOVLAND, M. & JUDD, A. G. 1988. *Seabed Pockmarks and Seepages.* Graham & Trotman, London, 293 pp.

JACOB, A. W. B., NEILSON, G. & WARD, V. 1983. A seismic event near the Hebrides Terrace Seamount. *Scottish Journal of Geology*, **19**, 287–296.

JONES, E. J. W., MITCHELL, J. G. & PERRY, R. G. 1986. Early Tertiary igneous activity west of the Outer Hebrides, Scotland – evidence from magnetic anomalies and dredged basaltic rocks. *Marine Geology*, **73**, 47–59.

KING, E. L., SEJRUP, H. P., HAFLIDASON, F., ELVERHOI, A. & AARSETH I. 1996. Quaternary seismic stratigraphy flow aprons, hemipelagic sediments, and large submarine slides. *Marine Geology*, **130**, 293–315.

LABERG, J. S. & VORREN, T O. 1995. Late Wechselian submarine debris flow deposits on the Bear Island Trough Mouth Fan. *Marine Geology*, **127**, 45–72.

LESLIE, A. 1993. Shallow Plio-Pleistocene contourites on the Hebrides Slope, northwest U.K. continental margin. *Sedimentary Geology*, **82**, 61–78.

LONG, D. 1993. *Geological characteristics and geohazards of the Hebridean and West Shetland Continental Shelf and Slope.* British Geological Survey Technical Report **WB/93/17C**.

MORGENSTERN, N. R. 1967. Submarine slumping and the initiation of turbidity currents. *In*: RICHARDS, A. F. (ed.). *Marine Geotechnique.* University of Illinois Press, Urbana, 189–220.

OVE ARUP & PARTNERS AND BGS 1993. *UK Continental Shelf Seismic Hazard.* Health & Safety Executive Report **46149**, 60 pp.

PAUL, M. A. & JOBSON, L. M. 1987. *On the geotechnical and acoustic properties of sediments from the British Continental Margin west of the Hebrides.*

Report to SERC (MTD). Department of Civil Engineering, Heriot–Watt University.

POTTER, P. 1989. *Marine slope instability on the continental slope of New Jersey.* Technical University of Nova Scotia, Internal Report **CE5990**.

STOKER, M. S. 1995. The influence of glacigenic sedimentation on slope-apron development on the continental margin off Northwest Britain. *In*: SCRUTTON, R. A., STOKER, M. S., SHIMMIELD, G. B. & TUDHOPE, A. W. (eds) *The Tectonics, Sedimentation and Palaeoceanography of the North Atlantic Region.* Geological Society, London, Special Publications, **90**, 159–177.

——, LESLIE, A. B. *et al.* 1994. A record of late Cenozoic stratigraphy, sedimentation and climate change from the Hebrides Slope, NE Atlantic Ocean. *Journal of the Geological Society, London*, **151**, 235–249.

STRACHAN, P. 1987. *Geotechnical data from vibrocores and gravity cores; Geikie and Lewis sheets.* British Geological Survey Marine Earth Sciences Research Programme Report Series, **87/16**.

—— & EVANS, D. 1991. A local deep-water failure on the Northwest Slope of the UK. *Scottish Journal of Geology*, **27**, 107–111.

TALBOT, L. A., PAUL, M. A. & STOKER, M. S. 1994. Geotechnical studies of a Plio-Pleistocene sedimentary sequence from the Hebrides Slope, NW UK Continental Margin. *Geo-Marine Letters*, **14**, 244–251.

TURBITT, T. (ed.) 1988. *Bulletin of British earthquakes 1986.* British Geological Survey Technical Report **WL/88/11**.

WHALLEY, W. B. 1978. Abnormally steep slopes on moraines constructed by valley glaciers. *In*: *The Engineering Behaviour of Glacial Material. Abstracts of a Conference Organised by the Midlands Soil Mechanic and Foundation Engineering Society*, April 1975, University of Birmingham. Geobooks.

Shallow geotechnical profiles, acoustic character and depositional history in glacially influenced sediments from the Hebrides and West Shetland Slopes

MICHAEL A. PAUL[1], LISA A. TALBOT[1] & MARTYN S. STOKER[2]

[1]*Department of Civil & Offshore Engineering, Heriot–Watt University, Edinburgh EH14 4AS, UK*

[2]*British Geological Survey, Murchison House, West Mains Road, Edinburgh EH9 3LA, UK*

Abstract: Surficial sediments from the Hebrides and West Shetland Slopes are mostly matrix-supported diamictons or laminated muds with subordinate sand. Those from the outer shelf and upper slope have water content and undrained shear strength profiles that suggest variable light to moderate overconsolidation, whereas those from the lower slope and basin floor have water content and undrained shear strength profiles that suggest normal consolidation. We believe that these geotechnical differences are genetic and reflect the processes operative in former glacial and glaciomarine palaeoenvironments along the continental margin: variably overconsolidated sediments are the products of marginal or proximal glaciomarine environments in which ice loading has occurred, whereas normally consolidated sediments are found in distal glaciomarine to basin plain settings. The depositional setting also controls the void index of the sediment via its sedimentary fabric and we find a sensible relationship between the compression curve of a sediment and its mode of deposition. It is usual to find that individual values lie scattered about this curve as a result of small variations in the packing. These variations correspond to minor sediment layers from which acoustic reflections may interfere constructively to give a multilayered acoustic texture, whose appearance is thus ultimately related to the magnitude of the scatter.

The continental margin off northwest Britain provides a well-studied example of a glacially constructed margin formed by glacial, glaciomarine and mass-flow processes, subsequently modified by bottom currents and hemipelagic sedimentation (Stoker 1995). The sea bed and immediate subsurface are thus composed of a variety of sediments, largely fine grained, whose geotechnical properties are related to depositional processes via the particle packing and sedimentary fabric. This paper will describe shallow geotechnical profiles from a number of contrasted depositional settings on the slope and outer shelf off northwest Scotland and will relate these profiles to the inferred depositional processes by which these sediments accumulated. The discussion will concentrate on the water content and shear strength of the sediments and will examine these in the context of the void index model proposed originally by Burland (1990). This model also provides an insight into the acoustic character of the sediments, which will also be briefly discussed.

Regional setting

There is considerable variation in physiography along the length of the Hebrides–West Shetland margin (Fig. 1). The Hebrides Slope changes in strike northwards from N- and NW-trending to NE-trending, whereas the strike of the West Shetland Slope is more uniformly NE-trending. West of the Outer Hebrides, the shelfbreak occurs at about 200 m, although the maximum change in slope angle occurs at about 600 m, at the Geikie Escarpment. West of the Shetlands the shelfbreak is located in water depths ranging from 120 to 250 m, although it may be as deep as 500 m on the Wyville-Thomson Ridge.

On the Hebrides Slope, average slope angles range from <1° to 4°, occasionally reaching 7° on the upper slope, although the Geikie Escarpment displays extreme slope angles of up to 26° (Strachan & Evans 1991). Slope angles on the West Shetland Slope are typically <1° on the upper slope, increasing to 1.5° to 2° in the mid-slope area, before decreasing towards the basin floor (Richards *et al.* 1987; Stoker 1990; Stoker *et al.* 1991; Damuth & Olson 1993; Howe *et al.* 1994), although the slopes are locally cut by gullies and channels (Kenyon 1987).

The geological framework of the study area consists of a Neogene sediment wedge which unconformably overlies tilted, eroded and locally deformed Palaeogene strata. On the outer shelf and slope, the Neogene wedge can be

PAUL, M. A., TALBOT, L. A. & STOKER, M. S. 1998. Shallow geotechnical profiles, acoustic character and depositional history in glacially influenced sediments from the Hebrides and West Shetland Slopes. *In*: STOKER, M. S., EVANS, D. & CRAMP, A. (eds) *Geological Process on Continental Margins: Sedimentation, Mass-Wasting and Stability*. Geological Society, London, Special Publications, **129**, 117–131.

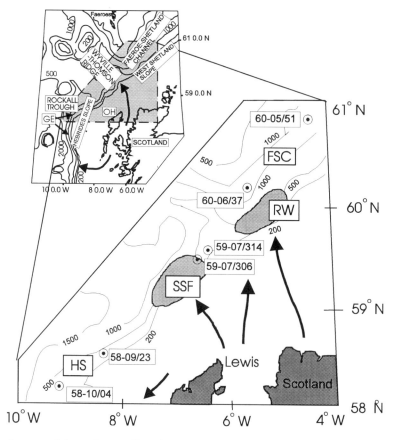

Fig. 1. Location map of the study area off northwestern Scotland showing the geographic areas and core sites referred to in the text. FSC, Faeroe–Shetland Channel; RW, Rona Wedge; SSF, Sula Sgeir Fan; HS, Hebrides Slope; GE, Geikie Escarpment; OH, Outer Hebrides. The arrows show the positions of major ice drainage paths.

divided into a laterally impersistent and eroded Miocene section (locally including upper Oligocene) unconformably overlain by a Plio-Pleistocene slope-apron with a Holocene veneer (Stoker 1995). However, in the deep-water basins of the Rockall Trough and the Faeroe–Shetland Channel, the Plio-Pleistocene appears to be conformable on the Miocene. Where the Miocene is absent, Plio-Pleistocene strata rest directly on the Palaeogene. In contrast, on the Geikie Escarpment lower Oligocene rocks locally crop out at the sea bed (Jones *et al.* 1986).

The sediments described in this paper are from the uppermost part of the Plio-Pleistocene slope-apron. They are predominantly glacigenic sediments deposited during the Devensian (Weichselian) glacial stage (Stoker 1995). On the Hebrides and West Shetland Slopes and in the Faeroe–Shetland Channel, the bulk of the cores penetrated into upper Devensian

glaciomarine sediments. Below about 400 m water depth, these sediments are commonly acoustically layered; above this depth, the slope sediments are often more weakly layered and display evidence of disturbance. The latter is due, in part, to iceberg scouring of the upper slope (Stoker *et al.* 1994). One of the cores studied here (59-07/314) is located on the outer part of the northern Hebrides Shelf and penetrated into older, early Devensian, sediments associated with a large submarine end-moraine (Stoker *et al.* 1994).

On the outer shelf, the near-surface sediment suite largely comprises diamictons, together with subordinate gravels and sands, of presumed subglacial to ice marginal origin, which occasionally form positive relief features (Stoker & Holmes 1991). On the upper slope the sediments comprise a variable suite of diamictons and muds believed to be largely of proximal glaciomarine origin whereas the middle to lower slope is dominated by muds and clast-poor diamictons of distal glaciomarine origin. Over much of the outer shelf and upper slope sandy sediments of Holocene age form a thin (<1 m) discontinuous veneer on the underlying glacial materials.

It is now recognized that the glacial sediments accumulated in a series of discrete depocentres, possibly related to cross-shelf transport paths determined by the positions of major ice streams (Stoker 1995), between which lay areas of sediment bypass in which accumulation was relatively attenuated. The Hebrides Slope, west of the Outer Hebrides, represents an area of bypass, flanked to the north by the Sula Sgeir Fan, which was a major depocentre. Farther north, the Wyville-Thompson Ridge marks a further area of bypass and the Rona Wedge a further, extended, depocentre. Despite the temporal variation, the depositional setting throughout the Devensian glacial stage has been one of grounded ice on the shelves and glaciomarine, ice-proximal to ice-distal sedimentation on the slopes. In the adjacent basins hemipelagic and ice-distal processes have operated. The deposits have been modified at times by bottom currents, which continued to be active during glacial intervals albeit at reduced strengths relative to the present-day, interglacial, current regime (Stoker et al. 1989; Akhurst 1991; Howe et al. 1994).

Sample collection and testing

The six cores described in this paper (Table 1) were collected in 1986–1987 by the British Geological Survey during their regional mapping of the UK continental shelf. Two cores have been studied respectively from the Sula Sgeir Fan, the Hebrides Slope and the Faeroe–Shetland Channel (Fig. 1). These areas represent three contrasted settings: the flank of a major fan, the outermost shelf and upper continental slope in an area of sediment bypass and the floor of a deepwater channel. All the cores were collected by the vibrocore method, in which a core barrel and liner of nominal inside diameter 90 mm are

driven into the sea bed to a depth of c. 6 m by a combination of sampler weight and vibratory action. Upon recovery, the cores were cut into 1 m sections, sealed and stored horizontally under refrigeration while awaiting testing. We note that some of the geotechnical profiles presented here show discontinuities at depths corresponding to the section breaks (i.e. at exact multiples of 1 m) and thus we discount these discontinuities as artefacts of the handling process.

We recognize that vibrocoring is not an ideal method for the collection of samples for geotechnical testing. The relatively thick wall of the core barrel and liner can cause sample disturbance and the vibratory motion can induce compaction in sandy or silty sediments. In the present study the authors have sought to minimize the effects of disturbance in four ways: by restricting the study to clayey sediments, in which the sampling process is largely undrained and so causes less overall change in water content than it does in sandy sediments; by collecting subsamples from the centre line of the core, where any disturbance is least (Baligh et al. 1987); by inspecting the core visually, after cleaning by osmotic knife, to avoid zones of obvious disturbance; and by analysing principally those geotechnical properties (grading, water content, plasticity and void index) that are insensitive to fabric disturbance.

In the following sections, the results of four basic geotechnical tests are presented. The grading of the samples was determined by conventional sieving and sedimentation methods. The water content (defined as the percentage of water per unit dry weight) was obtained by oven drying to 105°C. The liquid and plastic limits were determined respectively by cone penetrometer and hand rolling using the methods of BS1377 (Anon 1975). Undrained shear strength was determined using a 19 mm × 12 mm laboratory vane. From these data were calculated the liquidity index (the ratio of water content minus plastic limit to the liquid limit minus the plastic limit). The liquidity index expresses the relative plasticity of the sediment at its natural water

Table 1. *Details of cores*

Core no.	Location	Position	Water depth (m)	Recovered length (m)
58-09/23	Hebrides Slope	58°31.36′N, 08°22.44′W	300	5.96
58-10/04	Hebrides Slope	58°13.57′N, 09°15.13′W	353	6.00
59-07/306	Sula Sgeir Fan	59°24.00′N, 06°40.50′W	485	5.95
59-07/314	Sula Sgeir Fan	59°30.30′N, 06°27.60′W	284	5.80
60-05/51	Faeroe–Shetland Channel	60°43.24′N, 04°20.59′W	1115	5.66
60-06/37	Faeroe–Shetland Channel	60°15.54′N, 05°33.44′W	1142	5.73

content: at the liquid limit it is equal to unity and at the plastic limit to zero. There is a well-established correlation (Skempton & Northey 1953) between the liquidity index and the undrained shear strength of any clay soil, which is referred to in the sections below.

The void index of the sediment was also calculated from these data. This index (Burland 1990) is defined as the *in situ* void ratio minus the void ratio at 100 kPa effective normal stress, divided by the compression index of the remoulded sediment. This index is in effect a normalized void ratio, which relates the *in situ* void ratio to that expected from the fully remoulded sediment at the same effective stress. It is thus a measure of relative openness of the natural sediment fabric, because in the natural state a sediment always has a higher void index than when remoulded. In this study we have used values of compression index and

100 kPa void ratio calculated from the liquid limit in the manner described by Burland (1990). The *in situ* void ratio has been calculated from the water content using an assumed grain density of 2.70 Mg/m^3. A major advantage of this approach is that only the water content is susceptible to sample disturbance (above): the other parameters are measured on disturbed material in any event. It is hoped that the precautions described above will have minimized the effects of disturbance on the water content and hence on the results.

Geotechnical profiles

Sula Sgeir Fan

The Sula Sgeir Fan has been a major depocentre throughout the Plio-Pleistocene. The upper part of the fan succession consists of up to 200 m of

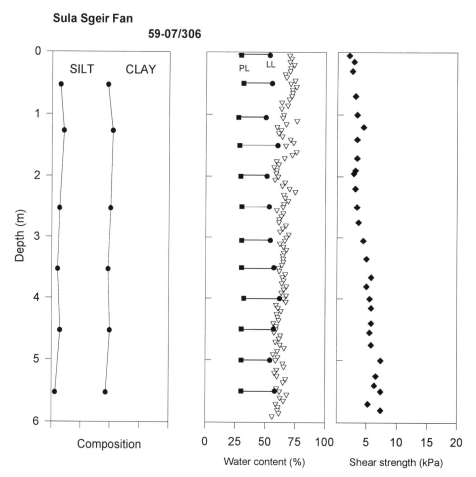

Fig. 2. Profiles of composition, water content and undrained shear strength measured on vibrocore 59-07/306 from the Sula Sgeir Fan.

glacigenic sediments of Anglian to mid-Devensian age, the bulk of which belongs to the Upper MacLeod Sequence (Stoker 1990: Stoker *et al.* 1993*a*). Seismic records reveal the glacigenic fan deposits to be composed of discrete, acoustically unstructured mass-flow sediment packages separated by thin, occasionally slope-wide, prograding clinoforms. The mass-flow packages are individually composed of fine-grained, matrix-supported diamicton with subordinate sand and mud layers and interpreted mainly as the products of glaciomarine mass flow or secondary slope instability (Baltzer *et al.* this volume). These packages merge upslope with an acoustically disordered facies (the MacDonald sequence) which occurs on the outer shelf and is associated with a rugose sea-bed morphology. The fan is partially overlain by sediments of the acoustically laminated MacAulay sequence, which forms a drape over the Upper MacLeod

sequence and is composed of uniform silty clays, interpreted as the product of distal glaciomarine sedimention during the mid- to late Devensian. The upper part of the fan and the adjacent shelf are heavily marked by iceberg scours. Cores 59-07/306 and 59-07/314 were studied from this area.

Core 59-07/306. This core (Fig. 2) was collected from the acoustically laminated MacAulay sequence in a water depth of 485 m, below the local limit of iceberg scouring. The sediment consisted of very uniform dark grey clayey silt (clay content 45–50%) which showed little downcore variation in lithology. The water content showed numerous small variations at the centimetre scale. Throughout the core the value was in excess of the liquid limit, falling from a liquidity index of around 1.5 near the top to around 1.1 near the base. The undrained shear strength was in consequence very low,

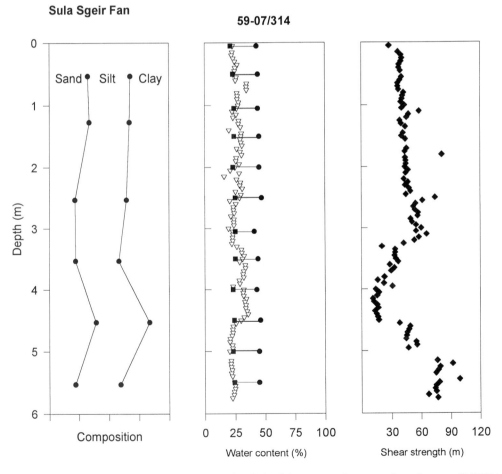

Fig. 3. Profiles of composition, water content and undrained shear strength measured on vibrocore 59-07/314 from the Sula Sgeir Fan.

rising from <4 kPa at the top of the core to around 8 kPa near the base.

Core 59-07/314. This core (Fig. 3) was raised from a depth of 284 m on the outer part of the northern Hebrides Shelf adjacent to the Sula Sgeir Fan. The core penetrated acoustically disordered sediments of the Macdonald sequence (Stoker *et al.* 1993*a*) associated with a submarine end-moraine of early Devensian age (Stoker *et al.* 1994). The sediment recovered was a dark grey–brown, silty, matrix-dominant diamicton containing clasts up to 100 mm in size, with subordinate sand pockets and clay layers. The consistency of the sediment was firm to stiff, with a water content close to the plastic limit, although there is some variability (for example, between 3.5 and 4.5 m) with zones of higher water content and lower strength which correlate broadly with

a change to a sandier lithology between these depths, although the spacing of the particle size observations precludes any more detailed analysis. The water content and shear strength profiles both suggest that the sediment is overconsolidated, as discussed below in terms of its void index.

Hebrides Slope, west of the Outer Hebrides

The Hebrides Slope succession in this area consists of sediments of the Upper MacLeod sequence. During the Anglian and Devensian glaciations this area was bypassed by the major sediment transport paths (see Fig. 1) leading to a reduced thickness (<100 m: Stoker *et al.* 1994) of glacigenic sediments in this area compared with the corresponding thickness on the Sula Sgeir Fan to the north (Stoker 1995). The

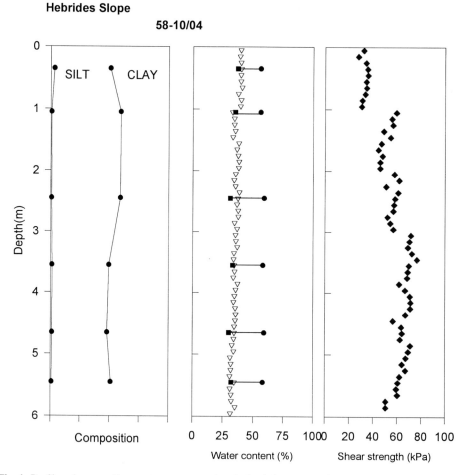

Fig. 4. Profiles of composition, water content and undrained shear strength measured on vibrocore 58-10/04 from the Hebrides Slope.

near-surface sediments have a well-layered, drape-like acoustic structure which in the upper slope becomes poorly layered to disordered, a change which appears to correlate with the maximum depth of iceberg disturbance. It is considered that the sediments originated by fall-out from suspension plumes and were subsequently modified by along-slope currents (Stoker *et al.* 1994). Cores from the disordered facies have yielded diamictons which range in consistency from firm to soft; those from the layered facies have mostly yielded silty clays of generally soft consistency (Evans & Strachan 1989). Cores 58-09/23 and 58-10/04 were studied from the area (Fig. 1).

Core 58-10/04. This core (Fig. 4) was collected from 353 m water depth on the upper Hebrides Slope, below the local shelf break, in an area

where the sea bed showed evidence of iceberg scour and the underlying seismic texture was disordered. It was composed of brown–grey silty clay with sporadic small clasts. In its top metre the core was markedly softer than the underlying material, although little compositional change was found. Below this depth the water content is relatively constant and lies close to the plastic limit throughout the profile, although with numerous minor fluctuations in water content of a few per cent (gross changes at depths of exactly 1 m intervals are discounted as artefacts). The undrained shear strength is generally high (40–80 kPa), as expected from a water content close to the plastic limit, other than in the topmost metre where the stength is around 20 kPa.

Core 58-10/23. This core (Fig. 5) was collected from a water depth of 325 m from the slope

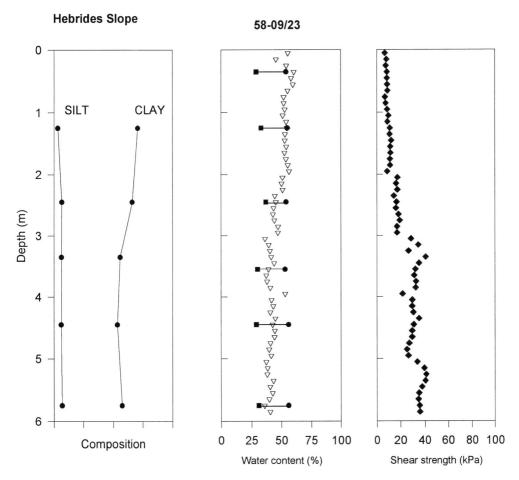

Fig. 5. Profiles of composition, water content and undrained shear strength measured on vibrocore 58-09/23 from the Hebrides Slope.

above the Geikie Escarpment, in an area where the underlying seismic texture was disordered to poorly stratified. The sediment, which was visually uniform with numerous monosulphide bands and streaks, consisted of a olive–grey clayey silt in which the proportion of clay increased in the top 3 m from around 25% to around 40%. The water content reduced throughout the profile, from near the liquid limit in the top 2 m to a liquidity index of around 0.5 at the base of the core. In the lower 3 m the water content reduced less slowly than in the upper 3 m, although with some variation at a scale of around 10 cm. Conspicuous breaks in the profile at exact 1 m intervals are discounted as artefacts. The undrained shear strength showed a corresponding rise from below 20 kPa at the top to around 40 kPa at the base: variations at the 10 cm scale appear correlated with similar variations in the water content.

Faeroe–Shetland Channel

The Faeroe–Shetland Channel represents a basin-plain setting in which sedimentation rates were generally low because of sediment capture on the higher slope. Two seismic sequences have been described from the area: (1) the Faeroe–Shetland Channel sequence, which shows a laminated acoustic texture, is associated with sediments of variable clay–sand lithofacies which show evidence of bioturbation, and is interpreted as a deep-water sequence of hemipelagic and bottom-current origin (Stoker *et al.* 1993*a*); (2) the Morrison sequence, which shows a transparent acoustic texture, is

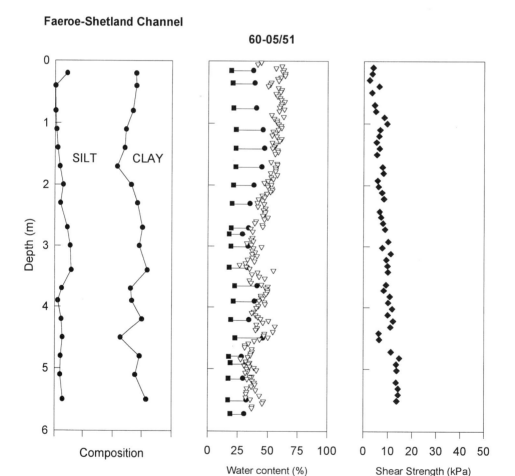

Fig. 6. Profiles of composition, water content and undrained shear strength measured on vibrocore 60-05/51 from the Faeroe–Shetland Channel.

Faeroe-Shetland Channel

60-06/37

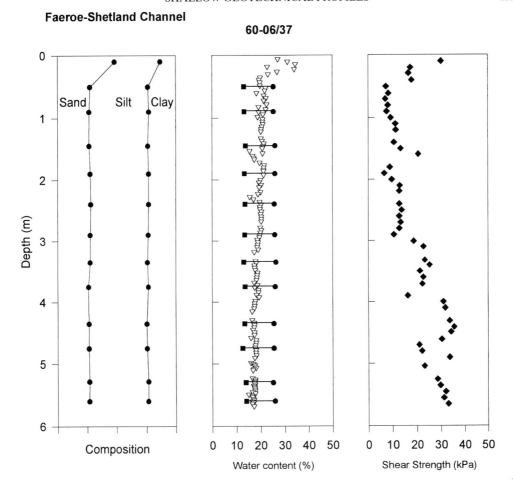

Fig. 7. Profiles of composition, water content and undrained shear strength measured on vibrocore 60-06/37 from the Faeroe–Shetland Channel.

associated with lensoid sediment packages of uniform sandy clay lithology, and is interpreted as a mass-flow deposit of glacigenic origin (Stoker *et al*. 1991). Cores 60-05/51 and 60-06/37 were studied from this area (Fig. 1).

Core 60-05/51. This core (Fig. 6) was collected from the floor of the Faeroe–Shetland Channel in a water depth of 1115 m, and penetrated the Faeroe–Shetland Channel sequence. The core contained four distinct lithofacies (mud, sandy mud, muddy sand, sandy silt), which have been described previously in this and related cores (Paul *et al.* 1993), and are similar to facies assemblages described by Akhurst (1991) from other cores in the Faeroe–Shetland Channel. Despite these facies changes the geotechnical profile is relatively simple: the water content lies above the liquid limit throughout the core, falling from

1.5 at the top to 1.25 near the base. In consequence, the undrained shear strength is low throughout the core, rising from around 5 kPa at the top to around 15 kPa at the base. Between 3 m and 5 m there is some variation as a result of facies changes, with the sandier facies showing a lower water content and higher undrained shear strength than the more clay-rich facies.

Core 60-06/37. This core (Fig. 7) was collected in the distal zone of the Rona Wedge in the Faeroe–Shetland Channel from a water depth of 1142 m. The sample site was located in an area underlain by sediments of the Morrison sequence, near the nose of an identifiable lensoid unit which showed a transparent acoustic texture. The top 50 cm of this core was composed of sandy silt, and had a correspondingly variable water content with values from 20 to 35%: below this depth the core consisted of uniform sandy mud with minor clasts and showed very little lithological variation. In consequence, the water content below 0.5 m varied only little with depth, falling from 20% to 15% from top to base of the core. As the sediment was of low plasticity, this represents a relatively large change in liquidity index (*c.* 0.5 to *c.* 0.25) and thus there is a corresponding increase in shear strength from 20 kPa to around 50 kPa. The low plasticity of the sediment causes the behaviour to be very sensitive to small variations in water content: an adjacent, lithologically similar core (60-06/34) with a water content around 12–15% has a shear strength around 100 kPa and showed evidence of preconsolidation (Paul *et al.* 1993).

Discussion

Relationship to depositional setting

The cores described above are drawn from three generalized settings: the outer shelf and upper slope, the middle slope to slope foot and the basin-plain. Each of these settings has been characterized by specific depositional processes which have influenced the near-surface geotechnical properties of the sediments within them.

The sediments on the outer shelf and upper slope are generally considered (Stoker 1988, Stoker & Holmes 1991; Stoker *et al.* 1993a) to have originated in a proximal glaciomarine environment and were probably periodically subjected to direct loading from grounded ice, tidewater shelves or iceberg turbation. We believe that this has introduced moderate (but variable) overconsolidation in some sediments (e.g. cores 58-10/04 and 59-07/314), as evidenced by their reduced water content (and hence liquidity index) and consequent increase in shear strength. However, we note that although core 58-09/23 was collected from a similar position on the upper slope, it is not similarly overconsolidated. This demonstrates that local palaeogeography may have exerted significant control on consolidation state near the shelfbreak and possibly implies a more ice-distal position at the time of deposition.

On the middle slope to slope foot, the sediments have in general been deposited by a combination of mass flow and suspension fall-out (Stoker *et al.* 1994), and in some cases may also have experienced alongslope current reworking (Kenyon 1987; Leslie 1993). Those considered here (cores 59-07/306 and 60-06/37) are thought to have been deposited by distal mass flows (60-06/37) or suspension fall-out (59-07/306). This has produced a sediment of high

water content and low shear strength, although in the case of 60-06/37 the very low plasticity has made the strength of the sediment abnormally sensitive to small changes in water content, as shown by the behaviour of other cores in the same area (Paul *et al.* 1993).

On the basin floor, the sediment has accumulated mainly by suspension fall-out, turbiditic run-out from sporadic debris flows, and bottom-current reworking (Stoker *et al.* 1991, 1993); pauses in sedimentation have allowed bioturbation to occur (Akhurst 1991). As a result, these sediments (e.g. core 60-05/51) commonly show greater lithofacies variation than those on the slope, which causes a corresponding variation in the water content and shear strength. However, the variation in liquidity index (the relative water content in relation to the liquid and plastic limits) is much less (Fig. 6: cf. Paul *et al.* 1993), which indicates that these variations are simply the result of lithologically controlled plasticity changes.

Void index model

Our results have shown that in all the settings considered, the water content profile of the sediment varies with inferred depositional process, but that this relationship can be obscured by lithological changes. It is thus useful to explore these relationships in the context of the void index of the sediments (Burland 1990), which provides a normalized framework within which to understand the one-dimensional compression of all natural sediments.

This model proposes that in terms of their void index, all naturally deposited, structured sediments follow a common compression curve termed the sediment compression line (SCL) when plotted against effective stress. Figure 8 outlines the essential features of the model, on which we have also superimposed details of the acoustic facies, which we shall discuss in the following section. Owing to small variations in fabric, a sequence of samples down any given profile (e.g. a in Fig. 8) will fall around the SCL with some scatter. Similarly, the model proposes that all remoulded sediments will follow another common compression line (the intrinsic compression line or ICL): samples down a profile will form a sequence such as b. Overconsolidated sediments fall along a rebound line below the ICL, as the *in situ* effective stress is lower than the past maximum stress under which the sediment fabric would have been derived: c in Fig. 6 shows a typical depth sequence.

Previous work (Talbot *et al.* 1994) has suggested that, as the position of a point relative to

SCHEMATIC PACKINGS AND ACOUSTIC FACIES
RESULTING FROM VARIOUS SEDIMENTATION PROCESSES

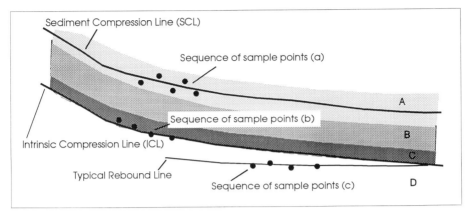

Fig. 8. Schematic void index diagram to show the major features discussed in the text. Areas A–D are approximate fields defined by the position of the sample points relative to the intrinsic and sediment compression lines. The lower panel shows the processes by which sediments within each field have been deposited and indicates the acoustic textures that these sediments commonly possess.

the ICL and SCL is a measure of the level of natural structure in the material, there is a correspondence between the position of a sample point on the void index diagram and the depositional process by which the sediment was formed. This relationship is shown schematically in Fig. 8. In general, sediments which plot around the SCL (area A) show an open, well-developed structure and have often originated by relatively slow deposition, perhaps also involving bioturbation. Sediments which plot in area B are more closely packed and in the case studied by Talbot *et al.* (1994) were deposited rapidly, with probable syndepositional failure and consequential fabric adjustment. Those which plot in area C have a closely packed structure following partial or complete remoulding

and appear characteristic of mass movement processes. We note that this finding agrees with the work of Audet (this volume), who reported analogous effects of mass movement processes on sediment properties in the Amazon fan. Finally, area D contains sample points from sediments that are overconsolidated: this may arise from a variety of processes which cause a post-depositional reduction in effective stress and in the present context are likely to include ice loading or erosion.

In Fig. 9 the void indices from the cores considered here have been plotted against the *in situ* effective stress (calculated from the sediment density assuming hydrostatic pore pressure). Each profile is shown as a connected sequence of points and it is clear that sediments which fall

Void Index Diagram

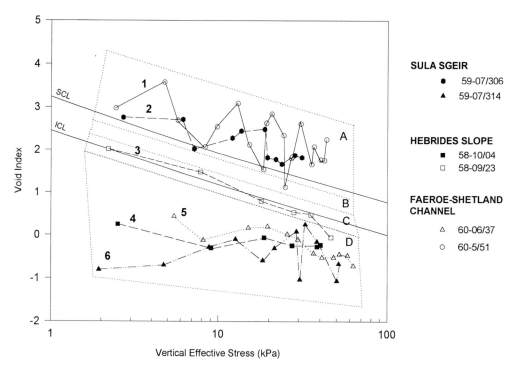

Fig. 9. Void index diagram for the vibrocores discussed in this paper, showing the relationship of the sample points to the fields indicated in Fig. 8. The lines connect sequences of sample points in order of increasing depth down the core.

onto the distinctive areas A–D of Fig. 8 are from differing depositional settings. The samples from cores 60-05/51 (sequence 1) and 59-07/306 (sequence 2) plot around or above the SCL (Fig. 9), indicating an ordered structure associated with an open fabric and high water content, although the level of structure (and hence the distance from the SCL) is variable. This behaviour is most obviously explained as the result of hemipelagic (60-05/51) or suspension fall-out (59-07/306) processes, assisted in the case of 60-05/51 by the action of bioturbation.

Core 58-09/23 (sequence 3), which is composed of soft, normally consolidated material, plots close to the ICL (Fig. 9). Because shear disturbance might be expected to cause the sample points to lie close to the ICL, as a result of destructuration by remoulding, and because this core was collected from a proximal glaciomarine setting, we argue that it represents sediments that have suffered downslope flow after release but have probably not been subject to ice loading. By contrast, cores 58-10/04 (sequence 4) and 59-07/314 (sequence 6) from the outer shelf and upper slope fall around a flatter rebound

line below the ICL (Fig. 9), a position which reflects a history of variable consolidation probably related to iceberg or ground ice disturbance. Rightward projection of the rebound line suggests that it will intersect the ICL at a relatively low effective stress of around 100–150 kPa, which, interestingly, thus provides an estimate of the past consolidation pressure which is independent of the interpretation of a conventional compression test. Investigators who have studied glacigenic sediments in analogous offshore settings have commented on the interpretation of past pressure in terms of depositional and post-depositional history (Saettem 1990; Saettem *et al.* 1992) and have argued that relatively low past pressures are not inconsistent with a subglacial origin, possibly because of subglacial drainage, which is generally agreed to play an important role in determining the subglacial effective stress (Boulton & Paul 1976; Boulton 1979; Alley *et al.* 1987).

The samples from core 60-06/37 (sequence 5) do not fall along a clear rebound curve, but instead lie around a curve parallel to the ICL, although displaced from it by a horizontal distance equivalent to about 100–200 kPa (Fig. 9). Although this might be the result of overconsolidation, the points do not fall along a single rebound line, which implies that different points in the profile have had differing stress histories. It is also possible, in view of the low plasticity of the sediment, that this core has suffered some differential mechanical compaction leading to a loss of water and a reduction in the void index. This could be a natural effect associated with a mass-flow origin or could may be a result of the vibrocore sampling process itself.

Relationship to acoustic texture

The results presented here, together with the earlier results of Talbot *et al.* (1994), indicate that there is a correspondence between the position of a sample relative to the ICL or SCL and the local acoustic texture of the unit from which it was collected. We would obviously not expect a perfect correspondence in all cases, for example in the case of overconsolidated units, and indeed the known variations in seismic facies interpretation on the Scottish shelf and slope (Stoker *et al.* 1993*b*) confirm this view. However, the broad principle that acoustically well-layered sediments yield data points around the SCL (Fig. 8: area A) and less well layered sediments yield data points in areas B–D seems established.

Samples from sediments associated with a layered acoustic texture are located around the SCL with considerable scatter (e.g. Fig. 9: cores 60-05/51 and 59-07/306). We believe that this scatter is fundamental to 'open' structures that plot around the SCL and that constructive acoustic interference from the minor density variations implied by this scatter generate the multilayered signature. Thus samples from any acoustically well-layered unit will yield a scatter of void index values which will fall in area A of Fig. 8. Although in the present study no samples have fallen within area B, our previous work (Talbot *et al.* 1994) suggests that this area is associated with weaker acoustic layering.

Cores 58-10/04, 58-09/23 and 59-07/314 are associated with units that show a disordered acoustic texture. Samples from these cores plot on (58-09/23) or below (58-10/04 and 59-07/314) the ICL in regions C and D of Fig. 9. This position is a direct result of their variable consolidation state and we propose that the disturbance that has created their disordered acoustic texture is also responsible for the variable state of their packing. Finally, 60-06/37 was collected from an acoustically transparent unit interpreted as the product of a mass flow. In this case

the significant feature is the relative lack of scatter between the sample points (sequence 5: Fig. 9): this implies a lack of internal acoustic contrasts in the sediment, probably as a result of fabric homogenization by the flow process, which will thus lack internal acoustic reflectors.

Conclusions

In the six vibrocores studied here the sediments have been found to be mostly matrix-supported diamictons or laminated muds with subordinate sands. The cores from the outer shelf and upper slope have provided sediments that are firm to stiff and whose water contents are close to or below the plastic limit throughout the profiles. As the cores are from the upper 6 m of the sea bed, this suggests that the material is lightly to moderately overconsolidated. Sediments from the lower slope and basin floor are soft to very soft and have water content close to or above the liquid limit throughout the profile, which suggests that the sediment is normally consolidated. Their undrained shear-strength profiles show similar variations, as they are controlled directly by the water content and follow a standard relationship to the liquidity index.

We consider that these geotechnical differences are genetic and reflect the processes operative in former glacial and glaciomarine palaeoenvironments along the continental margin. Sediments from the upper slope and shelf have formed in proximal glaciomarine to ice-front settings and have experienced iceberg turbation and possible ice loading. The extent of these effects appears to have varied from one core site to another. Sediments from the lower slope are largely the product of distal glaciomarine processes including suspension fall-out and mass flow, and on the basin floor are the products of bottom currents and hemipelagic sedimentation under conditions of lower sediment input. These have resulted in a soft material whose water content profile has been controlled only by the effective overburden stress.

The void index of the sediment shows a sensible relationship with the depositional setting. Those samples from the outer shelf and upper slope plot on or below the intrinsic compression line, which suggests a disordered or destructured fabric that may have resulted from post-depositional compression and flowage. The distance of the sample point below the ICL is a measure of the degree of overconsolidation. Those sediments from lower slope and basin floor settings plot around the sediment compression line characteristic of many marine sediments, which suggests a structured fabric with high pore space.

Thus the void indices provides a link between geotechnical character of the sediment and the processes by which it was deposited.

Fluctuations within the void index profile of a sediment also provide an explanation for the layered acoustic texture seen in many soft, muddy sediments. Changes in void index are the result of changes in packing and thus in bulk density: acoustic velocity is almost constant in soft sea-bed materials and so these changes will dominate the impedance profile. As such minor fluctuations are capable of generating layered signatures by acoustic interference, it follows that sediments which plot around the sediment compression line with the normal amount of scatter are likely to have a layered acoustic character. By contrast, sediments which plot with little scatter down their profile will have little impedance contrast and will show a faint to transparent acoustic character.

The authors thank M. Audet and C. F. Forsberg for their constructive review of the manuscript; B. Barras, H. Barras and G. Tulloch for their advice and assistance during the laboratory programme; and the British Geological Survey for access to seismic records and core material from the Hebrides and West Shetland Slopes. The financial support of MTD Ltd is gratefully acknowledged. The contribution of M. S. Stoker is published with the permission of the Director, British Geological Survey, NERC.

References

AKHURST, M. C. 1991. *Aspects of Late Quaternary Sedimentation in the Faeroe–Shetland Channel, Northwest UK Continental Margin*. British Geological Survey Technical Report **WB/91/2**.

ALLEY, R. B., BLANKENSHIP, D. D., BENTLEY, C. R. & ROONEY, S. T. 1987. Till beneath ice stream B2. Till deformation: evidence and implications. *Journal of Geophysical Research*, **92**, 8921–8929.

ANON 1975. *BS1377: Methods of Test for Soils for Civil Engineering Purposes*. British Standards Institution, London.

AUDET, D. M. 1998. Mechanical properties of terrigenous muds from levee systems on the Amazon Fan. *This volume.*

BALIGH, M. M., AZZOUZZ, A. S. & CHIN, C. T. 1987. Disturbance due to ideal tube sampling. *Journal of Geotechnical Engineering Division ASCE* **113**(GT7), 739–757.

BALTZER, A., HOLMES, R. & EVANS, D. 1998. Debris flows on the Sula Sgeir Fan, NW of Scotland. *This volume.*

BOULTON, G. S. 1979. Processes of glacier erosion on different substrata. *Journal of Glaciology*, **23**, 15–38.

—— & PAUL, M. A. 1976. Influence of genetic processes on some geotechnical properties of glacial

tills. *Quarterly Journal of Engineering Geology*, **9**, 159–193.

BURLAND, J. B. 1990. On the compressibility and shear strength of natural clays. *Geotechnique*, **40**, 329–378.

DAMUTH, J. E. & OLSON, H. C. 1993. Preliminary observations of Neogene-Quaternary depositional processes in the Faeroe-Shetland Channel revealed by high resolution seismic facies analysis. *In*: PARKER, J. R. (Ed.) *Petroleum Geology of Northwest Europe: Proceedings of the 4th Conference*. Geological Society, London, 1035–1046.

EVANS, D. & STRACHAN, P. 1989. *Geikie 58°N–10°W. Seabed sediments*. 1 : 250,000 map series, British Geological Survey.

HOWE, J. A., STOKER, M. S. & STOW, D. A. V. 1994. Late Cenozoic sediment drift complex, northeast Rockall Trough, North Atlantic. *Palaeoceanography*, **9**, 989–999.

JONES, E. J. W., PERRY, R. G. & WILD, J. L. 1986. Geology of the Hebridean margin of the Rockall Trough. *Proceedings of the Royal Society of Edinburgh, Section B*, **88**, 27–51.

KENYON, N. H. 1987. Mass-wasting features on the continental slope of northwest Europe. *Marine Geology*, **74**, 57–77.

LESLIE, A. B. 1993. Shallow Plio-Pleistocene contourites on the Hebrides Slope, northwest U.K. continental margin. *Sedimentary Geology*, **82**, 61–78.

PAUL, M. A., TALBOT, L. A. & STOKER, M. S. 1993. Geotechnical properties of sediments from the continental slope northwest of the British Isles. *Advances in Underwater Technology*, **28**, 77–106.

RICHARDS, P. C., RITCHIE, J. D. & THOMSON, A. R. 1987. Evolution of deep water climbing dunes in the Rockall Trough – Implications for overflow currents across the Wyville-Thomson Ridge in the (?)Late Miocene. *Marine Geology*, **76**, 177–183.

SAETTEM, J. 1990. Glaciotectonic forms and structures on the Norwegian continental shelf. *Norsk Geologisk Tidsskrift*, **70**, 81–94.

——, POOLE, D. A. R., ELLINGSEN, L. & SEJRUP, H. P. 1992. Glacial geology of outer Bjornoyrenna, southwestern Barents Sea. *Marine Geology*, **103**, 15–51.

SKEMPTON, A. W. & NORTHEY, R. D. 1953. The sensitivity of clays. *Geotechnique*, **3**, 30–53.

STOKER, M. S. 1988. Pleistocene ice proximal glaciomarine sediments in boreholes from the Hebrides shelf and Wyville-Thomson Ridge, NW UK continental shelf. *Scottish Journal of Geology*, **24**, 249–262.

—— 1990. Glacially-influenced sedimentation on the Hebridean Slope, northwestern United Kingdom continental margin. *In*: DOWDESELL, J. A. & SCOURSE, J. D. (Eds) *Glacimarine Environments: Processes and Sediments*. Geological Society, London, Special Publications, **53**, 349–362.

—— 1995. The influence of glacigenic sedimentation on slope-apron development on the continental margin off northwest Britain. *In*: SCRUTTON, R. A., STOKER, M. S., SHIMMIELD, G. B. & TUDHOPE, A. W. (Eds) *The Tectonics, Sedimentation and*

Palaeoceanography of the North Atlantic Region Geological Society, London, Special Publications, **90**, 159–177.

—— & HOLMES, R. 1991. Submarine end-moraines as indicators of Pleistocene ice-limits off northwest Britain. *Journal of the Geological Society, London*, **148**, 431–434.

——, HARLAND, R. & GRAHAM, D. K. 1991. Glacially influenced basin plain sedimentation in the southern Faeroe–Shetland Channel, northwest UK continental margin. *Marine Geology*, **100**, 189–199.

——, ——, MORTON, A. C. & GRAHAM, D. K. 1989. Late Quaternary stratigraphy of the northern Rockall Trough and Faeroe–Shetland Channel, northeast Atlantic Ocean. *Journal of Quaternary Science*, **4**, 211–222.

——, HITCHEN, K. & GRAHAM, C. C. 1993a. *United Kingdom offshore regional report: the geology of the Hebrides and West Shetland shelves, and adjacent deep water areas.* HMSO, London, for the British Geological Survey.

——, LESLIE, A. B., SCOTT, W. D., BRIDEN, J. C., HINE, N. M., HARLAND, R., WILKINSON, I. P., EVANS, D. & ARDUS, D. A. 1994. A record of late Cenozoic stratigraphy, sedimentation and climate change from the Hebrides Slope, NE Atlantic Ocean. *Journal of the Geological Society, London*, **151**, 235–249.

——, STEWART, F. S., PAUL, M. A. & LONG, D. 1993b. Problems associated with seismic facies analysis of Quaternary sediments on the Northern UK Continental Margin. *Advances in Underwater Technology*, **28**, 239–262.

STRACHAN, P. & EVANS, D. 1991. A local deep water sediment failure on the northwest slope of the UK. *Scottish Journal of Geology*, **27**, 107–111.

TALBOT, L. A., PAUL, M. A. & STOKER, M. S. 1994. Geotechnical and acoustic properties of Plio-Pleistocene sediments from the Hebrides Slope. *Geo-Marine Letters*, **14**, 244–251.

Mechanical properties of terrigenous muds from levee systems on the Amazon Fan

D. MARC AUDET

University of Oxford, Department of Earth Sciences, Parks Road, Oxford OX1 3PR, UK,
e-mail: marc@earth.ox.ac.uk

Abstract: The mechanical stability of channel levees may be an important control on channel avulsion and sediment distribution patterns in submarine fans. Levee stability depends on the shear strength of the levee sediments, which in turn depends on how the sediments are deposited and compacted. Long, continuous sections of levee sediments on the Amazon Fan were recently cored during Leg 155 of the Ocean Drilling Program. The porosity–depth profiles from Sites 939 and 944, situated on the levee of the Amazon Channel, were analysed using a theoretical compaction curve derived from soil mechanics theory. By fitting the theoretical compaction curve to porosity–depth data, estimates of the compression index and the void ratio at 100 kPa effective stress were determined. The compression index of the levee sediments was estimated to be 0.58–0.68. The mechanical parameters of the levee sediments are consistent with published geotechnical data for remoulded clays. Undrained shear strength data from Leg 155 were analysed using an infinite slope analysis to determine the slope angle for failure as a function of burial depth. The variations in the shear strength measurements were bounded by critical slope angles of 1–3°. The higher bound of 3° tends to be consistent with maximum slopes measured near the levee crest near Site 938 on the Blue channel. The lower bound of 1° appears to be consistent with the slope of the sediment deposit blanketing the slope away from the levee crest. The top 55 m of Site 941, sited in the Western Debris Flow, show a relatively undisturbed, normally compacted porosity profile, even though the sediments show soft-deformation features. In contrast, the core samples between 55 and 68 m show lower porosity values compared with the porosity profile from Site 939. This low-porosity zone was interpreted as a zone of reworked or sheared sediment that could be the failure surface of a translational slide.

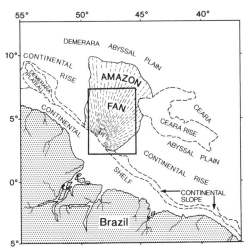

Fig. 1. Map showing the location of the Amazon Fan in relation to the physiographic features of the western equatorial Atlantic (modified from Damuth 1977).

The Amazon Fan is a deep-water depositional system located on the northeast Brazilian continental margin, extending from the continental shelf downslope for about 700 km to the Demerara and Ceara Abyssal Plains at depths >4600 m (Fig. 1). The morphology, sediments, structure, age and growth pattern of the Amazon Fan have been described in detail (Damuth & Kumar 1975; Kumar 1978; Milliman 1979; Damuth & Embley 1981; Damuth *et al.* 1988; Manley & Flood 1988; Flood *et al.* 1991). During Leg 155 of the Ocean Drilling Program (Flood *et al.* 1995a), long continuous sections of sediment were cored at 17 sites on the Amazon Fan, and sediment physical properties, such as porosity and undrained shear strength, were measured.

In terms of area, the Amazon Fan (3.3×10^5 km²) is the third largest in the world after the Bengal Fan (3×10^6 km²) and the Indus Fan (1×10^6 km²), and comparable in size to the Mississippi Fan (3×10^5 km²) (table 4, Reading & Richards 1994). Reading & Richards (1994) proposed a classification scheme for deep-water basin margins based on grain size and feeder system type. These four fans (Amazon, Bengal, Indus, Mississippi) are examples of point-source, mud-rich submarine fans and are characterized by elongate shapes and low slope gradients (0.2–18 m/km). Sediments tend to feed into the

AUDET, D. M. 1998. Mechanical properties of terrigenous muds from levee systems on the Amazon Fan   133
In: STOKER, M. S., EVANS, D & CRAMP, A. (eds) *Geological Processes on Continental Margins:*
Sedimentation, Mass-Wasting and Stability. Geological Society, London, Special Publications, **129**, 133–144.

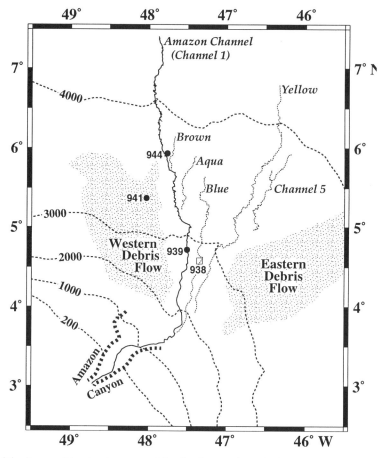

Fig. 2. Map of the Amazon Fan showing some of the distributary channels, including the Amazon Channel (Channel 1), which is the most recently active channel on the fan. Amazon Canyon, which feeds sediments to the channels, the Western and Eastern Debris Flows, and ODP Leg 155 Sites 939, 941, 944 are also shown. The square labelled 938 shows the location of profile C–D that passes through Site 938 (see Fig. 10). Channel 1 is the most recent channel. The older channels (in increasing order of relative age) are: Brown, Aqua, Blue, Yellow and Channel 5. The dashed contours show the bathymetry between 1000 and 4000 m. The 200 m contour represents the shelf break. (Adapted from fig. 9 of Flood *et al.* 1995*b*.)

fan through a mud-rich river delta and are transported downslope by low-density turbidity currents, often initiated by slumps and slides on the delta front.

The Amazon Fan has been developing since the Andean uplift (middle Miocene) when the Amazon River began to flow eastward into the Atlantic Ocean (Damuth & Kumar 1975). Terrigenous sediments, dominated by silts and clays, are derived from the Andes and from weathering within the Amazon Basin. Fan growth occurred primarily during glacial sealevel lowstands when the Amazon River was able to cross the emerged shelf and discharge sediments into the Amazon Canyon. These sediments were transported and distributed

downslope across the fan through a series of distributary channels (Fig. 2).

The fan architecture consists of two main features: (1) levee complexes built of overlapping channel–levee systems and (2) widespread debris-flow deposits (Manley & Flood 1988). The observed sediment sequences suggest cyclic depositional patterns that may have formed in response to glacio-eustatic sea-level fluctuations. During periods of falling sea level, sediments previously trapped on the shelf during high sea-level stands were carried directly into the deep sea. As a result, the increased sediment supply oversteepened the upper slopes, which induced large-scale sediment failure, resulting in large mass movements and turbidity flows.

These mass movements filled topographic lows and levelled pre-existing fan topography, thus creating the widespread basal transparent seismic units observed in the upper and middle fans. As sea level continued to fall, the Amazon River began to discharge directly into the deep sea, and upon reaching the shelf break, cut a sub-marine canyon, which acted as a point-source for turbidity flows that created the channel–levee systems. Once sea level rose, the sediments reaching the fan became finer grained, as the coarser-grained material was trapped in the aggrading river and on the shelf. During maximum highstand, such as the Holocene, all the river sediments were deposited on the river delta and upper shelf, and the fan became temporarily inactive (Damuth & Kumar 1975; Damuth & Flood 1985; Damuth et al. 1988).

Recent studies (Damuth et al. 1983a,b; Flood & Damuth 1987) of channel bifurcation patterns based on long-range side-scan sonar (GLORIA) suggest that only a single channel is active at any one time. Occasionally, the active channel levees fail through avulsion and the active channel changes course and causes sediments to be dis-tributed to a different region of the fan. Earlier in the fan's history, the sediments were distrib-uted in the eastern part of the fan through Channel 5. By a sequence of avulsions, the sedi-ment distribution pathway evolved into the Yellow channel (Fig. 2) and Channel 5 was aban-doned. The distributary channel continued to migrate westward, successively into the Blue, Aqua and Brown channels before reaching its current track as the Amazon Channel (Damuth et al. 1983b, 1988; Manley & Flood 1988).

Large parts of the upper and middle fan have also been affected by turbidity currents and debris flows that probably were initiated by slumping and sliding (Damuth & Embley 1981; Manley & Flood 1988). In addition to over-steepening, the migration of a gas hydrate phase boundary during sea-level fluctuation and diapiric activity may also be mechanisms for ini-tiating wide spread debris flow (McIver 1982; Manley & Flood 1988). Because of the presence of large debris flows in the Amazon Fan, and channel avulsion owing to levee breaching, it is worth while to study the slope stability of levee deposits. Submarine mass-wasting processes have been reviewed recently by Hampton et al. (1996), and by Mulder & Cochonat (1996).

The purpose of this paper is to study the com-paction and the shear strength of the sediments forming the levees of the main active channel of the Amazon Fan. Infinite slope analysis is uti-lized to assess whether the overburden (which depends on the compaction) and the measured

shear strength are consistent with slope measure-ments based on precision depth-recorder pro-files. This paper examines the porosity–depth profiles from three sites drilled during ODP Leg 155: Site 939 on the upper fan; Site 944 on the middle fan; Site 941 in the Western Debris Flow (Fig. 2). The porosity–depth profiles (Flood et al. 1995a) are used to determine the mechanical properties of the sediments in terms of C_c (com-pression index) and e_{100} (void ratio at 100 kPa effective stress) (i.e. compression parameters, Skempton 1944). Because C_c and e_{100} have been measured for many types of sediments, it is poss-ible to identify whether the sediments have been normally compacted, or alternatively, have undergone a more complicated stress history. Assuming an infinite slope analysis and actual measurements of the undrained shear strength, S_u, it is possible to assess whether slope measure-ments are consistent with mass movements triggered by sediment failure. This work comple-ments previous geotechnical studies related to assessing geohazard risk to submarine structures (e.g. Hampton 1989).

Soil mechanics and gravitational compaction

This section describes the basic theory for modelling gravitational compaction of normally pressured sediments. More details of the theor-etical derivation are given in an earlier paper (Audet 1995a).

Basic model

The sediments are modelled as a two-component system consisting of grains and pore fluid. The bulk density of the sediments, ρ, is given by

$$\rho = \phi\rho_f + (1 - \phi)\rho_g \qquad (1)$$

where ϕ is the porosity, ρ_g is the grain density, and ρ_f is the pore fluid density. In this paper, ρ_g and ρ_f are assumed to be constants.

For modelling compaction in sedimentary basins, it is often assumed that sediments deform only in the vertical direction, that is, one-dimen-sional compression, also known as K_0 conditions (Atkinson 1993). For this simplified state of stress, the force balance on a representative ele-mentary volume of sediment is

$$\frac{\partial \sigma_v}{\partial d} = + \rho g \qquad (2)$$

where σ_v is the total vertical stress (the over-burden), and g is the acceleration due to gravity, assumed to act in the positive d direction. The coordinate d represents the burial depth measured downward from the depositional surface. The pore fluid pressure, denoted by p_f, is assumed to be hydrostatic,

$$\frac{\partial p_{vf}}{\partial d} = + \rho_f g \qquad (3)$$

If the sediments were overpressured, then Darcy's Law would replace equation (3), leading to a nonlinear, diffusion equation for ϕ (Audet & McConnell 1992; Audet 1995b).

From soil mechanics theory, the sediments are assumed to obey Terzaghi's principle of effective stress (Terzaghi 1936) which states that the sediment strength depends on the effective stress, denoted by σ_v' and defined by

$$\sigma_v' = \sigma_v - p_f \qquad (4)$$

For a given lithology, σ_v' depends only on ϕ as discussed in the following section.

Sediment compressibility

The ability of sediments to sustain mechanical loading is controlled by the effective stress, which in turn depends on the sediment porosity (Skempton 1944, 1961, 1970; Atkinson 1993). Experimental measurements of the effective stress of clays and natural sediments can be found in the geotechnical engineering literature (see Burland 1990 for a review). In many cases, σ_v' and ϕ can be correlated by

$$\sigma_v' = \sigma_{100} \exp_{10} \left(\frac{e_{100} - e}{C_c} \right) \qquad (5)$$

where $e = \phi/(1 - \phi)$ is the void ratio. (Note: $\exp_{10} x = 10^x$.) The effective stress law given by equation (5) is based on observations that plots of e v. log σ_v' follow a straight line (Skempton 1944). Equation (5) is usually used to describe sediments subjected to uniform loading. Hysteresis effects caused by loading and subsequent unloading are not considered in this paper. The constant σ_{100} is the reference effective stress, usually taken to be 100 kPa (corresponding to about 10 m burial depth). The void ratio at $\sigma_v' = 100$ kPa is denoted by e_{100}. The sediment strength is characterized by C_c, a constant called the compression index. As the two parameters C_c and e_{100} characterize the effective stress, that is, mechanical compaction, the two parameters are referred to as the sediment mechanical parameters.

Equation (5) also applies to sediments recovered from deep-sea cores. To illustrate, Laughton (1957) measured the normal compression curves for two deep-sea sediment samples from the northeastern Atlantic: (1) Globigerina ooze cored at a water depth of 4713 m (54% CaCO$_3$); (2) terrigenous mud cored at a water depth of 2085 m (21% CaCO$_3$). The effective stress v. void ratio for the two samples are shown in Fig. 3. For effective stresses >1000 kPa, the compression data are linear on a semi-logarithmic plot. The negative slope of the best-fit line gives C_c and the intercept of the line at $\sigma_v' = 100$ kPa gives e_{100}. The mechanical parameters based on Laughton's (1957) data are: (1) $C_c = 0.677$, $e_{100} = 2.352$ for the ooze; (2) $C_c = 0.413$, $e_{100} = 1.289$ for the mud.

Correlation of mechanical parameters

Parasnis (1960) verified equation (5) for effective stress levels up to 82 MPa, corresponding to c. 8 km of burial depth. Parasnis observed that for carbonate sediments, the mechanical parameters C_c and e_{100} are linearly correlated according to

$$e_{100} = 3.37 \, C_c + 0.181 \qquad (6)$$

(see additional discussion in Audet 1995a)

Burland (1990) found an analogous correlation for remoulded clays:

$$e_{100} = 2.02 \, C_c + 0.287 \qquad (7)$$

Figure 4 shows equations (6) and (7) along with the original data from Parasnis (1960) and Burland (1990). The results for natural sediments are very different from those for

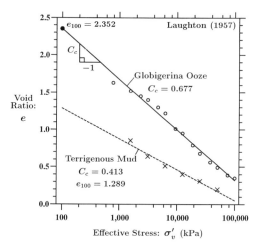

Fig. 3. The void ratio (e) as a function of the effective stress (σ_v') for Globigerina ooze and terrigenous mud from the northeastern Atlantic. The data are from the compression tests of Laughton (1957). The negative slope of the best-fit line of the semi-logarithmic plot is the compression index (C_c), and e_{100} is the void ratio of the line at an effective stress of 100 kPa.

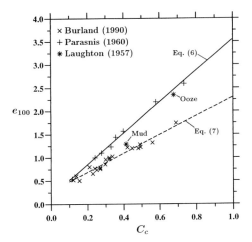

Fig. 4. The correlation of the mechanical parameters, C_c and e_{100}. The continuous line (equation (6)) is based on compression tests for carbonate sediments (Parasnis 1960), and the dashed line (equation (7)) is for remoulded clays (Burland 1990). Also shown are the mechanical parameters (from Fig. 3) for the Globigerina ooze and the terrigenous mud compression test data from Laughton (1957).

remoulded clays, which suggests that the fabric of naturally deposited sediments exerts a significant control on the effective stress. The results are consistent with the discussion by Burland (1990) concerning the sediment compression line for argillaceous sediments. The correlations by Parasnis and Burland are important because they demonstrate that the sediment mechanical parameters (C_c and e_{100}) span a relatively small range of values. The mechanical parameter correlations are useful because they can be used to verify the results from compression tests. The results from Fig. 3 for the northeastern Atlantic cores (Laughton 1957) are plotted in Fig. 4. The Globigerina ooze results are consistent with Parasnis' correlation for carbonate sediments, and the terrigenous mud parameters are consistent with Burland's correlation for remoulded clays. Thus, Laughton's compression tests are consistent with compression tests on a representative range of sediment types.

Theoretical compaction curve

Using equations (1)–(5), and assuming that the sedimentary column is homogeneous, which means that C_c, e_{100}, ρ_f and ρ_g are independent of burial depth d, it is possible to develop a theoretical relationship between porosity and burial depth (Parasnis 1960; Audet 1995a). The

soil mechanics based theoretical compaction curve is given by

$$\exp_{10}\left(\frac{e_{100}-e}{C_c}\right)\left(1+\frac{C_c}{\ln 10}+e\right) = \frac{d}{\mathcal{L}}+\mathcal{K} \quad (8)$$

where $\mathcal{L}$ and $\mathcal{K}$ are constants. The constant $\mathcal{L}$, defined by

$$\mathcal{L} = \frac{\sigma}{(\rho_g-\rho_f)g} \quad (9)$$

has dimensions of length and so it is simply called the length scale. The term $\mathcal{K}$ is dimensionless and is given by

$$\mathcal{K} = \left(\frac{\sigma_0}{\sigma_{100}}\right)\left(1+\frac{C_c}{\ln 10}+e_0\right) \quad (10)$$

where σ_0 is the effective stress at void ratio e_0 based on evaluating equation (5). Equation (10) arises as a constant of integration, and it can be evaluated by specifying the effective stress at $d = 0$, the depositional surface. Equation (10) has no special physical meaning. Equation (8) gives a concise relationship between burial depth d and the porosity ϕ (written in terms of the void ratio e), for the special case of one-dimensional compression of normally pressured sediments.

At $d = 0$, the sediments are loosely consolidated and have little strength, so the effective stress is expected to be zero. However, the effective stress law given by equation (5) is zero only when the void ratio is infinite, that is, $\phi = 1$. Therefore, to apply the theoretical profile, it is necessary to specify a non-zero but small value for σ_0. For shallow burial depth where there is negligible sediment compaction, the effective stress is given by

$$\sigma_v' \approx (\rho_g-\rho_f)(1-\phi_0)g\,d \quad (11)$$

Using $\rho_g = 2750$ kg/m^3 and $\rho_f = 1050$ kg/m^3 (Hamilton 1976), and $\phi_0 = 0.7$, a typical value for deep-sea sediments, an effective stress of 10 kPa corresponds to $d \approx 2$ m. Therefore, $d = 0$ in the model corresponds to some arbitrary, but shallow, burial depth at which the sediments take on cohesive, soil-like properties. As shown earlier (Audet 1995a), the theoretical porosity profiles are not too sensitive to the exact value of σ_0, which makes it possible to apply the model to real situations.

Compaction of levee deposits

General description of fan sediments

This brief description of Amazon Fan sediments is based on studies of short (<10 m) piston cores by Damuth & Kumar (1975) and Damuth

(1977). Detailed core descriptions of Leg 155 cores (Flood *et al.* 1995*a*) show that the types of sediments deeper in the fan are similar to those observed in the piston cores.

The top 30–65 cm of sediment on the fan consists of pelagic, foraminiferal clay, marl and ooze of Holocene age. Holocene sedimentation rates average about 4–6 cm/ka. The Pleistocene–Holocene boundary is commonly marked by an iron-rich crust 1–10 cm thick (McGeary & Damuth 1973). The underlying, pre-Holocene sediments consist mostly of dark grey to olive–grey hemipelagic clay with interbeds of redeposited silts and sands. The grey hemipelagic clay contains disseminated organic detritus, hydrotroilite and silt, plus pelagic components (planktonic foraminifera) which indicate that clay deposition was relatively continuous during glacial periods such as the Wisconsin (Damuth & Kumar 1975; Damuth 1977). The absence of hemipelagic clay from hills and seamounts on the continental rise and abyssal plains indicates that the terrigenous components were transported by gravity-driven mass flows instead of sedimenting from the upper water column (see Damuth *et al.* 1988 for discussion). Late Wisconsin and earlier glacial sedimentation rates are up to hundreds of cm/ka (Flood *et al.* 1995*a*).

The dominant sediment type consists of thin beds and laminae of silt, alternating with redeposited, turbiditic mud. Sand beds up to several metres thick tend to be found on the channel floors and in the lower fan lobe (Flood *et al.* 1995*a*). Particle sizes range from clay to fine gravel, but most beds are silt to medium sand (Damuth & Kumar 1975; Flood *et al.* 1991; Flood *et al.* 1995*a*). The sedimentary structures and the presence of displaced shallow-water fauna indicate that these beds were deposited by turbidity currents and related gravity-controlled flows. The silt–sand beds tend to be coarser, thicker and more abundant on the lower fan, suggesting that these coarse sediments bypassed the upper and middle fan via the distributary channels.

Lithostratigraphic description

This section describes the lithology and stratigraphy of holes 929B, 941B and 944A used in this study. The following summary is based on the extensive site descriptions available in the Initial Reports of Leg 155 (Flood *et al.* 1995*a*).

Site 939, Hole B.

Site 939 was sited about 1.0–1.5 km east of the crest of the eastern levee of the Amazon channel. At this location, the Amazon, Brown and Aqua channels coincide. Hole 939B penetrated to a depth of 102.7 m with 97% recovery. Two lithostratigraphic units, Units I and II, are distinguished on the basis of sediment texture and carbonate content.

The upper part of Unit I consists of burrowed, brown to greyish brown foraminifer–nannofossil clay. This sediment is underlain by a dark, reddish brown iron-rich crust (*c.* 0.5 cm thick). The iron-rich crust (0.38 m) is underlain by grey nannofossil-rich clay. The base of Unit I (0.68 m) is placed where the clay colour changes to a dark grey, which is associated with decreasing carbonate content. The carbonate content of Unit I decreases below the iron-rich crust and is about 3.6% near the base.

Unit II consists of terrigenous clay to silty clay of dark grey to very dark grey colour. The carbonate content is less than 3.0%. The sediment in this unit is stained by diagenetic hydrotroilite, resulting in irregular black patches and colour bands. Black, soft hydrotroilite concretions and micronodules (1–2 mm diameter) are common. Unit II is divided into five subunits based mainly on the frequency of silt laminae and beds.

Subunit IIA (0.68–23.00 m) consists of either clay or silty clay with intense to slight bioturbation. The top of Subunit IIB (23.00–44.23 m) is placed at the first silt laminae in Unit II. The sediment in Subunit IIB is moderately burrowed and/or colour-banded silty clay with thin laminae and beds of silt. Many of the silt layers have parallel or wavy lamination. Subunit IIC (44.23–66.17 m) consists mainly of moderately burrowed silty clay and has relatively few silt laminae or beds compared with Subunit IIB. Subunit IID (66.17–81.50 m) resembles Subunit IIB, except that the silt laminae and beds are more frequent. Subunit IIE (81.50–99.4 m) consists mainly of beds of structureless silty clay.

Site 944, hole A. Hole 944A, sited about 150 km downslope from Site 939, penetrated through 384 m of sediment with overall core recovery of 54%. However, in the section of interest to this study (top 114 m), core recovery was much higher (88%). Six lithological units are defined. Unit I (0–0.54 m) is a Holocene, bioturbated, foraminifer-rich nanofossil clay, with up to 38% carbonate. A dark greyish brown iron-rich crust (*c.* 1 cm thick) marks the base of Unit I. The upper part of Unit II (to 122 m) corresponds to the distal levee of the Amazon–Brown channel-levee system, and provides a downslope analogue to hole 939A located at the proximal portion of the Amazon–Brown system.

Unit II consists of olive grey to very dark grey terrigenous clay, silty clay, silt and sand. Similar to Site 939, Unit II sediment at Site 944 contains diagenetic hydrotroilite with similar black coloration patterns. Unit II is subdivided into four subunits based on frequency of silty clay, silt laminae and thin beds of sand.

Subunit IIA (0.54–15.38 m) contains moderately to heavily bioturbated silty clay. The base of Subunit IIA is placed at the uppermost occurrence of silt laminae. Subunit IIB (15.38–37.42 m) is characterized by silty clay that contains some silt laminae (one to three occurring per metre of core). There are some features of soft-sediment deformation, including folds. Subunit IIC (37.42–182.00 m) contains moderately bioturbated silty clay, along with silt laminae and thin beds of silt, very fine, fine and medium sand (5–30 layers per metre). The boundary between Subunits IIB and IIC is placed at the first thin bed that contains very fine sand. Beds of coarse and medium sand are found below 143 m. Subunit IID is similar to IIC. For brevity, details of the Units III–VI are omitted as these units are not studied in this paper.

Site 941, hole B. Site 941 is located on the western side of the Amazon Fan in a surficial debris-flow deposit that fills a depression between two levees. Four lithological units are recognized. Unit I (0–0.66 m) is a Holocene, bioturbated nanofossil–foraminifer clay, with about 30% carbonate, similar in lithology to Unit I at Sites 939 and 944. However, Site 941 does not contain the iron-rich diagenetic crust found at Sites 939 and 944.

The top of Unit II (0.66–3.0 m) consists of brown calcareous clay that grades in colour into an olive-grey to a greenish grey. The bottom of Unit II (3.0–6.66 m) consists of olive-grey to black homogeneous silty clay. No silt or sand laminae or beds are present, and parts of the unit are moderately bioturbated. Most of the sediment in Unit II is stained by diagenetic hydrotroilite.

Unit III (6.60–85.10 m) is a thick sequence of clay and silty clay of variegated colour ranging through shades of dark grey, dark olive grey, dark greenish grey and black. The sediment has been affected by soft-sediment failure, deformation and redeposition as evidenced by abrupt changes in lithology and colour. Discordant stratal relationships between laminae beds and lithologic contacts suggest that Unit III is composed of discrete, detached, commonly deformed blocks or clasts of variable size. Folds, indicative of deformation, occur on scales from centimetres to metres. Some small-scale folds are visible in the cores. Rare larger folds up to several metres in scale can be inferred from the attitudes of dipping beds in the cores. The unit is interpreted as a debris-flow or slump deposit, but the number of depositional events is uncertain. Unit IV (129.70–177.90 m) consists of very dark grey to dark olive-grey silty clay with numerous discrete laminae and beds of silt, similar to levee sequences cored at other site during Leg 155.

Analysis of porosity data

Site 939 (upper fan, water depth 2793 m) and Site 944 (middle fan, water depth 3701 m) sampled the levees of the Amazon Channel (Fig. 2), the most recently active channel of the Amazon Fan (Flood *et al.* 1995a). The levee deposits at both sites are >100 m thick, and both sites had good (88–97%), continuous core recovery; thus, they were good choices for this preliminary study.

Figure 5 shows the porosity–depth data from Site 939. The Advanced Hydraulic Piston Corer (APC) was used to core the sediments up to $d =$ 80 m. At depths greater than 80 m, the Extended Core Barrel (XCB) was used. There is a porosity anomaly at $d = 80$ m corresponding to the change in coring method. The porosity from the XCB sample is about 5% greater than the APC

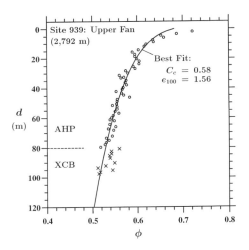

Fig. 5. The porosity (ϕ) v. burial depth (d) for the levee sediments from ODP Site 939 on the upper fan (2792 m water depth). ○, Data from samples cored using the Advanced Hydraulic Piston Corer (AHP); ✕, data from samples cored using the Extended Core Barrel device (XCB) (Shipboard Scientific Party 1995b). The continuous curve is the best fit of the theoretical compaction curve (equation (8)) to the APC data, using the values of $C_c = 0.58$ and $e_{100} = 1.56$.

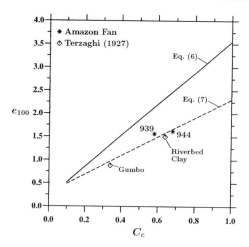

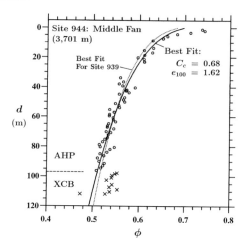

Fig. 6. A comparison of the mechanical parameters C_c and e_{100} derive from ϕ–d data for ODP Sites 939 and 944, compared with Parasnis' correlation (equation (6)) and Burland's correlation (equation (7)). Also shown are the mechanical parameters for Mississippi Gumbo and Mississippi Riverbed Clay measured by Terzaghi (1927).

Fig. 7. Porosity–depth data from the channel levee of ODP Site 944, on the middle fan (3701 m water depth) (Shipboard Scientific Party 1995d). The continuous curve is the best fit of equation (8) to the APC data from Site 944 (O), using the values of C_c and e_{100} shown. The dotted curve is the best fit to the APC data from Site 939 (from Fig. 5). The XCB data from Site 944 ($\times$) are shown for comparison.

sample. Because the XCB coring technique probably disturbs the core samples more than the APC technique, only porosity–depth data from APC cores were analysed in this paper. The XCB data are shown only for reference. The continuous curve in Fig. 5 represents the best non-linear fit of equation (8) to the APC data from Site 939 determined by adjusting the values of C_c and e_{100}. The effective stress at the depositional surface was taken to be $\sigma_0 - 10$ kPa. The densities used in equation (9), $\rho_g = 2710$ kg/m^3 and $\rho_f = 1024.5$ kg/m^3, are averaged values based on the index properties from Leg 155. The best estimates of the mechanical parameters for the Site 939 sediments are $C_c = 0.58$ and $e_{100} = 1.56$.

The compression parameters determined by analysing the Site 939 ϕ–d data are compared with earlier results from tri-axial or oedometer tests as shown in Fig. 6. Figure 6 shows the mechanical parameter correlations for carbonate sediments (equation (6)) and remoulded clays (equation (7)). Also shown are the compression properties of two mud samples from the Mississippi Delta, a Gumbo Clay and a Riverbed Clay, as measured by Terzaghi (1927). The mechanical parameters for the two Mississippi Delta samples are consistent with equation (7), which is based on Burland's data. The C_c–e_{100} values from Site 939 are also consistent with equation (7), which means that they are probably reasonable for the redeposited, terrigenous muds of the Amazon Fan.

Figure 7 shows the analysis of the porosity data from Site 944, located about 150 km downfan from Site 939. The XCB was used below $d = 97$ m, and there is a porosity offset of about 4% between APC and XCB data. The best fit of equation (8) for the data of Site 944 gives $C_c = 0.68$ and $e_{100} = 1.62$, which differs from the values determined from Site 939. However, for $d > 20$ m, the two best-fit curves do not differ that much with respect to the scatter of the data. The compression parameters for Site 944 are shown in Fig. 6 and lie close to equation (7). Both pairs of estimates (Site 939 and 944) are plausible, and given the scatter in the porosity data, the differences in C_c and e_{100} may not be significant. The consistency between Burland's data (i.e. equation (7)), compression tests on Mississippi Delta sediments (Terzaghi 1927) and the results shown in Fig. 6 for Sites 939 and 944 suggests that redeposited muds in submarine fans have mechanical properties similar to remoulded clays.

It should be noted that in both Figs 5 and 7 there is a discrepancy between the theoretical compaction curve and the porosity data in the first 5–10 m of burial. This may be caused by early diagenetic effects, such as chemical compaction (Wetzel 1989), that are not included in the simple mechanical compaction theory used to derive equation (8).

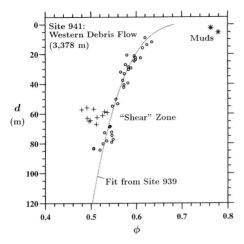

Fig. 8. The porosity–depth data from ODP Site 941 located in the Western Debris Flow (3378 m water depth) (Shipboard Scientific Party 1995c). All data are from APC cored samples. *, Two measurements of the pelagic mud overlying the debris-flow sediments. The crosses (+) highlight the reduction of porosity in the zone 55 m < d < 67 m, which is called 'Shear' Zone. The dotted curve is the best fit of equation (8) for the APC data from Site 939 (from Fig. 5).

Stability analysis of the Western Debris Flow

Porosity profile disturbances

Figure 8 shows the porosity profile from Site 941B in the Western Debris Flow (water depth 3378 m; Fig. 2). Site 941B was cored using the APC and the porosity data show less variance than those from Site 941A which was drilled in part using the XCB device (Flood *et al.* 1995a). The two shallowest points from Site 941B, labelled Muds in Fig. 8, are pelagic muds deposited after the debris flow event. At depths of about 55–70 m, the porosity shows an over-compacted zone, here labelled the 'Shear' Zone. The porosity profile shows a trend very similar to that from Site 939 as shown by the dotted line, the best fit of equation (8) from Fig. 5.

As described earlier, the sediments cored at Site 941B have undergone deformation and folding over various length scales. In spite of sediment reworking, the porosity profile in the top 55 m at Site 941B appears to be relatively undisturbed, as the porosity measurements are similar to those from Site 939. One possible explanation is that sediment can fail locally (stress and shear localization), thus creating clasts and blocks within which the sediment is not subjected to any large stresses that would

affect the porosity. Depending exactly on how samples were selected for measuring index properties, the full range of possible porosity variation resulting from soft-sediment failure may not necessarily be represented in all sections of a porosity profile. The porosity values in the 'Shear' Zone are overcompacted relative to the trend from Site 939. The lower porosity values may be caused by sediment reworking, if the zone represents the rupture or failure surface of some translational mass movement.

Infinite slope analysis

Mass-wasting processes such as slumps and debris flows may be important mechanisms for levelling the topography of the sea floor, and for this reason, slope stability is an important issue. The stability of sloping sediments depends on the undrained shear strength, S_u. Figure 9 shows S_u v. burial depth for Sites 939 and 944. The data from the two sites are similar in the first 20–30 m, and the S_u data show more scatter for $d > 40$ m.

As a preliminary analysis, one can consider the stability of an infinite layer of material on a slope of angle θ (Atkinson 1993). Let $\bar{\rho}$ be the average bulk density of a sediment layer of thickness d. Then, for a critical slope angle θ_c, the corresponding critical undrained shear strength, S_u^*, is given by

$$S_u^* = \frac{1}{2} \bar{\rho} g d \sin 2\theta_c \qquad (12)$$

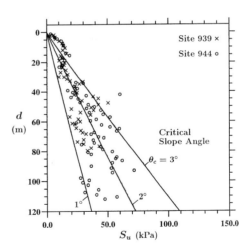

Fig. 9. The peak undrained shear strength (S_u) v. burial depth for ODP Sites 939 and 944. The shear strength was determined by the Shipboard Scientific Party of ODP Leg 155 using the ODP miniature vane shear device (Flood *et al.* 1995a). The three curves are based on the infinite slope stability analysis (equation (12)) for critical slope angles (θ_c) of 1°, 2°, and 3°.

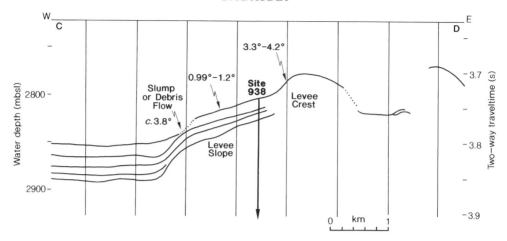

Fig. 10. A line drawing of a 3.5 kHz profile through Site 938 showing the western (left) levee of the abandoned Blue channel (JOIDES *Resolution*). The slope of the levee crest is about 3.3–4.2°, whereas the levee slope has an inclination of about 0.99–1.2°. The slump or debris-flow feature has a slope angle *c.* 3.8°. The location of profile C–D is shown in Fig. 2. (Adapted from fig. 2 of Shipboard Scientific Party 1995*a*.) mbsl, metres below sea level.

By assuming a value of θ_c, equation (12) can be used to determine S_u^* as a function of d using the theoretical compaction curve (equation (8)) to determine $\bar{\rho}$ (as $\bar{\rho}$ depends on the porosity profile). The critical shear strength for three values of θ_c are shown in Fig. 9. Most of the undrained shear strength measurements are bounded by the lines corresponding to $1° < \theta_c < 3°$. This relationship implies that most slopes with angles <1° are stable; whereas most slopes with angles >3° are unstable.

Damuth *et al.* (1988) measured apparent average slopes of levees at 60 locations on the upper and middle Amazon Fan. The measurements were based on 3.5 and 10 kHz precision depth recorder (PDR) profiles. All but two of the apparent average slope measurements ranged between 0.5° and 2.5°, and the average was 1.3°. Damuth *et al.* (1988) also measured the maximum apparent slopes near the crests of 22 of the steepest levees. Maximum crest slopes were in the range 1.2–9.0°, with an average of 3.5°. Maximum slope angles of >4.5° were found only in the upper fan.

Figure 10 shows a 3.5-kHz profile through the western levee of the abandoned Blue channel in the upper Amazon Fan, where Site 938 was located (Fig. 2). Slope angles are about 3.3–4.2° just down from the levee crest. Further down on the levee slope, angles are about 0.99–1.2°. There appears to be a slump or debris flow feature marking the end of the levee slope. The angle of the scar feature is about 3.8°. The levee morphology shown in Fig. 10 seems to support

the results of the infinite slope analysis shown in Fig. 9, that is, slope angles >3° are unstable. The section near the levee crest with angles *c.* 3° extends over a distance of about 400 m, whereas the levee slope section with the shallower angles *c.* 1° extends about 1.5 km. As larger slope angles tend to be more unstable, regions with slope angles >3° would tend to be less extensive than regions with the more stable, shallower angles <1°.

A possible interpretation is that the steep levee banks are unstable and prone to failure. When the levee banks fail, mass flows drape sediment over the levee slope at angles near the lower limit of stability predicted by the infinite slope analysis, i.e. *c.* 1°. As the sediment thickness develops on the levee slope, the edge of the slope furthest from the levee crest oversteepens, which leads to further sediment failure and transport of sediment further away from the channel–levee system, which may explain the features and slope angles seen in Fig. 10.

Discussion and conclusions

The analysis of porosity profiles from the levees of the main channel of the Amazon Fan suggests that these terrigenous sediments are normally compacted and have mechanical properties similar to that of remoulded clays. Sediments from Sites 939 and 944 show similar compaction trends, which suggests that both depositional environments are similar, even though the two sites are more than 150 km apart. This further implies that the sediment transport and depositional mechanisms behave similarly over large distances downfan.

Site 941A sampled sediments from the Western Debris Flow, a mass-transport deposit. The porosity profile shows a compaction trend similar to that of Site 939, except for an over-compacted zone around 55–70 m burial depth. This overcompacted zone is interpreted as a 'shear' zone where the sediments underwent shear failure and provided a slip surface that allowed the overlying 55 m of sediments to undergo translational sliding. The massive movement of the 55 m thick overlying sediment layer appears to have taken place without alter-ing the overall porosity structure of the layer, even though evidence (e.g. folds and abrupt changes in colour) suggests sediment defor-mation on length scales of centimetres to metres.

The measured undrained shear strengths of the sediments from Sites 939 and 944 are con-sistent with critical slope angles of 1–3°, a range which is consistent with apparent slope angles measured by Damuth *et al.* (1988). The analysis suggests that over length scales of >20 m, slope angles are limited by the intrinsic shear strength of the sediments. The infinite slope analysis shows a consistency between the porosity pro-files (i.e. overburden) and the undrained shear strength measurements, which supports the idea that slope instability could be the mechanism for breaching levees, thus leading to channel avul-sion and the creation of the distributary channel patterns observed in the fan. Furthermore, a similar slope instability mechanism could account for the translational slides that initiate mass movements downslope on the Amazon Fan (Damuth & Embley 1981).

In future work, data from the additional sites of ODP Leg 155 could be examined to deter-mine the range of C_c for levee muds. It may be possible to compare the C_c values based on ϕ–d profiles with actual measurements from oedometer tests. A more sophisticated slope stability analysis could be attempted to better understand the role of slope failure in control-ling the overall slope of the channel levee topog-raphy on the Amazon Fan.

I thank J. E. Damuth for his comments on an earlier version of the manuscript, and for pointing out line C–D from ODP Site 938. I also thank D. J. W. Piper and an anonymous referee for reviewing the manu-script and providing constructive comments. This work was supported by NERC Grant GR3/8992 (H7).

References

ATKINSON, J. 1993. *An Introduction to the Mechanics of Soils and Foundations*. McGraw–Hill, London.

AUDET, D. M. 1995a. Modelling of porosity evolution and mechanical compaction of calcareous sedi-ments. *Sedimentology*, **42**, 355–373.

—— 1995b. Mathematical modelling of gravitational compaction and clay dehydration in thick sedi-ment layers. *Geophysical Journal International*, **122**, 283–298.

—— & McCONNELL, J. D. C. 1992. Forward modelling of porosity and pore pressure evolution in sedi-mentary basins. *Basin Research*, **4**, 147–162.

BURLAND, J. B. 1990. On the compressibility and shear strength of natural clays. *Géotechnique*, **40**, 329–378.

DAMUTH, J. E. 1977. Late Quarternary sedimentation in the western equatorial Atlantic. *Geological Society of America Bulletin*, **88**, 695–710.

—— & EMBLEY, R. W. 1981. Mass-transport processes on Amazon Cone: Western Equatorial Atlantic. *American Association of Petroleum Geologists, Bulletin*, **65**, 629–643.

—— & FLOOD, R. D. 1985. Amazon Fan, Atlantic Ocean. *In*: BOUMA, A. H., NORMARK, W. R. & BARNES, N. E. (eds) *Submarine Fans and Turbidite Systems*. Springer, New York, 97–106.

—— & KUMAR, N. 1975. Amazon Cone: morphology, sediments, age, and growth pattern. *Geological Society of America Bulletin*, **86**, 863–878.

——, FLOOD, R. D., KOWSMANN, R. O., BELDERSON, R. H. & GORINI, M. A. 1988. Anatomy and growth pattern of Amazon deep-sea fan as revealed by Long-Range Side-Scan Sonar (GLORIA) and high-resolution seismic studies. *American Associ-cation of Petroleum Geologists, Bulletin*, **72**, 885–911.

——, KOLLA, V., FLOOD, R. D., KOWSMANN, R. O., MON-TEIRO, M. C., GORINI, M. A., PALMA, J. J. C. & BELDERSON, R. H. 1983a. Distributary channel meandering and bifurcation patterns on the Amazon deep-sea fan as revealed by long-range side-scan sonar (GLORIA). *Geology*, **11**, 94–98.

——, KOWSMANN, R. O., FLOOD, R. D., BELDERSON, R. H. & GORINI, M. A. 1983b. Age relationships of distributary channels on Amazon deep-sea fan: implications for fan growth patterns. *Geology*, **11**, 470–473.

FLOOD, R. D. & DAMUTH, J. E. 1987. Quantitative characteristics of sinuous distributary channels on the Amazon Deep-Sea Fan. *Geological Society of America Bulletin*, **98**, 728–738.

——, MANLEY, P. L., KOWSMANN, R. O., APPI, C. J. & PIRMEZ, C. 1991. Seismic facies and late Quater-nary growth of Amazon submarine fan. *In*: WEIMER, P. & LINK, M. H. (Eds) *Seismic Facies and Sedimentary Processes of Modern and Ancient Submarine Fans*. Springer, New York, 415–433.

——, PIPER, D. J. W., KLAUS, A. *et al.* 1995a. *Proceed-ings of the Ocean Drilling Program, Initial Reports*, **155**: Ocean Drilling Program, College Station, TX.

——, ——, & SHIPBOARD SCIENTIFIC PARTY 1995b. Introduction. In: FLOOD, R. D., PIPER, D. J. W., KLAUS, A. et al. Proceedings of the Ocean Drilling Program, Initial Reports, 155. Ocean Drilling Program, College Station, TX, 5–16.

HAMILTON, E. L. 1976. Variations of density and porosity with depth in deep-sea deposits. Journal of Sedimentary Petrology, 46, 280–300.

HAMPTON, M. A. 1989. Geotechnical properteis of sediment on the Kodiak continental shelf and upper slope, Gulf of Alaska. Marine Geotechnology, 8, 159–180.

——, LEE, H. J. & LOCAT, J. 1996. Submarine landslides. Review of Geophysics, 34, 33–59.

KUMAR, N. 1978. Sediment distribution in the western Atlantic off northern Brazil–structural controls and evolution. American Association of Petroleum Geologists, Bulletin, 62, 273–294.

LAUGHTON, A. S. 1957. Sound propagation in compacted ocean sediments. Geophysics, 22, 233–260.

MANLEY, P. L. & FLOOD, R. D. 1988. Cyclic sediment deposition within Amazon deep-sea fan. American Association of Petroleum Geologists, Bulletin, 72, 912–925.

MCGEARY, D. F. R. & DAMUTH, J. E. 1973. Postglacial iron-rich crusts in hemipelagic deep-sea sediments. Geological Society of America Bulletin, 84, 1201–1212.

MCIVER, R. D. 1982. Role of naturally occurring gas hydrates in sediment transport. American Association of Petroleum Geologists, Bulletin, 72, 912–925.

MILLIMAN, J. D. 1979. Morphology and structure of Amazon upper continental margin. American Association of Petroleum Geologists, Bulletin, 63, 934–950.

MULDER, T. & COCHONAT, P. 1996. Classification of offshore mass movements. Journal of Sedimentary Research, 66, 43–57.

PARASNIS, D. S. 1960. The compaction of sediments and its bearing on some geophysical problems. Geophysical Journal of the Royal Astronomical Society, 3, 1–28.

READING, H. G. & RICHARDS, M. 1994. Turbidite systems in deep-water basin margins classified by grain size and feeder system. American Association of Petroleum Geologists, Bulletin, 78, 792–822.

SHIPBOARD SCIENTIFIC PARTY 1995a. Site 938. In: FLOOD, R. D., PIPER, D. J. W., KLAUS, A. et al. (eds) Proceedings of the Ocean Drilling Program, Initial Reports, 155. Ocean Drilling Program, College Station, TX, 409–436.

—— 1995b. Site 939. In: FLOOD, R. D., PIPER, D. J. W., KLAUS, A. et al. (eds) Proceedings of the Ocean Drilling Program, Initial Reports, 155. Ocean Drilling Program, College Station, TX, 437–461.

—— 1995c. Site 941. In: FLOOD, R. D., PIPER, D. J. W., KLAUS, A. et al. (eds) Proceedings of the Ocean Drilling Program, Initial Reports, 155. Ocean Drilling Program, College Station, TX, 503–536.

—— 1995d. Site 944. In: FLOOD, R. D., PIPER, D. J. W., KLAUS, A. et al. (eds) Proceedings of the Ocean Drilling Program, Initial Reports, 155. Ocean Drilling Program, College Station, TX, 591–633.

SKEMPTON, A. W. 1944. Notes on the compressibility of clays. Quarterly Journal of the Geological Society, London, 100, 119–135.

—— 1961. Effective stress in soils, concrete and rocks. In: Pore Pressure and Suction in Soils, Butterworths, London, 4–16.

—— 1970. The consolidation of clays by gravitational compaction. Quarterly Journal of the Geological Society, London, 125, 373–411.

TERZAGHI, K. 1927. Principles of final soil classification. Public Roads, 8, 41.

—— 1936. The shearing resistance of saturated soil and the angle between the planes of shear. In: Proceedings of the 1st International SMFE Conference, Harvard, MA, 1, 54–56.

WETZEL, A. 1989. Influence of heat flow on ooze/chalk cementation: quantification from consolidation parameters in DSDP Sites 504 and 505 sediments. Journal of Sedimentary Petrology, 59, 539–547.

The Var submarine sedimentary system: understanding Holocene sediment delivery processes and their importance to the geological record

THIERRY MULDER[1], BRUNO SAVOYE[2], DAVID J. W. PIPER[3] & JAMES P. M. SYVITSKI[4]

[1]*Cardiff University of Wales, Department of Earth Sciences, PO Box 914, Cardiff CF1 3YE, UK*
Present address: Département de Géologie et Océanographie, CNRS URA 197, Université de Bordeaux I, Avenue des Facultés, 33405 Talence Cedex, France.
[2]*IFREMER, Géosciences Marines, BP 70, 29280 Plouzané, France*
[3]*Geological Survey of Canada–Atlantic, PO Box 1006, Dartmouth, Nova Scotia, B2Y 4A2, Canada*
[4]*Institute of Arctic & Alpine Research, University of Colorado at Boulder, 1560 30th Street, Campus Box 450, Boulder, CO 80309-0450, USA*

Abstract: The Var system extends off Nice in the Western Mediterranean. It comprises a river, a delta and a submarine valley leading to a deep-sea fan that together have been in operation since the Early Pliocene. The Var system is an area experiencing active sediment transport, where at least three major types of sediment transfer process are identified: hyperpycnal turbid plumes, surge-like turbidity currents generated by shallow failures induced by excess pore pressure during river flood periods, and by large earthquake-triggered slides. The last two processes might generate higher-density turbidity currents, but at different return intervals. Hydrological data, direct observations of the sea floor, geotechnical testing and numerical modelling confirm the very high frequency of these sediment transfer events. Some of the processes have catastrophic surge behaviour, others are continuous during periods of river flooding. In the latter case, all the sediment supplied to the vicinity of the river mouth is transferred seaward without or with only brief periods of deposition. The geological record of such continuous activity remains difficult to identify. The palaeo-events identified in sedimentary series are often widespread, high-magnitude events with return periods close to a millennium, i.e. usually beyond historical records. Normal 'background' processes provide only thin deposits that are not interpretable in the geological record.

In the past 30 a, offshore mass-wasting processes and other types of rapid sediment transfer have been widely studied. This international effort has been stimulated by the needs of the petroleum and other offshore industries, and national security concerns. As in other oceanographic disciplines, major advances were largely induced by the development of new submarine exploration tools and of new approaches such as remote sensing, *in situ* measurements, scaled laboratory experiments, numerical modelling and geographic information systems. These developments led to the multiplication of case study descriptions.

Early classifications of offshore mass-wasting processes arose in the 1970s (Middleton & Hampton 1973; Nardin *et al.* 1979; Lowe 1982). Whereas slumps and slides are similar to subaerial processes (Skempton & Hutchinson 1969),

these classifications underlined the existence of two main offshore processes: mass flows and turbidity currents. Mulder & Cochonat (1996) discussed the complexity of classification, as one single large event may experience several process changes from the failure event to the final deposition of the sediment gravity flow. The two main processes, mass flows and turbidity currents, can be further subdivided. For example, mass flows can be separated into debris flows, liquefied flows and fluidized flows. Differentiation of turbidity currents is usually based on the fluid flow density, high or low, although several authors such as Shanmugam (1996) and Mulder & Cochonat (1996) have criticized the use of the term 'high-density turbidity current': such flows may be compound with a dense bottom flow (Ravenne & Beghin 1983), which is actually not turbulent, and a low-density turbulent upper

MULDER, T., SAVOYE, B., PIPER, D. J. W. & SYVITSKI, J. P. M. 1998. The Var submarine sedimentary system: understanding Holocene sediment delivery processes and their importance to the geological record. *In*: STOKER, M. S., EVANS, D. & CRAMP, A. (eds) *Geological Processes on Continental Margins: Sedimentation, Mass-Wasting and Stability*. Geological Society, London, Special Publications, **129**, 145–166.

145

part. Recent flume experiments by G. Parker (St Anthony Falls Hydraulic Laboratory) show that turbidity currents may form from fast-moving debris flows (pers. comm.).

Sediment transfer from the land to a marine basin generally involves turbidity currents and mass-wasting, with two major processes predominating: (1) injection of suspended sediment at river mouths and (2), transformation of sediment failures into mass flows.

Hyperpycnal plumes can only be triggered from a river mouth if suspended sediment concentration is high enough compared with the receiving basin fluid density. This kind of turbidity current has long been associated with lakes (Forel 1885), but has only recently been introduced as a common marine phenomenon by Normark & Piper (1991). It has been shown to be a frequent feature of river mouths experiencing seasonally high concentrations of sediment (Wright *et al.* 1986). A world-wide study by Mulder & Syvitski (1995) of sediment discharge at river mouths confirmed the potential frequency of such plumes during river floods. The importance of this process is indicated by the interpretation of a 11 m thick turbidite as the deposit of a 11 day long hyperpycnal flow in Saguenay Fjord (Syvitski & Schafer 1996; Mulder *et al.* 1998).

In rivers, where sediment concentration is too low to create hyperpycnal plumes, low-density turbidity currents may occur as a result of the concentration of hypopycnal plumes by processes similar to density cascading described by Postma (1969). Sea-floor valleys seem to be necessary to concentrate and maintain such plumes (Normark & Piper 1991).

Mass-wasting processes are particularly well developed in both open-coast and bay-head deltas. The now classic studies of Prior & Coleman (1978, 1982) on the Mississippi Delta revealed a multitude of sedimentary processes: mud, silt and grain flows and low-density turbidity currents generated by oversteepening, excess pore pressure or wave reactivation (Prior *et al.* 1989). Similarly, fjords offer a large variety of mass-wasting processes (Prior *et al.* 1986, 1987; Syvitski & Farrow 1989), including shallow foreset failures occurring during high sedimentation periods described by Bornhold *et al.* (1994) in British Columbia fjords.

Published studies about recent mass-wasting processes rarely deal with the whole sedimentary system: the source area and trigger mechanisms, the travel path and depositional area. The 1929 Grand Banks event is one of the rare example of a relatively complete case study of a mass-wasting phenomenon (Piper *et al.* 1988). The Var submarine system is also relatively well known

by scientists and engineers dealing with mass-wasting processes, owing to its close relationship with the 1979 Nice Airport disaster.

The Var area provides the following points of interest for the study of sediment transfer from land to a deep basin: (1) it is fed by a well monitored short river that is forming an active delta. (2) It is located in a seismically active area. (3) Little sediment is trapped in shallow water, because of a very narrow continental shelf. (4) The head of the Var Canyon is directly connected to the Var River mouth. (5) The whole area from the head of the canyon to the distal deep-sea fan is relatively narrow, allowing extensive coverage by high-resolution geophysical and sampling surveys. (6) Several recent mass-wasting processes have been well documented: the 1979 airport slide, which evolved into a large submarine turbidity current that broke submarine cables in the basin, small superficial failures creating low-velocity and low-density turbidity currents as recorded by Gennesseaux *et al.* (1971) and hyperpycnal flows as recently examined in a study based on both hydrological and geological data (Mulder *et al.* 1997*b*).

The Var submarine sedimentary system gives us an opportunity to assess the relative occurrence and importance of the different types of sediment transfer processes. The questions addressed in this paper are: what is the quantitative importance of mass-wasting events in the overall sediment transport at the Var River mouth? How are these events preserved in the sedimentary record? What is the scale of sediment transfer events that produce extensive and discernible deposits within sedimentary series?

To address these questions we are going to study three phenomena: (1) the records of small-sized turbidity currents made by Gennesseaux *et al.* (1971), (2) the geological record of the 1979 event, and (3) the hydrological data that can be useful to predict the possible occurrence of hyperpycnal plumes during flooding periods. A particular flood that occurred in 1994 is also included in this study.

Geographical and geological setting

The Var River is 120 km long, draining a 2820 km^2 basin from its source (altitude 2352 m) to the city of Nice on the coast of the Baie des Anges, Mediterranean Sea (Fig. 1). It connects directly into a sinuous steep submarine canyon. The mean annual water discharge (50–55 m^3 s^{-1}) can be multiplied by tens during spring or autumn flash floods, as is the case for many Mediterranean rivers. For example, in November 1994, water discharge at the river mouth

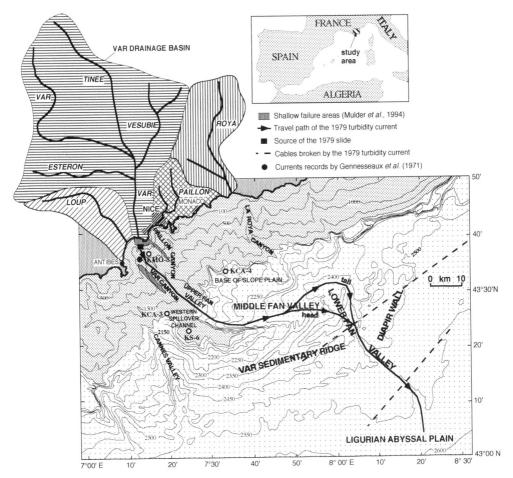

Fig. 1. Geographical setting showing the drainage basin of the Var River, the morphology of the submarine deep-sea fan, the travel path of the 1979 event, the current records made by Gennesseaux *et al.* (1971), the location of the KMO-8, KCA-3, KCA-4 and KS-6 cores and the potential shallow failure areas related to underconsolidated sediments (Mulder *et al.* 1994).

reached values close to 4000 m³ s⁻¹, i.e. 70–80 times larger than the average value. As the river crosses lithologies dominated by marls and silts, subaerial mudflows with concentration reaching 500 kg m⁻³ (Laurent 1971) are triggered in the drainage basin after strong rainfalls. Offshore, the Var has created a 20,000 km² deep-sea fan.

The Late Pleistocene fluvial deposits of the Var River are coarser than the modern deposits, indicating that the discharge was probably greater during this period (Dubar 1988). Early Holocene deposits older than 4 ka are finer than modern deposits (Dubar, pers. comm.). A gravelly delta prograded during the Late Pleistocene and this now forms a prominent prodeltaic depositional apron: the elevation of the top of the gravel surface suggests that the rise of sea level at about 11 ka stopped the progradation

and flooded the lower valley. Sea level stabilized close to its present position at about 6 ka (Dubar 1988; Savoye & Piper 1991).

Over the last seven years, five detailed oceanographic surveys have been conducted on the Var submarine system, providing observations all along the fan from the canyon head to the outer fan lobe. The fan is characterized by a single fan valley with abrupt bends separated by long straight reaches, and by an extremely asymmetric levee system (Savoye *et al.* 1993). On the southern side of the fan valley lies the prominent Var Sedimentary Ridge (Fig. 1). In contrast, no prominent levee exists on the northern side of the fan valley. The relief of the ridge above the channel bottom exceeds 300 m on its upslope part, but decreases to less than 50 m downstream. Sedimentation rates on the Var Sedimentary Ridge were high throughout the Late

Quaternary, even during sea-level highstands (Piper & Savoye 1993).

The modern Var fan has developed since the earliest Pliocene transgression (Savoye *et al.* 1993). Its morphology suggests that significant deposition has been produced by both sandy and muddy turbidity currents (Piper & Savoye 1993). Deep-water side-scan sonar imagery shows fresh gravel waves within the canyon and the upper fan valley (Piper & Savoye 1993). The continental slope off the Var Delta appears largely erosional. Evidence of recent erosion has been seen on chutes (gullies) that drain the upper slope down to the Var Canyon. Various pieces of the airport embankment have been found at the confluences between these chutes and the main Var Canyon (Savoye *et al.* 1988, 1995). Badlands features and anastomosing channels, similar to onshore alpine drainage patterns, have been observed at the canyon head suggesting the existence of frequent erosive density currents.

The record of mass-wasting events in the Var Canyon

In 1969, density currents were recorded *in situ* in the Var Canyon at 700 m water depth (Gennesseaux *et al.* 1971; Fig. 1). In the same period, Inman (1970) and Shepard *et al.* (1977) reported similar current recordings in La Jolla Canyon (California). The Var currents were typically 1–2 m thick, with maximum velocities greater than 1 m s^{-1}. Related sediment concentrations were estimated to be less than 1 kg m^{-3}. Although velocities varied over the 25 day long period of record, currents seemed to be a permanent phenomenon during the experiment period (Gennesseaux *et al.* 1971).

A mass-wasting event involving at least 8 × 10^6 m^3 of material occurred in 1979 (Gennesseaux *et al.* 1980; Malinverno *et al.* 1988). Although this event can be considered representative of large Holocene Var events, on a global scale it is small if compared with events involving 600–3000 km^3 of sediment on the Saharan and Norwegian margins (Embley & Jacobi 1977; Bugge *et al.* 1988). The 1979 Var event shares many features with the well-known Grand Banks event although it was not earthquake-triggered (Heezen & Ewing 1952; Kuenen 1952; Piper *et al.* 1985, 1988, 1992). On 16 October 1979, a large landslide occurred during landfilling operations related to the extension of Nice Airport. It affected both underconsolidated silty clays (later mapped by Cochonat *et al.* 1993) and some of the landfill aggregates. Modelling results using a visco-plastic approach (Mulder *et al.* 1997*a*) showed

that in accelerating and disintegrating along the steep offshore slopes, the slide transformed into a debris flow that ignited (Parker 1982) to produce a typical compound turbidity current with a dense lower part resulting directly from the debris flow, and a turbulent low-density upper plume. A piece of metal (probably part of a bulldozer carried by the 1979 event) was found during a submersible dive at 1400 m water depth on a 30 m high terrace in the Var Canyon, at a distance of 15 km from the failure area (Cochonat, pers. comm.). Recent numerical modelling (Piper & Savoye 1993; Mulder *et al.* 1997*a*) shows that velocities as high as 35–40 m s^{-1} may have been reached in the upper part of the canyon, decreasing to 5–10 m s^{-1} in the mid-fan valley. In the lower fan-valley, flow velocity ranged from 2 to 3 m s^{-1}, constrained by the timing of submarine cable failures. Flow thickness varied from 100 m to 250 m in the distal part of the mid-fan valley (Piper & Savoye 1993). At the surface, a tsunami affected the whole Baie des Anges. No earthquake was recorded during the time immediately preceding the event. Eleven people were killed onshore.

The first broken cable, located 100 km from the triggering area, was cut after 3 h 45 min after the initial failure; the second cable, situated at 120 km, was cut after 8 h (Fig. 1). The cable repair report indicates that the first cable break was not located exactly in the axis of the Var Fan Channel, but at about 5–6 km west. This suggests that the 1979 turbidity current spilled over the eastern end of the Var Sedimentary Ridge (Piper & Savoye 1993). The freshness of the flute-like scours on side-scan images at the eastern end of the levee (Malinverno *et al.* 1988; Savoye *et al.* 1988, 1995), the location of the first cable break and evidence from cores that the turbidite deposited by the 1979 event was the sandiest Holocene turbidite, suggest that the observed scours were eroded by the 1979 current.

Piper & Savoye (1993) mapped the surface sandy layer related to the 1979 event. Its thickness varies between 0.05 and 1 m and it extends over at least the first 120 km of the travel path, and probably more towards the deep basin. Using an average thickness of 0.1 m for the sandy layer and a surface of 1500 km^2 for the depositional area, the estimate of sand volume for the turbidite related to the 1979 event is 15 × 10^7 m^3. The sandy layer could have been provided from two sources: (1) the failure area and (2) erosion along the Var Channel floor, in a manner similar to that described by Piper & Aksu (1987) for the Grand Banks event. The estimated volume related to the 1979 event should be considered as a minimum, as the sand fraction represents only 10% of the sediments located on the upper slope, i.e. in the source area of the triggering slide.

Hydrology

Hydrological data and the occurrence of hyperpycnal events

Over a year, the Var River discharge curve is typically bimodal. Flash floods can occur both in spring and autumn, owing to snow melt and convective rainfall, respectively (Fig. 2). The average yearly discharge is 52.5 $m^3 s^{-1}$. Using the relationship

$$log Q_{flood} = -0.07 \, log \, A^2 + 0.865 \, log \, A \\ + 2.084 \text{ for } A < 10^6 \text{ km}^2, r^2 = 0.99 \quad (1)$$

relating the maximum discharge (Q_{flood}) to the basin area (A; Matthai 1990), the maximum discharge that can be produced by the modern Var River can be estimated as 17,200 $m^3 s^{-1}$.

On the other hand, Mulder & Syvitski (1995) showed that the following equation could be used to predict the minimum value of the rating coefficient b (the coefficient relating the suspended sediment discharge to the water discharge) to produce a hyperpycnal plume:

$$X = \left(\frac{Q_{flood}}{Q_{av}} \right)^b \frac{Cs_{av}}{Cc} \quad (2)$$

where Cs_{av} is the average suspended sediment concentration, Q_{av} the average annual discharge and Cc the critical suspended sediment concentration to produce an underflow. Cc is 42.7 kg m^{-3} for the Var River (Mulder & Syvitski 1995). For $X > 1$, a river may produce a hyperpycnal plume. Using equation (2), we calculated that the Var River is able to produce hyperpycnal turbidity currents as soon as b is greater than 0.67 (Fig. 3).

Data published by Laurent (1971) show that the rating coefficient measured for instantaneous discharge values varying from 24 to 411 $m^3 s^{-1}$ is 1.534 (Fig. 4):

$$Cs_{av} = 7.67 \times 10^{-4} \, Q_{av}^{-1.534} \quad (3)$$

therefore high enough to allow the river to produce underflows.

Using a value for b corresponding to monthly discharges (estimated to 1.7–1.75 by comparison with rivers of similar size and behaviour, see Mulder & Syvitski 1995), the average discharge value corresponds to sediment concentration of about 0.80–0.99 kg m^{-3} (mean annual value). Furthermore, the concentration values related to peak discharges during floods occurring after dry periods produce even larger rating coefficient (see discussion in Mulder *et al.* 1997*b*).

The calculation of the critical discharge to produce a hyperpycnal plume indicates a value close to 1200–1250 $m^3 s^{-1}$ for floods occurring in spring or after wet summers, with a suspended sediment concentration reaching at least 42.7 kg m^{-3} (Mulder & Syvitski 1995). The corresponding river flood return period is 4

MONTHLY DISCHARGE DURING YEAR 1971

SUSPENDED SEDIMENT LOAD (CALCULATED)

DISCHARGE (MEASURED)

} MEAN MONTHLY VALUES DURING PERIOD 1974-1994

CALCULATED ANNUAL SUSPENDED SEDIMENT LOAD : 663x10³ t

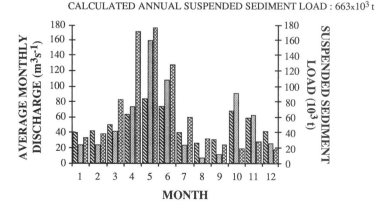

Fig. 2. Discharge during the year 1971 and average monthly discharge at the river mouth during the period 1974–1994. The corresponding sediment load is calculated using $b = 1.7$ and 1.75, and $a = 7.67 \times 10^{-4}$.

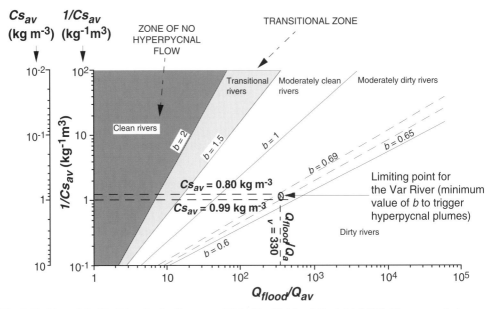

Fig. 3. Position of the Var River in the diagram published by Mulder & Syvitski (1995). Hyperpycnal plumes can be produced at the river mouth of the Var when *b* exceeds the limiting value 0.65–0.69. As the measured rating coefficient *b* is 1.534, the Var can produce hyperpycnal plumes. Cs_{av}, average annual suspended sediment concentration; Q_{flood}, discharge during flood; Q_{av}, average annual discharge.

years using instantaneous discharge values, and 21 years using daily discharge values. The critical discharge value might be as low as 620–750 $m^3 s^{-1}$ (with $1.65 \leq b \leq 1.7$) for floods following dry summer periods (Mulder *et al.* 1997*b*) and the return period could be substantially reduced

(2a for a short duration plume, 5a for a 24 h long plume).

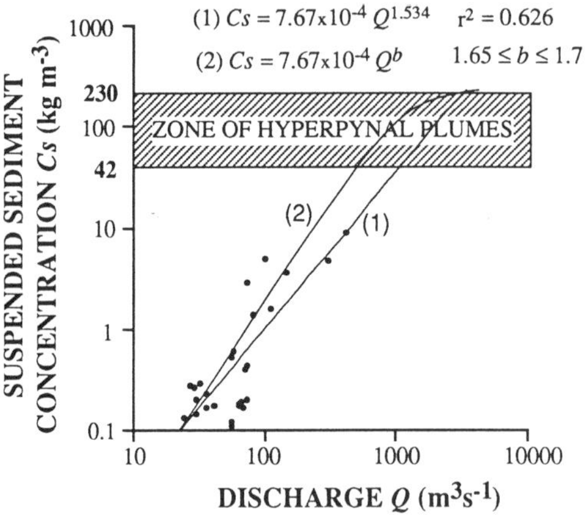

The Var sedimentary system therefore appears very active in terms of river supply of sediment to the deep basin, although river activity and delta construction have decreased strongly during the Holocene (Savoye & Piper 1991). The Var system involves a wide spectrum of sediment transport processes, larger than the variety experienced by larger rivers (cf. hyperpycnal events; Mulder & Syvitski 1995).

Fig. 4. Relationship between suspended sediment concentration and instantaneous discharge (rating curve) at the mouth of the Var River. Curve (1) has been calculated using data from Laurent (1971) and curve (2) corresponds to concentration after dry periods. *b*, rating coefficient; *Cs*, suspended sediment concentration; *Q*, discharge, *r*, regression coefficient (r^2, variance).

(1) $Cs = 7.67 \times 10^{-4} Q^{1.534}$ $r^2 = 0.626$
(2) $Cs = 7.67 \times 10^{-4} Q^b$ $1.65 \leq b \leq 1.7$

Data and interpretation

Bathymetric data down to –900 m have been provided by high-resolution multi-beam sonar Simrad EM1000. For greater water depths, the dataset has been completed using the Baie des Anges 1:25,000 map based on Seabeam data (Monti & Carré 1979). We also used SAR (deep-towed acoustic system) imagery and sea-floor pictures obtained from the CYANA submersible during the SAME cruise. During the SAME, MONICYA, PRENICE and CASANICE cruises, Kullenberg (piston) cores were collected for geotechnical and sedimentological purposes and further submersible dives were made. Hydrological data based on limnigraphs extending from 1974 to 1994 have been provided by the DIREN (Direction Régionale de l'Environnement) of the 'Région Provence–Alpes–Côte d'Azur'. Data for 1968 are provided by Laurent

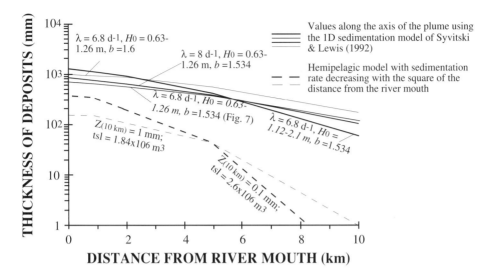

Fig. 5. Comparison of the gross sedimentation value at the Var River mouth during spring 1971 provided by two different models: (1) 1D plume sedimentation ($\lambda = 6.8$–8 d^{-1}; Syvitski & Lewis 1992); (2) hemipelagic plume sedimentation (sedimentation rate varies as the inverse square of the distance from the river mouth). tsl: total suspended sediment load.

(1971). From 1974 to 1976, and from 1985 to 1994, discharge values were measured at the river mouth. From 1975 to 1984, values were measured upstream (Pont de la Manda, drainage basin = 2790 km²).

Interpretation of the events recorded in 1971

Using the hydrologic data, we are going to try to provide an explanation for the turbid events that have been recorded by Gennesseaux *et al.* (1971).

The spring of 1971 was the rainiest spring of the 22 a in our dataset (Fig. 2). Daily values of discharge never exceeded 350 m³ s⁻¹ and peak instantaneous values of discharge at the river mouth estimated from daily discharge measured upstream at Usine de la Mescla (drainage basin = 1832 km²) reached values just below 600 m³ s⁻¹ over very short periods. Even using the minimum threshold value for the suspended sediment concentration needed to produce a hyperpycnal plume, such a plume could not have been maintained over many days, and consequently cannot explain the current records made by Gennesseaux *et al.* (1971). Another explanation for the recorded turbidity flows is slope failure as a result of rapid sediment accumulation. We have estimated the sedimentation rates just seaward of the Var River mouth during the 3 months of spring 1971. We used a 1D (centre line) version (Syvitski & Lewis 1992) of a model of sedimentation under a 2DH plume (a 2D model in three-dimensional

space with the third dimension averaged). In a 2DH model, the depth of the flow is averaged (Syvitski & Alcott 1995). We also compared the results with a basic hemipelagic plume model in which the sedimentation rate decreases as the square of distance from the river mouth (Fig. 5). The basic equation of the 1D model is (for distance from the river mouth less than 5.2 times the width of the river at the river mouth, i.e., 1300 m):

$$Z = \lambda H_0 C_0 \exp(-\lambda t) \qquad (4a)$$

Seaward, equation 11b of Syvitski *et al.* (1988) for a spreading plume should be used:

$$Z_{(x,y)} = \lambda H_0 C_0 \exp\left[-\lambda\left(1.76\,\frac{w_0}{u_0}+\frac{0.29x^{1.5}}{u_0 w_0^{0.5}}\right)\right] \quad (4b)$$

where Z is the sedimentation flux (in ML⁻² T⁻¹) assumed to be equal to the removal rate of sediment from the plume, λ is the removal rate, and H_0 and C_0 are the thickness and the concentration respectively of the turbidity current where sedimentation begins.

Total thickness of sediment deposited during the period of high discharge (19 March 1971 to 11 June 1971) is 700–900 mm ($b = 1.534$ and 1.6 respectively) calculated at the river mouth using the 1D plume model, using a high value of λ ($\lambda = 6.8$–8 d^{-1} corresponding to a medium silt) to simulate the possible change in the average grain size of the sediments deposited at the river mouth during this period of high discharge and $H_0 = 0.63$–1.26 m, and 160–370 mm for the basic hemipelagic sedimentation model (using the

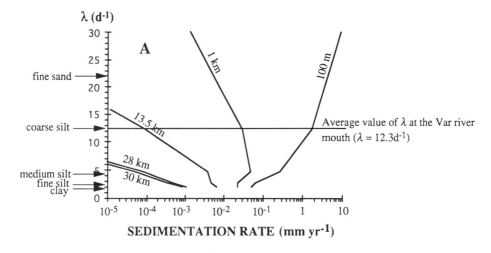

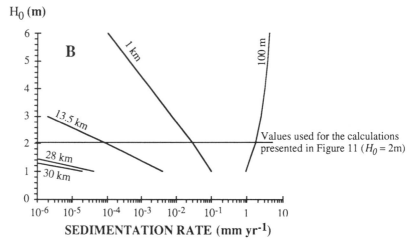

Fig. 6. Sensitivity analysis of the model for the deposition of a one-dimensional plume developed by Syvitski & Lewis (1992; after the 2DH model of Syvitski & Alcott 1995). Variation of calculated values of sedimentation rates with the removal rate, λ (**a**) and with the initial thickness of the flow, H_0 (**b**). The value of H_0 used to obtain results presented in Fig. 11 ($H_0 = 0.44$ m) and the average value of λ at the Var river mouth ($\lambda = 6.8$ d^{-1}) are indicated.

assumption that the sedimentation rate at 10 km is 1 and 0.1 mm, respectively). The theoretical daily sedimentation rate corresponding to the daily discharge during the same period (Fig. 7a and b) and at the river mouth averages just below 12(8) mm d^{-1} with peaks of sedimentation of up to 48(33) mm d^{-1} ($b = 1.6$, $\lambda = 6.8$ d^{-1}; in brackets, $b = 1.534$) over short periods. Sediment accumulation decreases rapidly seaward. At 2000 m and 10 000 m from the river mouth, mean daily sedimentation rates are less than 9.5(7) mm d^{-1} and 2(1.5) mm d^{-1}, respectively. Cumulative sedimentation curves are shown in Fig. 7c. We applied 1D consolidation theory (Terzaghi 1942) using a coefficient of consolidation equal to 2.8×10^{-7} to 2.34 at 10^{-7} m^2 s^{-1} at

1 and 2 m bsb, respectively (Mulder 1992) to calculate the excess pore pressure generated by such a fast sedimentation rate. Values of the critical slope leading to failure have been computed using an infinite slope analysis for undrained conditions that particularly fit this kind of event (Fig. 7d), and compared with the static critical slope (11.3°). The critical slope value drops a few degrees at the river mouth, where sediment supply is the highest. This reduction of the critical slope associated with a general over-steepening owing to the fast sediment supply can explain the triggering of shallow retrogressive failures in the area from the river mouth to the shelf edge, which corresponds to the area covered with

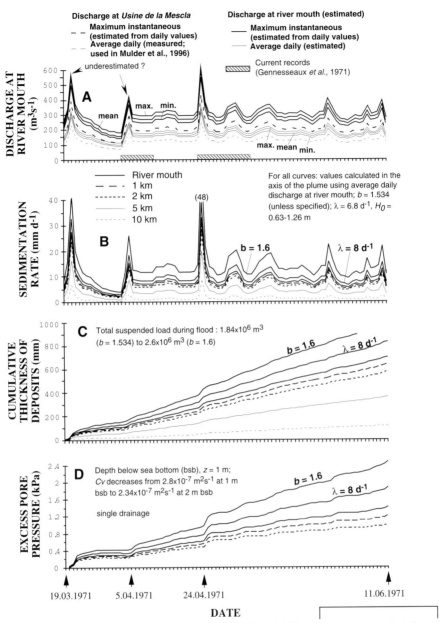

Fig. 7. Sedimentation and slope failure analysis during spring 1971. (**a**) Water discharge record at river mouth. Instantaneous values at the river mouth have been reconstituted using daily values measured upstream (usine de la Mescla). (**b**) Theoretical sedimentation rates calculated at the river mouth and at distances of 1000, 2000, 5000 and 10 000 m seaward using the centre line values of the 1D plume sedimentation model published by Syvitski & Lewis (1992), a flow thickness $H_0 = 0.63$–1.26 m and $\lambda = 6.8$–8 d^{-1} (corresponding to medium silt) to take into account the possible change in the average grain size of the sediment delivered at the river mouth during this high-discharge period. (**c**) Cumulative thicknesses of sedimentation over the same time period. Excess pore pressure (**d**) has been calculated using one-dimensional consolidation theory (Terzaghi 1942) with single drainage. Comparative simulations are made at the river mouth using $b = 1.6$ to take into account the possible change in b value that might occur because of the use of daily discharges instead of instantaneous discharges ($b = 1.534$). Discharge at rivermouth is estimated to be 1.25 (min) to 1.45 (max) times discharge at Usine de la Mescla. All calculations made with mean values: Q (river mouth) $= 1.35 \times Q$ (Mescla). $Q =$ discharge; bsb, below sea bottom.

underconsolidated sediment mapped by Cochonat *et al.* (1993). Mulder *et al.* (1994) calculated that the depth of a potential failure surface would be shallower than 2 m in these areas (the shear resistance is only 1.35 kPa at 1 m below the sea floor), which suggests that such surficial gravity processes can occur yearly (Mulder *et al.* 1996), and even several times a year.

The sedimentary impact of hyperpycnal events

The frequent occurrence of erosive bottom currents leading to the Var Canyon is indicated by the existence of badland-like erosional features that affect the submarine braided system down to 600 m water depth. The extension of the braided system at these water depths cannot be explained simply by the failures of gravel bars at the river mouth, as suggested by Gennesseaux (1962). Although hyperpycnal underflows can be considered to be less erosive than surge-like turbidity currents as they are less dense and exhibit lower velocities, their high frequency and their duration can lead to a substantial erosion of the bottom of the canyon and allow the badland system to be maintained.

The peak of precipitation related to the November 1994 flood occurred on 5 November. In 3 days, rainstorms provided 400 mm of water

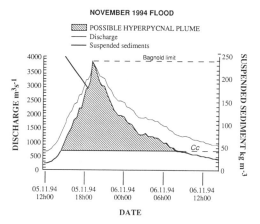

NOVEMBER 1994 FLOOD

Fig. 8. Discharge record during the 1994 flood (one value every 20 min, data from DIREN PACA) and calculated values of suspended sediment concentration using equation (3). Considering a critical value of 42.35 kg m^{-3} for suspended sediment concentration to produce hyperpycnal plumes, turbid plume may be produced during a period of almost 20 h. The critical concentration threshold to produce a hyperpycnal plume, Cc, and the Bagnold limit (9% volumetric concentration, i.e. 240 kg m^{-3}) are noted.

over southwest France (in Marseille, the mean monthly rainfall value for November is only 74 mm). All the rivers in the southwest part of France experienced exceptional discharge: the Durance, a tributary of the Rhône, reached 3300 m^3 s^{-1}, compared with a mean annual discharge of 20 m^3 s^{-1}. The Rhône discharge reached 9800 m^3 s^{-1}. It appears that rainstorms were more intense inland than near the coast. Bridges were washed out, mudflows, rockfall and rock avalanches were triggered, and six people were killed in France.

The Var River destroyed a part of the Nice–Digne railway, and Nice airport was closed because it was completely flooded by the river. Discharge at the river mouth reached a peak value close to 4000 m^3 s^{-1}. Accordingly, it is likely that a hyperpycnal plume was created for a few tens of hours. Figure 8 shows that flow concentration may have reached values close to the Bagnold limit above which a mud flow is created (a volume concentration of sediment within the plume of 9%, i.e. 240 kg m^{-3}; Bagnold 1962). This current may have transported 18 × 10^6 t of sediment over a period of 18 h, which represents 1.5 times the volume of sediment involved in the 1979 slide (Habib 1994). This value should also be compared with the 1.3–1.6 × 10^6 t mean annual suspended sediment load at the Var River mouth over the 22 a of our dataset. Using 1500 kg m^{-3} for the bulk density of freshly deposited sediments, a turbidite of few millimetres to a few centimentres thickness should have been deposited by this hyperpycnal plume in the upper and middle parts of the submarine valley, i.e. over an area covering *c.* 250 km^2.

The interpretation of the 1979 event

The 1979 event occurred in an area where sediment stability may have been affected by the nature of sediments and its relationship with sub-bottom hydrological processes, as well as human actions and the hydrological behaviour of the Var River (Fig. 9). The official investigation partially reported by Habib (1994) remains very unclear about what caused the slide, arguing about the inaccuracy of the time reported by the eye-witnesses. It is now clearly demonstrated that submarine landslides can produce tsunamis as large as those produced by earthquakes, for example, the subaerial Alika landslide produced a tsunami with related deposits found at altitudes up to 326 m (Moore & Moore 1984), but there is no published case of a tsunami causing a submarine landslide. As no earthquake was recorded in the study area immediately preceding the event, the tsunami was clearly caused by the slide. Because the slide

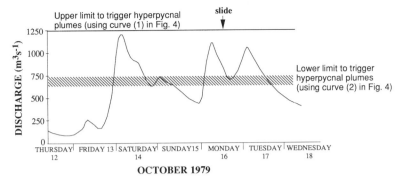

Fig. 9. Water discharge during the 5 day period including the triggering of the 1979 event (data from DIREN PACA).

occurred only at a few tens of metres from the shoreline, the tsunami would have reached the shore very soon after the slide occurred (at the most a few minutes), thus explaining the imprecision in the time sequence reconstructed from witnesses at various points along the coastline of the Baie des Anges. The question is: what caused the slide? Different speculative triggering scenarios have been proposed. They share the same background information, but differ in the interpretation of data. Discussing the respective 'weight' of each dataset could help to eliminate less probable scenarios.

Nature of sediments

The marine area surrounding Nice Airport is constituted by underconsolidated sediments (Cochonat *et al.* 1993). As floods produce hypopycnal plumes at least during the period in which discharge is less than the discharge threshold to produce underflows, we assumed that areas of underconsolidated sediment were covered by freshly deposited material during the beginning of the 1979 flood. Similar sediments accumulated in secondary channels merging with the main Var Canyon (Fig. 10). However, the main scar related to the 1979 event is located 1 km E of the Var Canyon (Fig. 10a; Habib 1994) in an area less affected by an increase of sediment supply at the river mouth than the head of the main channel. Moreover, the sediment supply during this period did not exceed a few millimetres per unit area at the most. This, even over the two days preceding the failure, is insufficient to generate an excess pore pressure high enough to trigger a deep failure. In addition, underconsolidated sediments are only a few metres thick, which is not consistent with the depth of failure (30–50 m) reported by Habib (1994).

Zones of low shear resistance deeper than a

few metres could be related to less cohesive sandy layers. Such layers are frequent in the Var Pleistocene deposits, as sediment deposited at the Var River mouth was coarser during this period (Dubar 1988). However, the top of the Pleistocene is about 80 m below the sea floor on the shelf, i.e. far below the maximum depth of failure reported by Habib (1994). Sandy layers occur in Holocene sediments but are generally less common than in Pleistocene deposits. Assuming that such a sandy layer was of large areal extent and located at 30–50 m below the sea floor, seepage and fluctuations of the top level of the aquifer related to the anomalous rainfall that occurred during the period preceding the slide might have produced an increase in pore pressure and the progressive liquefaction of the sediments. At this point, a slide failure might have occurred in a similar way to the triggering of subaerial mudslides after exceptional rainfall.

Anthropogenic action

During the period preceding the 1979 event, construction work on the airport might have destabilized the sediments settled below the airport. This anthropogenic action might have directly triggered a slide by overloading the edge of the airport with artificial material and by compaction. Landfilling operation took place in areas where original bathymetry varied from 10 to 15 m (Habib 1994) and thickness of landfill material reached 20–25 m in places. Water added through heavy rainfall may have also contributed to this overloading. Anthropogenic influence might also have acted in an indirect way, by creating a zone of weakness over a few tens of metres below the sea floor, for example, at a depth where a sedimentary discontinuity exists (e.g. a cohesionless sandy layer) and where stresses could have been accumulated,

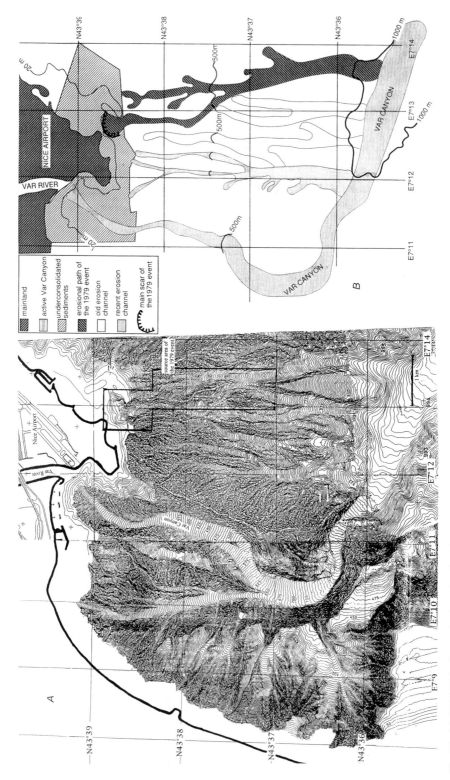

Fig. 10. (a) EM1000 bathymetric chart showing the source area of the 1979 event located just seaward of the Var River mouth and Nice Airport, and the evacuation channel down to its junction with the Var Canyon. Isobath interval is 5 m; grid step is 12.5 m. (b) Sketch map based on the EM1000 record. The channel purged by the 1979 event is shown in association with active or recently active erosional channels.

producing a direct failure. The process of failure is then similar to the natural process described in the preceding paragraph.

Behaviour of the Var River preceding the 1979 event

Similarly to the spring of 1971, October 1979 was a wet month. Maximum instantaneous discharge values at the river mouth were about 1200 m^3 s^{-1} (Fig. 9), i.e. just below the maximum limit required to trigger a hyperpycnal plume, but largely above the minimum limit (the mean daily discharge was 751 m^3 s^{-1} for 16 October). Over the period from 1974 to 1994, October 1979 exhibits the third top monthly discharge, Q_{av} (194 m^3 s^{-1}), just below November 1994 (Q_{av} = 276 m^3 s^{-1}) and October 1993 (Q_{av} = 207 m^3 s^{-1}).

According to the volume of water discharged during the early autumn of 1979, both small-sized hyperpycnal plumes and low-density turbidity currents related to shallow retrogressive foreset failures may have been triggered during the period preceding 16 October, and evacuated through the main Var Canyon. According to values published by Mulder *et al.* (1996, 1997*b*), hyperpycnal plumes were less than 10 h in duration and had a concentration of less than 100 kg m^{-3} at the river mouth. Mulder *et al.* (1997*b*) demonstrated that the main erosion caused by such a hyperpycnal plume occurs in less than 600 m water depth, although limited erosion could occur deeper, as sidewall slopes exceed 45°. This suggests that a mechanism involving retrogressive erosion from the main channel was unlikely to have been produced in 1979.

Considering the complete set of available data in our possession, it appears that the most likely cause of triggering of the 1979 event was an excessive loading at the head of the slope in a place where sediment had already a small shear resistance. The excessive rainfall at this period and the potential existence of small and short duration hyperpycnal plumes might have helped in the final triggering but it seems impossible to explain the 1979 slide using only these two factors as the main cause.

Whatever the origin of the loading was, natural, anthropogenic or both, the main part of the unstable mass began to move from the airport area and to rapidly disintegrate on the steep slopes. It transformed quickly into a liquefied flow (Habib 1994) and finally ignited to produce the huge turbidity current inferred by Piper & Savoye (1993). The slide is likely to have produced the observed tsunami. As is the case in other complex deltaic environments (e.g. Coleman & Garrison 1977; Prior & Coleman

1978, 1982), the cause of the 1979 event is probably not a single factor and it remains difficult to conclude if the event would have happened if one of these multiple causes had not been produced.

Earthquake-induced failures in the Baie des Anges

The 1979 event was not earthquake-induced. Nevertheless, the Baie des Anges, and more generally the Ligurian Sea, is considered an area of strong earthquake hazard. Few earthquakes with magnitude ranging from 4 to 6 have been recorded during the second half of the twentieth century (Rehault & Bethoux 1984). Horizontal ground accelerations ranging from 0.095 g to 0.26 g can be expected for earthquakes with return periods ranging from 100 to 1000 a, respectively (Mulder *et al.* 1994). Such earthquakes could purge the 20–30 m of surficial sediments in a similar manner to the 1979 event just seaward of Nice Airport, but over more extended areas because several channels could be purged simultaneously (Mulder 1992; Habib 1994). In that sense, the 1979 event may be a preview of what could happen at a larger scale in the case of an earthquake generating shaking intensity values in the Bay exceeding VII–VIII on Mercalli Modified Scale corresponding to horizontal acceleration values exceeding 0.13 g.

Discussion: The importance of rapid sediment transfer events in the geological record

We have described three recent rapid sediment transfer events in the Var Delta and the deep-sea fan. Two of them had a considerable impact on human life. The 1994 floods, which probably produced a hyperpycnal plume, affected a large part of the mainland of southwestern France and the 1979 event killed people through its related tsunami. But what is the impact of such events on the overall marine sedimentation and how do they affect the geologic record? What is the scale of the event required to produce a long-term discernible and extensive record?

To resolve these two major issues, we calculated the mean sedimentation rates seaward of the Var River mouth. We compared the centre line values of the one-dimensional plume model (Syvitski & Lewis 1992) with the model for hemipelagic sedimentation in which sedimentation rate decreases with the square of distance from the river mouth, using an average value of the flow thickness, H_0 = 0.3–0.58 m and a width

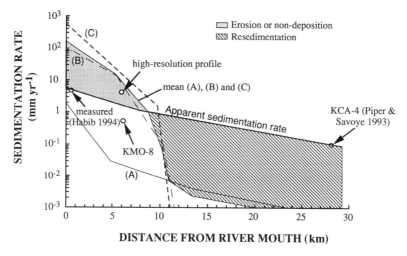

Fig. 11. Annual accumulation rate along a profile following the middle of the Var Canyon with its origin at the mouth of the river using mean total annual suspended load at the river mouth (tas). Curves (A) and (B), centre line values of the 1D model of Syvitski & Lewis (1992) with $\lambda = 6.8 \, d^{-1}$. Width of the Var River mouth is 250 m and the initial plume thickness is 0.3–0.58 m. Curves (C) and (D), using the model of hemipelagic sedimentation with the sedimentation rate varying as a inverse square function of distance from the river mouth. The average of (A), (B) (C) and (D) provides the theoretical sedimentation rate. On the other hand, the apparent sedimentation rate is defined as the interpolation between values provided by Habib (1994) and measurements on high-resolution profile and cores KMO-8 or KCA-4. The difference between mean (A), (B) (C) and (D) and the apparent sedimentation rate line provides an estimation of the missing sediments (eroded or non-deposited).

of the river mouth equal to 250 m. They provide theoretical maximum and minimum sedimentation rate at the river mouth, if the whole mean annual sediment load (1.32×10^6 t) was deposited and not removed (Fig. 11). Values vary between 63 mm a^{-1} and 54 cm a^{-1} at the river mouth, from 19 to 185 mm a^{-1} at a point 5 km seaward, and are less than 2×10^{-3} mm a^{-1} at a distance of 28 km from the river mouth. The one-dimensional plume model has been computed using $\lambda = 6.8 \, d^{-1}$ (medium silt).

If sedimentation rates measured on cores more than 17 km from the river mouth (core KCA-4 from Piper & Savoye 1993; Fig. 1) are larger than those calculated using hemipelagic models, then for distances closer to the river mouth, model sedimentation rates are larger than those published by Habib (1994) for a core collected in the neighbourhood of Nice Airport (4.8–5.9 mm a^{-1}). At this location, high sedimentation rates should be expected, as this area is covered with underconsolidated sediment (Cochonat *et al.* 1993). Core KMO-8, located at 6 km from the river mouth (water depth 820 m; Fig. 1), provided a radiocarbon age on a wood fragment of 3470 ± 80 BP at 184 cm below the sea floor, suggesting a maximum average sedimentation rate of 0.5 mm a^{-1}. A very high resolution seismic profile through core site KMO-8 reveals

a pebbly layer at 30 m below the sea floor. Assuming that these gravels were deposited during the late Pleistocene, the suggested value of the long-term sedimentation rate at this location would be 3 mm a^{-1}. These measured values are largely lower than the values provided by the two hemipelagic model that use the volume of sediment provided by the Var River. The question is: where are the sediments that are discharged yearly from the Var River mouth?

Our suggestion arising from the values of transported sediment reported in the preceding sections is that the major part of the sediment that should be deposited in the neighbourhood of the river mouth is either not deposited (when it is transported by hyperpycnal plumes), or deposited only for a short time period (a few weeks to a few years) before being evacuated towards the deep-sea fan and the abyssal plain. As a consequence, marine geological studies should be very careful with values of sedimentation rate measured in areas where gravity flow processes are active. One metre of sediment deposited during the Holocene might represent a real process of hemipelagic or pelagic deposition. It could also represent deposition between intermittent intense geological events and thus represent only an apparent sedimentation rate.

Using the value of sediment load for the

period 1974–1994, without taking into account the 1994 flood, and using 1250 m^3 s^{-1} for the discharge threshold to trigger a hyperpycnal plume, then 19–38% of the sediment that should have been deposited at the Var River mouth was not deposited, but was directly evacuated through hyperpycnal plumes. This hyperpycnal transport affects 55–63% of the theoretical sediment supply if the 1994 flood is taken into account. These values become higher using a lower discharge threshold for the occurrence of hyperpycnal events. Comparing the theoretical sedimentation rates in Fig. 11 and the values measured on cores, we calculated that long-term (i.e. ≥ 10 000 a) preservation of sediment deposited in the neighbourhood of the river mouth is less than 9% depending on the sedimentation model used. The remainder is deposited seaward of the middle fan valley, possibly with a large amount of reworking. Where it is redeposited, the apparent sedimentation rate would be higher than the theoretical sedimentation rate calculated using only hemipelagic and pelagic processes.

To answer the second question, i.e. the minimum size an event must reach to provide a significant deposit, wide and thick enough to be discernible in the geological record, we need to look at what is preserved in the sedimentary deposits.

Core KS-6, located on the levee bordering the right side of the upper and middle fan valley (Fig. 1) reveals 12–13 turbidites for the whole Holocene, i.e. one identified turbidite every 700–900 a. The top of the levee is located 350 m above the canyon floor. We know that the 1979 event did not spill over the levee. It also appears that retrogressive failure processes such as those inferred for spring 1971 can produce long duration continuous flows, but are unlikely to generate a turbidity current thick enough to reach the top of the levee. This suggests that only two processes can produce thick turbidity currents in the Var Canyon: (1) a large earthquake-triggered slide transforming into a turbidity current, according to a process similar to the one that occurred in 1979 but with a more effective increase of the turbidity current thickness, and (2) a hyperpycnal plume.

However, some characteristics of an ignitive earthquake-induced turbidity current are not consistent with overflow of the levee: basin experiments show that the dense flow at the base of a turbidity current is more likely to be deflected by any high topography rather than flow over it (Kneller et al. 1991; Alexander & Morris 1994; Kneller 1995). If high enough, a part of the plume might effectively separate from the rest of the main current, overflow the levee and create only a thin turbidite, as it is not sustained with the dense bottom flow. However, an earthquake might generate a large number of failures, 20–30 m thick (Mulder 1992; Mulder et al. 1994), which could then coalesce as independent turbid clouds, rather than a single deep-rooted flowslide, creating a huge turbidity current that can overflow the levee, in a similar way to those that occurred during the Pleistocene (Piper & Savoye 1993). As a consequence, turbidites cored on the top of the right levee could more probably be the result of the second process, hyperpycnal flows, as suggested by Mulder et al. (1997b).

The process leading to thick turbidity currents during hyperpycnal events is related to the nature of hyperpycnal events: they are processes occurring progressively as the concentration increases at the river mouth (Fig. 12). As discharge begins to increase at the river mouth, a hypopycnal plume is created (Fig. 12a) and persists during the whole flood. As suspended sediment concentration increases, part if not all of the plume begins to plunge (Fig. 12c). This stage corresponds to water discharges between 700 and 1200 m^3 s^{-1} at the Var River mouth. Shallow retrogressive failures may then be triggered simultaneously with the hyperpycnal process (Fig. 12b). In areas where slopes are high and the continental shelf is narrow, as in the Baie des Anges, a plunging plume at the river mouth may be stopped by a small density gradient in seawater at depth. When such a pycnocline is met, if speed is high enough (strongly dependent on the speed of the flow at the river mouth) the flow might continue to travel as a turbidity current, as shown by laboratory experiments (Rimoldi et al. 1996), or might spread as an intermediate mesopycnal plume (Fig. 12d). If any external forcing occurs and/or topography is encountered, this plume may concentrate and resume its plunging to join the main hyperpycnal flow according to a process inspired by density cascading (Postma 1969; Wilson & Roberts 1995). Past the plunging point (Wright et al. 1986), the hypopycnal plume tends to be reduced. Successive merging plumes may create a particularly thick turbidity current fed with sediments over a long duration.

Considering data published on other rivers over a period of time longer that 20 a, e.g. the Australian Burdekin River (Ward, 1978), the maximum discharge value calculated for a river (Matthai 1990) corresponds approximately to a return period in the range 1000–2000 a, i.e. the range of short-term climatic change that may affect the behaviour of the river. As a consequence, the hyperpycnal flows that may be at the origin of the turbidites detected at the top of the right levee could be the result of flood events

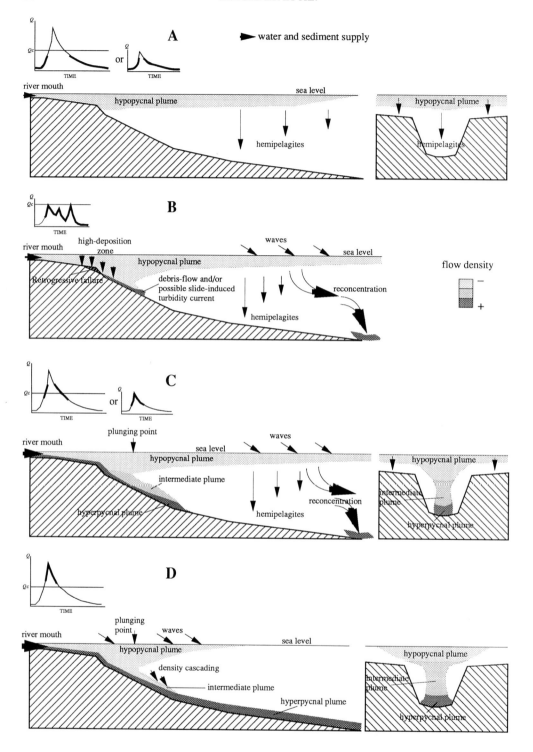

involving more than 10 000 m^3 s^{-1} at the Var River mouth, i.e. two to three times the maximum discharge measured in 1994. In that case, the 1994 flood would remain an event with a return period lower than 200 a. This suggests that usual *in situ* phenomena observed in marine sedimentology, although related to large catastrophic events such as the 1994 flood, are either not widely preserved in the long-term geological record (because they are completely eroded later), or are preserved in a very different way from high-magnitude events (Table 1).

In the Var, high-magnitude events are not important for the estimation of the daily flux but will be very important in the geological record, as they provide thick recognizable deposits with a large extension in the overall valley as well as on the levees where they have the best chance to be completely preserved and covered until the next high-magnitude event.

Low-magnitude events are the main part of the daily sediment flux. Sediments they involve are deposited only in the upper and middle parts of the fan valley (and not on the levees), and tend to be included in any larger-scale transport event through erosion, emphasizing our conclusion about the apparent sedimentation rates. Their importance is limited for the direct interpretation of the geological record, as their deposits are thin and sometimes indiscernible as the result of an individual event. These small steady-state deposits become strongly discontinuous because of erosion.

Conclusions

Three major sediment transfer processes are active during the present highstand in the Var System (Table 1): (1) large (generally) earthquake-induced ignitive turbidity currents in which we include the 1979 event although it was not earthquake-induced; (2) small-sized low-density turbidity currents related to retrogressive shallow failures generated by excess pore pressure owing to high sedimentation rates and

over-steepening occurring during floods on the narrow continental shelf and at the shelf break; (3) hyperpycnal plumes possibly associated with concentration phenomena similar to density cascading.

Among these three processes, only (1) and (3) can lead to significant deposits preserved over a long period of time. They lead to the deposition of turbidites that should be differentiated as follows (Table 1). Medium-sized failures and large earthquake-induced failures should include shelly fauna, as they have their source at the shelf break. Because the resulting high-density turbidity current produces strong erosion in the sandy channel floor, an extensive sand layer should be deposited on the fan, similar to that described by Piper & Savoye (1993). Hyperpycnal plumes could deliver a large amount of sediment carrying few marine fossils but instead freshwater and terrestrial organisms, and probably a higher organic matter content than turbidites resulting from submarine-triggered failures. As the flow concentration is lower than in a slide-generated turbidity current, erosion occurs preferentially on canyon walls and levees. This might explain the presence in core KCA-3 (Fig. 1) of both bark and resedimented pteropods with a considerable age (Piper & Savoye 1993).

In physical terms, processes (1) and (2) are typical unsteady flows (velocity is not constant with time at a given location within the flow) on the major part of their travel path. According to definitions published by Kneller (1995), they should be accumulative (velocity increases with distance) over a short distance after failure, depletive (velocity decreases with distance) in the Var Canyon, and then uniform (velocity is constant with distance). Deposition occurs only during the uniform steady phase of the travel path and leads to homogeneous well-sorted turbidites similar to the deposit related to the 1979 event (Piper & Savoye 1993). Before the uniform steady phase, flows are mainly erosional and bypassing dominates.

Fig. 12. Sedimentation processes occurring seaward of the Var River mouth during flooding periods. (**a**) During the rising limb of high-amplitude floods, and as long as water discharge, Q, at the river mouth is below the critical water discharge generating concentration high enough to produce a hyperpycnal plume, Q_c, or during low-amplitude floods, only a hypopycnal plume exists. (**b**) As the discharge increases during low-amplitude floods ($Q < Q_c$) in which high discharge values are maintained over a long period of time, sedimentation rate also increases, and shallow failures may be triggered at the river mouth and down to the shelf break, as happened in 1971. When $Q > Q_c$, the fluvial plume may begin to plunge, creating a hyperpycnal inflow (**c**). As concentration increase proceeds (**d**), the plume plunges deeper. Depending on both the existence of a pycnocline and the rate of concentration increase, intermediate plumes may be created. If concentration is high enough, a hyperpycnal plume can reach greater water depth located in the middle and lower parts of the fan valley (e.g. the 1994 flood). If any topography is met, concentration (density cascading) may create secondary plumes merging with the main hyperpycnal continuous current.

Table 1. *Features of the three main mass-wasting processes occurring in the Var Submarine Sedimentary System at the present time*

Nature of the event	Frequency	Thickness (T) and volume (V)	Sediment concentration	Velocity	Duration (from triggering to final deposition)	Erosion	Short-term deposition	Long-term deposition
Medium-sized slide and turbidity current, e.g. 1979	Unknown	Slide: $T = 20$–50 m; $V < 1$ km^3. Turbidity current: $T = 100$–300 m	>200 kg m^{-3} for the bottom part <100 kg m^{-3} for the upper part	≥30 m s^{-1} in the upper part of the canyon, 10–15 m s^{-1} in the lower part of the canyon, 2–3 m s^{-1} in the middle fan valley	<24 h	Important to very important (a few metres) on the main channel floor. Much sand is incorporated. May also include a shelly fauna from the shelf-break	Well-sorted, normally graded turbidite throughout the system. Extensive sandy tubidite in the fan valley	Well-sorted, normally graded turbidite in the fan valley
Earthquake-induced slide and turbidity current	≥100 a	Slide: $T = 20$–50 m; $V = 10$–100 km^3 Turbidity current: $T = 200$–500 m	Similar to above	Similar to above	24 h to a few days, depending on retrogression and earthquake aftershocks	Similar to above	Similar to above; a fine sandy turbidite may be deposited on the leveee	Similar to above. Possible thin fine turbidite on the levee
Shallow retrogressive failure on prodelta slope and related turbidity current, e.g 1971	Yearly	Slide: $T ≤2$ m; Turbidity current: $T ≤2$ m	≤2 kg m^{-3}	≤1–3 m s^{-1}	A few weeks with a variable intensity, depending on flood duration	Low and restricted above 600 m water depth	Normally graded turbidite in the upper and middle fan valleys	Sparse patchy turbidite in upper and middle valleys
High-frequency hyperpycnal plume, e.g. 1994	2–200 a	$T < 50$ m	42–230 kg m^{-3}	2–15 m s^{-1} in the canyon, ≤2 m s^{-1} in the fan valley	A few minutes to a few days, depending on flood duration	Affects hemipelagites deposited on channels floors/walls and levees. More important above 600 m water depth, reduced below	Inversely graded turbidite throughout fan valley, normally graded turbidite restricted to the upper and middle fan valleys. Possible high organic matter content, plant fragments and cont-inental fossils. High re-worked material content owing to the strong erosion of channels walls and levees	Sparse patchy turbidite in the upper and middle valleys

Table 1. *continued*

High-magnitude hyperpycnal plume	≥500 a	$T = 100–500$ m	$42–230$ kg m^{-3}	$2–20$ m s^{-1} in the canyon, $≤2$ m s^{-1} in the fan valley	A few days to a few weeks, depending on flood duration	Same as above	Same as above but the normally graded turbidite and inversely graded turbidite develops through-out the fan valley, on the right leveee and on western spillover channel. The inversely graded turbidite may be eroded in the upper fan valley during the rising limb	Normally graded turbidite on the right levee, western spill-over channel and on the fan valley, possibly capping an inversely graded turbidite

The physical evolution of hyperpycnal events is more complicated, as they are related to floods, i.e. to events in which intensity varies with time. Erosion and deposition by such events has been discussed in Mulder *et al.* (1998). During the rising limb, the plume is an unsteady waxing event (velocity increases with time at a given location), non-uniform on the steep slopes of the canyon, and then depletive seaward. An inversely graded turbidite is deposited and shifts seaward as discharge increases at the river mouth. During the falling limb of the hydrograph, the plume is a waning phenomenon (velocity decreases with time at a given location). A conventional Bouma-type (Bouma 1962) normally graded turbidite is then deposited and shifts landward as discharge slows down at the river mouth. During larger events, deposition may occur on levees, as such large-scale events are able to produce both huge low-density turbid clouds, a few hundreds of metres thick, and low-density hyperpycnal and intermediate plumes that can concentrate when high topography, such as a levee, is met. Only turbidity currents resulting from large failures (generally earthquake-triggered) cause tsunamis.

Only high-magnitude events related to processes (1) and (3), with a size resulting in thick deposits, are preserved in the sedimentary record as individual, continuous and recognizable bodies with an extent covering the whole Var system (Table 1). In this way, they constitute the best clues for the understanding of the geological record. The size of the 1979 event represents probably the lower limit for long-term preservation of an independent turbidite related to a single event over a large area. As process (2) and low-magnitude events related to processes (1) and (3) cannot generate deposition at the top of the levees and in the distal part of the basin, and produce only thin deposits on the canyon and fan valley floors, they will be rapidly eroded. A part of the deposited material will then be reworked by larger-scale sedimentary processes generated through (1) and (3) and only thin and indiscernible patchy turbidites might remain.

This study measures the importance of sediment transfer processes in an active delta and deep-sea fan system. It also highlights the relative importance of sedimentary processes and measurements in such active sedimentation areas. These concerns can be summed up as follows:

(1) Sedimentation rates measured in active sedimentation areas have no quantitative significance, except to underline and possibly quantify the magnitude of bypassing and erosional phenomena.

(2) As a direct consequence of (1), sedimentary processes in such areas should be interpreted in terms of fluxes (horizontal sediment transfer) rather than static processes (local accumulation rates).

(3) Long-term preservation of deposits related to large sediment transfer events in high-sedimentation environments includes only catastrophic high-magnitude events; in proximal areas, daily low-scale processes are erased from the record through erosion. In distal areas, their remains are either sparse and indiscernible from the rest of the surrounding sediments, or mixed with more recent deposits. Therefore, this suggests that processes measured or observed *in situ* at the present time and their related deposits are different in scale from those that are interpretable within sedimentary series.

The authors thank H. C. Jaubert from the Direction Régionale de l'Environnement (DIREN) Provence Alpes–Côte d'Azur, E. Evrard and P. Boissery of l'Agence de l'Eau de Marseille, and E. Astier of l'Agence de l'Eau Rhône–Méditerranée–Corse for their cooperation in the transmission of recent hydrological data. P. Cochonat and J. Alexander are kindly thanked for the time they spent in reviewing this paper. One of the authors (T. M.) has been funded by the MTR-MAST programme of the European Union.

References

ALEXANDER, J. & MORRIS, S. A. 1994. Observations on experimental, non-chanellized, high concentration turbidity currents and variations in deposition around obstacles. *Journal of Sedimentary Research*, **A64**, 899–909.

BAGNOLD, R. A. 1962. Auto-suspension of transported sediment: turbidity currents. *Proceedings of the Royal Society of London, Series A*, **265**, 315–319.

BORNHOLD, B. D., REN, P. & PRIOR, D. B. 1994. High-frequency turbidity currents in British Columbia fjords. *Geo-Marine Letters*, **14**(4), 238–243.

BOUMA, A. H. 1962. *Sedimentology of some Flysch Deposits: a Graphic Approach to Facies Interpretation*. Elsevier, Amsterdam.

BUGGE, T., BELDERSON, R. H. & KENYON, N. H. 1988. The Storegga Slide. *Philosophical Transactions of the Royal Society of London, Series A*, **325**, 357–388.

COCHONAT, P., DODD, L., BOURILLET, J.-F. & SAVOYE, B. 1993. Geotechnical characteristics and instability of submarine slope sediments, the Nice slope (N-W Mediterranean Sea). *Marine Georesources and Geotechnology*, **11**(2), 131–151.

COLEMAN, J. M. & GARRISON, L. E. 1977. Geological aspects of marine slope stability, Northwestern Gulf of Mexico. *Marine Geotechnology*, **2**, *Marine Slope Stability*, 9–44.

DUBAR, M. 1988. Niveaux marins étagés des Alpes-Maritimes: détritisme, tectonique et eustatisme sur le littoral méditerranéen au Pliocène supérieur et au Quaternaire. *Géologie Alpine, Mémoire hors série* **14**, 235–240.

EMBLEY, R. W. & JACOBI, R. D. 1977. Distribution and morphology of large submarine sediment slides and slumps on Atlantic Continental Margins. *Marine Geotechnology*, **2**, 205–229.

FOREL, F. A. 1885. Les ravins sous-lacustres des fleuves glaciaires. *Comptes Rendus de l'Académie des Sciences*, **101**(16), 725–728.

GENNESSEAUX, M. 1962. Une cause probable des écoulements turbides profonds dans le canyon sous-marin du Var (Alpes-Maritimes). *Comptes Rendus de l'Académie des Sciences, Paris*, **254**, 2038–2040.

——, GUIBOUT, M. & LACOMBE, H. 1971. Enregistrement de courants de turbidité dans la vallée sous-marine du Var (Alpes-Maritimes). *Comptes Rendus de l'Académie des Sciences, Paris, Série D*, **273**, 2456–2459.

——, MAUFFRET, A. & PAUTOT, G. 1980. Les glissements sous-marins de la pente continentale niçoise et la rupture de câbles en mer Ligure (Méditerranée Occidentale). *Comptes Rendus de l'Académie des Sciences, Paris, Série D*, **290**, 959–962.

HABIB, P. 1994. Aspects géotechniques de l'accident du nouveau port de Nice. *Revue Française de Géotechnique*, **65**, 3–15.

HEEZEN, B. C. & EWING, M. 1952. Turbidity currents and submarine slumps, and the 1929 Grand Banks earthquake. *American Journal of Science*, **250**, 849–873.

INMAN, D. L. 1970. Strong currents in submarine canyons. *Abstracts, Transactions of the American Geophysical Union*, **51**(4), 319.

KNELLER, B. 1995. Beyond the turbidite paradigm: physical models for deposition of turbidites and their implications for reservoir prediction. *In*: HARTLEY, A. J. & PROSSER, D. J. (eds) *Characterization of Deep Marine Clastic Systems*. Geological Society, London, Special Publications, **94**, 31–49.

——, EDWARDS, E., MCCAFFREY, W. & MOORE, R. 1991. Oblique reflection of turbidity currents. *Geology*, **19**, 250–252.

KUENEN, P. H. 1952. Estimated size of the Grand Banks turbidity current. *American Journal of Science*, **250**, 849–873.

LAURENT, R. 1971. *Charge solide en suspension et géochimie dans un fleuve côtier méditerranéen. Le Var (Alpes-Maritimes)*. PhD thesis, Université de Nice, 249 pp.

LOWE, D. R. 1982. Sediment gravity flows: II. Depositional models with special reference to the deposits of high-density turbidity currents. *Journal of Sedimentary Petrology*, **52**(1), 279–297.

MALINVERNO, A., RYAN, W. B. F., AUFFRET, G. A. & PAUTOT, G. 1988. Sonar images of the path of recent failure events on the continental margin off Nice, France. *Geological Society of America Special Paper*, **229**, 59–75.

MATTHAI, H. F. 1990. Floods. *In*: WOOLMAN, M. G. & RIGGS, H. C. (eds) *The Geology of North America, Vol. O1, Surface Water Hydrology*. Geological Society of America, Boulder, CO, 97–120.

MIDDLETON, G. V. & HAMPTON, M. A. 1973. Sediment gravity flows; mechanics of flow and deposition. *In*: MIDDLETON, G. V. & BOUMA, A. H. (eds) *Turbidity and Deep Water Sedimentation*, SEPM Pacific Section short course, Anaheim, 1–38.

MONTI, S. & CARRÉ, D. 1979. *Carte bathymétrique de la Baie des Anges, Nice, Côte d'Azur*, 1:25000. CNEXO, Brest.

MOORE, J. G. & MOORE, G. W. 1989. Deposit from a giant wave on the island of Lanai, Hawaiian Ridge. *Science*, **226**, 1312–1315.

MULDER, T. 1992. *Aspects géotechniques de la stabilité des marges continentales. Application à la Baie des Anges, Nice, France*. PhD thesis, INPL–ENSG Nancy, 264 pp.

—— & COCHONAT, P. 1996. Classification of offshore mass movements. *Journal of Sedimentary Research*, **66**(1), 43–57.

—— & SYVITSKI, J. P. M. 1995. Turbidity currents generated at river mouths during exceptional discharges to the world oceans. *Journal of Geology*, **103**, 285–299.

——, SAVOYE, B., SYVITSKI, J. P. M. & COCHONAT, P. 1996. Origine des courants de turbidité enregistrés à l'embouchure du Var en 1971. *Comptes Rendus de l'Académie des Sciences, Paris, Série IIa*, **322**(4), 301–307.

——, —— & —— 1997a. Numerical modelling of a mid-sized gravity flow: the 1979 Nice turbidity current (dynamics, processes, sediment budget and seafloor impact). *Sedimentology*, **44**, 305–326.

——, ——, —— & PARIZE, O. 1997b. Des courants de turbidité hyperpycnaux dans la tête du canyon du Var? Données hydrologiques et observations de terrain. *Oceanologica Acta*, **20**, 607–626.

——, —— & SKENE, K. I. 1998. Modeling of erosion and deposition by turbidity currents generated at river mouths. *Journal of Sedimentary Research*, **68**, 124–137.

——, TISOT, J.-P., COCHONAT, P. & BOURILLET, J.-F. 1994. Regional assessment of mass failure events in the Baie des Anges, Mediterranean Sea. *Marine Geology*, **122**, 29–45.

NARDIN, T. R., HEIN, F. J., GORSLINE, D. S. & EDWARDS, B. D. 1979. A review of mass movement processes, sediment and acoustic characteristics, and contrasts in slope and base-of-slope systems versus canyon-fan-basin floor systems, *Society of Economic Paleontologists and Mineralogists, Special Publication*, **27**, 61–73.

NORMARK, W. R. & PIPER, D. J. W. 1991. Initiation processes and flow evolution of turbidity currents: implications for the depositional record. *In*: OSBORNE, R. H. (ed.) *From Shoreline to Abyss*. Society of Economic Paleontologists and Mineralogists, Special Publication, **46**, 207–230.

PARKER, G. 1982. Conditions for the ignition of catastrophically erosive turbidity currents. *Marine Geology*, **46**, 307–327.

PIPER, D. J. W. & AKSU, A. E. 1987. The source and origin of the Grand Banks turbidity current inferred from sediments budgets. *Geo-Marine Letters*, **7**, 177–182.

—— & SAVOYE, B. 1993. Processes of late Quaternary

turbidity current flow and deposition on the Var deep-sea fan, north-west Mediterranean Sea. *Sedimentology*, **40**, 557–582.

——, COCHONAT, P., OLLIER, G., LE DREZEN, E., MORRISON, M. & BALTZER, A. 1992. Evolution progressive d'un glissement rotationnel en un courant de turbidité: cas du séisme de 1929 des Grands Bancs (Terre Neuve). *Comptes Rendus de l'Académie des Sciences, Paris, Série II*, **314**, 1057–1064.

——, SHOR, A. N., FARRE, J. A., O'CONNELL, S. & JACOBI, R. 1985. Sediment slides and turbidity currents on the Laurentian fan: side-scan sonar investigations near the epicenter of the 1929 Grand Banks earthquake. *Geology*, **13**, 538–541.

——, —— & HUGHES CLARKE, J. E. 1988. The 1929 Grand Banks earthquake, slump and turbidity current. *In*: CLIFTON, H. E. (ed.) *Sedimentologic Consequences of Convulsive Geologic Events*. Geological Society of America Special Paper **229**, 77–92.

POSTMA, H. 1969. Suspended matter in the marine environment. *In*: *Morning Review*. Lectures of the Second International Oceanographic Congress, Moscow, 1966, 213–219.

PRIOR, D. B. & COLEMAN, J. M. 1978. Disintegrating retrogressive landslides on very-low-angle subaqueous slopes Mississippi Delta. *Marine Geotechnology*, **3**(1), 37–60.

—— &—— 1982. Active slides and flows in underconsolidated marine sediments on the slope of Mississippi delta. *In*: SAXOV, S. & NIEUWENHUIS, J. K. (eds) *Marine Slides and Other Mass Movements*. Proceedings of a NATO Workshop, Algarve, Portugal, 15–21 December 1980, Series 4, **6**, 21–50.

——, BORNHOLD, B. D. & JOHNS, M. W. 1986. Active sand transport along a fjord bottom channel, Bute inlet, British Columbia. *Geology*, **14**, 581–584.

——,——, WISENAM, W. J., JR & LOWE, D. R. 1987. Turbidity current activity in a British Columbia fjord. *Science*, **237**, 1330-1333.

——, SUHAYDA, J. N., LU, N.-Z., BORNHOLD, B. D., KELLER, G. H., WISEMAN, W. J., WRIGHT, L. D. & YANG, Z.-S. 1989. Storm wave reactivation of a submarine landslide. *Nature*, **341**, 47–50.

RAVENNE, C. & BEGHIN, P. 1983. Apport des expériences en canal à l'interprétation sédimentologique des dépôts de cônes détritiques sous-marins. *Revue de l'Institut Français du Pétrole*, **38**(3), 279–297.

REHAULT, J.-P. & BETHOUX, N. 1984. Earthquake relocation in the Ligurian Sea (western Mediterranean): geological interpretation. *Marine Geology*, **55**, 429–445.

RIMOLDI, B., ALEXANDER, J. & MORRIS, S. A. 1996. Experimental turbidity currents entering density stratified water: analogues for turbidites in Mediterranean hypersaline basins. *Sedimentology*, **43**, 527–540.

SAVOYE, B. & PIPER, D.J.W. 1991. The Messinian event on the margin of the Mediterranean Sea in the Nice area, southern France. *Marine Geology*, **97**, 279–304.

——, COCHONAT, P., PIPER, D. J. W., NELSON, C. H., VOISSET, M. & DROZ, L. 1995. Late Quaternary turbidity current activity: the Var deep-sea fan case study (Northwestern Mediterranean). *In*: *IAS–16th Regional Meeting of Sedimentology–5ème Congrès Français de Sédimentologie–ASF, 24–26 April 1995, Aix-les-Bains, Book of abstracts*. Publication ASF, Paris, **22**, 132.

——, PIPER, D. J. W. & DROZ, L. 1993. Plio-Pleistocene evolution of the Var deep-sea fan off the French Riviera. *Marine and Petroleum Geology*, **10**, 550–571.

——, VOISSET, M., COCHONAT, P., AUFFRET, G. A., BOURILLET, J.-F., DROZ, L., OLLIER, G., COUTELLE, A., LE CANN, C., AUZENDE, J.-M. & ROBERT, S. 1988. 'Le canyon du Var': case study of a modern active proximal channel (abstract). *American Association of Petroleum Geologists, Bulletin*, **72**, 1023.

SHANMUGAM, G. 1996. High-density turbidity currents: are they sandy debris flows? *Journal of Sedimentary Research*, **66**(1), 1–10.

SHEPARD, F. P., MCLOUGHLIN, P. A., MARSHALL, N. F. & SULLIVAN, G. G. 1977. Current-meter recordings of low-speed turbidity currents. *Geology*, **5**, 297–301.

SKEMPTON, A. W. & HUTCHINSON, J. N. 1969. Stability of natural slopes and embankment foundations. State-of-the-art report. *In*: *Proceedings 7th International Conference*, SMFE, Mexico City, **2**, 291–339.

SYVITSKI, J. P. M. & ALCOTT, J. M. 1995. RIVER3: Simulation of river discharge and sediment transport. *Computers and Geosciences*, **21**(1), 89–151.

—— & FARROW, G. E. 1989. Fjord sedimentation as an analog for small hydrocarbon-bearing fan deltas. *In*: WHATELEY, M. K. G. & PICKERING, K. T. (eds) *Deltas: Sites and Traps for Fossil Fuels*. Geological Society, London, Special Publications, **41**, 21–43.

—— & LEWIS, A. G. 1992. The seasonal distribution of suspended particles and their iron and manganese loading in a glacial runoff fjord. *Geoscience Canada*, **19**(1), 13–20.

—— & SCHAFER, C. T. 1996. Evidence for an earthquake-triggered basin collapse in Saguenay Fjord, Canada. *Sedimentary Geology*, **104**, 127–153.

——, SMITH, J. N., CALABRESE, E. A. & BOUDREAU, B. P. 1988. Basin sedimentation and the growth of prograding deltas. *Journal of Geophysical Research*, **93**, 6895–6908.

TERZAGHI, K. 1942. *Theoretical Soil Mechanics*. John Wiley, New York, 510 pp.

WARD, R. 1978. *Floods. A Geographical Perspective. Focal Problem in Geography*. Macmillan, London, 244 pp.

WILSON, P. A. & ROBERTS, H. H. 1995. Density cascading: off-shelf sediment transport, evidence and implications, Bahama Banks. *Journal of Sedimentary Research*, **A65**(1), 45–56.

WRIGHT, L. D., YANG, Z.-S., BORNHOLD, B. D., KELLER, G. H., PRIOR, D. B. & WISENAM, W. J. JR 1986. Hyperpycnal plumes and plume fronts over the Huanghe (Yellow River) delta front. *Geo-Marine Letters*, **6**, 97–105.

Cenozoic changes in the sedimentary regime on the northeastern Faeroes margin

T. NIELSEN[1], TJ. C. E. VAN WEERING[2] & M. S. ANDERSEN[1]

[1]*The Geological Survey of Denmark and Greenland, Thoravej 8, 2400 Copenhagen, Denmark*

[2]*Netherlands Institute for Sea Research, P.O. Box 59, 1790 AB Den Burg, Texel, The Netherlands*

Abstract: An area on the northeastern Faeroes margin has been studied in detail using multichannel, high-resolution, seismic-reflection data which have revealed an early Tertiary basaltic basement overlain by an up to 2 km thick, wedge-shaped succession of sediments. Seismic facies analysis of the sedimentary succession has given rise to a sequence-stratigraphic interpretation of the slope. In addition, the study has provided information about changes in the sedimentary regime and variations in slope stability during deposition of the slope-apron caused by changes in sea level and the bottom-current regime.

The continental margin north of the Faeroe Islands has developed since the opening of the NE Atlantic c. 55 Ma ago (Talwani & Eldholm 1974). The shelf consists of Tertiary basalt with a thin cover of sediments (Andersen 1988), and the shelfbreak is not well defined. The average dip of the margin is less than 0.5° down to c. 1000 m water depth, below which the dip increases to 1–2°. At about 1500 m water depth a steep headwall up to 300 m high cuts the sea floor. Downslope from this headwall the regional dip is less than 1°, but slumped material gives rise to a rough sea-floor surface (Van Weering *et al.* this volume). The basement

underlying the margin consists of basalts extruded before or during magnetic chron C24R, i.e. 55 Ma ago (Talwani & Udintsev 1976; Smythe 1983; Waagstein 1988). The sedimentary succession overlying the basalt is wedge shaped, with a thickness range from a few tens of metres on top of the Faeroe Platform to more than 2 km in the mid-slope area.

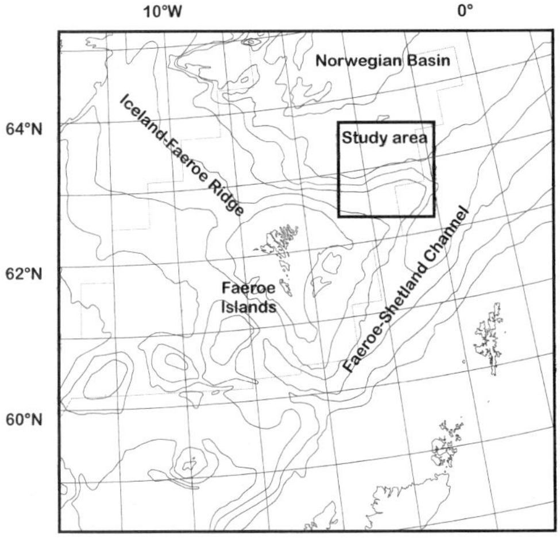

In 1991 and 1993 the Geological Survey of Denmark and Greenland (GEUS) and the Netherlands Institute for Sea Research (NIOZ) acquired more than 1000 km of seismic data on the eastern part of the northern Faeroes margin (Fig. 1). The study area transects subaerially extruded basalt on the Faeroe Platform, seaward-dipping reflectors in a transition zone underneath the slope, and oceanic basaltic basement in the Norwegian Basin. Thus, the area covers the entire continent to ocean transition zone.

Seismic stratigraphy

The database for the present study consists of high-resolution, multichannel seismic-reflection profiles acquired by GEUS and NIOZ in connection with the European MASTII programme ENAM (European North Atlantic Margin). An open survey grid comprises dip-lines with short, connecting cross-lines (Fig. 2). The resolution of the seismic data is c. 10 m. The penetration extends to the basaltic surface, which generally forms the acoustic basement. Based on detailed interpretation of the seismic data, nine seismic units and a slump deposit have been identified above the basaltic basement (Table 1 and Fig. 3).

On the seismic profiles the basaltic surface is characterized by a very-high-amplitude reflector of variable continuity and character. The

Fig. 1. Location of the study area, which is expanded in Fig. 2, and the bathymetric setting of the NE Atlantic Ocean (200 m contour and 500 m intervals thereafter).

NIELSEN, T., VAN WEERING, TJ. C. E. & ANDERSEN, M. S. 1998. Cenozoic changes in the sedimentary regime on the northeastern Faeroes margin. *In*: STOKER, M. S., EVANS, D. & CRAMP, A. (eds) *Geological Processes on Continental Margins: Sedimentation, Mass-Wasting and Stability*. Geological Society, London, Special Publications, **129**, 167–171.

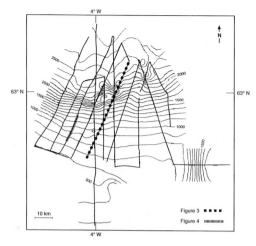

Fig. 2. Detailed bathymetry of the study area based on seismic-reflection profiles (contours in metres) showing locations of the seismic lines. Marked lines show profiles illustrated in Figs 3 and 4.

reflector dips northward and reaches a depth of 5500 m in the Norwegian Basin. The internal reflector character of the basalt allows its division into three zones (Andersen 1988): a zone with parallel reflectors on the Faeroe Platform, a zone with seaward-dipping reflectors in the slope area, and a zone with chaotic reflections in the basin-floor area.

The reflector marking the top of seismic unit 1 is faint and discontinuous, making the position of the upper boundary somewhat unclear. Below the upper part of the slope the internal configuration of the unit is characterized by subparallel reflections, which farther down the slope become disrupted and grade into a chaotic pattern in the midslope and basin-floor areas. This basinward change of the internal reflections is interpreted to be caused by repeated mass-movement events during deposition of the unit.

Following the definitions of Van Wagoner *et al.* (1988), seismic units 2–8 have been interpreted and classified as various systems tracts representing at least three depositional sequences (Table 1). The two lowermost sequences, termed sequences 1 and 2, are developed fully and remain partly preserved, whereas only the lowstand system tract remains from sequence 3. In the lower part of the slope, at *c.* 2000 ms two-way travel time (1500 m water depth), seismic units 4–8 are all cut at a major sediment failure headwall.

Seismic unit 9 is only present upslope of the major headwall. The base of unit 9 is defined by an unconformity which downslope becomes erosional and cuts deep into the underlying seismic units (Figs 3 and 4). Just above the erosional unconformity are a number of deposits characterized by a chaotic, semi-transparent reflection pattern and a mounded morphology, a

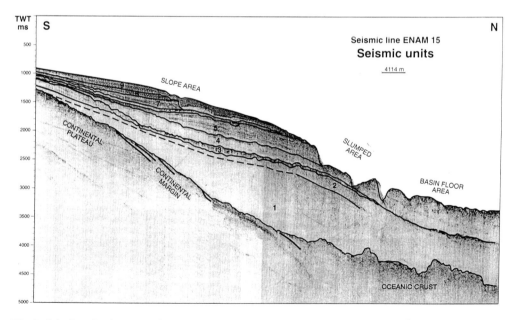

Fig. 3. Seismic-reflection profile (ENAM 15) across the continental margin north of the Faeroe Islands, and interpretation of the seismic units. (See Fig. 2 for location.)

Table 1. *Outline of the development history of the continental margin north of the Faeroe Islands based on the interpretation of the seismic-reflection profiles shown in Fig. 2*

Seismic unit	Seismic sequence	Depositional palaeoenvironment	Development history
Unit 9		Current regime Dominantly alongslope sedimentation	Norwegian Sea Deep Water current, changing with glacial–interglacial cycles Glacial erosion (upper slope)
Slump deposits		Unstable, sediments prone to gravity sliding	Initiation of glaciation? Onset of Norwegian Sea Deep Water current?
Unit 8	Sequence 3		
Unit 7			
Unit 6	Sequence 2	Quiescent and slowly changing Distribution dependent on sea-level changes	Low rate of differential subsidence Sea-level changes
Unit 5			
Unit 4		Dominantly downslope sedimentation	
Unit 3	Sequence 1		
Unit 2			
Unit 1		Unstable, sediments prone to gravity sliding	High rate of differential subsidence owing to thermal cooling

configuration typical of slump deposits (Macurda 1991). Their presence suggests that the erosional unconformity is closely related to mass movements. The upslope limit for this mass-movement erosion lies at about 1000 m water depth (14–1500 ms TWT). The erosional scar is partly buried (Fig. 4), but is still visible at the present sea floor, where it forms the landward flank of a contour-parallel channel of up to 70 m high and 50 km long (see Van Weering *et al.* this volume). Downslope from the erosional scar, above the slump deposits, the internal reflection pattern of seismic unit 9 is characterized by multiple, high-amplitude parallel to subparallel reflectors which display aggradational downlap onto the slump deposits. The overall configuration of the unit

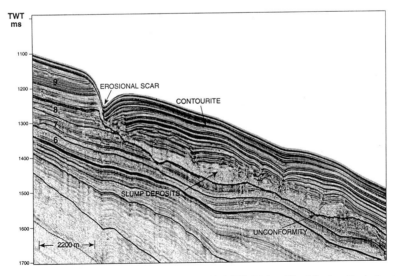

Fig. 4. Expanded section of part of seismic-reflection profile PE91-06 (see Fig. 2 for location). A seismic-reflection pattern indicative of contourite deposits is identified overlying slump deposits. The unconformity and the erosional scar are interpreted to have been formed by mass movements.

resembles that of a contourite as described by Stow (1994) and Wold (1994). The upslope termination of the contourite forms the basinward flank of the contour-parallel channel at the sea floor. The maximum thickness of the contourite is *c.* 100 m.

Discussion and conclusion

The Norwegian Basin is a classic thermal subsidence basin (Eldholm *et al.* 1989). As a result of differential subsidence, gradual steepening of the oceanward dip of the basement below the continental slope is predicted by thermal and tectonic modelling (McKenzie 1978). Steepening of the slope would be most dramatic during early cooling, but is expected to continue for at least 80 Ma after initiation of sea-floor spreading. The present dip of the basaltic basement within the study area is an indicator of the total amount of differential subsidence of the continental margin north of the Faeroe Islands since its formation (Smythe 1983; Andersen 1988). The greatest difference in the dip within the whole succession is that between the basement reflector and the upper boundary of seismic unit 1 (Fig. 3). This indicates that most of the differential subsidence in the area occurred during deposition of seismic unit 1. A rapid steepening of the slope angle may cause instability of the sediments, increasing the likelihood of gravity sliding. The mass movements interpreted for the downslope part of seismic unit 1 are therefore considered a result of rapid differential subsidence during the early development of the slope.

After deposition of seismic unit 1, the sediments on the continental slope became less prone to gravity sliding. The fact that seismic units 2–8 could be interpreted in a systematic sequence-stratigraphic framework indicates that sedimentation on this part of the continental margin off the Faeroe Islands has been continuous over a long period, and that the depositional pattern during this time was dominated by changes in relative sea level. It also appears that the upper and middle parts of the slope have been stable for a long period, as none of the seismic units here show signs of gravity movements. Thus, the slope became more stable and the depositional environment changed from mass-movement dominated to sea-level-controlled sedimentation. This scenario is in accord with subsidence models, which predict a reduction in the rate of differential subsidence with time. However, the change in subsidence rate is rather marked, and we speculate that it might be associated with the westward shift of

the sea-floor spreading axis, which occurred about 28 Ma ago (Nunns 1983).

It is not possible, based on the present database, to determine the timing and duration of the slope instability that led to the formation of the major headwall at the lower part of the slope and the lesser scar higher up the slope. Both these escarpments could be the result of the same mass-movement event, but another possibility is that the two were separate stages of instability. In the latter case, one stage affected only the lower part of the slope, leading to formation of the major headwall, and a second stage affected the sediments higher up on the slope. In either case, instability post-dating deposition of seismic unit 8 is clear although the controlling factor remains uncertain.

Deposition of contourites post-dating seismic unit 8 marks a shift in the sedimentary environment from dominantly downslope, sea-level-controlled to dominantly alongslope, current-controlled deposition, indicating a major change in the oceanographic regime of the Norwegian Basin. The slope instability could either be directly connected to the onset of this change, or alternatively, by the circumstances leading to it: either the subsidence of the Iceland–Faeroe Ridge or the initiation of glaciation of the northern hemisphere.

This work was carried out as part of EC MAST II contract MAS-CT93-0064.

References

ANDERSEN, M. S. 1988. Late Cretaceous and early Tertiary extension and volcanism around the Faeroe Islands. *In*: MORTON, A. C. & PARSON, L. M. (eds) *Early Tertiary Volcanism and the Opening of the NE Atlantic*. Geological Society, London, Special Publications, **39**, 115–122.

ELDHOLM, O., THIEDE, J. & TAYLOR, E. 1989. The Norwegian Continental Margin: tectonic, volcanic, and paleoenvironmental framework. *In*: ELDHOLM, O., THIEDE, J. *et al.* (eds) *Proceedings of the Ocean Drilling Program, Scientific Results*, **104**. Ocean Drilling Program, College Station, TX, 5–28.

MCKENZIE, D. 1978. Some remarks on sedimentary basins. *Earth and Planetary Science Letters*, **40**, 25–32.

MACURDA, B. D. 1991. Seismic stratigraphic and facies analysis of shallow water siliciclastics. *GeoQuest Training*.

NUNNS, A. G. 1983. Plate tectonic evolution of the Greenland–Scotland Ridge and surrounding regions. *In*: BOTT, M. H. P., SAXOV, S., TALWANI, M. & THIEDE, J. (eds) *Structure and Development of the Greenland–Scotland Ridge: New Methods and Concepts*. Plenum, New York, 11–30.

SMYTHE, D. K. 1983. Faeroe–Shetland Escarpment and continental margin north of the Fairways. *In*: BOTT, M. H. P., SAXOV, S., TALWANI, M. & THIEDE, J. (eds) *Structure and Development of the Greenland–Scotland Ridge: New Methods and Concepts*. Plenum, New York, 109–119.

STOW, D. A. V. 1994. Deep sea processes of sediment transport and deposition. *In*: PYE, K. (ed.) *Sediment Transport and Depositional Processes*. Blackwell, Oxford, 257–291.

TALWANI, M. & ELDHOLM, O. 1974. Margins of the Norwegian–Greenland Sea. *In*: BURK, C. A. & DRAKE, C. L. (eds) *The Geology of Continental Margins*, Springer-Verlag, NY, 361–374.

——, UDINTSEV, G. *et al*. 1976. *Initial Reports of the Deep Sea Drilling Project*, **38**. US Government Printing Office, Washington, DC, 1256 pp.

VAN WAGONER, J. C. *et al*. 1988. An overview of the fundamentals of sequence stratigraphy and key definitions. *In*: WILGUS, C. K. *et al*. (eds) *Sea-Level Changes – an Integrated Approach*. Society of Economic Palaeontologists and Mineralogists Special Publication, **42**, 39–45.

VAN WEERING, TJ. C. E., NIELSEN, T., KENYON, N. H., AKENTIEVA, K. & KUIJPERS, A. H. 1998. Large submarine slides at the NE Faeroe continental margin. *This volume*.

WAAGSTEIN, R. 1988. Structure, composition and age of the Faeroe basalt plateau. *In*: PARSON, L. M. & MORTON, A. C. (eds) *Early Tertiary Volcanism and the Opening of the NE Atlantic*. Geological Society, London, Special Publications, **39**, 225–238.

WOLD, C. N. 1994. Cenozoic sediment accumulation on drifts in the northern North Atlantic. *Paleoceanography*, **9**, 917–941.

The Southeast Greenland glaciated margin: 3D stratal architecture of shelf and deep sea

LENE CLAUSEN

Geological Museum, University of Copenhagen, Øster Voldgade 5–7, DK-1350
Copenhagen K, Denmark

Abstract: A seismic stratigraphic study of the Upper Neogene and Pleistocene shelf and deep-sea deposits of Southeast Greenland is presented. The depositional environment was controlled by glacial, turbidity current, and contour current processes. Below the mid- and outer shelf a Tertiary sedimentary wedge, as much as 1.7 km thick, is resting on a seaward-dipping volcanic acoustic basement. A lower marine part is separated by a regional uncon-formity, SB-A, from two seismic sequences of early Pliocene to Pleistocene prograding clinoforms. SB-A is surmised to represent a seaward-dipping palaeo-relief surface, accen-tuated by modest submarine erosion related to a sea-level lowering. The overlying two Sequences x and y reflect shelf-edge progradation in front of grounded ice during periods of maximum glaciation. The topset beds of the youngest sequence, possibly of Pleistocene age, are characterized by six or more sub-horizontal, landward-dipping internal unconfor-mities, taken to reflect individual glacial cycles of grounded ice erosion. The present-day shelf is overdeepened, landward dipping, and traversed by troughs. These features are characteristic of a shelf sculptured by grounded ice. The position of the transverse troughs, reflecting the positions of former icestreams, seems to have been more or less permanent since the late Neogene. The shelf progradation has been higher in front of the troughs than in front of adjacent banks, resulting in the development of trough-mouth fans (TMF). The strata continue seawards below the shelf break across a crustal monoclinal flexure to the deep sea. Three seismic sequences have been identified within the late Miocene and Pleisto-cene deep-sea sedimentary succession on the continental rise. They are bounded by uncon-formities of mid-Pliocene, early Pliocene, and (?) late Miocene ages. The oldest sequence is interpreted as basin floor fans, probably linked to a sea-level lowstand. The two youngest sequences were deposited from interplay of contour and turbidity currents during glacial periods, and they form large sediment ridges on the sea floor, oriented almost perpendic-ular to the continental rise, nucleated on the older sequence. Inter-ridge channel–levee complexes recognized within the youngest sequence were deposited from turbidity cur-rents. The major turbidity channels are prolongations of the transverse shelf trough–trough-mouth fan systems. A moat zone, at the base of the continental slope, reflects the position of strong modern contour currents. Appreciable contourite deposition did not commence before the early Pliocene, and was probably linked to glaciation of the shelf and onset or intensification of the Labrador Sea Water deep current.

The study area is located at the Southeast Green-land continental margin in the western part of the Irminger Basin (Fig. 1). The study is based on multichannel reflection seismic data acquired by the North Atlantic Project D (NAD) and North Atlantic Rifted Margin (NARM) Projects (Larsen 1985, 1990; Larsen *et al.* 1993). The main objective of the present paper is to describe the stratal architecture of the late Neogene and Pleistocene sedimentary successions from the shelf to the deep sea and to discuss the deposi-tional and erosional processes involved in their formation. Interpretations are based on seismic stratigraphic analysis combined with sea-floor morphology and drilling data (ODP Leg 152 borehole data; Larsen *et al.* 1994*b*).

The study area has previously been investi-gated by Sommerhoff (1973, 1979) and Johnson *et al.* (1975), who studied the shelf sea-floor topography. They concluded that the shelf and continental slope bottom morphology is glacially formed, but modified by turbidity current and contour current erosional processes. The bathymetry and sediment cover of the shelf north of the study area were investigated by Larsen (1980), and a tentative seismic stratigra-phy for the Cenozoic sedimentary succession of the Southeast Greenland shelf and rise was out-lined by Larsen (1985, 1990). Preliminary interpretations of the Neogene development of the shelf were published by Larsen *et al.* (1994*a*) and Lykke-Andersen (1998). It is documented in these studies that the Greenland continent has experienced glaciation back to the late Miocene (7 Ma) and that the shelf has prograded considerably because of glacial advances across the shelf. Prograding clinoforms beneath the shelf off Kangerdlugssuaq immediately north of

CLAUSEN, L. 1998. The Southeast Greenland glaciated margin: 3D stratal architecture of shelf and deep sea. *In*: STOKER, M., EVANS, D. & CRAMP, A. (eds) *Geological Processes on Continental Margins: Sedimentation, Mass-Wasting and Stability.* Geological Society, London, Special Publications, **129**, 173–203.

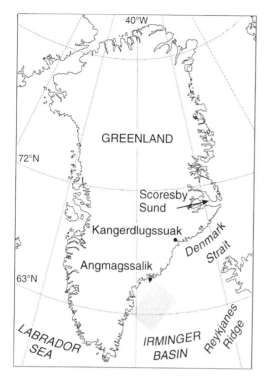

Fig. 1. Location of the Irminger Basin, bounded northwards by the Denmark Strait Sill forming part of the Faroe–Iceland–Greenland Ridge and eastwards by the Reykjanes Ridge. The Irminger Sea connects southwards to the Labrador Sea. The study area (enlarged in Fig. 2) is indicated by fine stipple.

the study area were also interpreted as glacially formed (Larsen 1994). A general review of the Quaternary geology of the Greenland shelves has been given by Funder & Larsen (1989).

Major deep-sea contourite deposits were previously recognized only in the northeastern Irminger Basin (the Snorri Drift; see Shor & Poore 1979), and south of Cape Farewell (the Eirik Ridge; see Johnson & Schneider 1969; Jones *et al.* 1970; Ruddiman 1972; Egloff & Johnson 1975). Recently, a number of smaller sediment ridges were recognized along the continental rise adjacent to the study area (Clausen 1998, who refers to the ridges as elongate drifts). These are separated from the steep continental slope by contour-current-eroded moats. The main results of that work are summarized below for the purpose of linking the depositional systems of the shelf and the deep sea.

Seismic data

The seismic grid covers an area extending from the mid-shelf to the upper continental rise (Fig.

2). It comprises *c.* 4000 km of high-resolution, relatively shallow seismic profiles and *c.* 1500 km of conventional seismic profiles, referred to as deep seismic profiles. Technical details have been given by Clausen (1998). Because of the usage of a short streamer the deeper parts of the high-resolution lines are masked by sea-bed multiples. The vertical resolution is estimated at 5 m in the shallow part of these profiles.

The thickness of seismic units beneath the shelf is estimated from seismic stacking velocities. Below the continental slope thicknesses of seismic sequences are based on the preliminary two-way travel time v. depth relationship at Site 918 (Larsen *et al.* 1994*b*) calibrated to seismic velocities adopted from the GGU 1981 seismic survey (Larsen 1990). For water a seismic velocity of 1480 m/s has been used.

Morphology of the Southeast Greenland margin

The shelf along the northwestern Irminger Basin is about 150–200 km wide, but tapers to about 50 km in the central and southern parts of the study area (Fig. 3). The water depth on the reverse-sloping shelf generally ranges between 200 and 400 m. East–west-trending troughs, 400–1000 m deep, crosscut the shelf. Each trough is flanked at the outer shelf by bank areas covered by about 200 m of water, slightly deepening shorewards.

The present study is concerned with the middle and outer shelf parts of the Skjoldungen, Ilertakajik, and Gyldenløves Troughs (Figs 3 and 4). They extend landwards and appear to be prolongations of fjords, but in the absence of nearshore data exact relations are unresolved. The closely spaced troughs are 8–25 km wide and between 600 and 900 m deep at the midshelf position. The larger Sermilik Trough, located 100 km north of Gyldenløves Trough, is 50 km wide and as much as 900 m deep, (Clausen in press). The shelf troughs deepen significantly from the outer shelf and shorewards, with a sea floor dipping as much as 2.7°. The troughs are separated by shallow bank areas, namely, Fylkir Bank, Skjoldungen Bank, Møsting Bank, and Heimland Bank ('Heimland Rücke' of Sommerhoff 1979). The last is considerably larger than the other banks, and shows a much more complex morphology with highs and lows. Each shallow bank area is subdivided by a central bathymetric low area, about 250 m deep. The outer shelf is generally smooth, in contrast to the inner rugged shelf. The outer parts of the Skjoldungen and the Ilertakajik

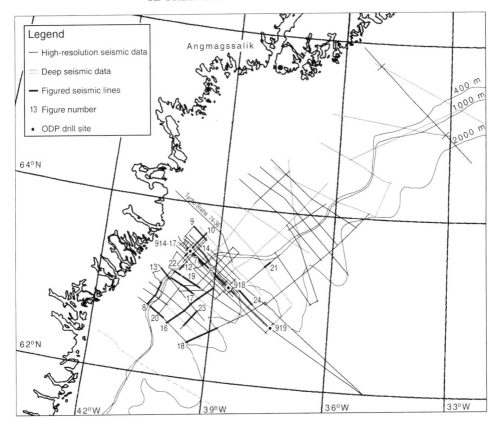

Fig. 2. Grid of multichannel seismic data in the study area. Figured line segments are highlighted and ODP drill sites are indicated. The dashed line represents a deep seismic line of the ICE Survey (Larsen *et al.* 1995). A simplified bathymetry is indicated with the 400 m isobath marking the position of the shelf break.

troughs show, however, a fairly rugged bottom topography (see, e.g. Fig. 8, below). Small-scale furrows, about 5–10 m deep, formed by drifting icebergs, occur down to a water depth of 350 m (Mienert *et al.* 1992). The shelf edge occurs at about 300 m water depth seaward of the bank areas, increasing to 350–400 m water depth in front of the transverse shelf troughs.

The continental margin south of 65°N is characterized by a well-defined continental slope dipping 6–13° (locally as much as 17°), passing abruptly into a moat zone and a gently dipping continental rise. A prominent feature of the continental slope is bathymetric bulges, located in front of the transverse shelf troughs, referred to as trough-mouth fans (TMF) (Vorren *et al.* 1989) (Fig. 4). Apart from spoon-shaped scours the slope is generally smooth and there are no incised submarine canyons. On the upper continental rise five sediment ridges (I–V, Fig. 3) are separated by interjacent downslope turbidity channels, as much as 150 m deep and 1–3 km wide, originating at the base of the

trough-mouth fans. The sediment ridges are about 30 km across, some 40–60 km long, and reach up to 400 m above the surrounding sea floor. They are offset slightly to the south of the neighbouring trough-mouth fans and are oriented almost perpendicular to the continental slope, from which they are separated by a zone of contour parallel channels (moats). This 15–25 km wide zone is characterized by sediment mounds and interjacent moats, and lies at the base of the slope at a water depth of about 1500 m (varying between 1000 and 1700 m) (see Clausen (1998) for further details).

Oceanographic currents of the Irminger Basin

The Irminger Sea is characterized by strong surface and deep oceanic currents. The former comprise the Irminger Current and the East Greenland Current, affecting the upper 200 m of the water column. The Irminger Current is

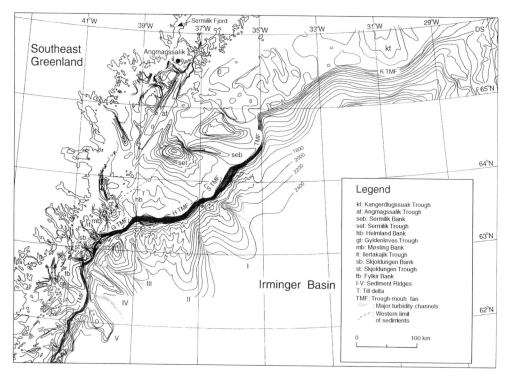

Fig. 3. Bathymetric map of the Southeast Greenland margin compiled from all data available (see also Fig. 4). Shelf troughs and bank areas are labelled with letters, whereas the deep-sea sediment ridges are referred to by Roman numbers. The Denmark Strait is indicated by DS (northeastern corner). Isobath intervals are 50 m on the shelf and 100 m in the deep sea. Note the steepness of the continental slope south of 65°N.

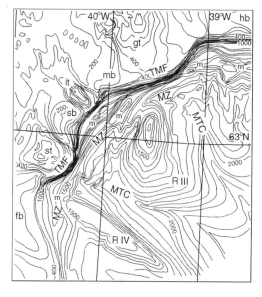

derived from the northern branch of the North Atlantic Current, and carries relatively warm (3.5–4.0°C) salty water northwards along the Reykjanes Ridge. At the Denmark Strait it turns southwards and flows along the Southeast Greenland coast. Some of this water rounds the southern tip of Greenland and continues to the Labrador Sea, whereas a second portion remains in the Irminger Sea and recirculates in cyclonical gyres (Fig. 5a) (Buch 1990). The East

Fig. 4. Enlarged segment of the bathymetric map (Fig. 3), showing distribution of discussed morphological elements: transverse troughs (gt, it, st) and bank areas (hb, mb, sb, fb) of the shelf, trough-mouth fans (TMF) of the continental slope, moat zone (MZ) with mounds (m) at the base of the continental slope, and sediment ridges (numbered I–V), separated by major turbidity channels (MTC) with associated subsidiary channels of the upper continental rise.

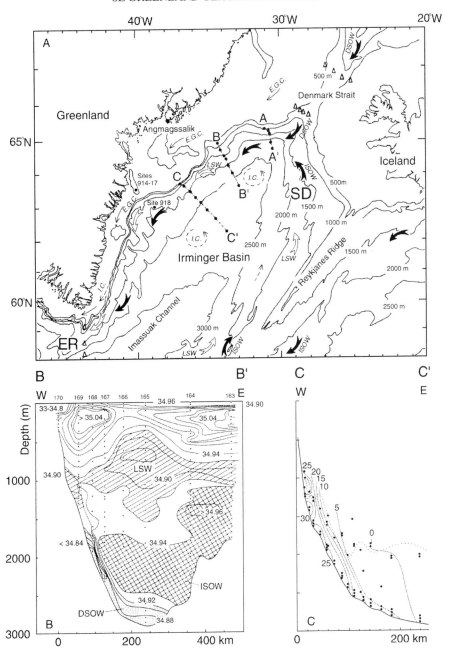

Fig. 5. Map showing surface and deep currents in the Irminger Sea (modified from Dickson & Brown 1994). (a) The near-surface Irminger Current (I.C.) and East Greenland Currents (E.G.C.) are indicated by thin dashed and unbroken arrows, respectively. The pathways of the Denmark Strait Overflow Water (DSOW) and the Iceland–Scotland Overflow Water (ISOW) are indicated by filled arrows; open arrows show the Labrador Sea Water (LSW). Transects A–A′, B–B′, and C–C′ are long-term current meter arrays, of which two are shown in (b) and (c). Locations of ODP drill sites 914–919, the Eirik Ridge (ER), and the Snorri Drift (SD) are indicated. (b) Main water bodies forming the North Atlantic Deep Water current (NADW) in transect B–B′ (a). Abbreviations as in (a). (c) Mean velocities (cm/s) of the deep-sea currents in transect C–C′ (a). Note that the LSW constitutes the high-velocity zone at 800–1500 m (compare b). The dotted line represents the σθ = 27.8 isopycnal.

Greenland Current carries water of predominantly arctic origin (<0°C) that passes through the Fram Strait between Greenland and Spitsbergen and flows southwards along the Greenland shelf carrying great amounts of sea ice and icebergs along with it (Swift & Aagaard 1981). The current velocity at the surface is as much as 0.5 m/s, decreasing to zero below 200 m depth, and the transport totals 2–3 Sverdrups ($1Sv=10^6$ m^3/s) (Buch 1990). The current rounds the southern tip of Greenland and joins the Baffin Current off the northeast coast of Canada. The East Greenland Current (or a precursor) is considered to have been established about 4 Ma ago (Bohrmann et al. 1990).

The deep current in the study area, the North Atlantic Deep Water (NADW), comprises three main water bodies (Dickson et al. 1990) (Fig. 5a and b). At Dohrn Bank (transect A–A′, Fig. 5a), the Denmark Strait Overflow Water (DSOW) totals 5.6 Sv, but at some point around 64°N latitude the deep transport increases rapidly, as a result of merging with other NADW components, of which the cold Labrador Sea Water (LSW) may be the most important contributor (Dickson et al. 1990; Dickson & Brown 1994). The core of the overflow current, characterized by mean current speeds just above 25 cm/s, is positioned at 1600–2200 m depth (Fig. 5c). Two other high-velocity zones are located at the base of the continental slope at a depth of 800–1500 m (Dickson et al. 1990; Dickson & Brown 1994), which corresponds to the level of the LSW. The onset of the Iceland–Scotland Overflow Water (ISOW) may be dated at around 11–13 Ma (Bohrmann et al. 1990), coinciding with or slightly pre-dating the onset of the NADW (see Larsen et al. 1994b).

Seismic characteristics of the shelf

The mid- and outer shelf of Southeast Greenland is characterized by a post-Paleocene sedimentary wedge, c. 1.7 km thick, resting on volcanic basement (Larsen 1990; Larsen et al. 1994b). The upper part of the succession consists of seaward-dipping clinoforms that show increasing angle of dip toward the present-day shelf edge (Fig. 6). The sedimentary package has been drilled at a mid-shelf position close to where the volcanic basement subcrops the sea floor (Fig. 3) and dating is thus available for the oldest dipping strata and for the youngest flatlying sediments capping the shelf (see section 'Correlation with ODP Leg 152').

The greater part of the prograding clinoform succession, which is the main subject of this paper, overlies a major unconformity, SB-A, and

is divided into two seismic stratigraphic sequences. The oldest, Sequence x, is characterized by a tangential oblique progradational pattern and the youngest, Sequence y, is characterized by a complex sigmoid-oblique reflection pattern (see Mitchum et al. 1977). The correlative deep-sea sedimentary succession is summarized below.

Sequence x

The regional unconformity SB-A separates a sedimentary package, Sequence x, denoted by steeply eastward-dipping strata, from an older, more southward and gently dipping succession (Figs 6 and 7). It is truncated by the landward-dipping unconformity SB-B in the inner part of the Gyldenløves Shelf Trough. SB-A slants about 2–2.5° (locally as much as 6°) towards the east below the mid-shelf, thence flattens beneath the outer shelf at a depth of up to 900 m. The unconformity dips eastwards again under the continental slope. It appears to roughly correlate with unconformity SB2, identified below the continental rise (Fig. 6). SB-A, which is a downlap surface, is most prominent where it is steeply dipping, i.e. below the mid-shelf and the flanks of the present-day shelf troughs, where it also clearly truncates reflections below (Figs 8 and 9). Few details of the unconformity can be distinguished below the outer shelf and continental slope, where it is discernible only on the few deep seismic lines available, owing to strong sea-bed multiple interference on the high-resolution seismic profiles.

The upper reflection terminations within Sequence x are generally difficult to recognize because of masking by bubble pulses, but clinoforms under the mid-shelf area tend to maintain parallelism as they terminate abruptly against SB-B, indicating erosional truncation. Clinoforms further seawards tangentially approach SB-B, indicating toplap relations (Fig. 7).

Sequence x, consisting of tangential oblique clinoforms, shows in the early part a gradual change or shift from moderately (as much as 5–7°) to steeply dipping (as much as 10°) foreset beds. High-amplitude high-continuity reflections, showing truncation and downlap relations, define small-scale intra-sequence units, easiest to recognize in the early, moderately dipping part (Fig. 7).

A wedge of small-scale dipping reflections is encountered within the shallow part of one of these intra-sequence units (Fig. 9). The wedge is about 2 km long and consists of steeply dipping clinoforms (about 8°), 75 m high, prograding towards the northeast (i.e. almost perpendicular

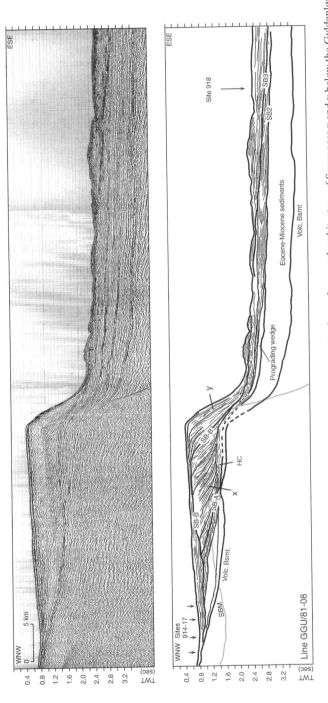

Fig. 6. Deep seismic profile 81-08 (for location. see Fig. 2). The interpreted line drawing shows the stratal architecture of Sequences x and y below the Gyldenløves Shelf Trough and Sequences β and γ below the adjacent inter-ridge area. The major unconformities (SB-A, SB-B, SB2, and SB3) and ODP drill sites are indicated. Sea-bed multiples are labelled SBM. TWT, Two-way travel time.

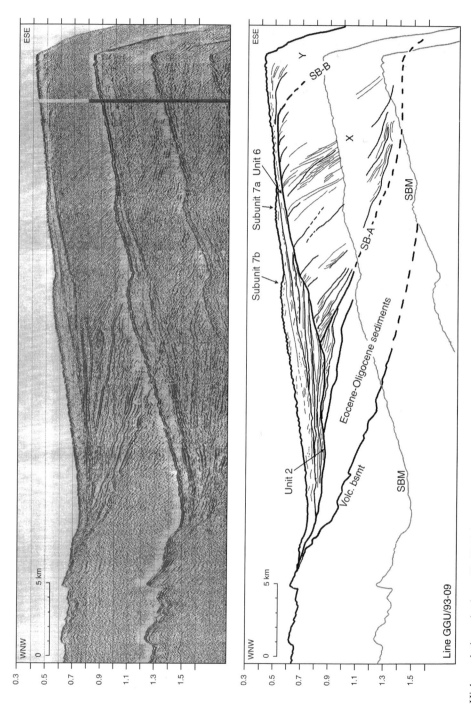

Fig. 7. High-resolution seismic profile 93-09 (for location, see Figs 2 and 15). The interpreted line drawing illustrates the seismic units recognized within Sequence y below the Gyldenløves Shelf Trough. Subunits 7a and b terminate seaward as ramps, and unit 2 shows small-scale seaward-dipping faint clinoforms. Note also slide scars at the uppermost slope. A parallel profile is illustrated in Fig. 14.

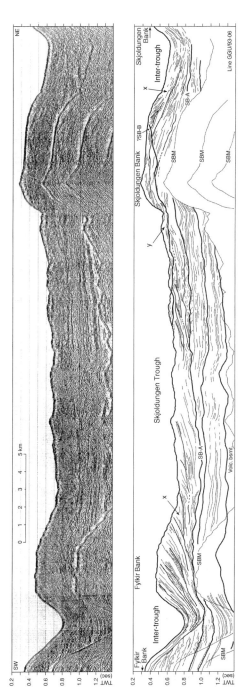

Fig. 8. High-resolution seismic profile 93-06 oriented approximately parallel with the shelf edge (for location, see Fig. 2). The interpreted line drawing shows the stratal architecture of Sequence x below the Skjoldungen Trough and the adjacent inter-trough areas. Note the very rugged sea-floor relief of the shelf trough.

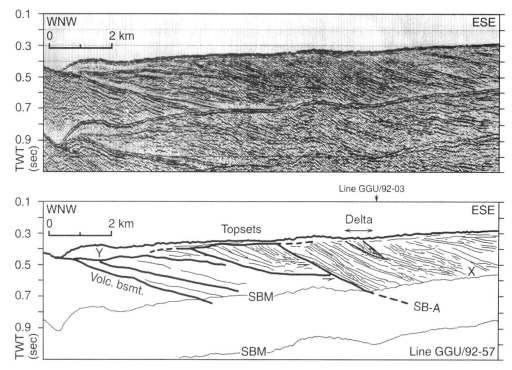

Fig. 9. High-resolution seismic profile 92-57 along the northern flank of the Gyldenløves Trough (for location, see Fig. 2). The interpreted line drawing illustrates the stratal architecture of Sequence x. Unconformity SB-A shows erosional truncation of gently dipping reflections below. Small-scale seaward-dipping clinoforms are interpreted as delta deposits (see text for discussion). The crossing line 92-03 is shown in Fig. 10.

to the main direction of shelf progradation) onto a clinoform showing synform configuration (Fig. 10).

Below the outer Gyldenløves Shelf Trough the oldest part of Sequence x connects to hummocky clinoform reflection configuration downdip of the prograding clinoforms (Figs 6 and 11), but their temporal relations cannot be resolved. The hummocky clinoform unit, about 150 m thick, is restricted to a relatively small area, about 5 km long and 8–9 km wide. Analogous facies have not been recognized beneath other shelf troughs in the area.

The main part of Sequence x consists of steep, densely spaced clinoforms that either downlap onto the hummocky clinoform unit or terminate shortly above it. Vestiges of an intra-sequence unit, like those recognizable in the oldest part, are occasionally discernible (especially near the base and the top of the clinoforms), but cannot consistently be followed. However, even though individual units cannot be mapped out, the overall architectural pattern is evident from seismic profiles located parallel to the shelf edge, revealing a systematic variation in dip direction

of the clinoforms relative to the sea-floor morphology (Fig. 8). Thus clinoforms below transverse shelf troughs appear as sub-horizontal strata that progressively dip away from the trough axis to meet in a synform configuration below the bank areas (inter-troughs).

The middle part of Sequence x is characterized by large-scale contortion (see Fig. 13, below). The deformation is centred below the outer Ilertakajik Shelf Trough, but can be followed to clinoforms under the southeastern part of the Gyldenløves Shelf Trough.

The total time-thickness of Sequences x and y has been mapped in the central and southern regions of the study area, revealing three elongate depocentres (Fig. 11), each located in front of a transverse shelf trough.

Sequence y

Unconformity SB-B at the base of Sequence y is a flat-lying or gently landward-dipping erosional surface that truncates reflections below (Figs 6 and 7). It passes into a conformity within the package of clinoforms beneath the outer shelf,

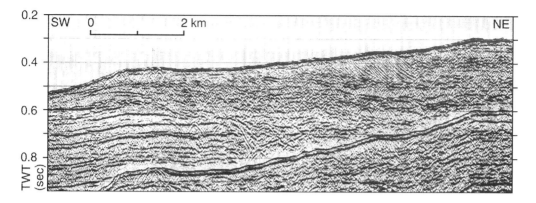

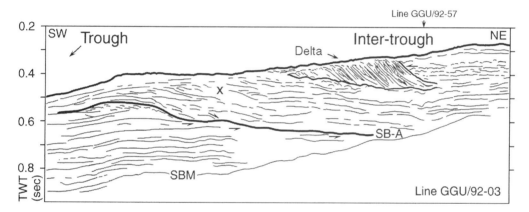

Fig. 10. High-resolution seismic profile 92-03 across the northern flank of the Gyldenløves Trough and adjacent inter-trough area (for location, see Fig. 2). The interpreted line illustrates the stratal architecture of Sequence x. Unconformity SB-A shows erosional truncation of southward gently dipping reflections below. Small-scale seaward-dipping clinoforms are interpreted as delta deposits (see text for discussion). The crossing line 92-57 is shown in Fig. 9.

but further downdip merges with a downlap surface immediately overlying the hummocky clinoform unit of Sequence x. Further offshore the amalgamated downlap surface comes close to the sea floor at the slope–rise transition and thence joins unconformity SB3 below the continental rise (Fig. 6).

SB-B outlines a buried depression below the Gyldenløves Shelf Trough, and can be followed northwards across the Heimland Bank to the Sermilik Shelf Trough, where it is truncated by the sea floor along the flank of the Sermilik Bank. Similarly, it is truncated by the sea floor along the northern flank of the Møsting Bank just south of the Gyldenløves Shelf Trough. The unconformity is absent in the Ilertakajik and Skjoldungen shelf troughs, or it may occur immediately below the sea floor (see Fig. 13, below). Below the southern flank of the

Skjoldungen Bank a major unconformity, buried beneath c. 200 ms of flat-lying strata, may represent SB-B (Fig. 8).

In the area south of the Gyldenløves Shelf Trough, with the Skjoldungen Bank as a possible exception, it is not possible to recognize the correlative conformity matching SB-B within the package of seaward-dipping clinoforms.

The shelf part of Sequence y consists of subhorizontal strata, up to 220 m thick, thinning towards the shelf break (e.g. Fig. 7). The succession passes into steep clinoforms approaching the present-day shelf break. Below the Heimland Bank and the Sermilik Trough the succession includes a number of overlapping buried shallow troughs (Clausen in press).

At least seven intra-sequence units, numbered 1–7 in ascending order, have been identified within the buried depression below the

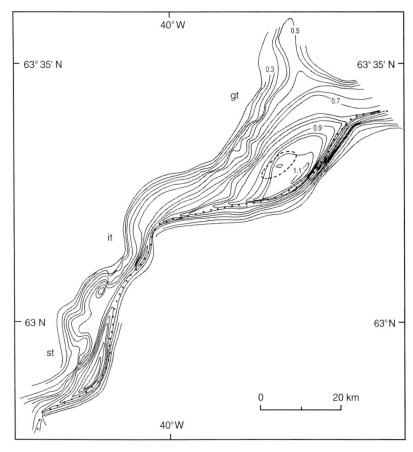

Fig. 11. Map showing the total time-thickness of Sequences x and y. Three depocentres are located in front of the Gyldenløves Trough (gt), Ilertakajik Trough (it), and the Skjoldungen Trough (st). The present-day shelf break is indicated by a dotted line and the hummocky clinoform unit at the base of Sequence x is marked by a dashed line. Contour interval is 100 ms.

Gyldenløves Shelf Trough area. They are separated by sub-horizontal to landward-dipping relatively smooth high-amplitude reflections that truncate reflections below. The lowermost five units, which are restricted to the deepest, outer part of the depression, are stacked in a southwestwards direction. The units show but few and discontinuous, faint internal reflections, and are generally seismically transparent. Rather indistinct small-scale seaward-dipping reflections have however been observed within unit 2 (Fig. 7).

Unit 6 is the only unit that can be followed from the mid- to the outer shelf, where it passes into steeply dipping clinoforms. Diffractions caused by iceberg scours in the sea floor obscure the internal seismic reflection pattern, but faint discontinuous sub-horizontal and landward-dipping reflections can occasionally be discerned below the outer shelf. Small-scale

seaward-dipping faint reflections have in one instance been observed to downlap onto the SB-B at or near the palaeo-shelf break (see Fig. 14, below). In the topset–foreset transition zone the toplapping and offlapping reflections are rather faint and slightly chaotic, but further downslope they become more distinct and exhibit high continuity. In some areas this reflection pattern passes into hummocky clinoforms at the base of the slope (Fig. 12).

Unit 7 is restricted to the Gyldenløves Shelf Trough, where it extends landwards about 5 km from the shelf edge to where the volcanic basement crops out on the sea floor. It wedges out laterally towards the adjacent bank areas (Fig. 15), but is believed to continue as a thin veneer draping the landward side of the Heimland Bank. Diffractions in general mask the internal seismic configuration, as in unit 6, but a faint semi-continuous landward-dipping reflection

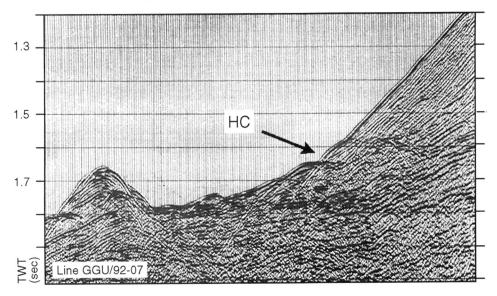

Fig. 12. Detail of high-resolution seismic profile 92-07 across the continental slope–rise intersection (for location, see Fig. 2). The line segment shows foreset strata of Sequence γ passing into hummocky clinoforms (HC), interpreted as debris flow lobes.

divides unit 7 into two subunits (*a* and *b* in Fig. 7). These subunits terminate seawards as ramps dipping up to 4°. A few nebulous reflections within subunit *a* dip parallel or subparallel to the ramp surface and downlap onto the high-amplitude, smooth basal reflection (Fig. 14). Subunit *b* is thinner and less extensive than subunit *a*, and the associated ramp is moderately to poorly developed.

Seismic characteristics of the continental rise

Three seismic sequences (α, β, and γ, in ascending stratigraphical order) have been recognized within the upper Neogene to Quaternary deep-sea sedimentary succession (Figs 16 and 17). The sequences are bounded by gently dipping regional unconformities, referred to as SB1, SB2, and SB3, respectively. Only the youngest Sequence γ, which is bounded upwards by the sea floor, forms a continuous unit throughout the study area. All sequences exhibit external mound forms (*sensu* Mitchum *et al.* 1977) that together make up the sediment ridges, but the internal reflection configuration differs (see Clausen (1998) for a more comprehensive description).

Sequence α

This sequence appears to be present in all ridges, but is thin in ridges I and II. It onlaps the lower

boundary (SB1) toward the continental slope and downlaps basinward. Externally, Sequence α displays symmetric mound forms, as much as 550 m thick, that makes up the nuclei of ridges III–V (Fig. 16). The internal reflection configuration is semi-chaotic with abundant diffractions (Fig. 18e). Locally this facies passes into a mounded-chaotic facies pattern (*sensu* Myers & Piper 1988) (Fig. 18d). Four units, 40–170 m thick, can be identified within the sequence ('subunits' of Clausen 1998). They are separated by internal discontinuities, characterized by high-amplitude hyperbolic diffractions (Fig. 18e).

Sequence β

This sequence, which is recognized within sediment ridges II–V but may also be present as a thin erosional remnant in ridge I, shows both onlap and downlap relations with the basal unconformity SB2. It displays external mound forms, asymmetric both in cross-section (Fig. 16) as well as longitudinally (Fig. 17). The mounded parts are as much as 450 m thick, but thin considerably across the crest of α (Fig. 16). Internally, the sequence is characterized by a mounded-stratified seismic facies (*sensu* Myers & Piper 1988), where low- to moderate-amplitude subparallel reflections converge toward the continental slope and toward the northeastern flank of the sediment ridges (Figs 16, 17, 18a

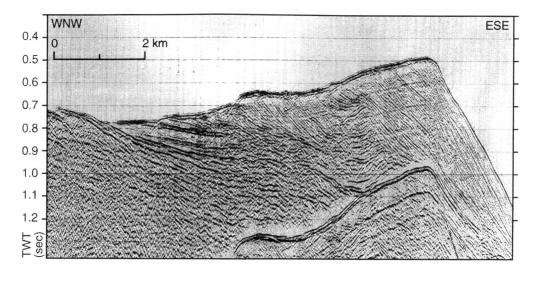

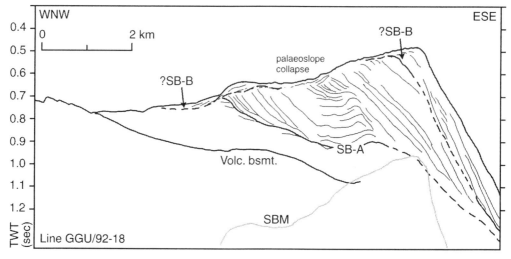

Fig. 13. High-resolution seismic profile 92-18 along the axis of the Ilertakajik Trough (for location, see Fig. 2). The interpreted line drawing illustrates the stratal architecture of Sequence x. Deformation of the clinoforms is interpreted as large-scale palaeoslope failure. Small-scale clinoforms are developed above the collapse structure. Unconformity SB-A shows erosional truncation of gently dipping reflectors below. The upper unconformity, SB-B, may be preserved immediately beneath the sea floor.

and c). Within ridges II and IV this facies is furthermore characterized by a well-developed cyclic pattern, comprising about ten units, each being about 30–70 m thick (Fig. 18c). The mounded-stratified facies passes into a migrating-wave seismic reflection configuration (*sensu* Mitchum *et al.* 1977) on the southwestern flanks of the sediment ridges (Fig. 16). A buried moat, correlative with the base of Sequence β, is identified below the present-day moat (Figs 17 and 19). A distinct prograding wedge is also

recognized on one seismic line at the base of Sequence β (Fig. 6).

Sequence γ

This sequence is present within all five sediment ridges in the study area. It has an external mound shape within the ridges, but also constitutes trough and basin fill (*sensu* Mitchum *et al.* 1977) in the bathymetric lows between the ridges (Fig. 16). The mounded forms are as

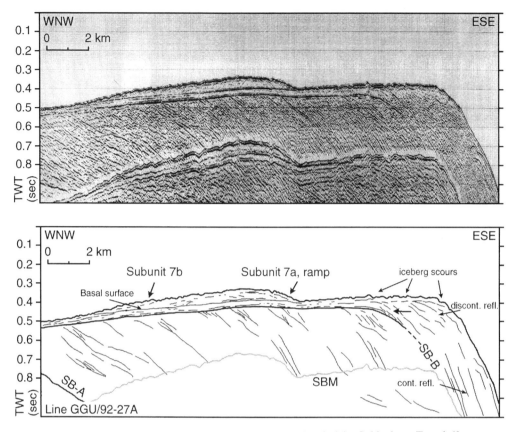

Fig. 14. High-resolution seismic profile 92-27A across the seaward end of the Gyldenløves Trough (for location, see Figs 2 and 15). The interpreted line drawing illustrates the topset strata of Sequence y. Subunit 7a shows a well-developed ramp. Internal indistinct reflections, dipping parallel to the ramp surface, downlap onto a high-amplitude smooth reflection. The sea bed, however, is characterized by intensive iceberg scouring. Subunit 7b is thin and exhibits a poorly developed ramp. Small-scale clinoforms within unit 6 downlap onto the SB-B unconformity at the palaeoshelf break. A parallel profile is illustrated in Fig. 7.

much as 460 m thick, but thin to 75–150 m across the crest of the underlying Sequence β.

A migrating- and climbing-wave seismic reflection configuration characterizes the mound forms. Relatively steep waves are developed on the southwestern and northwestern sides of the ridges (Fig. 18b). The wave crests on the southwestern flanks are subparallel to the ridges and are thus approximately perpendicular to the modern contour current flow. Irregular, flat waves are identified on the northwestern flanks and across the crests of the ridges. Toward the continental slope the reflections of the mounded form are either truncated by the present-day moat towards the northwest (Fig. 20), or converge and show a buried moat (Fig. 19).

Within the moat zones mounds are reflected by the sea-floor relief. Smaller mounds, whose bases are level with the surrounding sea floor, exhibit a relatively simple internal structure with parallel and subparallel reflections (Fig. 21). Large mounds, extending down to several hundreds of metres below the sea floor, show an internal aggrading and migrating-wave reflection configuration (Fig. 22). Some mounds, dissected by up to 150–200 m deep furrows, display irregular shapes (Fig. 22).

Sequence γ is characterized by subparallel and parallel reflection configurations within the trough or basin fill between the ridges. Buried troughs, about 10–15 km wide, are located beneath the modern major sea-floor channels north of ridges III and IV (Fig. 23). The fill of the basin southwest of ridge II contains numerous buried subsidiary channels, which are located beneath or adjacent to the channels exposed on the sea floor (Fig. 24). There is evidence of

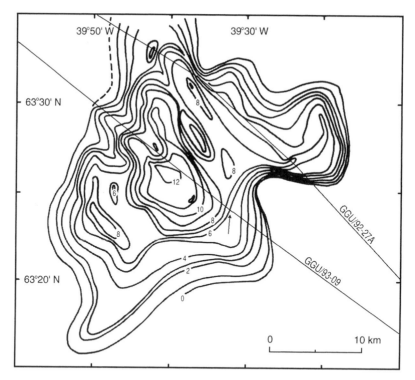

Fig. 15. Map showing the time-thickness of unit 7 (Sequence y) below the Gyldenløves Trough (for location, see Fig. 2). Contour interval is 10 ms. Subunit 7a is characterized by two lobe-shaped ramp terminations. Subunit 7b is less extensive and shows ramp termination only in the central part of the trough (indicated by an arrow). The locations of seismic lines 93-09 (Fig. 7) and 92-27A (Fig. 14) are indicated.

lateral channel migration, generally towards the northeast. The associated levees appear acoustically well layered and consist of regular, semi-continuous, parallel reflections of medium to high amplitude (Figs 23 and 24). Both recent and buried levees are characterized by numerous internal low-angle unconformities.

Correlation with ODP Leg 152

The seismic units and sequences can be correlated with ODP Leg 152 sites 914–919 (Larsen *et al.* 1994*b*). Of these, the shelf drill sites 914–917 penetrated seismic units 2, 3, 7 and possibly 4 of Sequence y, in addition to Oligocene and Eocene sediments. Sequence x is not present at the shelf drill sites. At Site 914 gravel of presumed Quaternary age with no matrix preserved was recovered from levels corresponding to seismic units 2–?4 (lithologic unit 1C). It is followed by a compacted diamictite (lithologic unit 1B), about 1 m thick, dated as Quaternary (for bio- and lithostratigraphy, see Larsen *et al.* 1994*b*); this interval may correspond to the upper part of seismic subunit 7b. A 5 m thick

succession of glaciomarine mud (lithologic unit 1A) immediately below the sea floor may correspond to the uppermost part of subunit 7b. In this succession two cooling periods and an intervening interglacial period are inferred from benthic foraminifer assemblages (Larsen *et al.* 1994*b*). The upper 1.5 m (core 152-914A-1H-1 in Larsen *et al.* 1994*b*), which includes the youngest glacial interval, is correlated with oxygen isotope stages younger than stage 4 (80 ka), whereas the uppermost veneer consisting of sandy mud, about 30 cm thick, is interpreted as post-glacial sediments. At Site 916 five intervals of washed gravel (lithologic unit 1C) seem to match levels in seismic subunit 7a. Three intervals of compacted diamicton succeeded by a level of washed gravel (lithological units 1B and 1A) correspond to seismic subunit 7b. The compacted diamicton (lithological unit 1B) was interpreted as a lodgement till and the gravel intervals (lithological units 1A and 1C) as ice-rafted dropstones washed out from glaciomarine mud (Larsen *et al.* 1994*b*). Alternatively, the washed gravels could originate from deformed till deposits, an interpretation that would be in

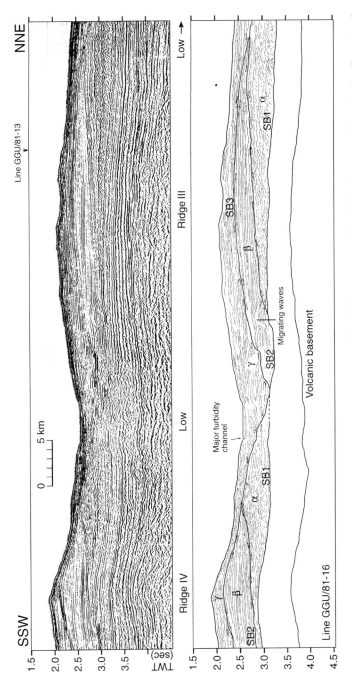

Fig. 16. Deep seismic profile 81-16 crossing sediment ridges III–IV on the continental rise (for location, see Fig. 2). The interpreted line drawing illustrates the morphology and internal architecture of the ridges and the interjacent channel–levee system. The 30–40 km wide and 50–60 km long ridges display a relief of as much as 450 m. The ridges comprise three seismic sequences (α, β, and γ) separated by unconformities SB1-3. The crossing line 81-13 is shown in Fig. 17; a detail of Sequence α (right-hand side, ridge III) is illustrated in Fig. 18e.

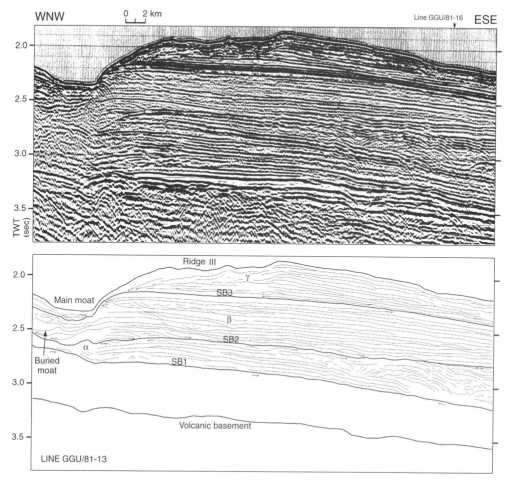

Fig. 17. Deep seismic profile 81-13 oriented parallel to the crest of sediment ridge III (for location, see Fig. 2). The interpreted line drawing illustrates the morphology and internal architecture of ridge III and the moat at the base of the continental slope. Sequences α, β, and γ and their bounding unconformities SB1, SB2, and SB3 are outlined.

agreement with the seismic data. The crude biostratigraphy does not constrain the age of the till-delta deposits (unit 7), and it remains unclear whether the latest grounded ice on the Southeast Greenland Shelf was Weichselian, Saalian or pre-Saalian.

There is no direct stratigraphic control on Sequence x or the remaining part of Sequence y. However, unconformity SB-A correlates, as far as can be established, with unconformity SB2 of the deep sea, which provides a tie to ODP Site 918. Hence the early part of Sequence x is

Fig. 18. Seismic facies configurations of the sediment ridges illustrated by high-resolution seismic profile 93-02 (a–d, ridge IV) and detail of deep seismic profile 81-16 (e, ridge III) (for location, see Fig. 2). (a) Crossing line showing spatial relations of the three sequences within ridge IV. White frames indicate location of enlargements (b)–(d). (b) Upslope migrating seismic-wave reflection configuration characteristic of Sequence γ (wavelength 1.5–3.5 km, height 10–40 m, and apparent angle of climb 1.0–1.5°). (c) Mounded-stratified reflection configuration characteristic of Sequence β. Low- to moderate-amplitude subparallel reflections converge rapidly to the north and west. (d) Mounded-chaotic facies pattern characteristic of Sequence α. (e) High-amplitude hyperbolic diffractions characteristic of Sequence α. The white bars indicate the four intra-sequence units recognized. A larger segment of line 81-16 is shown in Fig. 16.

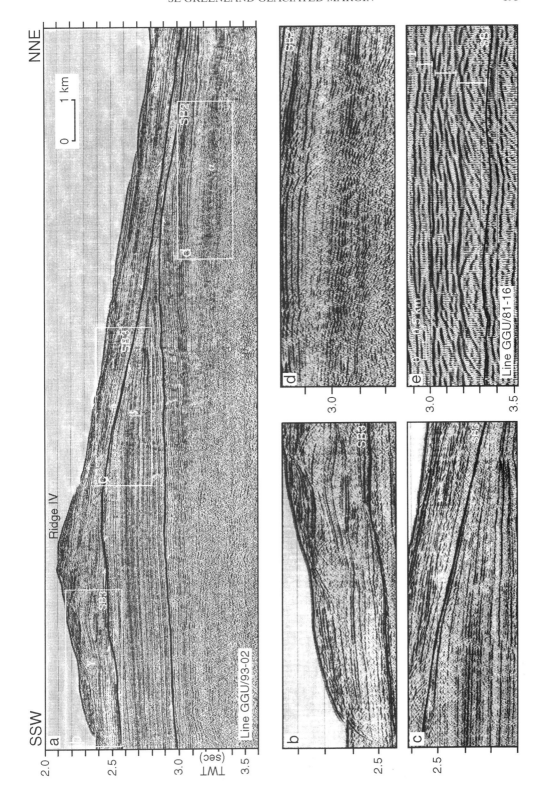

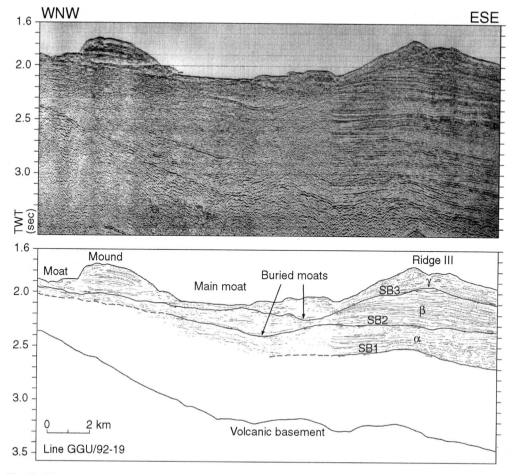

Fig. 19. High-resolution seismic profile 92-19 across the moat zone (for location, see Fig. 2). The interpreted line drawing illustrates that the main moat is developed as a shallow depression (compare Fig. 20), flanked basinwards by sediment ridge III, comprising Sequences α, β, and γ. High-amplitude reflections below the modern moat are interpreted as buried moats. A large mound and a smaller moat are located toward the continental slope. The elongate mound parallels the slope and exhibits an asymmetric cross-section similar to that of the sediment ridges.

probably of early Pliocene age, whereas the main part of Sequence x plus Sequence y, which appears to correlate with Sequence γ, is of mid-Pliocene to Pleistocene age.

Site 919, situated on the lower continental rise in the seaward end of the study area, penetrated 147 m of Sequence γ. Site 918, located on the upper continental rise, penetrated Sequences γ (channel–levee complex) and β. The drill site, however, is situated between two ridges and thus the internal reflection configuration of the ridges cannot be compared with the borehole data. Sequence boundary SB3 is of mid-Pliocene age and located immediately above the base of lithological unit IC, which marks a significant increase in the sedimentation rate (Larsen *et al.*

1994*b*). Sequence boundary SB2 is of early Pliocene age, corresponding to a level somewhat above the first occurrence of dropstones (Larsen *et al.* 1994). Sequence α is not present at the drill site, but sequence boundary SB1 overlies an unconformity that at Site 918 is positioned at the middle–upper Miocene boundary. Hence Sequence α is late Miocene to early Pliocene in age, and thus postdates the mid-Miocene onset of the NADW (Larsen *et al.* 1994*b*).

Interpretations

Interpretations of the youngest Sequences y and γ have generally been used as templates to infer the sedimentary processes involved in forming

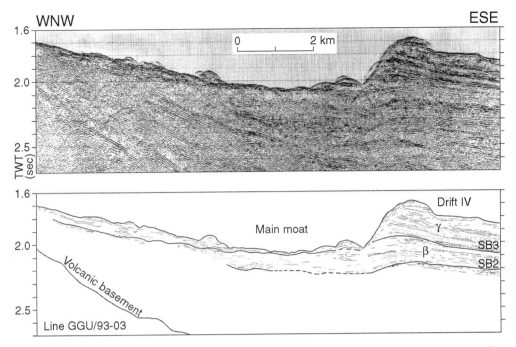

Fig. 20. High-resolution seismic profile 93-03 across the moat zone (for location, see Fig. 2). The interpreted line drawing illustrates a deeply entrenched main moat with rather small mounds (compare Fig. 19). The internal structure of the mounds is obscured by hyperbolic diffractions.

the underlying sequences, and for this reason the units are dealt with in reverse stratigraphical order. This approach is employed as the primary external architecture of the youngest units is more intact, and the internal structure is far better resolved by the high-resolution seismic data.

The seismic facies analysis is inspired by the work of Mitchum *et al.* (1977) and Myers & Piper (1988), whose terminology is adopted.

Shelf and slope: Sequences x and y

In the shelf area, unconformity SB-B generally truncates the underlying clinoforms and represents one or possibly several major erosional events. Its landward-dipping orientation and trough-like geometry, especially below the Gyldenløves Trough, are taken as evidence for erosion by grounded ice. The origin of the older unconformity SB-A is not clear, but no features suggestive of glacial erosion have been recognized (for further discussion, see next section).

Sequence y, showing little aggradation and strong progradation, is similar to the type IA sequence recognized below the Antarctic margins (Cooper *et al.* 1991), and considered deposited in association with multiple grounded

ice events on the shelf (see also Larsen 1994; Larsen *et al.* 1994*a*).

The six high-amplitude reflections each truncating reflections below, identified within the succession of topset beds of Sequence y, are also interpreted as surfaces eroded by grounded ice. Their rather smooth appearance suggests that they have not been furrowed by icebergs like the present-day sea bed. Some of the seismic units between these erosional surfaces are rather thin and show either transparent or faint and discontinuous sub-horizontal reflections (Fig. 7), and may represent deformed basal till layers like those observed at the base of ice stream B in the Ross Ice Shelf (Alley *et al.* 1987, 1989; Blankenship *et al.* 1987). The indistinct sub-horizontal reflections and the associated small-scale clinoforms within the otherwise seismic transparent units 2, 6, and 7 are interpreted as topset and foreset beds of small-scale delta-like features (Fig. 7). The facies configuration is suggested to represent stacked successions of deformed till layers deposited beneath grounded icestreams, and associated grounding line 'till deltas' following Alley *et al.* (1987, 1989). Unit 7 is actually considered to represent two such till deltas with more or less intact surface morphology, an interpretation strongly corroborated by the

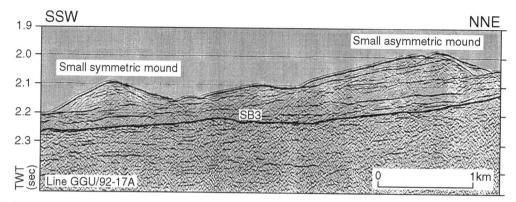

Fig. 21. High-resolution seismic profile 92-17A across the moat zone (for location, see Fig. 2). The line shows small, seismically well-layered mounds exhibiting symmetric and asymmetric cross-sections.

presence of small-scale prograding clinoforms within subunit a. Unit 7 apparently signals two 'conveyor belt' grounding-line advances, during which sediments at the upstream end of the delta were recycled to be deposited at the downstream end as foreset beds (compare Powell 1984). The indistinctness and/or lack of reflections within subunit b and parts of subunit a may indicate a very homogeneous sediment supply, as suggested for the 'subglacial delta' described by Larter & Vanneste (1995).

Another delta-like morphological element is observed on the landward side of the Heimland Bank, on the basis of a single channel seismic

profile (Tycho Brahe line 74-36, Fig. 2). If this structure is part of the same system as unit 7, the grounding-line delta extends for more than 30 km. Similar morphological features have been described from the shelves off the Antarctic Peninsula ('subglacial deltas' of Vanneste & Larter 1995), Wilkes Land ('lateral moraines' of Barnes 1987; Domack, 1987; Eittreim *et al.* 1995), and Prydz Bay ('grounding line moraine' of O'Brien 1994). The last two are considered to be more than 50 km long.

The discontinuous reflections characterizing the shallow part of the large-scale clinoforms is taken to represent small-scale slide scars of the

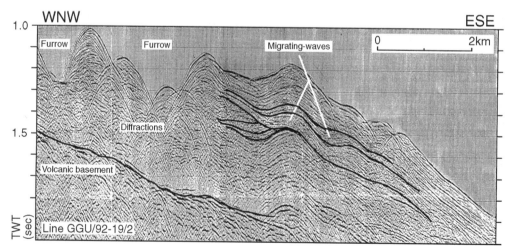

Fig. 22. Detail of high-resolution seismic profile 92-19/2 across the moat zone (for location, see Fig. 2). The line segment crosses a relatively large mound displaying aggrading reflection configuration with numerous internal unconformities. The thicker upslope limbs suggest some upslope migration (indicated by white lines). The upslope part of the mound is dissected by 150–200 m deep erosional features and hyperbolic diffractions obscure the reflection image.

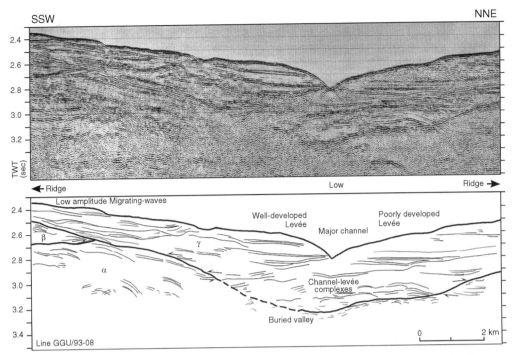

Fig. 23. High-resolution seismic profile 93-08 across channel–levee complexes on the continental rise (for location, see Fig. 2). The interpreted line drawing shows a buried valley, *c.* 10 km wide, underlying a modern turbidity channel, and filled in by channel–levee deposits representing Sequence γ. Note the asymmetric pattern of the levees flanking the modern turbidity channel (see also Fig. 24).

palaeo-shelf edge and upper slope. Such features are seen in many places at the present-day shelf edge. The more continuous reflections of the slope are interpreted as stacked, steeply dipping debris flow deposits. Some extend to the base of the slope as snout-shaped units (hummocky clinoform facies).

The shelf prograded along its entire length, but strongest so in front of the shelf troughs (Fig. 11). The elongate shape of the depocentres reflects deposition from multiple 'line sources' (cf. Haugland *et al.* 1985; Kuvaas & Kristoffersen 1990; Larter & Cunningham 1993). In the central and southern part of the study area the positions of depocentres seem to have been more or less permanent during progradation of the shelf (Fig. 8), whereas they shifted laterally in the Heimland and Sermilik area (Clausen in press).

Sequence x differs from Sequence y by lacking topset beds. It is, however, considered likely that such beds were removed by glacial erosion during formation of SB-B. Except from the most landward part, the preserved foreset segments of Sequence x are exceedingly similar to that of Sequence y, comprising internal small-scale clinoforms and steeply dipping intra-sequence

unconformities. Hence the main part of Sequence x is likely to have been deposited by the same processes as outlined for Sequence y.

The oldest clinoforms of Sequence x differ by showing a lower seaward dip, more prominent intra-sequence unconformities and conspicuous internal small-scale dipping reflections. The latter are interpreted as representing delta deposits (Figs 9 and 10). The steep, well-defined foreset beds are not typical for a till delta (Alley *et al.* 1989; Larter & Vanneste 1995; Vanneste & Larter 1995), but rather point to a well-sorted sediment. Besides, the delta appears to be laterally restricted, indicating an origin from a point-source, and the deposit is suggested to represent a grounding-line fan formed by melt-water (compare Powell, 1984, 1990); see also discussion below.

The transition from low- to high-dipping large-scale clinoforms is gradual, suggesting a general shift in the depositional environment from glaciomarine and meltwater dominated to subglacial conditions.

The hummocky clinoform unit at the base of Sequence x is tentatively interpreted as a fan deposit consisting of channelized turbidites

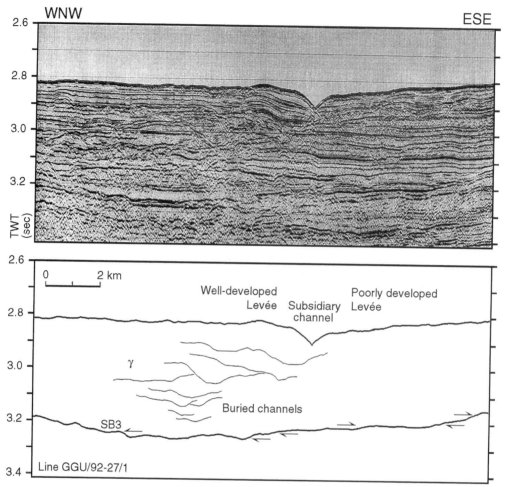

Fig. 24. High-resolution seismic profile 92-27/1 across channel–levee complexes on the continental rise (for location, see Fig. 2). The interpreted line drawing illustrates lateral migration of buried channels located beneath a modern turbidity channel. Note that the western levee of the modern channel is better developed than the eastern counterpart. This asymmetric pattern suggests preferential deposition on the western levee as a result of the effect of the Coriolis Force.

(cf. Macurda 1993). Individual channels cannot be resolved, but the facies may reflect rapid deposition of sandy sediments along the channel axis that were less subject to compaction. The deposit is restricted to a depression with a width-to-depth ratio of about 60:1, which may represent a topographic low at the end of a canyon.

Continental rise: Sequences γ, β, and α

Sequence γ is interpreted as sediments deposited mainly from turbidity and contour current processes. The inter-ridge trough and basin fill represents aggrading channel–levee complexes within buried valleys. The levees are characterized by abundant internal unconformities,

indicative of repeated overbank deposition with intermittent nondeposition and/or erosion. This feature and the orientation of the valleys strongly suggest a turbidity current origin, sourced from the trough-mouth fans (Clausen 1998).

The moat zone, comprising alongslope channels interspersed with mounds, is the most compelling evidence for contour current erosion and deposition on the continental rise. Buried moats are indicated by high-amplitude, convex-up reflections beneath the main moat (Fig. 19). The complex larger mounds within the moat zone consist of aggrading and/or upslope migrating sediment waves (Fig. 22). Seawards, smaller mounds imperceptibly merge with the levees flanking the turbidity channels. The orientation of prominent furrows in some of the large mounds (see Fig. 22) is indeterminable, and thus it cannot be decided whether downslope or alongslope erosional processes are responsible for their formation.

Within the sediment ridges the migrating-wave seismic reflection configurations of sequence γ are interpreted as upslope migrating sediment waves generated mainly by contour currents, although turbidity current may also have contributed (see Clausen 1998, and references therein). It is envisaged that sediment transported down the main turbidity channels was picked up by the contour current and thence largely bypassed the northwestern flanks as low-amplitude sediment waves, to be deposited on the lee side as steep sediment waves.

The increase in seismic facies complexity from Sequence β to γ is inferred to reflect a shift from comparatively uniform to more fluctuating depositional conditions. The internal reflection configuration and the asymmetric outer geometry of Sequence β indicate a sediment supply from the northwest (along the base of the continental slope) and from the northeast (along major turbidity channels), as in Sequence γ. The convergence of strata toward the moat zone and below the interjacent buried valleys suggests sediment bypass across the northwestern and northeastern sides of the ridges with preferential deposition on the southeastern and southwestern sides. A buried moat matching the base of Sequence β provides support of contour current activity, conceivably associated with deposition in the tranquil zone basinwards of the current core (Clausen 1998). By comparison with Sequence γ it is also envisaged that major turbidity channels, each acting as a sediment conduit, were present along the northeastern flanks of the ridges, but they are not preserved in the record. Changes in sediment supply are possibly responsible for the cyclic seismic pattern of Sequence β.

By analogy with Sequence γ the migrating-wave seismic reflection configuration is interpreted as sediment waves formed on the lee side of the ridge by a southward-flowing contour current.

Although Sequence α makes up the nucleus of the sediment ridges III–V it is not believed to have been deposited by contour currents. The lack of buried moats, the internal facies distribution, and the symmetric external shape point to a more likely origin as basin floor fans. Small slump-related faults or minor channels may be responsible for the diffractions characterizing the top of each unit within the sequence (Fig. 18e). Basin floor fans are normally associated with sea-level lowstands, and Sequence α may

reflect the second-order late Miocene lowstand indicated by Haq et al. (1987). The idea could even be entertained that the four intra-sequence units match the four third-order cycles during that period (cf. Haq et al. 1987), but more precise dating is required to investigate this notion.

Profile 93-19 (not shown), crossing ridges I and II, indicates that a succession, presumably representing Sequence α, actually is present within these ridges; it is as much as 200 ms thick. However, no spatial relations can be deduced from the sparse data.

Discussion and conclusions

Numerous features of Sequence y, notably the shelf bathymetry including reverse-slope profile, the till deltas, the broad trough-shaped, landward-dipping erosion surfaces, and the location of elongate depocentres in front of transverse shelf troughs, indicate that deposition on the Southeast Greenland margin during the Pleistocene was controlled by grounded ice events. The close resemblance of the foreset beds in Sequence x to those of sequence y makes it likely that most if not all of the clinoforms of the former were deposited under broadly similar conditions, i.e. related to shelf-edge progradation in front of grounded ice, suggesting periodic glaciations on the shelf since the early Pliocene. The likely correlation with the deep-sea Sequences γ and β, characterized by high sedimentation rates, diamictites and dropstones (Larsen et al. 1994a), corroborates this interpretation. It is stressed that progradation took place in front of both bank areas and shelf troughs, producing an essentially uniform architecture that reflects dissimilar rates of progradation rather than different depositional styles. This pattern is consistent with the line source model (e.g. Haugland et al. 1985; Kuvaas & Kristoffersen 1990; Bartek et al. 1991; Cooper et al. 1991; Hiscott & Aksu 1994).

The oldest clinoforms of Sequence x differ by showing a lower seaward dip, more prominent intra-sequence unconformities and, locally, conspicuous internal small-scale dipping reflections. It is uncertain what these differences signify in terms of depositional environment, but they might reflect melt-water dominated glacial conditions during the initial phase of the early Pliocene cooling period. The internal small-scale clinoforms are interpreted as representing delta deposits (Figs 9 and 10). The delta is suggested to represent a submarine grounding-line fan formed by melt-water, but a glacio-fluvial origin cannot be excluded. The tangential terminations of the upper reflections in the outer part of the delta indicate that little erosion has occurred. Thus the top of the delta roughly equates to the palaeo-depositional base level, which at the present day is situated at about 230 m below sea level. If the delta is of glacio-fluviatile origin a thermal subsidence rate of at least 50 m/Ma during the last 3–4 Ma must be invoked, otherwise the delta appears to be beyond reach of subaerial conditions, even during sea-level lowstands. This hypothetical figure is significantly higher than the poorly constrained subsidence rate calculated by Larsen *et al.* (1994*b*, pp. 174–175), supporting the inference of a submarine environment. The small delta is difficult to pick out on the crossing dip-line (Fig. 9), and similar deposits elsewhere in the area may have escaped notice in the absence of intersecting strike-lines.

The origin of unconformity SB-A is uncertain. Analogous buried unconformities at the Antarctic margins were ascribed by Cooper *et al.* (1993) and Eittreim *et al.* (1995) to grounded ice erosion. These authors suggested that initial advance(s) of grounded ice sheet(s), capable of eroding up to 500 m of strata, produced these basal unconformities over which the later glacial sequences prograded. However, their interpretation addresses only the overdeepening of the palaeoshelf and does not account for the associated glacial deposition that inevitably must have taken place further offshore. Thus their initial 'erosional phase' differs fundamentally from the erosional–depositional scenario inferred for the overlying glacial sequences, characterized by extensive shelf progradation. In the present case the absence of noticeable shelf progradation matching unconformity SB-A is taken to suggest that it represents a gently seaward-dipping palaeorelief surface modified by erosion, locally as much as 100 m. Either or both the hummocky clinoform unit of Sequence x and the prograding wedge of Sequence beta may represent the corresponding basin-floor fan deposition. Accordingly, it may have been brought about by a lowering of the base level, plausibly a drop of sea level. A potential candidate is the *c.* 100 m lowering of the sea level at 3.8 Ma (cf. Haq *et al.* 1987), and it is obviously tempting to suggest a connection between the sea-level lowering and the first glacial advance across the shelf, but this issue is beyond the scope of the present paper.

Unconformity SB-B, interpreted as formed by glacial erosion, marks a shift from progradation to progradation combined with aggradation. After its formation, more than 200 m of topset beds, representing several glacial depositional cycles, were preserved on the shelf. A corresponding development has been observed offshore Kangerdluggsuaq and Scoresby Sund (Larsen 1994; Vanneste *et al.* 1995), but dating is not available for these areas (for location, see Fig. 1). Broadly similar changes in stratal architecture have also been described from the Home Bay transverse trough in the Baffin Bay (Hiscott & Aksu 1994) and from the Bear Island Trough, northwest Norway (Vorren *et al.* 1991; Sættem *et al.* 1992), both dated as early to mid-Pleistocene. The global nature of this shift in stratal architecture has been suggested to reflect modifications of icestream dynamics, possibly related to climatic changes and glacio-eustasy (Bartek *et al.* 1991; Vanneste *et al.* 1994).

Aggradation on the Southeast Greenland Shelf is restricted to trough areas (Gyldenløves and Sermilik Troughs as well as buried troughs below the Heimland Bank), strongly indicating that deposition on the shelf was intimately linked to the icestreams. The topset strata wedge out toward the surrounding bank areas (Sermilik and Møsting Banks). The shallow banks are ascribed to reduced erosional capacity of thin, slow-flowing ice, flanking fast-flowing thick icestreams during the advancing phase. The lack of topset beds below the bank areas (and some of the troughs) reflects nondeposition during retreat of the ice (compare discussion of the till deformation model in the preceding section). Flat-lying strata occur, however, on the Heimland and Skjoldungen Banks, but actually these successions were probably deposited within shelf troughs (Clausen in press). The unconformity at the base of the flat-lying strata below the Skjoldungen Bank (surmised to represent SB-B) is slightly concave-up, and is interpreted as the base of a former shelf trough, which partly has been eroded away by the younger and still preserved trough located slightly to the south.

The outer bank areas were suggested to represent terminal moraines by Sommerhof (1973), an interpretation later cited by Funder & Larsen (1989, p. 778). There is little question that the glaciers did reach the shelf edge, but the outer bank areas cannot *per se* be classified as terminal moraines.

Deposition in troughs and non-deposition on banks is an observation at variance with the interpretations of Cooper *et al.* (1991) and Eittreim *et al.* (1995), based on data from the Eastern Ross Sea, Prydz Bay and Wilkes Land Shelf. They inferred that extensive outer shelf banks formed by aggradation of topset strata, whereas trough areas were considered sedimentary bypass zones. This fundamental difference bears further investigation.

The structure of the wide Heimland Bank is complex compared with other banks, and at least one buried low-relief trough appears present, suggesting that a former pathway for icestreams was located here (matching deep-sea deposits are discussed below). The shift in location of shelf troughs in the northern part of the study area is unlike the situation further south (central and southern part of the study area), where the positions of shelf troughs were relatively fixed (Clausen in press).

The small-scale clinoforms below the ramp surface of unit 7 in Sequence y downdip on a smooth reflection, which is taken as evidence that till-delta progradation took place immediately after the surface had been smoothened by erosion and compaction by a grounded ice. It is envisaged that subunits a and b reflect till deltas deposited by two readvances of the icestream grounding line during the same waning stage. The underlying stacked succession of landward-dipping strata might consist of reworked late stage till deltas and basal tills. The presence of till deltas has been suggested to indicate deposition by low profile fast-flowing grounded ice (Vanneste & Larter 1995); such icestreams would have markedly reduced erosional capacity, thereby allowing for the preservation of older topset beds. No precise biostratigraphic dates are available for these deltas, representing the latest glaciation event.

The very top of unit 7 comprises Holocene glaciomarine deposits (ODP Leg 152; Larsen *et al.* 1994*b*). Similar intervals are likely to have formed during interglacial and perhaps especially during late deglacial periods also in the past, but there is no seismic evidence for glaciomarine deposition within the topset segment of Sequence y. Hence such deposits or units are either below seismic resolution (i.e. less than 10 m thick) or have been remoulded into deformed basal tills.

Till deltas deposited at the shelf edge when grounded ice reached the palaeoshelf edge were subsequently subjected to downslope redeposition. Modern scars in the uppermost part of the continental slope show that slumps and slides currently are involved in the initial downslope sediment transport. The discontinuous reflection facies pattern that characterizes the upper part of some clinoforms (Fig. 14) may represent similar scars near the palaeoshelf edge. The absence of canyons (and palaeocanyons), and the scarcity of gullies at the upper slope suggest that unchannelized debris flows probably were the main process by which the slope prograded. Debris flow lobes at the downdip end of some of the clinoforms (Fig. 12) support this interpretation.

The unusually steep and smooth continental slope in the study area is considered to be related to erosion by strong contour currents, whereas the nature and supply of sediment appear to be of secondary importance (Clausen 1998). The circumstance that the continental slope steepens significantly south of latitude 65°N, where the deep-water transport rapidly increases because of entrainment of the Labrador Sea Water (Fig. 5a), corroborates the importance of contour current erosion. The circumstance that the steepest slope segments are developed at the trough-mouth fans (Figs 3 and 4) may simply reflect that these protruding features are subject to stronger erosion by contour currents. By analogy with modern conditions, the steeply dipping palaeoslope (clinoforms) of Sequences x and y is ascribed to erosion and maintenance by strong contour currents.

The position of the moat zone coincides with the location of the high-velocity core of the present-day LSW contour current. It is inferred that the erosion was (and is?) associated with deposition in the adjacent basinward, comparatively tranquil zone, resulting in the formation of sediment ridges (Fig. 25).

Analogues to the modern ridge–moat and channel–levee complexes of the continental rise are preserved also in the subsurface, indicating that this depositional scenario became established at about the early Pliocene.

The buried valleys and major turbidity channels are without exception adjoining present or former shelf trough–trough-mouth-fan systems, indicating that the downslope sediment transport on the continental rise was linked to depositional processes on the shelf. It is envisaged that small-scale slope-failure, particularly in front of the shelf troughs, where sediment supply was highest, led to debris flows on the slope, eventually passing into turbidity currents on the continental rise (Fig. 25). This scenario implies that turbidity-current activity was intensified at times when grounded ice reached the shelf edge because of the associated high or enhanced clastic supply to the slope.

The location, morphology and architecture of the sediment ridges strongly indicate that they are genetically associated with the downslope processes. The ridges are all located immediately south of major turbidity channels, and it appears that the channelized turbidity flows were affected by both the southward-flowing contour current and the Coriolis force, resulting in preferential deposition on the southwestern overbank.

The Ilertakajik Shelf Trough system apparently lacks a corresponding sediment ridge. However, it is conceivable that Ridge III, showing an atypical arcuate shape (Fig. 4), at

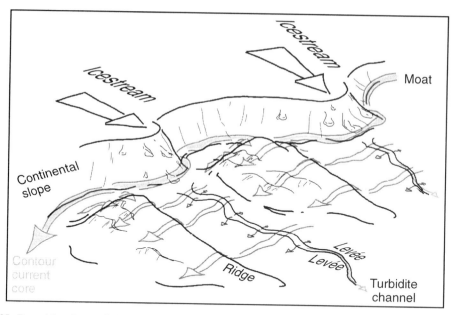

Fig. 25. Depositional scenario for the deep sea during glaciation showing the link between glacial, turbidity and contour current deposition. High sedimentation rates in front of icestreams (large arrows) caused sediment failures and debris flows on the continental slope, in turn generating turbidity currents on the continental rise. The turbidites were subsequently winnowed by contour currents (large grey arrows) and redeposited on large ridges. Sediment waves developed on the lee side of the ridges. A moat zone formed along the base of the continental slope below the high-velocity contour current core.

some stage was sourced also from this system. The associated major turbidity channel, which now exhibits an unusual southwards bend, may have been deflected to its present course, where it became a tributary to the Skjoldungen major turbidity channel, perhaps as a result of lower turbidity current activity in front of the Ilertakajik TMF. The rather insignificant size of the modern Ilertakajik Shelf Trough system is consistent with the conjecture that this system at some stage became low-active or was even abandoned.

The large-scale palaeoslope collapse located below the Ilertakajik Shelf Trough (Fig. 13) attests that one or perhaps several major failures occurred, which probably had an impact on the deep-sea sedimentary environment, too. However, whether or not the slope breakdown was connected to the abandonment of the system remains obscure.

The age relations and architecture of Sequences γ and β of the sediment ridges match those of Units 1 and 2 of the Eirik Ridge in the Labrador Sea (cf. Arthur *et al.* 1989). The onset of contour current deposition in the western Irminger Basin is comparatively late viewed in a North Atlantic context (Clausen 1998), but coincides with a generally intensified drift deposition

throughout the region (cf. Miller & Tucholke 1983). The older Sequence α displays no obvious features indicative of contour current activity, and is interpreted as consisting of basin-floor fans. It thus appears that the early phases of NADW, established in the mid-Miocene (Larsen *et al.* 1994*b*), were not associated with local contour current related deposition, which raises the question of which factor or set of circumstances controlled the initiation of deposition on the ridges.

There is a remarkable coincidence between onset of profoundly increased shelf progradation and start of deposition from contour currents, pointing to sediment availability as one determinant factor. It is conceivable that the greatly enhanced sedimentary supply to the shelf-slope environment, associated with the first period of grounded ice to the shelf break, also generated intensified turbidity current activity across the continental rise, as noted above. The large amounts of sediment carried in suspension via the turbidity channels to the deep sea may then have been winnowed or directly picked up by the contour current and eventually deposited as sediment ridges (Fig. 25). However, changes in the deep-sea currents may have been a contributing factor, as Sequence α does not exhibit features characteristic of contour current processes. It is of course possible that the currents were incapable of reworking unchannelized mass-flows, or maybe deposition did not commence until these fans provided a lee-effect favourable for accumulation of winnowed sediment, but these restraints are not considered likely. The erosional event ('R2') before construction of the Eirik Ridge (Arthur *et al.* 1989) was interpreted by Bohrmann *et al.* (1990) as reflecting a current intensification possibly linked with large-scale changes in the deep-sea current pattern related to the closure of the Panama Strait. Hence the onset or intensification of the LSW in the Irminger Basin may be an important contributing factor facilitating contour current related deposition in the study area (and at the Eirik Ridge). As discussed above, the unusual steepness of the SE Greenland margin is considered a result of the strong contour current (primarily the LSW) in the area. The progressive steepening of the clinoform foreset beds in the Pliocene package of strata could be another indication of contour current intensification.

This work is part of a PhD study funded by the University of Copenhagen and supervised by S. Funder (Geological Museum, University of Copenhagen) and H. C. Larsen (Danish Lithosphere Centre). The high-resolution seismic surveys were financed by the Danish Natural Science Research Council, and the seismic data were processed by S. Berendt Marstal and H. Lykke-Andersen, Geological Institute, University of Aarhus. The seismic lines were made available by the Geological Survey of Greenland (now merged with the Geological Survey of Denmark). C. Marcussen (Geological Survey of Greenland) is thanked for technical assistance. A. T. Nielsen (Geological Museum, University of Copenhagen) is gratefully thanked for continued support. The initial manuscript benefited from comments by D. A. V. Stow and A. R. Viana.

References

ALLEY, R. B., BLANKENSHIP, D. D., BENTLEY, C. R. & ROONEY, S. T. 1987. Till beneath ice stream B, 3, Till deformation: Evidence and implications. *Journal of Geophysical Research*, **92**, 8921–8929.
—, —, — & — 1989. Sedimentation beneath ice shelves – the view from ice stream B. *Marine Geology*, **85**, 101–120.
ARTHUR, M. A., SRIVASTAVA, S. P., KAMINSKI, M., JARRARD, R. & OSLER, J. 1989. Seismic stratigraphy and history of deep circulation and sediment drift development in Baffin Bay and the Labrador Sea. *In*: SRIVASTAVA, S. A. *et al.* (eds) *Proceedings of the Ocean Drilling Program, Scientific Results*, **105**, Ocean Drilling Program, College Station, TX, 957–988.

BARNES, P. W. 1987. Morphologic studies of the Wilkes Land Continental Shelf, Antarctica–glacial and iceberg effect. *In*: EITTREIM, S. L. & HAMPTON, M. A. (eds) *The Antarctic Continental Margin: Geology and Geophysics of Offshore Wilkes Land*. CPCEMR Earth Science Series, **5**, 175–194.
BARTEK, L. R., VAIL, P. R., ANDERSON, J. B., EMMET, P. A. & WU, S. 1991. Effect of Cenozoic ice sheet fluctuations in Antarctica on the stratigraphic signature of the Neogene. *Journal of Geophysical Research*, **96**, 6753–6778.
BLANKENSHIP, D. D., BENTLEY, C. R., ROONEY, S. T. & ALLEY, R. B. 1987. Till beneath ice stream B, 1, Properties derived from seismic travel times. *Journal of Geophysical Research*, **92**, 8903–8911.
BOHRMANN, G., HEINRICH, R. & THIEDE, J. 1990. Miocene to Quaternary paleoceanography in the northern North Atlantic: variability in carbonate and biogenic opal accumulation. *In*: BLEIL, U. & THIEDE, J. (eds) *Geological History of the Polar Oceans: Arctic versus Antarctic*. Nato ASI Series C, **308**, 647–675.
BUCH, E. 1990. *The Physical Oceanography of the Greenland Waters*. Royal Danish Administration of Navigation and Hydrography. Open-File Report.
CLAUSEN, L. 1998. Late Neogene and Quaternary sedimentation on the continental slope and upper rise offshore Southeast Greenland: interplay of contour and turbidity processes. *In*: LARSEN, H. C., SAUNDERS, A. D. *et al.* (eds) *Proceedings of the Ocean Drilling Program, Scientific Results*, **152**, Ocean Drilling Program, College Station, TX, in press.
—— Plio-Pleistocene seismic stratigraphy of the southeast Greenland Shelf: Outline of a glacial depositional model. *Sedimentary Geology*, in press.
COOPER, A. K., BARRETT, P. J., HINZ, K., TRAUBE, V., LEITCHENKOV, G. & STAGG, H. M. J. 1991. Cenozoic prograding sequences of the Antarctic continental margin: a record of glacio-eustatic and tectonic events. *Marine Geology*, **102**, 175–213.
——, EITTREIM, S., TEN BRINK, U. & ZAYATZ, I. 1993. Cenozoic glacial sequences of the Antarctic continental margin as recorders of Antarctic ice sheet fluctuations. The Antarctic paleoenvironment: a perspective on global change. *Antarctic Research Series*, **60**, 75–89.
DICKSON, R. R. & BROWN, J. 1994. The production of North Atlantic Deep Water: sources, rates, and pathways. *Journal of Geophysical Research*, **99**, 12,319–12,341.
——, GMITROWICZ, E. M. & WATSON, A. J. 1990. Deep water renewal in the northern North Atlantic. *Nature*, **344**, 848–850.
DOMACK, E. W. 1987. Preliminary stratigraphy for a portion of the Wilkes Land Continental Shelf, Antarctica: evidence from till provenance. *In*: EITTREIM, S. L. & HAMPTON, M. A. (eds) *The Antarctic Continental Margin: Geology and Geophysics of Offshore Wilkes Land*. CPCEMR Earth Science Series, **5**, 195–203.

EGLOFF, J. & JOHNSON, G. L. 1975. Morphology and structure of the Southern Labrador Sea. *Canadian Journal of Earth Sciences*, **12**, 2111–2133.

EITTREIM, S. L., COOPER, A. K. & WANNESSON, J. 1995. Seismic stratigraphic evidence of ice-sheet advances on the Wilkes Land margin of Antarctica. *Sedimentary Geology*, **96**, 131–156.

FUNDER, S. & LARSEN, H. C. 1989. Quaternary geology of the shelves adjacent to Greenland. *In*: FULTON, R. J. (ed.) *Quaternary Geology of Canada and Greenland*. Geology of Canada, Geological Survey of Canada, **1**, 769–772.

HAQ, B. U., HARDENBOL, J., VAIL, P. R. AND 14 OTHERS 1987. Mesozoic–Cenozoic cycle chart. Version 3.1A. *Insert in*: WILGUS, C. K. *et al.* (eds) *Sea-level Changes: an Integrated Approach*. Society of Economic Paleontologists and Mineralogists, Special Publication, **42**.

HAUGLAND, K., KRISTOFFERSEN, Y. & VELDE, A. 1985. Seismic investigations in the Weddell Sea Embayment. *Tectonophysics*, **114**, 293–313.

HISCOTT, R. N. & AKSU, A. E. 1994. Submarine debris flows and continental slope evolution in front of Quaternary ice sheets, Baffin Bay, Canadian Arctic. *Bulletin, American Association of Petroleum Geologists*, **78**, 445–460.

JOHNSON, G. L. & SCHNEIDER, E. D. 1969. Depositional ridges in the North Atlantic. *Earth and Planetary Science Letters*, **6**, 416–422.

——, SOMMERHOFF, G. & EGLOFF, J. 1975. Structure and morphology of the west Reykjanes Basin and the southeast Greenland continental margin. *Marine Geology*, **18**, 175–196.

JONES, E. J. W., EWING, M., EWING, J. & EITTREIM, S. L. 1970. Influence of Norwegian Sea overflow water on sedimentation in the northern North Atlantic and Labrador Sea. *Journal of Geophysical Research*, **75**, 1655–1680.

KUVAAS, B. & KRISTOFFERSEN, Y. 1990. The seismic stratigraphy of glacigenic sediments along the Weddell Sea margin. *In*: COOPER, A. K. & WEBB, P. N. (eds) *International Workshop on Antarctic Offshore Acoustic Stratigraphy (ANTOSTRAT): Overview and Extended Abstracts*. US Geological Survey Open-File Report, **90-309**, 171–176.

LARSEN, B. 1980. A marine geophysical survey of the continental shelf of East Greenland 60°–70°N – project DANA 79. *Rapport Grønlands geologiske Undersøgelse*, **100**, 94–98.

—— 1994. Morphology and seismic stratigraphy of the East Greenland Shelf in the Denmark Strait area compared to the Prydz Bay, Antarctica. *Terra Antarctica*, **1**, 427–430.

LARSEN, H. C. 1985. Petroleum geological assessment of the East Greenland Shelf. *Grønlands geologiske Undersøgelse Open-File Report*, **99**, 78, plus appendix.

—— 1990. The East Greenland Shelf. *In*: GRANTZ, A., JOHNSON, L. & SWEENEY, J. F. (eds) *The Arctic Ocean region. The Geology of North America*, **L**. Geological Society of America, Boulder, CO, 185–210.

——, BROOKS, C. K., HOPPER, J. R., DAHL-JENSEN, T.,

PEDERSEN, A. K., NIELSEN, T. F. D. & FIELD PARTIES 1995. The Tertiary opening of the North Atlantic: DLC investigations along the east coast of Greenland. *Rapport Grønlands geologiske Undersøgelse*, **165**, 106–115.

——, SAUNDERS, A. D., CLIFT, P. D., BEGET, J., WEI, W., SPEZZAFERRI, S. & ODP LEG 152 SCIENTIFIC PARTY 1994a. Seven million years of glaciation in Greenland. *Science*, **264**, 952–955.

——, ——, —— & LEG 152 SHIPBOARD SCIENTIFIC PARTY 1994b. *Proceedings of the Ocean Drilling Program, Initial Reports*, **152**. Ocean Drilling Program, College Station, TX.

——, LARSEN, L. M., LYKKE-ANDERSEN, H., ODP LEG 152 SHIPBOARD PARTY, MARCUSSEN, C. & CLAUSEN, L. 1993. ODP activities on the South-East Greenland margin: Leg 152 drilling and continued site surveying. *Rapport Grønlands geologiske Undersøgelse*, **160**, 73–79.

LARTER, R. D. & CUNNINGHAM, A. P. 1993. The depositional pattern and distribution of glacial–interglacial sequences on the Antarctic Peninsula Pacific margin. *Marine Geology*, **109**, 203–219.

—— & VANNESTE, L. E. 1995. Relict subglacial deltas on the Antarctic Peninsula outer shelf. *Geology*, **23**, 33–36.

LYKKE-ANDERSEN, H. 1998. Neogene–Quaternary depositional history of the East Greenland shelf in the vicinity of Leg 152 shelf sites. *In*: LARSEN, H. C., SAUNDERS, A. D. & LEG 152 SHIPBOARD SCIENTIFIC PARTY (eds) *Proceedings of the Ocean Drilling Program, Scientific Results*, **152**. Ocean Drilling Program, College Station, TX, in press.

MACURDA, B. 1993. *Seismic Facies Analysis*. Petroleum Information Corporation, Houston, TX.

MIENERT, J., ANDREWS, J. T. & MILLIMAN, J. D. 1992. The East Greenland continental margin (65°N) since the last deglaciation: changes in seafloor properties and ocean circulation. *Marine Geology*, **106**, 217–238.

MILLER, K. G. & TUCHOLKE, B. E. 1983. Development of Cenozoic abyssal circulation south of the Greenland–Scotland Ridge. *In*: BOTT, M. H. P., SAXOV, M., TALWANI, M. & THIEDE, J. (eds) *Structure and Development of the Greenland–Scotland Ridge*. Plenum, New York, 549–589.

MITCHUM, R. M., JR, VAIL, P. R. & SANGREE, J. B. 1977. Part Six: stratigraphic interpretation of seismic reflection patterns in depositional sequences. *In*: PAYTON, C. E. (ed.) *Seismic Stratigraphy – Applications to Hydrocarbon Exploration*. American Association of Petroleum Geologists, Tulsa, OK, 117–133.

MYERS, R. A. & PIPER, D. J. W. 1988. Seismic stratigraphy of late Cenozoic sediments in the northern Labrador Sea: a history of bottom circulation and glaciation. *Canadian Journal of Earth Sciences*, **25**, 2059–2074.

O'BRIEN, P. 1994. Morphology and late glacial history of Prydz Bay, Antarctica, based on echo sounder data. *Terra Antarctica*, **1**, 403–404.

POWELL, R. D. 1984. Glacimarine processes and the inductive lithofacies modelling of the ice shelf and tidewater glacier sediments based on Quaternary examples. *Marine Geology*, **57**, 1–52.

—— 1990. Glacimarine processes at the grounding-line fans and their growth to ice-contact deltas. *In*: DOWDESWELL, J. A. & SCOURSE, J. D. (eds) *Glacimarine Environments: Processes and Sediments*. Geological Society, London, Special Publications, **53**, 53–73.

RUDDIMAN, W. F. 1972. Sediment distribution on the Reykjanes Ridge: seismic evidence. *Geological Society of America Bulletin*, **83**, 2039–2062.

SÆTTEM, J., POOL, D. A. R., ELLINGSEN, L. & SEJRUP, H. P. 1992. Glacial geology of outer Bjørnøyrenna, southwestern Barents Sea. *Marine Geology*, **103**, 15–51.

SHOR, A. N. & POORE, R. Z. 1979. Bottom currents and the ice rafting in the North Atlantic: interpretation of the Neogene depositional environments of Leg 49 cores. *In*: LUYENDYK, B. P., CANN, J. R. *et al.* (eds) *Deep Sea Drilling Project, Initial Reports* **49**. US Government Printing Office, Washington, DC, 859–872.

SOMMERHOFF, G. 1973. Formenschatz und morphologische Gliederung des südostgrönländischen Schelfgebietes und Kontinentalabhanges. *'Meteor' Forschungsergebnisse, Reihe C*, **15**, 55.

—— 1979. Submarine glazial übertiefte Täler vor Südgrönland. *Eiszeitalter und Gegenwart*, **29**, 201–213.

SWIFT, J. H. & AAGAARD, K. 1981. Seasonal transitions and water mass formation in the Iceland and Greenland Seas. *Deep-Sea Research*, **28A**, 1107–1129.

VANNESTE, L. E. & LARTER, R. D. 1995. Deep-tow boomer survey on the Antarctic Peninsula Pacific Margin: an investigation of the morphology and acoustic characteristics of late Quaternary sedimentary deposits on the outer continental shelf and upper slope. *Geology and Seismic Stratigraphy of the Antarctic Margin*. Antarctic Research Series, **68**, 97–121.

——, BART, P., DE BATIST, M., MILLER, H. & THEILEN, F. 1994. A comparison of reflection seismic geometries on three different glacial margins. *Terra Antarctica*, **1**, 437–439.

——, UENZELMANN-NEBEN, G. & MILLER, H. 1995. Seismic evidence for long-term history of glaciation on central east Greenland shelf south of Scoresby sund. *Geo-Marine Letters*, **15**, 63–70.

VORREN, T. O., LEBESBYE, E., ANDREASSEN, K. & LARSEN, K.-B. 1989. Glacigenic sediments on passive continental margin as exemplified by the Barents Sea. *Marine Geology*, **85**, 251–272.

——, RICHARDSEN, G., KNUTSEN, S.-M. & HENRIKSEN, E. 1990. Cenozoic erosion and sedimentation in the western Barents Sea. *Marine and Petroleum Geology*, **8**, 317–340.

Recent geological processes in the Central Bransfield Basin (Western Antarctic Peninsula)

G. ERCILLA[1,2], J. BARAZA[1,2], B. ALONSO[1,2] & M. CANALS[1,3]

[1]UAGM, Unidad Asociada de Geociencias Marinas, CSIC-UB, Barcelona, Spain

[2]CSIC, Instituto de Ciencias del Mar, Paseo Juan de Borbón, s/n, 08039 Barcelona, Spain

[3]Universidad de Barcelona, Departament de Geología Dinámica, Geofísica y Paleontología, 08028, Barcelona, Spain

Abstract: Acoustic character of the near-surface geological features in the Central Bransfield Basin, located between the northwestern Antarctic Peninsula and the South Shetland Islands, are products of recent sedimentary, volcanic and oceanographic processes. A platform, between 250 and 750 m water depth, is characterized by a highly rough surface with prolonged and hyperbolic reflections, which are interpreted as a result of ice-sheet grounding that occurred during the last glacial advance and iceberg scouring. Locally, this surface is draped by thin, stratified sediments deposited from suspension during the present interglacial stage. The upper slope shows prolonged and hyperbolic reflections without sub-bottom returns that form an irregular surface resembling slide scars and slumped material. Futher downslope, sediments become stratified, and individual, chaotic or semi-transparent slump masses occur locally, within the eroded valleys and in inter-valley areas. Basin sediments are defined by stratified, semi-transparent and chaotic deposits that correspond to a variety of mass-transport, volcanic and oceanographic processes. Mass-transport and volcanic processes develop slump masses, turbidite deposits and volcanic edifices of varying characteristics. The activity of slope-parallel bottom currents builds sediment drifts between the volcanic edifices and the base of the Antarctic Peninsula slope.

The Bransfield Basin is a young and active rift basin (Saunders & Tarney 1984; Lawver *et al.* 1995) located in the northwestern tip of the Antarctic Peninsula, between the Antarctic Peninsula and the South Shetland Islands (Fig. 1). Extension in the Bransfield Strait, which would have begun between 4 and 1.3 Ma (Roach 1978; Barker & Dalziel 1983), is thought to be accommodated by convergence at the South Shetland Trench, which lies to the northwest of the South Shetland Islands (Pelayo & Wiens 1989). Several geological studies confirm the existence of volcanic (Saunders & Tarney 1984; Gracia *et al.* 1996) and hydrothermal activity (Han & Suess 1987; Schlosser *et al.* 1988), seismicity (Pelayo & Wiens 1989), negative gravimetry anomalies (Davey 1972) and positive magnetic anomalies (Garret & Storay 1987). In spite of its complex geological framework, the Bransfield Basin is an interesting area for the study of recent geological processes because it is one of the very few places where a young passive margin and a large glacial sediment input coexist. Several studies on the seismic stratigraphy of this basin have been made (Jeffers 1988; Anderson & Molnia 1989; Gambôa & Maldonado 1990; Jeffers & Anderson 1990; Banfield 1994) to determine the processes and controlling factors of its sedimentary evolution. However, although these studies have revealed interesting depositional patterns, ultra-high-resolution seismic studies of the most recent processes are still lacking.

The Bransfield Basin is divided into three sub-basins: eastern, central and western, separated by volcanic ridges. The Central Bransfield Basin, which is the focus of the present paper, was surveyed by RV *Hespérides* in 1993. The following analysis is mainly based on over 1000 km of ultra-high-resolution seismic profiles acquired with the Bentech TOPAS (TOpographic PArametric Sonar) system, and the new bathymetric mosaic and side-scan sonar records obtained by means of the Simrad EM12/EM1000 multibeam echosounders (Fig. 1). The TOPAS system is a hull-mounted sea-bed and sub-bottom echosounder which operates using the non-linear acoustic properties of the water (Dybedal & Bøe 1994). The system transmits two (primary) acoustic signals of high frequency that interact within the water column generating a secondary (interference) signal of low frequency, and thus provides a very good vertical resolution within the upper 50–80 m of the sediment column at any water depth (Webb 1993). An analysis of the acoustic character and surface and sub-bottom features of the physiographic provinces off the shelf was made to determine the most recent active geological processes in the Central Bransfield Basin.

ERCILLA, G., BARAZA, J., ALONSO, B. & CANALS, M. 1998. Recent geological processes in the Central Bransfield Basin (Western Antarctic Peninsula). *In*: STOKER, M. S., EVANS, D. & CRAMPS, A. (eds) *Geological Processes on Continental Margins: Sedimentation, Mass-Wasting and Stability.* Geological Society, London, Special Publications, **129**, 205–216.

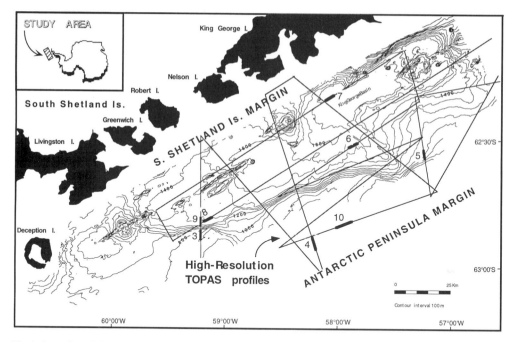

Fig. 1. Location of the study area showing the new bathymetry data obtained with the EM12/EM1000 multibeam echosounders (from Canals *et al.* 1994), and the location of TOPAS profiles. The numbers refer to profiles illustrated in Figs 3–10.

Physiography

The Central Bransfield Basin is an asymmetric, ENE–WSW-trending basin 230 km long, 125 km wide and with a maximum depth of 1950 m. It is bounded by two physiographically distinct margins. Its eastern margin (the Antarctic Peninsula margin) is wide and has an undulating shape in plan view, whereas the western margin (South Shetland Island margin) is narrow and has a linear trend only interrupted by several incised valleys (Figs 1 and 2). On the Antarctic Peninsula margin, three physiographic provinces have been defined basinward of the shelf: platform, slope and basin (Jeffers & Anderson 1990). The platform extends from 250 to about 750 m water depth and is 50–60 km wide, its lower boundary being clearly marked by a break in slope. The seismic profiles studied in the present paper only cover the outer part of the platform, below 400 m water depth. The slope on the Antarctic Peninsula margin extends down to 1400 m water depth and shows a variable width (from 10 to >25 km), and an undulating trend. On the margin of the South Shetland Islands the shelf is very narrow and it has not been studied here, as the available seismic records only extend up to the upper slope. The South Shetland slope is very narrow (generally <25 km) and extends seaward from

100 m to 1400–1800 m water depth. The basin has a relatively flat surface clearly differentiated from both the eastern and western slopes and has a step-like profile characterized by four bathymetric levels separated by scarps (Gracia *et al.* 1996) (Fig. 2).

Acoustic character

Seven types of acoustic character have been differentiated (Figs 3–9): (1) prolonged; (2) hyperbolic; (3) stratified; (4) chaotic; (5) semi-transparent; (6) stratified–prolonged; (7) stratified–hyperbolic. The prolonged acoustic character (1) defines an irregular or planar distinct surface of high amplitude, generally without showing penetration of the acoustic signal (Figs 5 and 9). However, where some penetration of the acoustic signal occasionally exists, structureless sediments up to 15–20 ms thick can be identified (Fig. 5). The hyperbolic acoustic character (2) is defined by single and multiple hyperbolas of high-reflectivity with their apices generally elevated above the surrounding sea floor. The distribution of the multiple hyperbolas is irregular, and they overlap with sub-bottom reflections (up to 15 ms of penetration) that are also characterized by irregular, but less reflective hyperbolas (Fig. 4). The

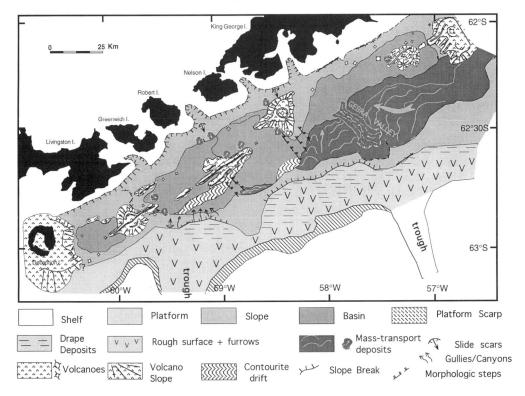

Fig. 2. Location of the surface geological features identified on the platform, slope and basin of the Central Bransfield Basin. Shelf province and areas that are not included on the new bathymetry map (Fig. 1) are taken from Jeffers & Anderson (1990).

single hyperbolic reflections are generally associated with the prolonged acoustic character (Fig. 3). The stratified acoustic character (3) defines a sequence of up to 100 ms thick with individual reflections of low and high amplitude and high lateral continuity showing parallel, divergent-fill and onlap-fill configurations (Figs 5–9). The chaotic acoustic character (4) is composed of disorganized, contorted and disrupted reflections of high amplitude, sometimes showing traces of the original bedding (Figs 5 and 9). The semi-transparent acoustic character (5) is characterized by the near-absence of reflections (Figs 6 and 7). The chaotic and semi-transparent reflections form mound-like or lenticular bodies of a few kilometres length and a variable thickness (up to 25 ms) that occasionally interrupt the stratified reflections (Figs 5 and 6). The stratified–prolonged (6) and stratified–hyperbolic (7) acoustic characters result from the combination of two types. They appear as up to 20 ms thick, parallel stratified reflections that locally drape an irregular surface defined by prolonged or hyperbolic reflections (Fig. 3).

Near-surface geological features

Based on the bathymetric map, TOPAS profiles and sonographs, seven major types of surface and sub-bottom geological features were identified from platform to basin: (1) highly rough surface; (2) undisturbed stratified deposits; (3) furrows; (4) mass-transport deposits; (5) valleys; (6) volcanoes and volcanic flows; (7) contourite drifts.

The highly rough surface (1) is identified on the platform from 400 to about 750 m water depth (Figs 2 and 3). It is characterized by relief of different scales varying from a few metres to tens of metres, and is acoustically characterized by prolonged and hyperbolic reflections.

The undisturbed stratified deposits (2) are identified on the most external areas of the platform, on the lower slope of the Antarctic Peninsula margin, and locally in the basin where the deposits are not affected by erosive or deformational features. On the platform, the undisturbed deposits are acoustically defined by parallel, stratified–prolonged or stratified–hyperbolic reflections. This type of deposit mimics the underlying

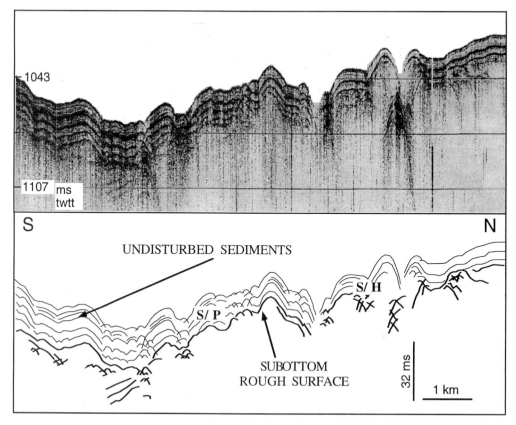

Fig. 3. TOPAS profile and line drawing showing the acoustic character and surface (undisturbed sediments) and sub-bottom (rough surface) features that characterize the outer platform of the Antarctic Peninsula margin. S/P, Stratified–prolonged reflections; S/H, stratified–hyperbolic reflections; twtt, two-way travel time. Location of profile is shown in Fig. 1.

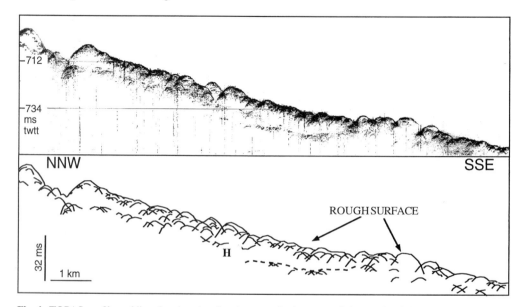

Fig. 4. TOPAS profile and line drawing showing the acoustic character of the proximal platform in the Antarctic Peninsula. H, Hyperbolic reflections. Location of profile is shown in Fig. 1.

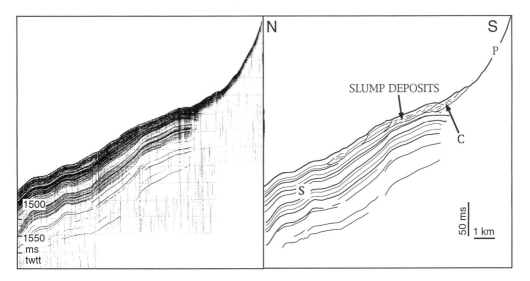

Fig. 5. TOPAS profile and line drawing showing the acoustic character and the geological features (undisturbed sediments and slumps) identified on the Antarctic Peninsula slope. S, Stratified reflections; C, chaotic reflections; P, prolonged reflection. Location of profile is shown in Fig. 1.

topography of the highly rough surface (Fig. 3). On the slope and basin, such deposits appear as packages of up to 100 ms thick, parallel or sub-parallel stratified reflections with alternating low and high amplitudes, which develop a planar sea-floor surface (Figs 5–8). In the basin, these reflections may also show a divergent-fill configuration, and are occasionally interrupted by zones of acoustic blanking with well-defined lateral boundaries and diffuse upper boundaries (Fig. 7). These acoustic anomalies may be caused by the presence of interstitial gas within the sediments, or by hydrothermal processes related to the nearby volcanoes.

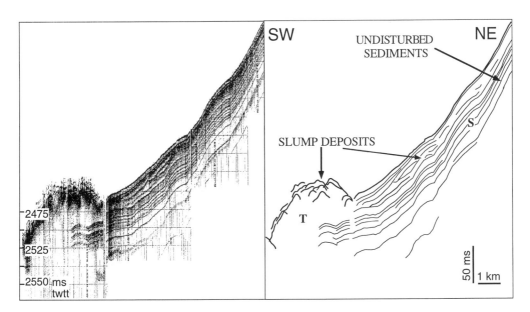

Fig. 6. TOPAS profile and line drawing illustrating the geological features (undisturbed sediments and slumps) that characterize the slope. T, Transparent acoustic character; S, stratified reflections. Location of profile is shown in Fig. 1.

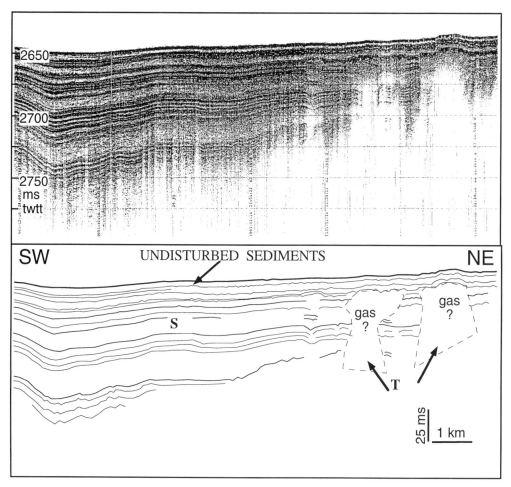

Fig. 7. TOPAS profile and line drawing showing the undisturbed sediments that fill the basin and their acoustic character. S, Stratified reflections; T, transparent acoustic character. Location of profile is shown in Fig. 1.

The furrows (3), which have been identified on sonographs, are from a few kilometres to tens of kilometres long and a few metres wide, and show a totally chaotic orientation (Fig. 10). These furrows are clearly observed down to about 600–700 m water depth on the platform (Fig. 2), where they produce an irregular surface characterized on TOPAS records by multiple hyperbolic reflections (Fig. 4).

The mass-transport deposits (4) mostly comprise slumps that are especially identified at the base of the steep slopes of the Antarctic Peninsula and the South Shetland Island margins, and also in the basin associated with the largest volcanic edifices (Fig. 2). The slumps on the Antarctic Peninsula slope appear as acoustically transparent, chaotic or disrupted stratified deposits of up to 75 ms thick, which usually form

a mound-like, positive relief with an irregular, strongly reflective surface (Figs 5 and 6). Other slumps are located adjacent to scarps with relief of up to 18 m at the sea floor. In this case, the slumped sediments consist of packages of stratified sediments that terminate abruptly against the failure plane, and are only slightly deformed as a result of the downslope movement, retaining most of their original structure. The area affected by slumping can be variable, ranging from tens to several hundreds of square kilometres (Fig. 2). On the slope of the South Shetland Islands and in the basin the slumps are smaller (up to tens of square kilometres). The sediments are totally disrupted, and are acoustically defined by transparent and chaotic reflections with an irregular surface of high reflectivity (Fig. 2). Upslope of these slumps there occur

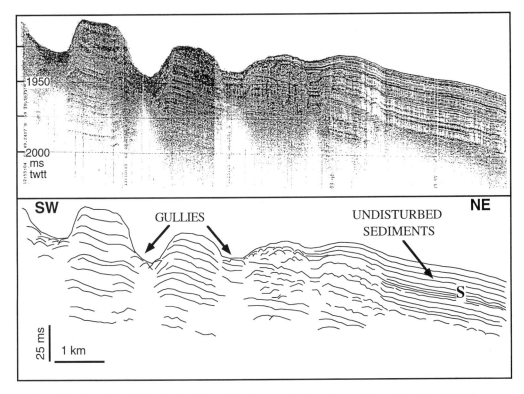

Fig. 8. TOPAS profile and line drawing illustrating the acoustic character and the gullies incised into the undisturbed sediments of the basin. S, stratified reflections. Location of profile is shown in Fig. 1.

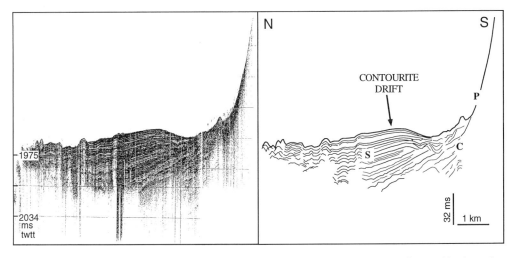

Fig. 9. TOPAS profile and line drawing showing the acoustic character identified in the slope and basin, and the contourite drift developed in the basin at the foot of the slope. S, Stratified reflections; C, chaotic reflections; P, prolonged reflection. Location of profile is shown in Fig. 1.

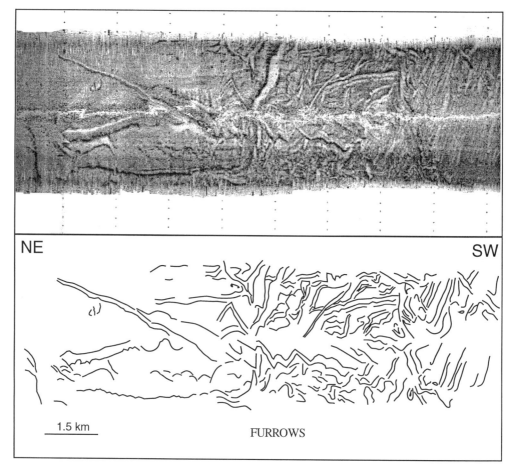

Fig. 10. Sonograph from the EM1000 multibeam echosounder showing the furrows identified in the proximal platform of the Antarctic Peninsula margin. Location of sonograph is shown in Fig. 1.

well-defined scarps which are characterized by hyperbolic reflections and probably result from the slump processes.

The studied valleys (5) are represented by gullies and by the large Gebra Valley eroded in the slope of the Antarctic Peninsula margin (Fig. 2). A group of valleys eroded in the South Shetland Island upper slope (Fig. 2) were not surveyed in the present cruise, and are drawn only from the literature (Jeffers & Anderson 1990). The gullies are located in the southwestern slope sector of the Antarctic Peninsula margin, and appear as a group of several parallel incisions up to 10–15 km long, crossing the slope and reaching the basin floor. There, the gullies display a U-shaped cross-section, are up to 2 km wide with a 40 m relief, and erode the surrounding undisturbed stratified deposits (Fig. 8). The Gebra Valley has a SE–NW-trending course and extends from 1200 m water depth down to the

basin, with a total length of 30 km (Canals *et al.* 1994). The head of this valley has an amphitheatre shape and a smooth, U-shaped cross-section 10 km wide. The entire length of the valley is partially filled with sediments slumped from its walls (Fig. 2). These slumped sediments are acoustically characterized by chaotic and disrupted stratified reflections, and form an irregular sea floor of high reflectivity.

Seven major volcanic edifices (6) and several smaller ones have been identified in the basin (Gracia *et al.* 1996) (Fig. 2). The largest volcanoes are aligned parallel to the axis of the basin, reach heights of up to 400 m, and display varied morphologies: conical, semicircular, crescentic and elongated. The outer surface of these edifices is characterized by prolonged reflections of strong reflectivity, resulting from the presence of hard volcanic material or coarse-grained debris that prevents the penetration of the acoustic signal. The volcanic surface occasionally displays hyperbolic reflections that are probably related to slumps and the presence of slide scars. Individual mounds or lenticular bodies a few kilometres long and up to 35 ms thick are identified in the basin intercalated between the volcanoes. These depositional mounds are formed by semi-transparent acoustic facies that interrupt the surrounding undisturbed sediment and are interpreted as volcanic-flow deposits.

The contourite drifts (7) appear as depositional mounds associated with moats located adjacent to the Antarctic Peninsula slope and to some of the volcanic highs in the basin (Fig. 2). The mounds are up to 13 m high with an asymmetric cross-section and the steeper slope oriented towards the moat. These mounds are internally defined by stratified deposits that converge and are truncated towards the moat. Some of the reflections show downlap terminations over ancient drift deposits. However, the moats have U-shaped cross-sections about 7 m deep and 1.5 km wide, and are acoustically defined by an irregular surface of high reflectivity that shows their erosive nature. Internally, they are composed of acoustically chaotic deposits in which it is locally possible to identify erosional down-cutting features filled with onlapping sediments (Fig. 9).

Geological significance

The geological interpretation of the different types of acoustic character and the surface and sub-bottom features described above provides a partial insight into the recent geological processes that have been active in the platform, slope and basin of the Central Bransfield Basin. The geological processes that could have been responsible for the present near-surface structure, stratigraphy and physiography of the Central Bransfield Basin include glacial-ice erosion, iceberg scouring, hemipelagic settling from suspensions and nepheloid layers, mass-wasting, volcanic eruptions and bottom-current processes. Evidence that supports the occurrence of these processes will be discussed in the following paragraphs.

Platform

The platform of the Antarctic Peninsula margin seems to show the effects of both erosional (glacial-ice erosion and iceberg scouring) and depositional processes (hemipelagic sedimentation from suspension and nepheloid layers). With respect to the erosional processes, stratigraphic and sedimentological studies (Jeffers & Anderson 1990; Banfield 1994) indicate that during the last glaciation, ice-shelves and glaciers extended across the entire platform, scouring and grounding on the sea floor to water depths close to 750 m. Our data suggest that the highly rough surface that characterizes the platform could represent the activity of erosive processes produced by the ice-sheet advance that occurred during the last glaciation (Fig. 2). The high acoustic reflectivity and the prolonged or hyperbolic reflections that characterize this surface may be caused by the presence of glacial residual deposits, mainly composed of coarse-grained sediments, which prevent the further penetration of the acoustic signal. In fact, coarse gravel and sand lags, classified as residual glacial-marine, have been sampled on this platform (Anderson et al. 1983; Banfield 1994). On the other hand, the erosive action of the wandering icebergs seems to have affected the platform, and apparently remains active today (Figs 2 and 10). This process could be responsible for the furrows observed on sonographs along the platform, which we interpret as iceberg plough marks. The presence of these iceberg plough marks has also been described in adjacent areas (Pudsey et al. 1994). The considerable seaward extension (down to 600–700 m water depth) of the iceberg plough marks on the Bransfield Basin suggests that erosional activity of icebergs probably started during the early stages of the interglacial period when ice-shelves and glaciers began to retreat and break up, increasing the number of icebergs and favouring their calving activity (Josenhans & Fader 1989; Heinrich 1990; Pudsey et al. 1994; Larter & Vanneste 1995). The icebergs interact with the sea floor, reworking the surficial deposits and creating an irregular surface that is acoustically characterized by multiple, irregular overlapping hyperbolas (Fig. 3). This fact has been established by correlation of the sonographs with the TOPAS profiles. In contrast, the areas less affected by iceberg scouring show a sea-floor surface characterized by prolonged and single hyperbolic reflections.

With respect to depositional processes on the platform, hemipelagic settling from suspensions and nepheloid layers seems to have occurred during interglacial stages such as the present, when glaciers retreat into fjords. These processes would be responsible for the formation of the parallel stratified deposits described on the outer platform (Figs 2 and 3). The blanket-like seismic expression of this facies that drapes the irregularities of the highly rough surface (Fig. 3) suggests the occurrence of hemipelagic settling. These undisturbed sediments are only identified on the outer platform, and their limited distribution could be related to the shelf circulation (Mandelli & Burkholder 1978; Stinger 1987) that would initiate the winnowing of fine sediment (Anderson *et al.* 1983), and/or to the intense iceberg scouring that affects the most proximal sectors of the platform.

Slope

Mass-wasting processes and products are mainly identified along the steep slopes of the Antarctic Peninsula and the South Shetland Island margins, on normal slopes and within the valleys (Fig. 2). In normal slope sectors, individual slump masses of a relatively small scale (a few hundreds of square metres) occur on the very steep slope (average gradient 12°) of the South Shetland Islands. In contrast, large volumes (hundreds of square kilometres) of slumped sediments characterize the normal slope and inter-valley areas of the Antarctic Peninsula margin. There, the slumped sediment masses are mainly identified downslope and close to areas with a basinward convex bathymetry (Figs 1 and 2). By comparison with similar bathymetric features in other glacial environments (Stoker 1990; Vorren *et al.* 1990; Kuvaas & Kristoffersen 1991), we think that these probably represent trough-mouth fans, in which the development of instability processes is known to be common. We suggest that the growth of these fans would be related to the presence of deep troughs incised in the shelf (in this case also in the platform, Fig. 2), and would take place mainly during glacial stages, when the glacigenic sediments are deposited on the slope, become unstable and move downslope because of the action of gravity. Mass-wasting processes also appear to be relevant within the valleys eroded in the slope, such as the Gebra Valley, as can be deduced from the amphitheatre shape of its head and the characteristics of the deposits that cover the valley floor. There, the slumped material forms an irregular surface with prolonged and hyperbolic reflections that resemble a blocky sea floor (Fig. 2). The occurrence of instabilities or mass-wasting processes along the valleys was probably most pronounced during glacial stages, when sediment supply increased in deep environments and sedimentation occurred on steeper sea floor, i.e. on the slope. The relevance of these processes during the glacial periods has been also described in the Bransfield Basin by Jeffers & Anderson (1990).

Basin

The basin sectors mostly show the activity of mass-transport, volcanic, and bottom-current-related processes. Mass-transport processes result in the downslope movement of the numerous slumps that have occurred on the slope, and in the channelized flow of sediment through the Gebra Valley and other small valleys (Fig. 2). The recent activity of the gullies as conduits for sediment flow to the basin is shown not only because they reach the basin floor, but also by their acoustic nature, which reflects erosive processes (Fig. 8). Many of the slumps initiated on the slope reach the basin floor as well-defined masses of disturbed sediment, but others may have evolved to gravity flows and then developed turbidity currents. Therefore, we suggest that the high continuity and strong reflectivity of the undisturbed and stratified basin deposits could be interpreted as turbidites (Figs 7 and 8). The divergent-fill configuration displayed by these deposits in the deepest sector of the basin (Fig. 7) suggests that turbidite deposition was contemporaneous with the deepening of the basin.

The distribution of volcanic edifices aligned parallel to the NE–SW axis of the basin is perpendicular to the NW–SE opening direction, and their different shapes may indicate successive evolutionary stages (Gracia *et al.* 1996) (Fig. 2). The activity of these volcanoes may have triggered gravitative processes represented by volcanic-flow mounds and slumps identified on their steep flanks and at their bases. The volcanic-flow mounds have very sharp boundaries and interrupt the sedimentary continuity of the undisturbed sediments. Their volcanic nature is suggested by the almost total absorption of the acoustic signal, which prevents the recognition of the underlying deposits. The volcanic activity has also developed turbidites (Yoon *et al.* 1994), as is revealed by the presence of graded volcaniclastic sediments deposited from high-density currents (Anderson & Molnia 1989).

Finally, bottom-current-related processes along the basin seem to have also been relevant for the recent geological history of the Central Bransfield Basin, although they have never been cited before in the scientific literature. These processes would be responsible for the development of the contourite drifts, although we cannot establish if these drifts are relict or modern sedimentary features. With respect to the present-day oceanography, studies in this area refer to surficial circulation (Mandelli & Burkholder 1978; Stinger 1987, among others). A very recent study (Calafat *et al.* 1996) demonstrates the presence of tidal movements observed down to 1250 m water depth. This study indicates a direction of tidal wave propagation towards the west with velocities between 1.7 and 46 cm/s. Based on sedimentary features displayed by the contourite drifts of the Central Bransfield Basin, we can only indicate that some type of bottom current would be responsible for the local development of contourite drifts at the bottom of the Antarctic Peninsula slope and around some of the volcanic highs located in the southwestern sector of the basin (Fig. 2). The narrow passages existing in this sector between the steep lower slope of the Antarctic Peninsula and the volcanic edifices have caused restrictions to the general circulation, and the passages have acted as corridors that accelerate the bottom currents, locally erode the sea-floor sediments and develop moats by lateral drift accumulation (Fig. 9). The contourite drifts are characterized by parallel or sigmoidal stratified facies, and locally show scour-and-fill features and erosional unconformities, probably related to variations in bottom-current speed and the subsequent lateral migration of the erosional and the depositional sectors within the drift (Fig. 9).

This paper has benefited from reviews by R. Bøe and R. Larter. The study was funded by the Spanish CICYT (Project ref. ANT-1008/93-CO3-02). We thank C. F. Victor Quiroga and the officers and crew of the RV BIO *Hespérides*, for their assistance during the GEBRA 93 cruise.

References

ANDERSON, J. B. & MOLNIA, B. F. 1989. *Glacial–Marine Sedimentation*. Short Course In Geology, 28th International Congress, American Geophysical Union, Washington, DC, 127 pp.

——, BRAKE, C. F., DOMACK, E. W., MYERS, N. C. & WRIGHT, R. 1983. Development of a polar glacial–marine sedimentation model from Antarctic Quaternary deposits and glaciological information. *In*: MOLNIA, B. (ed.) *Glacial Marine Sedimentology*. Plenum, New York, 233–264.

BANFIELD, L.A. 1994. *Seismic facies investigation of the Late Cenozoic glacial history of the Bransfield Basin, Antarctica*. MA thesis, Rice University, Houston, TX, 184 pp.

BARKER, P. F. & DALZIEL, I. W. D. 1983. Progress in geodynamics in the Scotia Arc region. *American Geophysical Union Geodynamic Series*, **9**, 137–170.

CALAFAT, A. M., CANALS, M. & DURRIEU DE MADRON, X. 1996. Registro de temperaturas a largo plazo del agua profunda del Estrecho de Bransfield. *VI Simposio Español de Estudios Antárticos*, Miraflores de la Sierra, Madrid, 89.

CANALS, M., ACOSTA, J., BARAZA, J., BART, P., CALAFAT, A. M., CASAMOR, J. L., DE BATIST, M., ERCILLA, G.,

FARRÁN, M., FRANCÉS, G., GRACIA, E., RAMOS-GUERRERO, E., SANZ, J. L., SORRIBAS, J. L. & TASSONE, A. 1994. La Cuenca Central de Bransfield (NW de la Península Antártica): primeros resultados de la campaña GEBRA-93. *Geogaceta*, **16**, 132–135.

DAVEY, F. J. 1972. Marine gravity measurements in Bransfield Strait and adjacent areas. *In*: ADIE, R. J. (ed.) *Antarctic Geology and Geophysics*. Universitetsforlaget, Oslo, 39–46.

DYBEDAL, J. & BØE, R. 1994. Ultra high resolution sub-bottom profiling for detection of thin layers and objects. *Oceans '94*. Osates, Brest, France.

GAMBÔA, L. A. P. & MALDONADO, P. R. 1990. Geophysical investigations in the Bransfield Strait and the Bellingshausen Sea – Antarctica. *In*: ST JOHN, B. (ed.) *Antarctica as an Exploration Frontier – Hydrocarbon Potential, Geology, and Hazards*. American Association of Petroleum Geologists, Studies in Geology, **31**, 127–141.

GARRET, S. W. & STORAY, B. C. 1987. Lithospheric extension on the Antarctic Peninsula during Cenozoic subduction. *In*: COWARD, M. P., DEWEY, J. F. & HANCOCK (eds) *Continental Extensional Tectonic*. Geological Society, London, Special Publications, **28**, 419–431.

GRACIA, E., CANALS, M., FARRAN, M., PRIETO M.J., SORRIBAS, J. & GEBRA TEAM 1996. Morphostructure and evolution of the Central and Eastern Bransfield Basins (NW Antarctic Peninsula). *Marine Geophysical Research*, **18**, 429–448.

HAN, M.W. & SUESS, E. 1987. Lateral migration of pore fluids through sediments of an active back-arc basin, Bransfield Strait, Antarctica. *American Geophysical Union EOS Transactions*, **68**(50), 1769.

HEINRICH, R. 1990. Cycles, rhythms, and events in Quaternary Arctic and Antarctic glaciomarine deposits. *In*: BLEIL, U. & THIEDE, J. (eds) *Geological History of the Polar Oceans: Arctic versus Antarctic*. Kluwer Academic, Dordrecht, 213–244.

JEFFERS, J. D. 1988. *Tectonics and sedimentary evolution of the Bransfield Basin, Antarctica*. MA thesis, Rice University, Houston, TX.

—— & ANDERSON, J. B. 1990. Sequence stratigraphy of the Bransfield Basin, Antarctica: implications for tectonic history and hydrocarbon potential. *In*: ST JOHN, B. (ed.) *Geological Evolution of Antarctica*. Cambridge University Press, Cambridge, 481–485.

JOSENHANS, H. W. & FADER, G. B. J. 1989. A comparison of models of glacial sedimentation along eastern Canadian margin. *Marine Geology*, **85**, 273–300.

KUVAAS, B. & KRISTOFFERSEN, Y. 1991. The Crary Fan: a trough-mouth fan on the Weddell Sea continental margin, Antarctica. *Marine Geology*, **97**, 345–362.

LARTER, R. D. & VANNESTE, L. E. 1995. Relict subglacial deltas on the Antarctic Peninsula outer shelf. *Geology*, **23**(1), 33–36.

LAWVER, L. A., KELLER, R. A., FISK, M. R. & STRELIN, J. 1995. Bransfield Strait, Antarctic Peninsula: active extension behind a dead arc. *In*: TAYLOR, B. (ed.)

Back-Arc Basins: Tectonics and Magmatism. Plenum, New York, 315–342.

MANDELLI, E. F. & BURKHOLDER, P. R. 1978. Primary productivity in the Gerlache and Bransfield Straits of Antarctica. *Journal of Marine Research*, **24**, 15–27.

PELAYO, A. M. & WIENS, D. A. 1989. Seismotectonics and relative plate motions in the Scotia Sea region. *Journal of Geophysical Research*, **94**, 7239–7320.

PUDSEY, C. J., BARKER, P. F. & LARTER, P. R. 1994. Ice sheet retreat from the Antarctic Peninsula shelf. *Continental Shelf Research*, **14**, 1647–1675.

ROACH, P. J. 1978. The nature of back-arc extension in Bransfield Strait. *Geophysical Journal of the Royal Astronomical Society*, **53**, 165.

SAUNDERS, A. D. & TARNEY, J. 1984. Geochemical characteristics of basaltic volcanism within back-arc basins. *In*: KOKELAAR, B. P. & HOWELLS, M. F. (eds) *Marginal Basin Geology: Volcanic and Associated Sedimentary and Tectonic Structures in Modern and Ancient Marginal Basins.* Geological Society, London, Special Publications, **16**, 59–76.

SCHLOSSER, P., SUESS, E., BAYER, R. & RHEIN, M. 1988. ³He in the Bransfield Strait waters: Indication for local injection from back-arc rifting. *Deep-Sea Research*, **35**, 1919–1935.

STINGER, J. K. 1987. Terrigenous, biogenic, and volcaniclastic sedimentation patterns of the Bransfield Strait and bays of the northern Antarctic Peninsula: implications for Quaternary glacial history. PhD dissertation, Rice University, Houston, TX.

STOKER, M. S. 1990. Glacially-influenced sedimentation on the Hebridean slope, northwestern United Kingdom continental margin. *In*: DOWDESWELL, J. A. & SCOURSE, J. D. (eds) *Glacimarine Environments: Processes and Sediments.* Geological Society, London, Special Publications, **53**, 349–362.

VORREN, T. O., LEBESBYE, E. & LARSEN K. B. 1990. Geometry and genesis of the glaciagenic sediments in the southern Barents Sea. *In*: DOWDESWELL, J. A. & SCOURSE, J. D. (eds) *Glacimarine Environments: Processes and Sediments.* Geological Society, London, Special Publications, **53**, 269–288.

WEBB, D. L. 1993. Seabed and sub-seabed mapping using a parametric system. *Hydrographic Journal*, **68**, 5–13.

YOON, H. I., HAN, M. W., PARK, B. K., OH, J. K. & CHANG, S. K. 1994. Depositional environment of near-surface sediments, King George Basin, Bransfield Basin, Antarctica. *Geo-Marine Letters*, **14**, 1–9.

Seismic characterization of Palaeogene depositional sequences: northeastern Rockall Trough

PAUL D. EGERTON

British Geological Survey, Murchison House, West Mains Road, Edinburgh EH9 3LA,
UK e-mail: p.egerton@bgs.ac.uk.

Abstract: Palaeogene depositional geometries and sedimentation patterns along the Hebridean margin in the northeastern Rockall Trough have been strongly influenced by the interaction of complex structure and bathymetric morphology. An evaluation of the seismo-stratigraphy of this margin has been undertaken using multichannel seismic-reflection data. Analysis of these profiles reveals that the late Palaeocene volcanic unit displays well-developed oblique to sigmoidal clinoforms and internal seismic facies variations indicative of westward progradation into a marine basin. This volcanogenic unit is overlain by a lower to middle Eocene seismic unit characterized by moderate- to high-amplitude downlapping reflection configurations and is interpreted to represent a shelf-margin deltaic system. A prominent upper Eocene to lower Oligocene unconformity surface that truncates the shelf-margin wedge can be identified and correlated regionally along the Hebridean margin. This significant unconformity surface is onlapped by upslope accreting high-amplitude continuous reflectors with complex mounded geometries which are interpreted as deep-water contourite drifts of middle to upper Oligocene age. Distinct upper Eocene to lower Oligocene point-sourced fans are preferentially developed along the southern margin of the Wyville-Thomson Ridge. These seismic packages are inferred to have been deposited during a lowstand in relative sea level induced by compressional uplift of the Wyville-Thomson Ridge, and have wide-ranging implications for understanding the distribution of sand-prone fans in general around the western and eastern margins of the Rockall Trough.

The northeastern sector of the Rockall Trough (Fig. 1) has been the subject of relatively few recent published analyses of the Tertiary seismic stratigraphy and seismic facies (Jones *et al.* 1986b; Earle *et al.* 1989; Stoker *et al.* 1993). At present, no unified seismo-stratigraphical framework exists within the public domain and correlation of reflectors has relied upon sparse exploration well control and limited Deep Sea Drilling Project (DSDP) sites located around the northeast Atlantic margin. The release of commercial, multichannel seismic-reflection data across this margin is providing an insight into the depositional processes and the distribution of sediment packages related to phases of Eocene–Oligocene tectonic activity. The aim of this paper is to describe the seismic character of Palaeogene sequences within the West Lewis Basin and adjacent northeast Rockall Trough, and to highlight the principal reflector geometries developed in this area. This study is based on interpretation of a grid of 2D, migrated, multichannel seismic-reflection profiles, supplemented with additional information from exploration wells (164/25-2, 164/25-1z and 154/3-1). The seismo-stratigraphical analysis was undertaken within the British Geological Survey (BGS) regional mapping programme of the Sula Sgeir area (Egerton & Stoker in press) and forms part of a wider attempt to understand and characterize Palaeogene evolution along the margins of the Rockall Trough.

Regional structural setting

The structural configuration of the extensional basins along the Hebridean margin (Fig. 1) has been influenced by a combination of tectonic events ranging from Caledonian low-angle thrusting (Snyder 1990; Stein & Blundell 1990) to major phases of extensional tectonics during the late Mesozoic and early Tertiary (Earle *et al.* 1989; Musgrove & Mitchener 1995, 1996). Compressional deformation during late Eocene to Oligocene times resulted from plate reorganization and intraplate stresses (Boldreel & Andersen 1993; Knott *et al.* 1993; Doré & Lundin 1996). Palaeogene sedimentation patterns along the Hebridean margin have been controlled by the combination of Eocene extensional subsidence related to the onset of sea-floor spreading in the mid-Atlantic, and subsequent major pulses of compression–transpression during the Oligocene (Doré & Lundin 1996). Inversion of normal faults resulted in the generation of compressional folds (ramp anticlines) such as the Wyville-Thomson and Ymir Ridges in the northeast Rockall Trough. The development of these

EGERTON, P. D. 1998. Seismic characterization of Palaeogene depositional sequences: northeastern Rockall Trough. *In:* STOKER, M. S., EVANS, D. & CRAMP, A. (eds) *Geological Processes on Continental Margins: Sedimentation, Mass-Wasting and Stability.* Geological Society, London, Special Publications, **129**, 217–228.

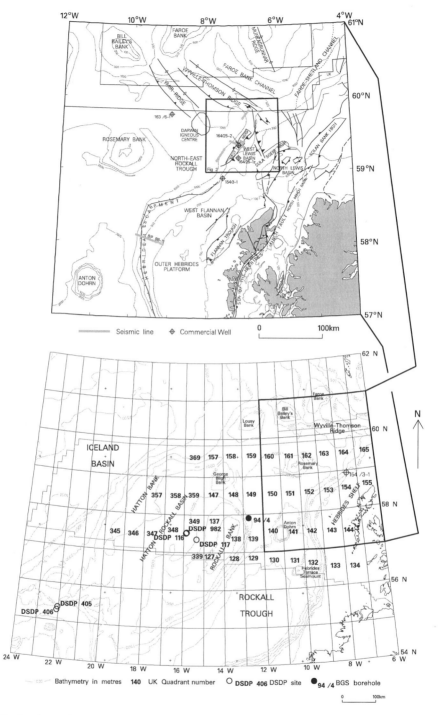

Fig. 1. Regional overview of the Rockall Plateau and Trough showing the position of key exploration wells and DSDP borehole sites referred to in this paper. The detailed map of the northeastern Rockall Trough and adjacent Hebridean shelf margin highlights the bathymetry, distribution of major fault trends and structural features south of the Wyville-Thomson Ridge.

structures during the early Oligocene has subsequently controlled both the distribution and geometry of flanking Tertiary sediments. Complex bathymetric and palaeoceanographic circulation patterns were also directly influenced by the development of these compressional structures during Oligocene to Miocene times (Boldreel & Andersen 1995).

Seismic-stratigraphical framework

Understanding of the Tertiary evolution of the Hebridean margin and flanking deep-water Rockall Trough is in an early stage of development because of the sparse well control and lack of closely spaced seismic coverage. Dating and correlation of key seismic reflectors across the Hebridean margin and westwards into the Rockall Trough has been aided by the release of four exploration wells (163/6-1, 154/3-1, 164/25-1,1z and 164/25-2) drilled between 1980 and 1992. These wells allow some degree of regional correlation of seismic reflectors. Early studies of DSDP sites 116 and 117 in the Hatton–Rockall Basin (Roberts, 1975) dated a prominent reflector (R4) as late Eocene in age. This reflector is a significant angular unconformity truncating the underlying strata, but attempts to correlate this reflector across into the Rockall Trough were initially problematic.

Masson & Kidd (1986), using evidence from DSDP site 610 in the southern Rockall Trough, presented a seismic stratigraphy which essentially modified the previous interpretation of Roberts (1975) and Roberts *et al.* (1981). Bull & Masson (1996) have identified four principal Cenozoic reflectors in an area to the south of Edoras Bank, southwestern Rockall Plateau; these reflectors have been linked to DSDP sites 405 and 406 and are important for constraining the stratigraphical evolution of these frontier basins. Recently, significant volumes of high-resolution and seismic-reflection data have been acquired by the BGS as part of the Rockall Continental Margin Project. Interpretation of these data has allowed the establishment for the first time of a unified stratigraphical framework for the mid–late Cenozoic strata of the Rockall margin (Stoker this volume).

Although no seismo-stratigraphical framework currently exists for the Tertiary sediments of the Hebridean margin, recent offshore geological mapping (Egerton & Stoker in press) of quadrants 164 and 165 in the northeastern part of the Rockall Trough has provided new information on the distribution and geometry of Palaeogene units in the West Lewis Basin adjacent to the Wyville-Thomson Ridge. The

results of this offshore reconnaissance mapping provide the basis for this paper.

Correlation of seismic reflectors

This study has correlated the reflectors imaged on the multichannel seismic lines with exploration wells 164/25-1,1z in the West Lewis Basin and 164/25-2 located on the West Lewis Ridge (Fig. 2). These wells provide an important constraint on the age and lithological variations within individual seismic units. The key reflectors identified in this study are presented in Fig. 3 tied to the two exploration wells in quadrants 164 and 165. Extrapolation of chronostratigraphically significant surfaces within the West Lewis Basin and adjacent northeastern Rockall Trough allows a seismo-stratigraphical framework to be constructed. Exploration well 154/3-1 penetrated over 900 m of basaltic volcanic succession with interbedded sediments reaching total depth in Lewisian gneisses. The location of this well allows direct comparison between the prograding seismic facies and the geophysical composite log, and additional information on the lithologies is provided from the core chippings.

A number of shallow BGS boreholes positioned on the Hebrides Shelf also give information on the Palaeocene and Eocene units. On the western side of the West Lewis Basin, the middle to upper Eocene unit forms a thick 900–1000 m prograding shelf-margin wedge. Shallow drilling to the north of the Sula Sgeir High (BGS borehole 90/2) has proved the presence of shallow-marine deltaic–paralic sediments associated with the landward part of this downlapping wedge. The Palaeocene volcanic succession has been penetrated in BGS borehole 85/7, which recovered fine-grained amygdaloidal N type mid-ocean ridge basalts (MORB) dated as Danian to Thanetian in age (Stoker *et al.* 1988, 1993).

Seismic character of the upper Palaeocene to mid-Miocene succession

Middle–upper Palaeocene seismic unit

The top of this seismic unit is characterized by a distinct high- to very-high-amplitude reflector which in some places significantly degrades the seismic resolution of the pre-Tertiary geology. In a number of locations the horizon shows indications of post-depositional faulting and internal facies variations. Parallel, prograding and hummocky seismic facies units can be distinguished in the volcanic sequence; this is taken to indicate

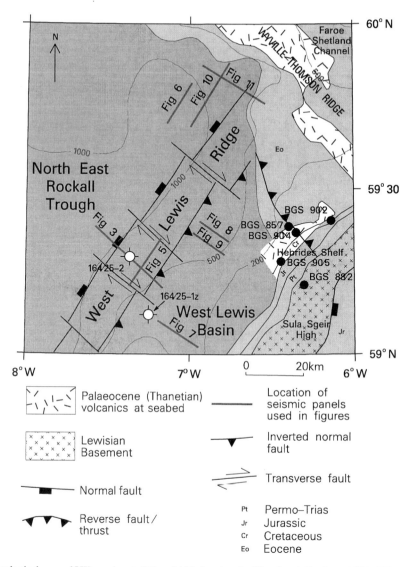

Fig. 2. Geological map of UK quadrants 164 and 165 showing the West Lewis Basin and NE–SW-trending West Lewis basement ridge (after Egerton & Stoker in press). The locations of the illustrated seismic panels relative to exploration wells 164/25-2 and 164/25-1z are shown.

lateral changes from subaerial to sub-aqueous environments. Across the Wyville-Thomson Ridge the volcanic rocks are interpreted to crop out at or near to the sea bed.

On high-resolution shallow seismic profiles the volcanic rocks are characterized by irregular, high-amplitude reflection configurations forming a complex bathymetric relief. Wood *et al.* (1988) described a suite of primary depositional scarps and terraces in the upper Palaeocene volcanic

unit fringing the margins of the Rosemary Bank igneous centre. These features were interpreted as representing the location of a palaeoshoreline where vertically stacked subaerially extruded basalt flow-fronts reached a marine environment and solidified at the boundary, forming a steep escarpment with flanking volcanic debris apron. Multichannel seismic profiles across the Hebridean margin, tied to well 154/3-1, reveal the presence of large-scale (700 ms two-way

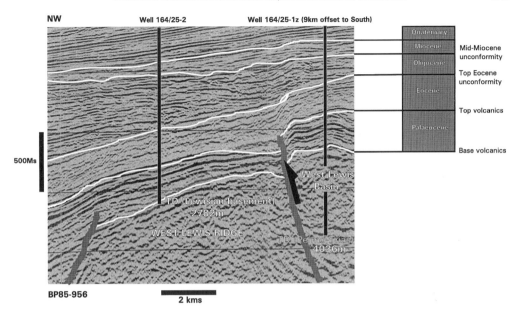

Fig. 3. Simplified seismic tie to exploration wells 164/25-2 on the West Lewis Ridge and well 164/25-1z in the adjacent West Lewis depocentre. An inverted normal fault bounding the eastern side of the West Lewis Ridge has generated an inversion anticline.

travel time (twt)) downlapping clinoforms (Fig. 4) within the volcanic succession (Andersen *et al.* 1997).

Well 154/3-1 penetrated over 900 m of basaltic volcanic rocks and thin interbedded sediments with a marked change in the gamma response between the upper and lower parts of the unit. Overall, the prograding sequence can be subdivided into a series of basinward-stepping packages bounded by unconformities or their correlative conformities (Fig. 4).

These prograding units have been interpreted to represent the product of pulses of extrusive volcanic activity along the margins of a marine basin interacting with changes in relative sea level. Based on comparisons with modern and ancient field analogues in Hawaii and western Greenland (Moore *et al.* 1973; Pedersen *et al.* 1996), the clinoforms are predicted to be largely composed of basaltic hyaloclastic breccias passing upwards and landwards into parallel-layered coherent lava flows. Small-scale

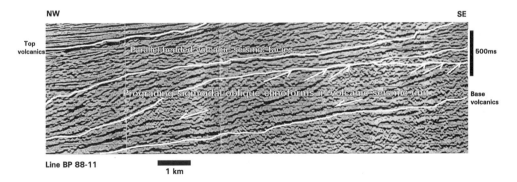

Fig. 4. Multichannel seismic-reflection profile (BP 88-11) across the outer Hebrides Platform (Fig. 1) showing oblique and sigmoidal reflectors in a prograding seismic unit, with an overlying parallel- layered seismic unit. Correlation from well 154/3-1 located to the northeast of this profile indicates that the prograding package represents interbedded hyaloclastic breccias and basaltic lava flows of late Palaeocene age. These units are bounded by erosional unconformities suggesting changes in relative sea level.

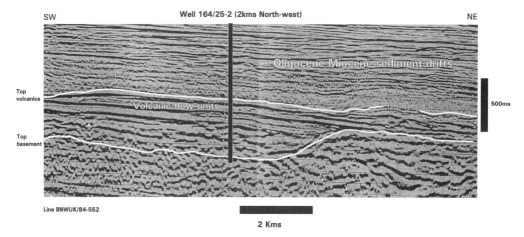

Line BNWUK/84-552

2 Kms

Fig. 5. Structurally controlled, incised valley fill on the crest of the West Lewis Ridge. The internal fill is interpreted to represent volcanic flow units in the basal and upper parts of the valley interbedded with sedimentary units. The volcanic seismic units thin onto the flanks of the valley with marginal levees developed at areas of overspill.

variations in the individual clinoforms are beyond the level of seismic resolution. Seismic data shot parallel to the depositional slope of the Hebridean margin indicate the presence of possible palaeovalleys which have been infilled with the prograding basaltic lava flows (Fig. 5). It is suggested that pre-existing structural lineaments may have acted as feeder conduits focusing the basaltic flows into the zone of low topography. The seismic resolution in the West Lewis Basin is in one example sufficient to image interleaved volcanic and sedimentary packages along the margins of these valleys. Overspill of the volcanic flow units is indicated by the development of marginal levees flanking the valley features (Fig. 5). Evidence from exploration well 164/25-2 indicates the presence of late Palaeocene shallow-marine or coastal clastic deposits interbedded with basaltic

volcanic rocks on the crest of the West Lewis Ridge.

Lower–middle Eocene seismic unit

This depositional unit is manifested on the seismic-reflection profiles as acoustically well-layered, moderate- to high-amplitude reflectors overlying a complex, hummocky, high-amplitude event representing the top of the volcanic units. In a number of locations along the Hebridean margin, the lower–middle Eocene seismic package appears to become acoustically transparent or displays very low amplitudes.

This dramatic amplitude reduction may be an indication of a change in the overall lithology between the lower and middle Oligocene and the relatively homogeneous nature of the lower

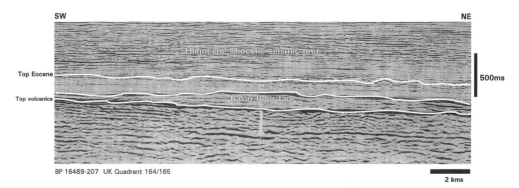

BP 16489-207 UK Quadrant 164/165

2 kms

Fig. 6. Multichannel seismic-reflection profile showing an upper Palaeocene–lower Eocene basin-floor fan interpreted to be sourced from the southern margin of the Wyville-Thomson Ridge.

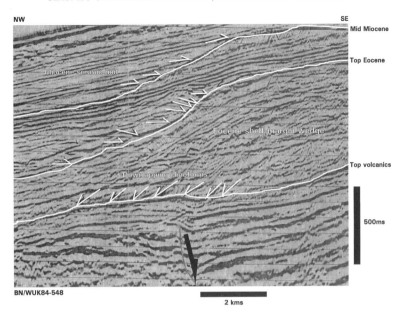

Fig. 7. Multichannel seismic-reflection profile BN/WUK84-548 located in quadrants 164 and 165. A thick (>800 ms) Eocene shelf-margin wedge can be observed downlapping onto the uppermost Palaeocene volcanic units. The wedge is truncated by a major unconformity at uppermost Eocene level, which is onlapped by a mounded sediment–drift sequence of Oligocene to early Miocene age. Normal faulting may be observed affecting the underlying Palaeocene volcanic unit.

Eocene in this location. Discrete, potentially sand-prone basin-floor fans displaying moderate-amplitude discontinuous reflectors have been imaged in lateral association with onlapping moderate- to low-amplitude reflectors (Fig. 6). The fans display acoustic thicknesses of 200–250 ms in this location and probably represent the distal portions of basin-margin prograding wedges developed along the margin of the northeastern Rockall Trough.

Middle–upper Eocene seismic package

This depositional package is characterized by prograding clinoforms (Fig. 7) which downlap onto the low-amplitude parallel-bedded reflectors of the lower Eocene. In a number of locations the middle–upper Eocene package can be observed downlapping directly onto the uppermost Palaeocene volcanic horizon. The lower Eocene deposits are interpreted to be thin or absent in these areas. The clinoforms lose expression to the west into the basin and pass laterally into a low- to moderate-amplitude, discontinuous, reflection character.

A major regional unconformity characterizes the top of this depositional package, forming a prominent high-amplitude continuous reflector which shows evidence of erosional truncation of

underlying middle–upper Eocene reflectors. This reflector has been tied to well 164/25-2 located on the crest of the West Lewis Ridge. The unconformity surface displays a relatively complex morphology, especially in the southern and central parts of the West Lewis Basin and west of the Hebridean Escarpment, where evidence of channelling and incision by alongslope current activity has been observed on the seismic profiles (Fig. 8).

The unconformity becomes increasingly difficult to map in proximity to the Wyville-Thomson Ridge, where compressional deformation structures obscure the reflector morphologies (Boldreel & Andersen 1993). This upper Eocene–lower Oligocene unconformity surface has been recognized throughout the western and eastern margins of the Rockall Trough and the Hatton–Rockall Basin (Boldreel & Anderson 1994; Stoker this volume), the southeastern margin of the Hatton–Rockall Basin (Roberts 1975), and more recently south of Edoras Bank (Bull & Masson 1996). Vanneste *et al.* (1995) also recognized a distinct erosional unconformity in the Bill Bailey's Bank and Lousy Bank area to the southwest of the Faeroe Islands which has been correlated with the middle–upper Eocene reflector in the Rockall Trough.

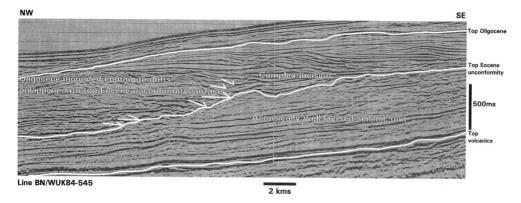

NW   SE

Top Oligocene

Top Eocene
unconformity

Oligocene mounded contourite drifts   Complex incision
onlapping onto top Eocene unconformity surface

500ms

Acoustically well-layered seismic unit

Top
volcanics

Line BN/WUK84-545

2 kms

Fig. 8. The complex morphology of the uppermost Eocene unconformity surface is apparent on this multichannel profile. The incision is interpreted to represent alongslope current activity as evidenced by the presence of mounded Oligocene contourite drifts directly overlying the unconformity surface.

Upper Eocene–lower Oligocene seismic unit

A distinct change in seismic character occurs above the upper Eocene unconformity surface with the widespread development of onlapping and upslope-accreting high-amplitude reflectors (Fig. 9). Mounded and sheeted geometries occur in proximity to the Wyville-Thomson Ridge with complex patterns of drift development. Significant fan geometries have been identified on the multichannel seismic profiles south of the Wyville-Thomson Ridge.

Line BN-WUK84-551 (Fig. 10) shows a very well-developed convex-upwards seismic package characterized by internally downlapping high- to moderate-amplitude reflectors. The package downlaps onto acoustically well-layered lower–middle Eocene events and thins towards the southwest. The wedge has an acoustic thickness ranging from 250 ms to a maximum of 400

ms adjacent to the ridge. Seismic profile WM-90-332, trending NW–SE (Fig. 11), shows a distinct wedge development prograding to the WNW and restricted to the base of the palaeoslope to the south of the compressional fault bounding the Wyville-Thomson Ridge.

The upper Eocene–lower Oligocene seismic unit is overlain by a high-amplitude unconformity surface which is onlapped by upslope-accreting middle–upper Oligocene sediment drifts. Similar wedge-shaped seismic facies units have been recognized and drilled BGS borehole 94/4 along the western margin of the Rockall Trough adjacent to Rockall Bank (Stoker this volume). To the south of the Edoras Bank margin, a large carbonate-dominated submarine fan has been revealed by DSDP drilling and multichannel seismic acquisition (Roberts et al. 1984).

BGS shallow borehole 90/2 located to the south of the Wyville-Thomson Ridge on the

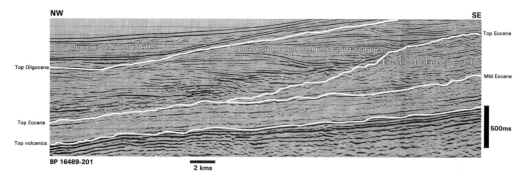

NW   SE

Top Eocene

Miocene sediment drifts   Oligocene mounded sediment drift complex

Top Oligocene   Eocene shelf-margin wedge

Mid Eocene

Top Eocene

500ms

Top volcanics

BP 16489-201

2 kms

Fig. 9. Multichannel seismic-reflection profile showing an Oligocene mounded drift-unit overlying an Eocene shelf-margin wedge in the West Lewis Basin.

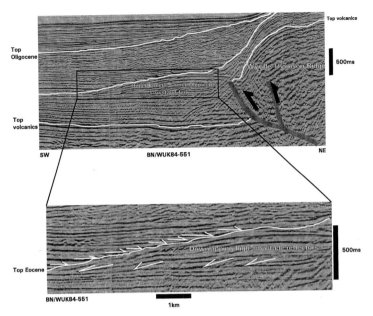

Fig. 10. Multichannel seismic-reflection profile across the southern flank of the Wyville-Thomson Ridge. A well-developed downlapping seismic package is present at the base of slope, which is interpreted to represent a point-sourced fan deposited during a relative lowstand in sea level. The lowstand fan is onlapped by upslope-accreting contourite drifts.

Hebrides Shelf penetrated the upper part of a prograding wedge, revealing the presence of fine- to coarse-grained, poorly sorted, bioclastic sandstone. The sandstone comprised fragmented shell debris and a poorly sorted siliciclastic component (Stevenson 1990). These sediments have been interpreted as having been deposited by mass-flow processes that moved the carbonate sands downslope. Because of the restricted areal extent of the lowstand fans imaged on seismic, it seems likely that the sediment was point sourced and potentially channelled through canyons incised into the western flank of the Wyville-Thomson Ridge. The timing of fan development may be related to pulses of compressional uplift which affected the whole region during the late Eocene–early Oligocene (Boldreel & Andersen 1993; Doré & Lundin 1996).

Middle–upper Oligocene seismic unit

This interval is characterized by very complex drift development, with acoustically well-layered mounded and sheetlike geometries (Stoker *et al.* 1993; Stoker this volume). These complicated internal features are interpreted to represent sediment drift and contourite development along the Hebridean slope during this time.

Adjacent to the Wyville-Thomson Ridge the upper Oligocene package is affected by rotational faulting, which is commonly clearly imaged on high-resolution and multichannel seismic profiles. Upslope-accreting contourite drifts have developed overlying the listric fault planes (Baltzer *et al.* this volume). The top of this distinct seismic unit is marked by a very-high-amplitude reflector which truncates the underlying seismic reflectors. This erosional unconformity is well developed above the Darwin Igneous Centre, where there is a marked angular discontinuity and erosional truncation of the underlying reflectors.

Along the margins of the West Lewis Basin the unconformity displays a progressive westward incision with evidence of complex basal channelling adjacent to the Wyville-Thomson Ridge. Above the uppermost Oligocene reflector, downlapping high-amplitude continuous reflectors dominate the seismic character. Evidence from well 164/25-2 (Fig. 2) indicates that the middle–upper Miocene package overlying the West Lewis Ridge is mud dominated whereas the sequence below the unconformity is characterized by interbedded shallow-marine sandstones and mudstones with moderate degrees of bioturbation.

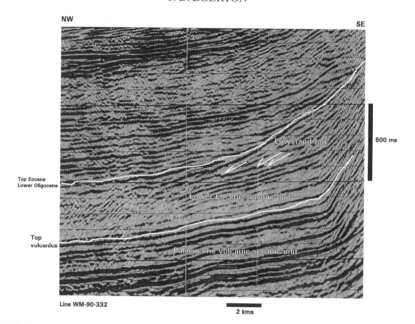

Fig. 11. Multichannel seismic profile along the southern flank of the Wyville-Thomson Ridge showing the wedge-shaped geometry of a late Eocene–early Oligocene lowstand fan. This profile is perpendicular to Fig. 10, illustrating the external architecture of the seismic package.

Spatial variations in stratigraphical thickness

The stratigraphical thickness variations in the middle–upper Eocene seismic unit reflect contemporaneous normal faulting and associated subsidence along the Hebridean margin during the Eocene. Distinct progradational delta lobes can be recognized along the shelf-slope margin comprising an 800–1000 m thick accumulation in the south of the area and a relatively isolated fan adjacent to the Wyville-Thomson Ridge that attains 800–900 m in thickness.

Subsurface mapping reveals that there is a significant reduction in the thickness of the middle–upper Eocene seismic unit in the western part of the West Lewis Basin and across into the northeastern Rockall Trough (Egerton & Stoker in press). This transition probably represents the distal apron of the prograding delta passing into deep-marine pelagic–hemipelagic sediments. It is thought likely that extensional subsidence along a westerly dipping normal fault in part controlled the accumulation of the shelf-margin wedge.

Discussion

Progradational geometries within the volcanic succession have been imaged along the Hebridean margin together with the presence of palaeovalleys acting as feeder channels for the volcanogenic material into the basin. The seismic reflection data show that the thickness and distribution of the prograding volcanogenic wedge may have been influenced by pre-existing fault-controlled palaeovalleys. Cenozoic compressional deformation affecting the Rockall Trough and Norwegian Sea (Doré & Lundin 1996; Doré et al. 1997) was extensively focused along NW–SE pre-existing basement trends, therefore the orientation and preferential location of these conduits may be strongly influenced by these basement lineaments.

The overlying lower Eocene shelf-margin wedge is extensively developed along the Hebridean margin. Extensional subsidence of the basin margin accumulated in excess of 900 m of deltaic–paralic sediments during this time. Adjacent to the Wyville-Thomson Ridge, point-sourced fans have been imaged on seismic profiles; these seismic packages have been interpreted to represent mass-flow input possibly resulting from a lowstand of relative sea level induced by pulses of late Eocene to early Oligocene compression which affected the North Atlantic region and Greenland Sea (Doré & Lundin 1996; Doré et al. 1997).

The seismic identification of a discrete upper Eocene–lower Oligocene fan geometry along the southwestern flank of the Wyville-Thomson Ridge provides evidence to suggest tectonic uplift and erosion of the structurally generated topography or bathymetry. Pulses of transpression controlled by a sinistral strike-slip fault structure bounding the southern part of the ridge (Egerton & Stoker in press) may have induced a relative lowstand in sea level. Incision of the shelf and re-sedimentation of the resultant sediments into the flanking basin as mass-flow deposits is thought to be the likely depositional process. The occurrence of the upper Eocene–lower Oligocene fan provides important constraints on the timing of uplift and deformation on the Hebridean margin. However, the lack of direct borehole or exploration well evidence penetrating this seismic unit necessarily limits the interpretation of the internal composition of the seismic package.

Comparisons can be made between the lowstand wedge identified on the western margin of the Rockall Trough and the Edoras Bank margin (Roberts *et al.* 1984; Stoker this volume). It is thought likely that lowstand fans of a similar nature and areal extent to the example identified around the southern flank of the Wyville-Thomson Ridge are present adjacent to compressional structures in the Faeroe Basin and Mid-Norwegian margin. This has important implications for hydrocarbon exploration and play identification in the frontier acreage of the NE Atlantic. The seismic characteristics of Palaeogene depositional packages along the Hebridean margin and northeast Rockall Trough reflect the interaction of tectonics and complex bathymetry resulting from distinct phases of fault-generated subsidence and uplift.

Listric faulting which has been identified (Baltzer this volume) affecting the upper part of the Oligocene sequence adjacent to the Wyville-Thomson Ridge may have been generated by seismic events resulting in sediment instability and downslope mobilization.

The incision recognized within the lower–middle Miocene seismic unit may be explained by the concentration and focusing of bottom currents in proximity to the steep bathymetric profile of the Wyville-Thomson Ridge (Boldreel & Andersen 1995; Stoker this volume). It is clear that Tertiary sedimentation patterns and processes along the northeastern margin of the Rockall Trough have been strongly influenced by the complex structural history of the area combined with relative changes in sea level.

I would like to express my thanks to B. Mitchener & BP Exploration Operating Company Ltd (Atlantic Frontier Programme) for access to extensive seismic data and exploration well information from the northeastern Rockall Trough. Thanks are due to D. Evans, K. Hitchen and referees Tj. van Veering and D. Masson for their comprehensive and constructive reviews of the manuscript. This paper is published with the permission of the Director of the British Geological Survey (NERC).

References

ANDERSEN, M. S., EGERTON, P. D., HITCHEN, K. & BOLDREEL, L. O. 1997. Seismic facies analysis of volcanic successions: offshore UK and the Faroes. In preparation.

BALTZER, A., HOLMES, R. & EVANS, D. 1998. Debris flows on the Sula Sgeir Fan, NW of Scotland. *This volume.*

BOLDREEL, L. O. & ANDERSON, M. S. 1993. Late Paleocene to Miocene compression in the Faeroe–Rockall Area. *In:* PARKER, J. R. (ed.) *Petroleum Geology of NW Europe: Proceedings of the 4th Conference,* Geological Society, London, 1025–1034.

—— & —— 1994. Tertiary development of the Faeroe–Rockall Plateau based on reflection seismic data. *Bulletin of the Geological Society of Denmark,* **41,** 162–180.

—— & —— 1995. The relationship between the distribution of Tertiary sediments, tectonic processes and deep-water circulation around the Faeroe Islands. *In:* SCRUTTON, R. A., STOKER, M. S., SHIMMIELD, G. B. & TUDHOPE, A. W. (eds) *The Tectonics, Sedimentation and Palaeoceanography of the North Atlantic Region.* Geological Society, London, Special Publications, **90,** 145–158.

BULL, J. M., & MASSON, D. G. 1996. The southern margin of the Rockall Plateau: stratigraphy, Tertiary volcanism and plate tectonic evolution. *Journal of the Geological Society, London,* **153,** 601–612.

DORÉ, A. G. & LUNDIN, E. R. 1996. Cenozoic compressional structures on the north-east Atlantic margin: nature, origin and potential significance for hydrocarbon exploration. *Petroleum Geoscience,* **2,** 299–311.

——, LUNDIN, E. R., FICHLER, C. & OLESEN, O. 1997. Patterns of basement structure and reactivation along the NE Atlantic margin. *Journal of the Geological Society, London,* **154,** 85–92.

EARLE, M. M., JANKOWSKI, E. J. & VANN, I. R. 1989. Structural and stratigraphic evolution of the Faeroe–Shetland Channel and Northern Rockall Trough. *American Association of Petroleum Geologists, Memoir,* **46,** 461–469.

EGERTON, P. D. & STOKER, M. S. *Sula Sgeir, 59N–08W, Solid Geology, 1:250,000 map series,* 2nd edition (revised). Ordnance Survey, Southampton, for the British Geological Survey, Natural Environment Research Council, in press.

JONES, E. J. W., PERRY, R. G. & WILD, J. L. 1986. Geology of the Hebridean margin of the Rockall

Trough. *Proceedings of the Royal Society of Edinburgh, Section B*, **88**, 27–51.

KNOTT, S. D., BURCHELL, M. T., JOLLEY, E. W. & FRASER, A. J. 1993. Mesozoic to Cenozoic plate reconstructions of the North Atlantic and hydrocarbon plays of the Atlantic margins. *In*: PARKER, J. R. (ed.) *Petroleum Geology of Northwest Europe: Proceedings of the 4th Conference*. Geological Society, London, 953–974.

MASSON, D. G. & KIDD, R. B. 1986. Revised Tertiary seismic stratigraphy of the southern Rockall Trough. *In*: RUDDIMAN, W. F., KIDD, R. B. *et al.* (eds) *Initial Reports of the Deep Sea Drilling Project*, **94**. US Government Printing Office, Washington, DC, 1117–1126.

MOORE, J. G., PHILLIPS, R. L., GRIGG, G. R. W., PETERSON, D. W. & SWANSON, D. A. 1973. Flow of lava into the sea, 1969–1971, Kilauea volcano, Hawaii. *Geological Society of America Bulletin*, **84**, 537–546.

MUSGROVE, F. W. & MITCHENER, B. 1995. A structural history of the Rockall Trough, west of Britain. *EAGE 57th Conference and Technical Exhibition* (Abstract).

—— & —— 1996. Analysis of the pre-Tertiary rifting history of the Rockall Trough. *Petroleum Geoscience*, **2**, 353–360.

PEDERSEN, A. K., LARSEN, L. M., PEDERSEN, G. K. & DUEHOLM, K. S. 1996. Filling and plugging of a marine basin by volcanic rocks: the Tunoqqu Member of the Lower Tertiary Vaigat Formation on Nuussuaq, central West Greenland. *Grønlands Geologiske Undersøgelse Bulletin*, **171**, 5–28.

ROBERTS, D. G. 1975. Marine geology of the Rockall Plateau and Trough. *Philosophical Transactions of the Royal Society of London, Series A*, **278**, 447–509.

——, BACKMAN, J., MORTON, A. C., MURRAY, J. W. & KEENE, J. B. 1984. Evolution of volcanic rifted margins: synthesis of Leg 81 results on the west margin of Rockall Plateau. *In*: ROBERTS, D. G., SCHNITKER *et al.* (eds) *Initial Reports of the Deep Sea Drilling Project*, **81**. US Government Printing Office, Washington, DC.

——, MASSON, D. G. & MILES, P. R. 1981. Age and structure of the southern Rockall Trough: new

evidence. *Earth and Planetary Science Letters*, **52**, 115–128.

SNYDER, D. B. 1990. The Moine Thrust in the BIRPS data set. *Journal of the Geological Society, London*, **147**, 81–86.

STEIN, A. M. & BLUNDELL, D. J. 1990. Geological inheritance and crustal dynamics of the northwest Scottish continental shelf. *Tectonophysics*, **173**, 455–467.

STEVENSON, A. G. 1990. *British Geological Survey shallow drilling programme 1990. Final Geological Report*. BGS Technical report **WB/90/47C**.

STOKER, M. S. 1998. Sediment-drift development on the continental margin off NW Britain. *This volume*.

—— 1997. Mid- to late Cenozoic sedimentation on the continental margin off north-west Britain. *Journal of the Geological Society, London*, **154**, 509–515.

——, HITCHEN, K. & GRAHAM, C. C. 1993. *United Kingdom offshore regional report: the geology of the Hebrides and West Shetland shelves and adjacent deep-water areas*. HMSO, London, for the British Geological Survey.

——, MORTON, A. C., EVANS, D., HUGHES, M. J., HARLAND, R. & GRAHAM, D. K. 1988. Early Tertiary basalts and tuffaceous sandstones from the Hebrides shelf and Wyville-Thomson Ridge, NE Atlantic. *In*: MORTON, A. C. & PARSON, L. M. (eds) *Early Tertiary Volcanism and the Opening of the NE Atlantic*. Geological Society, London, Special Publications, **39**, 271–282.

VANNESTE, K., HENRIET, J.-P., POSEWANG, J. & THEILEN, F. 1995. Seismic stratigraphy of the Bill Bailey and Lousy Bank area: implications for subsidence history. *In*: SCRUTTON, R. A., STOKER, M. S., SHIMMIELD, G. B. & TUDHOPE, A. W. (eds) *The Tectonics, Sedimentation and Palaeoceanography of the North Atlantic Region*. Geological Society, London, Special Publications, **90**, 125–139.

WOOD, M. V., HALL, J. & DOODY, J. J. 1988. Distribution of early Tertiary lavas in the NE Rockall Trough. *In*: MORTON, A. C. & PARSON, L. M. (eds) *Early Tertiary Volcanism and the Opening of the NE Atlantic*. Geological Society, London, Special Publications, **39**, 283–292.

Sediment-drift development on the continental margin off NW Britain

M. S. STOKER

British Geological Survey, Murchison House, West Mains Road, Edinburgh EH9 3LA,
UK

Abstract: The long-term record of sediment-drift development on the continental margin off NW Britain was investigated using seismic-reflection data calibrated with boreholes and short cores. These data indicate that the onset of bottom currents on the NW British margin occurred in the early late Eocene following a major phase of subsidence in the Rockall Trough and Hatton–Rockall Basin. Seismic reflection profiles reveal the downwarped and eroded surface of lower upper Eocene and older strata onlapped by middle to upper Cenozoic sediments. The latter consist predominantly of deep-marine contourites preserved both as mounded, elongate, and broad, sheeted, sediment drifts with associated sediment waves. Seismic-stratigraphical analysis of the drift succession has revealed a broad two-stage depositional history. A vigorous bottom-current regime in the late Eocene to mid-Miocene interval was accompanied by significant lateral migration of sediment by upslope accretion onto the flanks of the basins, including the construction of the bulk of the Feni Ridge. Mid-Miocene tectonism modified the palaeoceanographic and sedimentary regime. In the ensuing mid-Miocene to Holocene interval, drift accumulation persisted in the axial and eastern part of the Rockall Trough whereas a largely erosional regime prevailed on its western margin. The overall locus of drift sedimentation shifted westwards onto the Rockall Plateau.

Sediment drifts and associated bedforms, such as sediment waves, are a major depositional product of bottom-current activity in deep-marine environments. In the NE Atlantic, for example, it has been estimated that such deposits form about 20% of the total volume of Cenozoic sediments preserved between the Charlie Gibbs Fracture Zone and the Greenland–Scotland Ridge (Wold 1994). The most characteristic drifts are the mounded, elongate drifts of open basin systems, which lie parallel to continental margins, and may have a relief of several hundred metres above the sea floor. Less well known are the extensive broad-domed to flat-lying sheeted drifts that form basin-plain fills in more topographically enclosed basins (Faugères *et al.* 1993; Howe *et al.* 1994; Stoker *et al.* 1998). The composition of drifts tends to reflect the regional setting; in the open ocean, biogenic contourites of typically pelagic aspect are common (Kidd & Hill 1986), whereas an increased abundance of terrigenous material occurs in contourites adjacent to continental margins (Stow 1986).

The development of sediment drifts can be used as an indicator of the palaeoceanographic history of a region. The long-term record, especially, may provide information about the distribution and timing of large-scale sedimentary changes related to changes in bottom-current circulation, as expressed by unconformities and changing styles of deposition, together with the factors responsible for the changes. Plate-tectonic processes, in particular, may result in changes in basin geometry and palaeobathymetry, thereby modifying palaeoceanographic circulation and deep-water sedimentation patterns (Cloetingh *et al.* 1990).

The impact of plate reorganization on sediment-drift development is particularly evident on the continental margin off NW Britain. This is an area of complex bathymetry flanking the oceanic Iceland Basin (Fig. 1). The continent–ocean boundary is located on the western flank of Hatton Bank (White 1992). This structural configuration largely reflects late Mesozoic–early Cenozoic continental rifting which was related to the processes that led to the opening of the NE Atlantic Ocean. The initiation of sea-floor spreading in the NE Atlantic would have placed the continental margin off NW Britain in a buffer zone between the spreading ridge to the west and the Africa–Europe convergence zone to the southeast (Knott *et al.* 1993), thus making it susceptible to any changes in the intra-plate stress field. Intra-plate tectonism, linked to intermittent tectonic phases, directly affected the evolution of the continental margin by creating its present morphological expression during a major, mid-Cenozoic, basin-subsidence event which, in turn, led to the onset of bottom currents (Stoker 1997).

STOKER, M. S. 1998. Sediment-drift development on the continental margin off NW Britain. *In*: STOKER, 229
M. S., EVANS, D. & CRAMP, A. (eds) *Geological Processes on Continental Margins: Sedimentation, Mass-Wasting and Stability*. Geological Society, London, Special Publications, **129**, 229–254.

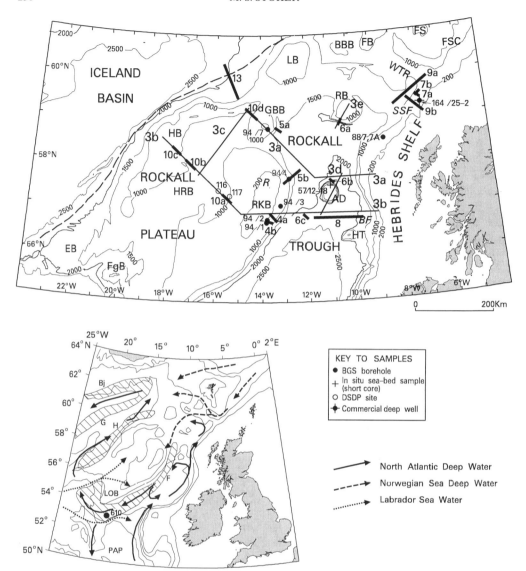

Fig. 1. Bathymetric setting of the continental margin off NW Britain (contours in metres), showing locations of sections and seismic profiles illustrated in Figs 3–10 and 13, and sample sites referred to in figures and text. Dashed line indicates approximate position of the continent–ocean boundary. Small map shows present-day bottom-circulation pattern (derived from a variety of sources: see text for references) and major, well-known, sediment drifts (diagonal rule). FSC, Faeroe–Shetland Channel; FS, Faeroe Shelf; FB, Faeroe Bank; BBB, Bill Bailey's Bank; LB, Lousy Bank; WTR, Wyville-Thomson Ridge; RB, Rosemary Bank; AD, Anton Dohrn; HT, Hebrides Terrace; RkB, Rockall Bank; HRB, Hatton–Rockall Basin; HB, Hatton Bank; GBB, George Bligh Bank; EB, Edoras Bank; FgB, Fangorn Bank; LoB, Lousy Bank; BF, Barra Fan; SSF, Sula Sgeir Fan; R, Rockall islet; F, Feni Ridge; H, Hatton Drift; G, Gardar Drift; Bj, Bjorn Drift; PAP, Porcupine Abyssal Plain.

The aim of this paper is to outline the long-term record of sediment-drift development on the continental margin off NW Britain, specifically the Rockall Trough and Hatton–Rockall Basin, and the tectono-oceanographic changes controlling their formation and evolution.

Emphasis is placed on the large-scale sedimentary response to changes in basin geometry and palaeocirculation identified from the regional seismic expression of sediment-drift deposits preserved within the middle to upper Cenozoic succession. Both the Rockall Trough

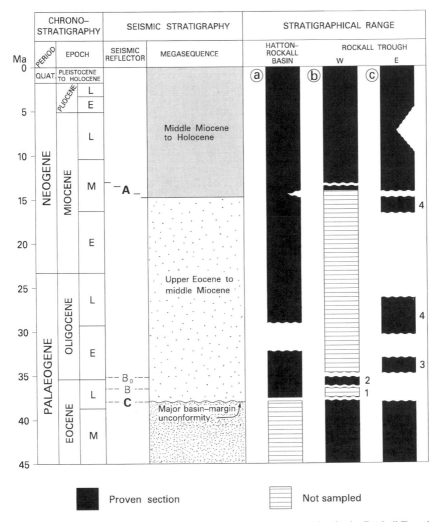

Fig. 2. Stratigraphical-range chart of the middle and upper Cenozoic succession in the Rockall Trough and Hatton–Rockall Basin. Information derived from: (**a**) DSDP sites 116 and 117 (Laughton *et al.* 1972; Berggren & Schnitker 1983); (**b**) BGS boreholes and short cores (see text for details); (**c**) BGS borehole 88/7,7A (Stoker *et al.* 1994), well 164/25-2, and Jones *et al.* (1986). Numbered sections (1–4) refer to upper Eocene to middle Miocene component sequences identified in Rockall Trough (see Fig. 3). The timescale is from Harland *et al.* (1990).

and the Hatton–Rockall Basin were open to the NE Atlantic Ocean throughout the mid- to late Cenozoic interval, and thus were sensitive to changes in bottom-current regime. From this, it is possible to reconstruct the gross depositional history of drift development, demonstrate the changing seismic character and sediment-body geometry of the drifts, and place specific phases of deposition and palaeocirculation within a tectono-stratigraphical framework.

The study is based on the extensive geophysical and geological database collected by the British Geological Survey (BGS) as part of its reconnaissance mapping programme of the United Kingdom continental margin, which is at present focused on the western frontier basins in the Rockall region as part of a joint BGS–oil company initiative (see acknowledgements). These data include high-resolution seismic-reflection profiles, comprising 2 × 40 and 4 × 40 cu. in. airgun, 1 and 2 kJ sparker and deep-tow boomer, together with available commercial deep-seismic data. Geological calibration of the seismic stratigraphy was achieved with BGS boreholes and short cores, combined with other published material such as that of the Deep Sea

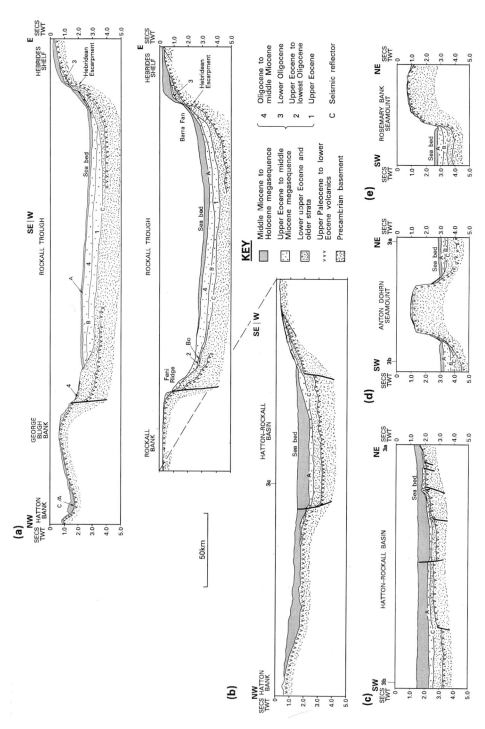

Fig. 3. Interpreted geoseismic cross-sections across the continental margin off NW Britain, focusing on the middle to upper Cenozoic stratigraphy (see text for details). Based on BGS and Mobil high-resolution and deep-seismic data acquired on behalf of the 'Rockall' continental margin consortium (see acknowledgements). Sections located in Fig. 1.

Drilling Project (DSDP) and available commercial well data. The geometry and stratigraphical range of the middle to upper Cenozoic succession in the area of study is depicted in Figs 2 and 3. Stratigraphical information is based on a number of key sample sites which are shown in Fig. 1. For Cenozoic correlation purposes, the calcareous nanoplankton zonal scheme of Martini (1971) has been utilized: the prefix 'NP' applies to the Palaeogene zones, and 'NN' to the Neogene–Quaternary zones.

Regional setting

Physiography

The continental margin off NW Britain is a wide, passive margin extending up to 800 km from the Scottish mainland (Fig. 1). Morphologically, it can be divided into an inner margin consisting of the Hebrides Shelf and an outer margin comprising the Rockall Plateau and adjacent smaller banks (e.g. Lousy Bank, Bill Bailey's Bank), which together form relatively shallow platform areas, separated by the deeper-water Rockall Trough. The latter is a major NE–SW-trending basin bounded to the northeast by the Wyville-Thomson Ridge, but deepening to the southwest where it opens out into the Porcupine Abyssal Plain. Water depths in the trough increase from 1000 m in the northeast to 4000 m in the southwest. The width of the trough (at the 1000 m isobath) is between 200 and 250 km. In the central and northern part of the trough, the continuity of the basin floor is interrupted by the Rosemary Bank, and Anton Dohrn and Hebrides Terrace seamounts.

The Rockall Plateau consists of a series of relatively shallow banks disposed around the deeper-water Hatton–Rockall Basin. The shallowest banks are Rockall Bank, Hatton Bank and George Bligh Bank, which fringe the northern half of the plateau; a number of lesser banks, such as Lorien, Fangorn and Edoras banks, mark the southern and western elevations of the plateau (Roberts 1975). Rockall Island, on Rockall Bank, is the sole subaerial expression of the plateau. The Hatton–Rockall Basin is generally shallower (<1500 m water depth) than the Rockall Trough, but in common with the trough it deepens and opens out to the southwest. The two basins are linked by two narrow channels (5–20 km wide at 1000 m isobath) either side of George Bligh Bank. Similarly, channels cut between the northern edge of Rockall Plateau, Lousy Bank and Bill Bailey's Bank incise the outer margin and link the Rockall Trough directly to the Iceland Basin.

One of the best known and major NE Atlantic sediment drifts, the Feni Ridge, is located on the western margin of the central and southern Rockall Trough, and can be traced for about 700 km along the eastern slope of Rockall Bank. Before this study, the Feni Ridge was regarded as the oldest drift on the continental margin, inferred to have started accumulating at about the Eocene–Oligocene boundary (Masson & Kidd 1986; Wold 1994). This drift stands out because of its giant elongate morphology. However, as will be demonstrated in this paper, it represents only a part of a more complex arrangement, not previously synthesized, of sediment-drift deposits preserved on the NW British margin.

Structure and stratigraphy

The Rockall Trough and Hatton–Rockall Basin are the main expressions of continental rifting across the margin. The locations of the major basin-bounding faults, where known, are indicated in Fig. 3. The structure of the eastern margin of the Rockall Trough remains ambiguous owing to the masking effect, on seismic profiles, of the thick basalt pile forming the Hebridean Escarpment, although recent interpretations favour an easterly dipping bounding fault (Musgrove & Mitchener 1996; England & Hobbs 1997).

The deep bathymetric expression of the Rockall Trough and Hatton–Rockall Basin reflects the underlying structure of the continental margin off NW Britain. The crust beneath these basins was extensively stretched and thinned during a phase of Early Cretaceous rifting associated with an abortive attempt at continental break-up between Greenland and NW Europe along the axis of the Rockall Trough (Roberts et al. 1988; Shannon et al. 1994; Musgrove & Mitchener 1996). During the Late Cretaceous and Palaeogene, the continental margin underwent differential post-rift subsidence which resulted in the overall deepening of the Rockall Trough and Hatton–Rockall Basin. Since the earliest-Eocene onset of sea-floor spreading west of Rockall Plateau (Roberts et al. 1984), this subsidence may have been linked, at least in part, to intra-plate tectonism as a consequence of the plate tectonic setting of the margin relative to the NE Atlantic spreading ridge and the Pyrenean–Alpine orogenic belt (Knott et al. 1993).

The present morphological expression of the continental margin is a mid- to late Cenozoic phenomenon, initiated by a major phase of rapid subsidence in the Rockall Trough and

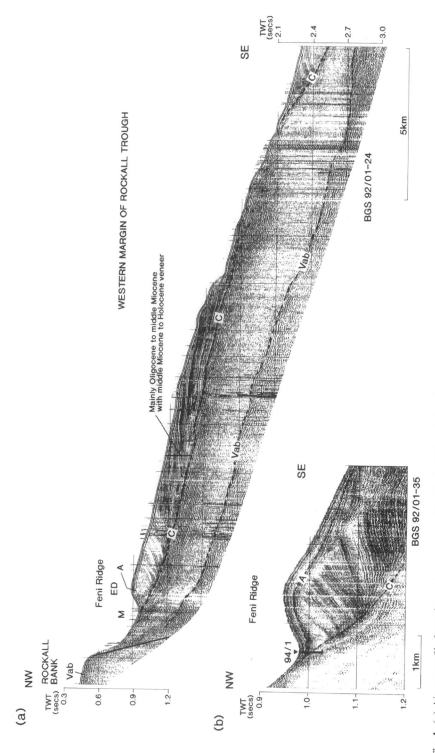

Fig. 4. (**a**) Airgun profile from the western margin of the Rockall Trough showing the seismic characteristics of the Feni Ridge, its modified geometry (by erosion), and progressive upslope migration across the early late Eocene unconformity surface. (**b**) Sparker profile showing detail of lower, upslope-accreting unit, and upper, draped unit where tested by BGS borehole 94/1. Vab, Volcanic acoustic basement (upper Paleocene to lower Eocene volcanics); C, early late Eocene unconformity; A, mid-Miocene unconformity; ED, elongate drift; M, moat; TWT, two-way time. Location of profiles is shown in Fig. 1. Estimates of scale on the seismic profiles are calculated by assuming the velocity of sound in water is 1.45 km s⁻¹, and in sediments is from 1.55 to 1.8 km s⁻¹, based on increasing compaction with depth (Hamilton 1985).

Hatton–Rockall Basin (Stoker 1997). This led to a deepening of the basins. On seismic-reflection profiles, this basin-subsidence event is manifest in the form of a widespread, deep-water unconformity which is particularly enhanced at the margin of the basins and adjacent to the axial seamounts of Rosemary Bank and Anton Dohrn, where the downwarped and eroded surface of lower upper Eocene (reflector C) and older strata is onlapped by middle to upper Cenozoic sediments (Fig. 3). This boundary has also been imaged on seismic profiles from the NW Rockall Trough adjacent to Lousy and Bill Bailey's banks (Vanneste *et al.* 1995). The unconformity is broadly equivalent to reflector R4 as originally defined by Roberts (1975) from the Rockall Plateau, and may also correlate with the 'brown' reflector of Masson & Kidd (1986) from the southern Rockall Trough, and reflector III of Bull & Masson (1996) from the southern margin of the Rockall Plateau.

On the western flank of the Rockall Trough, BGS borehole 94/1 (Fig. 4b) penetrated below the unconformity surface and recovered shallow-marine mudstones and sandstones of early late Eocene (NP16–18) age. Rocks of similar age (NP16–19) have also been proved by BGS boreholes 94/2 and 94/3 (Rockall Bank), and 94/7 (George Bligh Bank) where the unconformity surface occurs at or near to the sea bed (Figs 1 and 3), and in well 164/25-2 on the eastern flank of the basin (Fig. 1) (Egerton this volume). In the Hatton–Rockall Basin, DSDP site 116 terminated in upper Eocene (NP19) limestones at about the level of reflector R4 (Laughton *et al.* 1972; Berggren & Schnitker 1983), although, from inspection of the seismic data, it is unclear whether site 116 actually penetrated below reflector R4. However, a late Eocene age is inferred as the most probable age for the unconformity surface on the basis of the data from the Rockall Trough, together with confirmation of a late Eocene age for the equivalent reflector III south of Rockall Plateau (Bull & Masson 1996).

Regional stratigraphical evidence (Fig. 2) suggests that the oldest rocks overlying the unconformity are also of late Eocene age. This implies that the unconformity developed after a major and rapid phase of basinal subsidence, and resulted in a change in sedimentation style. Although borehole 94/1 only penetrated the feather-edge of the overlying sediments in the Rockall Trough (Fig. 4b), the recovery of fine-grained, deep-marine limestones resting directly on the upper Eocene shallow marine clastic deposits demonstrates a marked deepening of the basin. This subsidence is also apparent from

the change in geometry of the pre- and post-unconformity sequences, corresponding to a change from a shelf-margin to basin infill to a predominantly ponded-basin infill (Fig. 3). Supporting evidence for substantial late Eocene subsidence is provided by BGS short core 57-12/18, which proved a middle to upper Eocene (NP17-19) nearshore conglomerate from the top of Anton Dohrn seamount (Fig. 1), which was sampled at a current water depth of 705 m. From these data, the origin of the unconformity is attributed to submarine erosion by deep-water bottom currents following basinal subsidence, and marks the onset of bottom-current activity in the Rockall Trough and Hatton–Rockall Basin.

The middle to upper Cenozoic succession overlying the early late Eocene unconformity consists of two megasequences established as late Eocene to mid-Miocene (NP19 to NN5–7), and mid-Miocene to Holocene (NN5–7 to NN21) in age (Fig. 2). The boundary separating these megasequences (reflector A) is a regional unconformity surface that is locally angular. It is present both in the Rockall Trough and Hatton–Rockall Basin and forms the base of the Barra Fan on the Hebrides Slope (Fig. 3; see also Fig. 9, below). This boundary may correlate with NE Atlantic reflector R2 of Miller & Tucholke (1983), the 'green' reflector of Masson & Kidd (1986) from the southern Rockall Trough, and reflector II of Bull & Masson (1996) from the southern margin of the Rockall Plateau. Other basinal reflectors and component sequences within the upper Eocene to middle Miocene megasequence of the Rockall Trough, which have been assigned ages on the basis of core data, are indicated in Figs 2 and 3. No similar subdivision of the equivalent package in the Hatton–Rockall Basin has yet been established. On the Hebridean margin, the Plio-Pleistocene subdivision of the middle Miocene to Holocene megasequence has been documented elsewhere (Stoker *et al.* 1993) and is beyond the scope of this paper.

Present bottom-water circulation

The gross nature of bottom-water circulation around the continental margin off NW Britain has not changed substantially since mid-Miocene time (Miller & Tucholke 1983; Stow & Holbrook 1984). The stability of the oceanographic regime may be related to the formation of North Atlantic Deep Water (NADW) at about 12 Ma (Blanc *et al.* 1980). At the present day, NADW is a mixture of water from several sources which collectively influence the physical oceanography of the continental margin (Fig. 1). The most important bottom-water component is the cold (–0.5°C), southwest-flowing, Norwegian Sea Deep Water (NSDW) which enters the Rockall Trough from the north across the western part of the Wyville-Thomson Ridge, where peak flow velocities between 40 and 101 cm s^{-1} have been measured (Ellett & Roberts 1973; Dickson & Kidd 1986). In the Rockall Trough, NSDW entrains Atlantic Water from intermediate levels, mixing with Labrador Sea Water (LSW) as it descends to greater depths in a southward flow along the western margin of the trough. Additionally, a contribution of Antarctic Bottom Water (AABW) is further incorporated within an anticlockwise gyre developed between Porcupine and Rockall banks (Stow & Holbrook 1984; Dickson & Kidd 1986). At the southern end of the Rockall Trough, NADW flows westwards along the southern margin of the Rockall Plateau before turning to the northeast along the edge of Hatton Bank (Fig. 1). Peak flow velocities between 27 and 39 cm s^{-1} have been measured at the southwest end of the Rockall Trough (Dickson & Kidd 1986), whereas velocities between 12 and 23 cm s^{-1} have been recorded from the flank of Hatton Bank (McCave et al. 1980). Northeast-flowing LSW may also enter the Hatton–Rockall Basin from the south (Dickson & Kidd 1986).

Above the NADW, the eastern margin of the Rockall Trough is influenced by a northward-flowing slope-current down to a depth of 1000 m on the Hebrides Slope (Booth & Ellett 1983; Kenyon 1986). The origin of this slope-current is disputed: some elements of warmer saline Mediterranean Deep Water (MDW) have been detected (Harvey 1982; Hill & Mitchelson-Jacob 1993), and some entrainment of the deeper NADW is possible. Termed North-East Atlantic Water (NEAW), this slope-current continues north at water depths of less than 1000 m until it is deflected and accelerated to the west by the Wyville-Thomson Ridge. Peak maximum flows between 15 and 25 cm s^{-1} have been reported from the Hebrides Slope (Howe & Humphery 1995).

Regional seismic expression of sediment-drift deposits

The gross geometry of the mid- to late Cenozoic megasequences, as depicted in Fig. 3, highlights the angular nature of the early late Eocene unconformity (C) and the predominantly ponded, basinward-thickening character of much of the overlying middle to upper Cenozoic drift-dominated succession. This megasequence geometry, together with an internal reflection configuration dominated by onlapping, hummocky to wavy reflectors (detailed below), is deemed to be characteristic of basinal sediment packages derived largely through bottom-current processes, as it is unlikely that any other process could have controlled the development of these packages in such a temporal and spatial manner (Mountain & Tucholke 1985). It should be noted that the megasequences are not composed wholly of drift deposits. On the eastern flank of the Rockall Trough, lower Oligocene reefal sediments (depicted as sequence 3 in Figs 2 and 3) have been reported from the Hebrides Shelf (Jones et al. 1986; Stoker 1997). There is also evidence for localized downslope input into the Rockall Trough, including a lowstand fan adjacent to Rockall Bank (sequence 2 in Figs 2 and 3; see also Fig. 5), but most significantly associated with the mid-Miocene to Holocene progradation of the Hebridean shelf-margin, incorporating a major glacigenic input during the Pleistocene (Stoker 1995, 1997). The latter has resulted in a marked asymmetry in the geometry of the mid-Miocene to Holocene megasequence in the Rockall Trough, in contrast to the equivalent section in the Hatton–Rockall Basin, which is more symmetric in basinal geometry and more wholly related to bottom-current processes.

A number of different sediment-drift styles are preserved within the megasequences, including single- to multi-crested, elongate, drifts, broad, flat-lying to gently domed, sheeted drifts, and fields of associated sediment waves. The variable seismic expression and relations of the drift deposits are described below. Because of the large size of the study area and the contrast in drift preservation and association between the Rockall Trough and Hatton–Rockall Basin, the two basins will be described separately.

Rockall Trough

Western Rockall Trough, Feni Ridge. The Feni Ridge sediment drift displays a classic elongate, mounded geometry with a well-developed moat separating the crest of the drift from Rockall Bank (Fig. 4). The crest of the drift can be traced the length of Rockall Bank (Masson & Kidd 1986), and in the study area has a sea-bed relief of up to 45 m relative to the adjacent moat. Seismic-reflection profiles have revealed a well-layered internal acoustic signature which displays significant upslope accretion over a distance of about 15–18 km. Individual reflections display either onlapping or downlapping

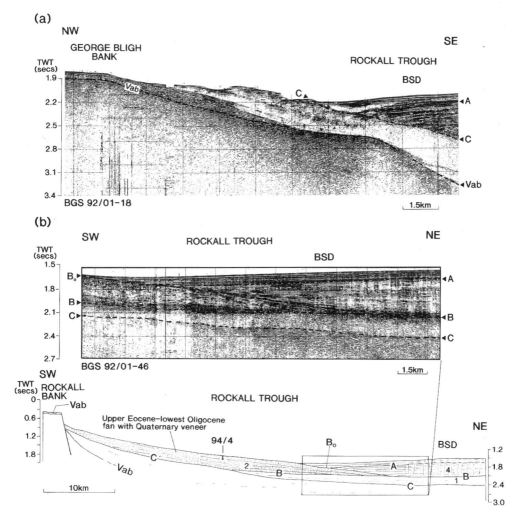

Fig. 5. Airgun profiles showing the seismic characteristics and morphology of broad sheeted-drift deposits in the NW Rockall Trough adjacent to (**a**) George Bligh Bank, and (**b**) NE Rockall Bank, and their progressive westward onlap and pinch-out onto the basin margin. In (**a**) the pre-sediment-drift strata have been eroded and sculpted by bottom currents into an irregular sea-bed topography. In (**b**), a lowstand fan (stippled), calibrated by BGS borehole 94/4, is interbedded with the drift deposits; the stratigraphical relationships are clarified in the associated line drawing. Abbreviations as in Fig. 4 except: B, late Eocene reflector; B_0, earliest Oligocene reflector; BSD, broad, sheeted drift. Component sequences 1, 2 and 4 as in Figs 2 and 3. Location of profiles is shown in Fig. 1.

reflector terminations onto the early late Eocene basin-margin unconformity surface (C) as the evolving drift-crest migrated upslope (Fig. 4). The Feni Ridge loses expression north of Rockall Bank, where a different style of drift accumulation is preserved (see below) (Fig. 5).

Regional stratigraphical information (Figs 2 and 3) indicates that the main phase of drift accumulation and upslope migration occurred during the Oligocene to mid-Miocene interval, and that since the mid-Miocene the western

margin of the Rockall Trough, in the study area, has been largely subjected to erosion by bottom currents. This erosive activity may have removed a significant portion of the drift which, as illustrated in Fig. 4a, has become locally detached from the main basinal accumulation of contourite deposits exposing older, pre-sediment-drift strata at the sea bed. Alternatively, some of the erosion may have occurred during the deposition of the Oligo-Miocene succession. The possibility that the present disposition of these drift deposits represents the synchronous accumulation of two separate drifts deposited at different depths on the slope cannot be discounted. Their separation may be an original feature caused by the increased slope angle in this zone focusing strong bottom currents on this part of the slope, thereby locally restricting sedimentation.

Boreholes and short cores have proved a post-middle Miocene sea-bed veneer blanketing the drift. This veneer increases in thickness southwards, coincident with an increasing water depth, where a drape up to 16 m thick is preserved (Fig. 4b). The upslope feather-edge of the drift, including both the upslope-accreting unit and the overlying drape, was penetrated at this location by BGS borehole 94/1, which recovered fine-grained biocalcarenites and sporadic gravel of mid-Miocene and younger age. The age and character of the sediments are totally unlike anything preserved on Rockall Bank, and taken collectively the evidence for bottom-current sedimentation is overwhelming, refuting at least one earlier hypothesis (Roberts 1972) that these sediments represent a major slump from the eastern side of Rockall Bank.

The upper draped unit continues to thicken farther southwards (Jansen et al. 1995), and DSDP site 610, about 400 km south of the study area (Fig. 1), proved in excess of 600 m of middle Miocene to Holocene nanofossil ooze interbedded with clays in the upper part of the section (Kidd & Hill 1986). It would appear that whereas the western side of the north central Rockall Trough has undergone net erosion since the mid-Miocene, drift accumulation has persisted farther south adjacent to Rockall Bank. This pattern of sedimentation may have been controlled to some extent by the anticlockwise loop of NADW circulating in the central and southern Rockall Trough (Fig. 1) (Dickson & Kidd 1986).

Northwest Rockall Trough. North of Rockall Bank, contourite deposits display a sheeted style of drift accumulation (Fig. 5). Along the northern edge of Rockall Bank and adjacent to George Bligh Bank, the acoustically well-layered drift deposits onlap the NW margin of the Rockall Trough, but upslope accretion of sediment has not occurred to the same degree as is associated with the Feni Ridge to the south. The layered internal reflection configuration varies from flat-lying to wavy and hummocky in expression, although this reflection character is locally accentuated on seismic profiles by small-scale faulting which is locally common in the basinal sediments (Fig. 5b). A broad, partly

erosional, depression in the sea bed, ranging from about 1.5 to 15 km wide and several tens of metres deep, occurs within the present-day zone of basin-margin pinch-out of the basinal drift deposits. Bottom-current-derived sediment preserved on the flank of the George Bligh Bank is very thin and has little or no seismic expression.

In the basin, the upper Eocene to middle Miocene megasequence can be divided into several components consisting of sheeted-drift sequences of late Eocene and Oligocene to mid-Miocene ages (sequences 1 and 4 in Figs 2 and 3) partially separated by a lowstand fan (sequence 2) which downlaps onto the older drift deposits but is onlapped by the younger sediments. The fan, proved in BGS borehole 94/4, which recovered coarse-grained, massive, mass-flow calcarenites, is of late Eocene to earliest Oligocene age. These stratigraphical relationships, as demonstrated in Fig. 5b, suggest that drift deposition has been restricted to the basin since the late Eocene onset of bottom-current activity, whereas the northwest margin of the basin has been extensively reworked and/or eroded. On the eastern flank of George Bligh Bank, long-lived erosive bottom currents have sculpted an irregular sea-bed relief cut into pre-upper Eocene strata (Fig. 5a). Short cores indicate that this irregular surface is mantled by a gravelly veneer of mixed siliciclastic–carbonate composition. Borehole 94/4 further indicates extensive reworking of the lowstand fan deposits adjacent to the northeast corner of Rockall Bank, particularly during the Quaternary.

North central Rockall Trough. In the axial part of the Rockall Trough, the drift succession is preserved largely as an extensive sheetform, basin-plain fill (Fig. 3). However, marked onlap and downlap of the basinal succession occurs onto the early late Eocene unconformity (C) on the flanks of the axial seamounts (Fig. 6a). The internal reflection character of the succession is highly variable, ranging from a well-layered, flat-lying to wavy, continuous, reflection configuration to a more discontinuous, commonly jumbled, hummocky to waveform character. The latter is particularly characteristic of the upper Eocene to middle Miocene megasequence, and is associated with large-scale, buried, sediment waves, up to 40 m high with a wavelength of 2–3 km (Fig. 6b and 6c). Internal reflectors within the sediment-wave packages typically display a progradational style indicative of a migrating waveform (Fig. 6b). These buried sediment waves developed during the same time interval (Oligocene to mid-Miocene) as the major upslope accretion associated with

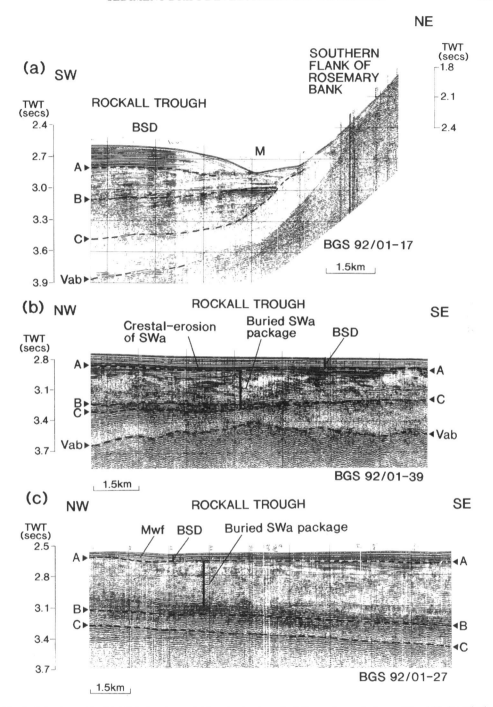

Fig. 6. Seismic profiles highlighting the variable seismic characteristics and morphology of the drift deposits in the north central Rockall Trough. (**a**) Airgun profile showing broad, sheeted drifts and a wide, deep, moat developed around Rosemary Bank. (**b**) and (**c**) Airgun profiles north and southwest, respectively, of Anton Dohrn showing large-scale, buried, sediment waves preserved in the upper Eocene to middle Miocene strata (C–A interval), buried beneath the predominantly flat-lying middle Miocene to Holocene (A to sea bed) deposits. The crestal region of the buried waves has been locally eroded by the mid-Miocene (A) unconformity. Abbreviations as in Figs 4 and 5 except: SWa, sediment waves; Mwf, migrating waveform. Location of profiles is shown in Fig. 1.

the formation of the Feni Ridge. The tops of the sediment waves are locally eroded by the mid-Miocene unconformity (Fig. 6b). The sediment waves are best developed in the vicinity of the Anton Dohrn seamount; elsewhere in the basin, including adjacent to the Rosemary Bank seamount, a more flat-lying to undulatory configuration characterizes this megasequence.

The upper part of the succession (middle Miocene to Holocene) is predominantly well layered and typically forms a drape on the floor of the north central Rockall Trough. However, subtle undulations on the sea bed are locally associated with migrating waveforms which have a similar wavelength to, but a much lower relief than, the underlying buried waves (Fig. 6c). Moreover, erosive bottom currents have scoured around the base of the seamounts, particularly Rosemary Bank where the sheeted-drift deposits abruptly downlap onto the underlying sediments, and are separated from the adjacent seamount by a wide, deep (180 m), moated area (Fig. 6a).

Eastern Rockall Trough. Onlapping, mounded and sheeted drifts with sediment waves (Figs 7 and 8) are developed throughout the middle to upper Cenozoic succession along the entire eastern margin of the Rockall Trough northwest of Britain (Stoker *et al.* 1993). However, in contrast to the western margin, the drift succession is commonly buried beneath, or interbedded with, locally thick accumulations of mass-flow deposits derived from the Hebridean margin. Whereas the Feni Ridge sediments are exposed on the western side of the basin, the equivalent deposits on the eastern margin are wholly buried (Fig. 3). Although some downslope input into the Rockall Trough from the Rockall Plateau (Fig. 5b) occurred during the late Eocene to mid-Miocene interval, it was not until the mid-Miocene that a significant shelf-margin input was initiated, associated with the development of the Barra and Sula Sgeir fans (Fig. 1). The fans are dominated by thick packages of acoustically structureless debris flows which interdigitate at their distal ends with layered drift deposits; however, the latter tend to be overwhelmed by downslope sediments higher on the fans (Figs 8b and 9) (Stoker *et al.* 1993). Beyond the limits of the fans, drifts are well preserved and strongly influence the sea-bed morphology.

The contrast between alongslope and downslope sedimentation is well demonstrated in the northeast corner of the Rockall Trough, adjacent to the Wyville-Thomson Ridge. In this area, the most coherent build-up of sediment drifts has occurred since the mid-Miocene,

although evidence of erosion and probable buried sediment waves in the underlying sediments has been reported (Stoker *et al.* 1993). Post-mid-Miocene bottom-current activity is expressed in the form of an acoustically well-layered sediment-drift complex, comprising a single- to multi-crested, mounded, elongate drift, a broad sheeted drift, and fields of sediment waves, all of which are exposed at the sea bed north of the Sula Sgeir Fan (Fig. 8a) (Howe *et al.* 1994; Stoker *et al.* 1998).

The mounded, elongate drift is up to 300 m thick and up to 20 km wide, with a maximum relief above the sea bed of 150 m. It displays a gradual upslope migration (Figs 7 and 8a). The drift axis has been traced along the base of the Wyville-Thomson Ridge and Hebrides Slope for up to 60 km. Locally, the axis of the drift splits over relatively short distances into several contemporaneously active crestal regions (Fig. 7). Inspection of the seismic profile in Fig. 7b suggests that the local development of the drift axis has also varied considerably with time, from a single- to multi-crested form. Well-developed sediment waves, up to 20 m high and 2 km wide, are locally preserved on the basinward-side of the elongate drift (Fig. 7a). Their internal geometry displays alternating asymmetric and sinusoidal units indicative, respectively, of active upslope migration and passive suspension draping (Howe 1996).

The elongate drift is separated from the Wyville-Thomson Ridge and Hebrides Slope by an erosional moat, up to 6 km wide and 100 m deep relative to the adjacent drift crest (Figs 7a and 8a). Small moat-related drifts, 30–100 m thick and 1–4 km wide, are occasionally developed within the moated area (Fig. 8a) (Stoker *et al.* 1998).

The broad, sheeted drift occupies a large part of the basin floor west of the elongate drift (Fig. 8a). It is up to 490 m thick, is several tens of kilometres wide, and displays a broad domal relief of up to 60 m above the general basin floor. Low-angle downlap of internal reflectors to the northeast reflects a slopeward migration of the broad crestal region by up to 10 km during the late Cenozoic. On the southwest flank of the sheeted drift, the acoustically well-layered drift sediments pass laterally into a large field of sediment waves which extend over an area of 550 km², and form a sediment package up to 200 m thick (Howe *et al.* 1994; Stoker *et al.* 1998). The internal configuration of the waves changes upwards from a basal set of climbing waves into an upper unit of sinusoidal waves, some of which have wave heights of 18 m and wavelengths of over 1 km. The slopeward migration of the

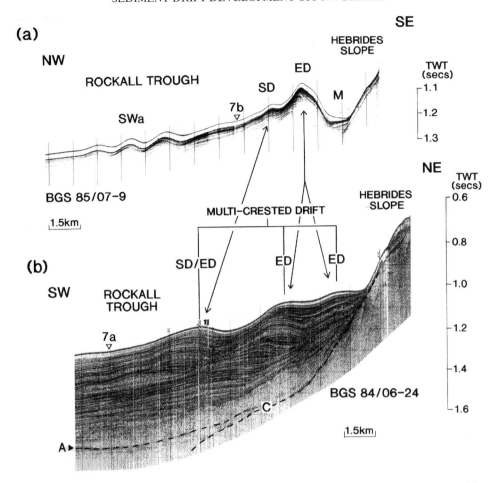

Fig. 7. Seismic profiles highlighting the seismic characteristics and morphology of drift deposits in the NE Rockall Trough. (**a**) Deep-tow boomer profile and (**b**) sparker profile showing extensive mounding and waveform development in the middle Miocene to Holocene section. These adjacent, and intersecting, seismic lines demonstrate the along-strike development from a single elongate drift, with a smaller subsidiary drift in (**a**), into three separate, but contemporaneously active drifts of broadly equal dimension in (**b**). The drifts and associated sediment waves have a distinct sea-bed expression. Abbreviations as in Figs 4–6 except: SD, subsidiary drift. Location of profiles is shown in Fig. 1.

sediment waves is consistent with the migration direction of the drift complex.

As the sediment-drift complex is traced southwards into the area of the Sula Sgeir Fan, the bulk of the complex becomes buried beneath the fan deposits. In Fig. 8a, the sediment-wave field is partially overlain by a debris flow at the distal edge of the Sula Sgeir Fan. Farther in towards the Hebrides Slope the drift sediments are buried by up to 400 m of mass-flow deposits of the Sula Sgeir Fan (Fig. 8b); these are stratigraphically equivalent to the upslope-migrating drift deposits associated with the mounded, elongate drift which has a sea-bed expression (Figs 7 and 8a) beyond the northern limit of the fan.

The glacigenic component of the Sula Sgeir Fan is highlighted in Fig. 8b, and illustrates the degree of sediment input onto the fan within the last 0.5 Ma associated with extensive mid- to late Pleistocene glaciation of the Hebrides Shelf (Stoker 1995). Such a high sediment input has clearly overwhelmed the normal background sedimentation on this part of the Hebrides Slope. However, a recent study (Stoker *et al.* 1994) from the upper Hebrides Slope, south of the Sula Sgeir Fan, demonstrated that bottom-current activity prevailed throughout the entire phase of mid-Miocene to Holocene shelf-margin progradation, even during glacial stages albeit probably at reduced strength. It seems likely

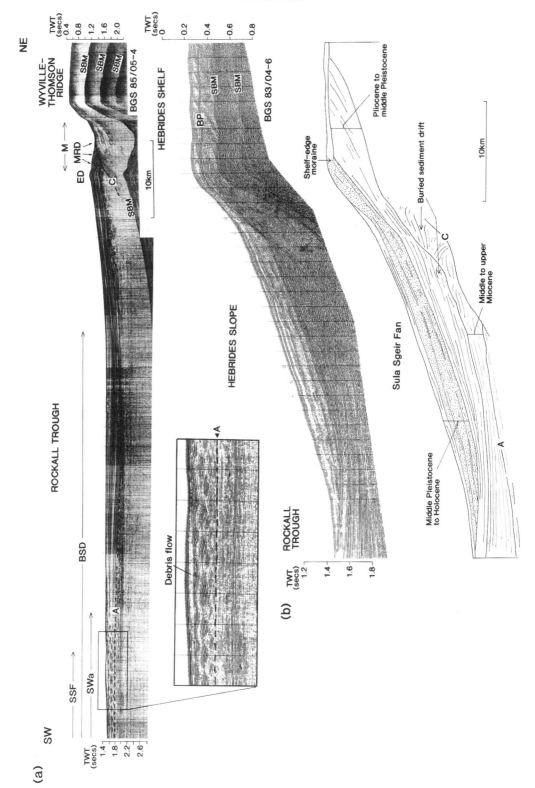

that the slope-wide clinoforms separating the glacigenic mass-flow packages (Fig. 8b) may represent thin units of deep-marine sediments deposited during relatively quiescent intervals of deposition on the slope. Interbedded units of downslope and deep-water sediments are also observed on seismic profiles farther south at the distal edge of the Barra Fan (Fig. 9). The acoustically layered deep-water deposits can be traced into the lower parts of the fan where they are preserved as semi-continuous reflections within an otherwise chaotic, mass-flow succession. This concept is supported by core data from the sea-bed veneer overlying the deposits of both fans which recovered interbedded hemipelagites, contourites and thin-bedded turbidites of latest Pleistocene to earliest Holocene age (Stoker *et al.* 1989; Howe *et al.* 1994; Howe 1995, 1996). Sea-bed photographs and current-meter data from along the Hebrides Slope further confirm bottom-current activity at the present day (Howe & Humphery 1995).

Hatton–Rockall Basin

The middle to upper Cenozoic succession in the Hatton–Rockall Basin has been little influenced by downslope processes, and bottom-current processes have dominated throughout this interval. DSDP site 116 (Fig. 1) proved a predominance of calcareous ooze, becoming chalky in the upper Eocene to lower Oligocene section, and with interbeds of dropstone muds in the upper Pliocene to Holocene sediments (Davies & Laughton 1972). In contrast to the Rockall Trough, where the bulk of the drift succession accumulated during the late Eocene to mid-Miocene interval, bottom-current deposition in the Hatton–Rockall Basin has increased since the mid-Miocene (Fig. 3). Although the middle to upper Cenozoic fill of the Hatton–Rockall Basin displays an overall symmetry, the component megasequences reflect a more asymmetric spatial development of the basin-fill.

On seismic profiles, the upper Eocene to middle Miocene succession is best developed and thickest (up to 400 m) on the eastern flank of the basin, thinning markedly to the west and

also along the axis of the basin to the north (Fig. 3). In the east, the buried drift displays a mounded, elongate geometry whose acoustic character is uniformly hummocky to wavy with extensive waveform development displaying upslope migration onto the western flank of Rockall Bank (Fig. 10a). The waveforms are up to 50 m high and range from about 0.5 km to several kilometres wide. Where internal reflectors are imaged, they show a consistent upslope progradation. Locally, the succession has been eroded, and the erosional hollows subsequently infilled. A highly dynamic depositional regime is envisaged, with a vigorous bottom circulation. To the west and north, the sediments display a more sheeted geometry with flat-lying, continuous reflections terminating at the basin margin by low-angle onlap.

The overlying middle Miocene to Holocene succession reflects a shift in depocentre to the central and western part of the basin, with extensive onlap onto the eastern side of Hatton Bank (Fig. 3). Within the basin, the drift deposits display a predominantly flat-lying, sheeted geometry, although an undulatory reflection character and small erosional hollows are locally developed. However, the reflection character is further complicated by faults which locally extend up to the sea bed, producing an undulatory sea-bed morphology (Fig. 10b). On the eastern side of the basin, the mid-Miocene unconformity is well imaged on seismic profiles, preserved as a marked onlap surface (Fig. 10a). Here also, the sheeted drift becomes slightly mounded on the edge of the basin, and erosion at the sea bed has sculpted an irregular surface tens of metres in relief. The middle Miocene to Holocene sediments extend onto Rockall Bank largely as a sheet-like veneer, and it seems probable that the underlying older drift deposits may be locally exposed at the sea bed.

On the western flank of the Hatton–Rockall Basin, the internal character of the middle Miocene to Holocene succession becomes more complex west of the basin-bounding fault. Although the sediments retain a sheeted external geometry there is an increased amount of internal erosion and sediment-wave

Fig. 8. (**a**) Airgun profile showing the seismic characteristics and morphology of the elongate and moat-related drifts adjacent to the Wyville-Thomson Ridge, together with the gently domed, broad, sheeted drift and associated sediment waves in the NE Rockall Trough. Inset shows buried sediment-wave field with upward change from asymmetric to sinusoidal waves, overlain by a debris flow package at the distal edge of the Sula Sgeir Fan. (**b**) Airgun profile and interpreted line drawing across the Sula Sgeir Fan showing mounded drift deposits buried beneath the fan adjacent to the Hebrides Slope. The glacial component (mainly debris flows) of the fan is stippled, and the basinward limit of these deposits is clearly defined. Abbreviations as in Figs 4–7 except: MRD, moat-related drifts; SSF, Sula Sgeir Fan; SBM, sea-bed multiple; BP, bubble pulse. Location of profiles is shown in Fig. 1.

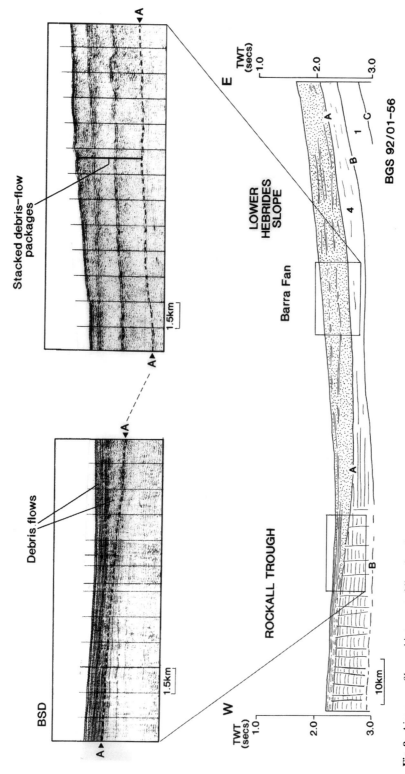

Fig. 9. Airgun profiles and interpreted line drawing across the lower part of the Barra Fan showing the contrasting seismic characteristics of, and the stratigraphical relationships between, the debris flows of the fan (stippled) and the basinal drift deposits, and the angular nature of the mid-Miocene unconformity (reflector A). Abbreviations as in Figs 4–8. Component sequences 1 and 4 as in Figs 2 and 3. Location of profile is shown in Fig. 1.

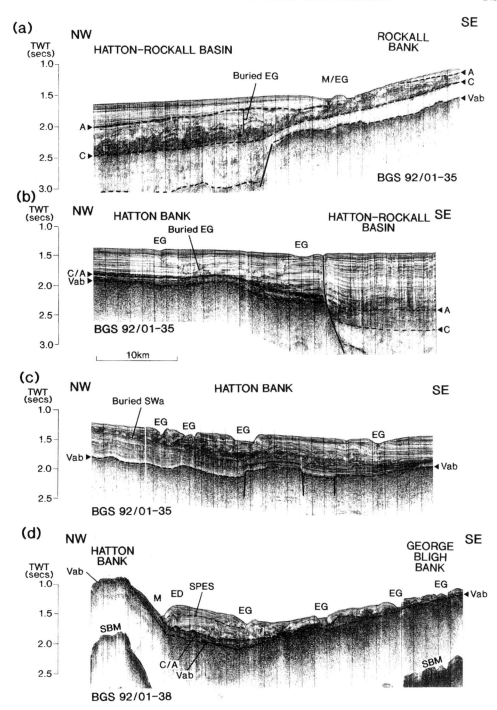

Fig. 10. Airgun profiles from the Rockall Plateau showing the variable seismic characteristics and morphology of the drift deposits on (**a**) the eastern flank of the Hatton–Rockall Basin; (**b**) the western flank of the Hatton–Rockall Basin; (**c**) the eastern flank of Hatton Bank; and (**d**) within the narrow channel between George Bligh and Hatton banks. Abbreviations as in Figs 4–9 except: C/A, multiple unconformity surface; EG, erosional gully; SPES, sub-planar erosion surface. Horizontal scale same in all sections. Location of profiles is shown in Fig. 1.

development (Figs 10b and 10c). Palaeo-erosional gullies ranging from a few tens of metres to about 70 m in depth, and up to 2–3 km wide, and subsequently infilled, are preserved within the succession. These probably represent palaeo-scour zones through which strong currents flowed during their formation. A number of open gullies are still present at the sea bed. Internal reflections within the infilled gullies typically display lateral progradation, with both westerly and easterly infill directions noted. Sediment waves are preserved at discrete levels throughout the succession, and are generally a few tens of metres high and range from 0.5 to 1.5 km in width. Not all discontinuities within the drifted section are scours; in the flat-lying lower part of the section, low-angle onlaps (Fig. 10c) reveal more subtle changes in sedimentation patterns.

In the narrow channel between the George Bligh and Hatton banks the only drift deposits preserved are of mid-Miocene to Holocene age. This channel is asymmetric in profile, with the steeper slope on Hatton Bank (Fig. 10d). Although the drift deposits exhibit onlap onto both banks, the thickest accumulation (up to 350 m) is preserved adjacent to the steeper eastern slope of Hatton Bank in the form of a mounded drift with a well-developed moat. The internal structure of the drift is locally complex with evidence of several phases of erosion, manifest both as discrete scours and drift-wide erosional surfaces. The present-day surface of the drift is undulatory, owing, in part, to erosion which has formed gullies several tens of metres deep and up to 1.5 km wide. Locally, the erosion has almost cut through the entire drift succession. On the shallower eastern slope of the channel (the western flank of George Bligh Bank) the drifted section has a sheeted geometry, but internally it comprises discrete packages of upslope-migrating sediment waves, up to 90 m thick and 2 km in width (Fig. 10d), locally eroded along channels or gullies some of which remain open at the present day. Near the top of the George Bligh Bank, a more uniform sheeted

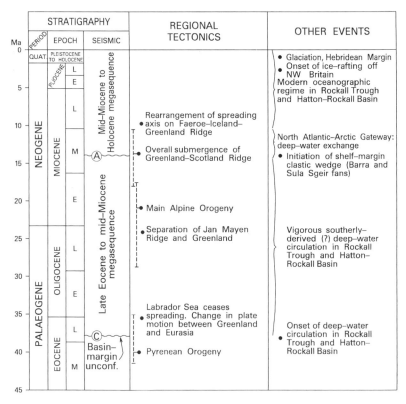

Fig. 11. Mid- to late Cenozoic tectono-stratigraphical framework for the continental margin off NW Britain. Regional tectonic and other events derived from a variety of sources (see text for references). Timescale from Harland *et al.* (1990).

character is developed with low-angle onlap onto the bank.

Sediment-drift development – discussion

Any attempt to understand the long-term record of bottom-current activity and sediment-drift development on the continental margin off NW Britain during the mid- to late Cenozoic interval has to take into consideration its regional tectonic setting (Fig. 11). Megasequence development and regional unconformities (megasequence boundaries) tend to reflect major phases of basin evolution, often a response to tectonism which modified sedimentation patterns and palaeoceanographic circulation. The regional unconformities (reflectors C and A) are submarine erosion surfaces largely cut by deep-water currents, and imply significant palaeoceanographic responses to tectonic events during the early late Eocene and the mid-Miocene. In the following discussion, the tectonostratigraphical framework provides the basis for evaluating the factors controlling (1) the initiation of bottom-current activity on the continental margin, and (2) the depositional history and palaeogeography of the mid- to late Cenozoic interval. Gross environment maps for the late Eocene to mid-Miocene, and mid-Miocene to Holocene intervals are presented in Fig. 12. For completeness, information from beyond the study area is included where relevant to the discussion.

Initiation of bottom-current activity on the continental margin off NW Britain

Sediment-drift development on the continental margin off NW Britain occurred in several stages commencing with the initiation of bottom-current activity within the Rockall Trough and Hatton–Rockall Basin during the late Eocene (Fig. 11). The timing of the onset of bottom-current activity is well constrained by the seismic stratigraphy, and on seismic profiles this event is represented by the basin-wide, early late Eocene, deep-water unconformity (megasequence boundary, reflector C). On the flanks of both basins, this unconformity is a markedly angular boundary which is commonly onlapped by the middle to upper Cenozoic drift deposits.

There is evidence for bottom-current activity around the periphery of the continental margin since the early Eocene, notably in the Charlie Gibbs Fracture Zone (south of the Rockall Trough) (Scrutton & Stow 1984) and on the western flank of the Rockall Plateau, although no significant deposition is thought to have occurred in the latter area (Stow & Holbrook 1984). This widespread, largely non-depositional event through much of the Eocene was interpreted as probably being related to global changes in bottom-water circulation, which were most likely to be linked to Antarctic cooling and the development of strong AABW flow throughout the Atlantic, and perhaps also to the formation of cold LSW (Shor & Poore 1979; Schnitker 1980; Stow 1982). Before the early late Eocene, there is no evidence for deep-water circulation within the UK sector of the Rockall Trough and Hatton–Rockall Basin, which may reflect their relatively shallow setting.

The incursion of bottom waters into the Rockall Trough and the Hatton–Rockall Basin may have been a response to a change in the shape and geometry of the basins. One effect of increased basinal subsidence is a change in the water circulation pattern within a basin (Cloetingh et al. 1990). The basin-subsidence event which led to the creation of the early late Eocene unconformity (C) may reflect a change in the intra-plate stress regime across the margin possibly triggered by regional tectonic events. It is perhaps no coincidence that the timing of the early late Eocene subsidence in the Rockall Trough and Hatton–Rockall Basin was broadly coeval with the culmination of the Pyrenean orogeny to the southeast (Knott et al. 1993), and with a major reorganization of spreading rate and direction in the North Atlantic to the west possibly caused by the cessation of sea-floor spreading in the Labrador Sea (Fig. 11) (Nunns 1983; Hinz et al. 1993). The erosional character of the early late Eocene unconformity around the basin margins may have been caused by the onset of vigorous bottom-current activity, the strength of which may be due to the new pathways funnelling the currents. The eroded basin-floor material may have formed a major source of sediment at least in the early stages of drift construction.

Late Eocene to mid-Miocene (Fig. 12a)

This interval was characterized by a vigorous, deep-water bottom-current circulation and significant lateral migration of sediment by upslope accretion onto the flanks of both the Rockall Trough and Hatton–Rockall Basin. An anticlockwise, southerly derived, palaeocirculation pattern is here inferred for the trough on the evidence of drift accretion on both flanks of the basin, a characteristic not previously described in this area for sediments of this age. This palaeocirculation pattern contrasts with that described by earlier workers (e.g. Miller & Tucholke 1983;

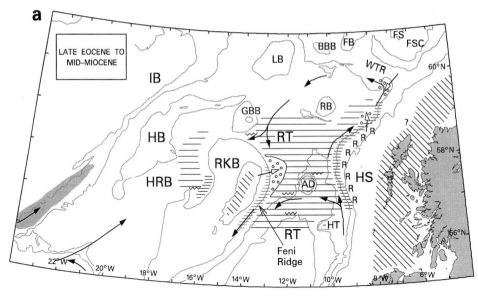

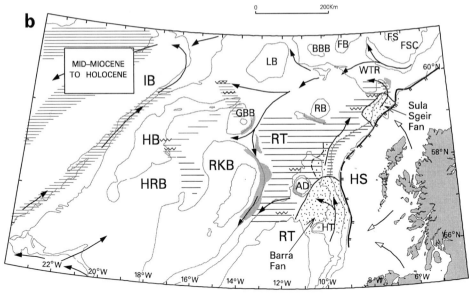

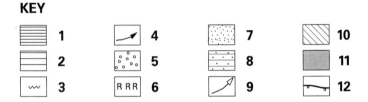

KEY

Heinrich *et al.* 1989), who have tended to assume that all of the northern North Atlantic sediment drifts, including the Feni Ridge, formed in response to a northerly source of deep water overflowing the Greenland–Scotland Ridge. Although an anti-clockwise gyre may work together with a northerly source to enhance the circulation pattern, the presence of the latter, at this time, remains ambiguous. In the Rockall Trough, a southerly derivation is consistent with recent suggestions that bottom-current activity north of the Greenland–Scotland Ridge has essentially been a Neogene–Quaternary phenomenon, and that deep-water exchange between the Arctic and North Atlantic basins was not fully established until the mid- or even the late Miocene (Eldholm 1990; Boldreel & Andersen 1995). Moreover, had deep northern-water existed in the late Eocene to mid-Miocene, palaeobathymetric reconstructions suggest that the southeast end of the Greenland–Scotland Ridge (Wyville-Thomson Ridge and Faeroe–Shetland Channel area) would have been too shallow to allow the passage of large quantities of dense outflow into the Rockall Trough (Wold 1994, 1995).

Limited downslope input into the Rockall Trough occurred adjacent to the northeast Rockall Bank. Upper Eocene–lowest Oligocene mass-flow deposits proved in BGS borehole 94/4 were derived from Rockall Bank and redeposited as a lowstand fan on the northwest flank of the basin. A similar early Oligocene input occurred on the northern Hebrides Slope and possibly adjacent to the Wyville-Thomson Ridge (Egerton this volume). An early Oligocene reefal development which fringed the eastern margin of the Rockall Trough was largely detached from the deep-water basin.

The dataset is too limited to allow detailed palaeocirculation patterns to be established for the Hatton–Rockall Basin, although the mounded, asymmetric geometry of the sediments on the eastern flank of the basin may be due to northward-flowing bottom currents. Wold (1994) has suggested that the Rockall Plateau could have been a local dense-water source during the early Oligocene, contributing to the early development of the Feni Ridge.

However, it remains uncertain how this would have affected the regional circulation pattern both on the Plateau and in the Rockall Trough, as the model fails to consider the development of the stratigraphically equivalent drifts in the Hatton–Rockall Basin and on the eastern flank of the Rockall Trough. On the western flank of the Rockall Plateau, bordering the Iceland Basin, strong bottom currents prevented any significant deposition during this interval (Stow & Holbrook 1984).

Mid-Miocene to Holocene (Fig. 12b)

The sedimentary regime across the NW British margin underwent a significant change during this interval in response to: (1) mid-Miocene tectonism; (2) deep-water exchange between the North Atlantic and Arctic oceans; and (3) Hebridean margin progradation (Fig. 11).

(1) The late early Miocene culmination of the Alpine orogeny, together with further changes in the sea-floor spreading geometry which are known to have resulted in compressional deformation in the Faeroes region (Boldreel & Andersen 1993), may be responsible for a change in the intra-plate stress regime. Across the NW British margin, the two main effects of this tectonism were the subsidence of the Greenland–Scotland Ridge (including the Wyville-Thomson Ridge) (Miller & Wright 1989), and the possible uplift of the Hebridean margin (Anderton *et al.* 1979; Stoker 1997). The former led to the onset of deep-water exchange between North Atlantic and Arctic watermasses (Stow & Holbrook 1984; Jansen *et al.* 1995), whereas the latter initiated the growth of the shelf-margin clastic wedge along the eastern flank of the Rockall Trough. The main basinal expression of these events is preserved as the mid-Miocene regional unconformity surface (megasequence boundary, reflector A).

(2) Deep-water exchange between the North Atlantic and Arctic oceans culminated in the establishment of the generalized circulation pattern of modern bottom currents (Fig. 1). This resulted in a change in the deep-water sedimentation pattern not only on the NW British margin but in the North Atlantic in general. On

---

Fig. 12. Gross depositional environment maps summarizing the main depositional systems existing during the development of the mid- to late Cenozoic megasequences. Key to symbols: 1, mounded, elongate sediment drift; 2, sheeted sediment drift; 3, sediment waves; 4, bottom-current circulation, inferred in (**a**), present-day in (**b**); 5, late Eocene–earliest Oligocene lowstand fan; 6, early Oligocene reef; 7, mid-Miocene to Holocene fans; 8, interbedded drift and downslope deposits; 9, sediment-source areas; 10, inferred land areas; 11, areas of extensive bottom-current erosion; 12, maximum extent of mid- to late Pleistocene ice-sheets. Abbreviations as in Fig. 1 except: HS, Hebrides Shelf; RT, Rockall Trough; IB, Iceland Basin. Present-day contours (as in Fig. 1) superimposed on maps to act as an approximate guide to the palaeo-morphology of the margin.

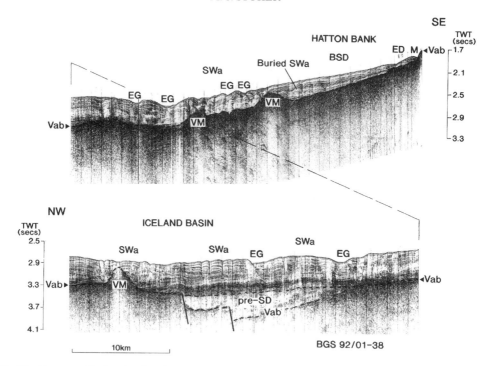

Fig. 13. Airgun profile showing the seismic characteristics and morphology of the middle Miocene to Holocene sediment-drift deposits on the northwest flank of the Rockall Plateau, which form the northern end of Hatton Drift. Several volcanic mounds and a small half-graben of early Palaeogene age are buried beneath the drifts. The sea-bed topography is strongly influenced by the bottom-current activity. Abbreviations as in Figs 4–10 except: pre-SD, pre-sediment-drift deposits (half-graben infill); VM, volcanic mound. Location of profile is shown in Fig. 1.

the NW British margin, sediment drifts and waves continued to develop along the axis and eastern margin of the Rockall Trough, but the western flank of the basin suffered erosion and appears to have been more a zone of sediment export than drift accumulation. Much of this sediment may have accumulated in the southern Rockall Trough where the Feni Ridge continued to develop (Kidd & Hill 1986). In the area of study, by far the greatest thickness of middle Miocene to Holocene bottom-current deposits accumulated on the Rockall Plateau (Fig. 3), with sheeted drifts, complex waveforms and erosional features preserved in the Hatton–Rockall Basin and on Hatton Bank. In places such as the channel between George Bligh Bank and Hatton Bank (Fig. 10d), the complexity of the succession probably reflects a vigorous, complex, circulation pattern accentuated or enhanced by the local topography, with currents focused and strengthened within the narrow channel. Strong bottom currents may persist at the present day either actively eroding or maintaining the sea-bed gullies.

This overall westward shift in net accumulation across the continental margin coincides with the growth of several other major sediment drifts in the NE Atlantic, such as the Hatton (on the western flank of Rockall Plateau) and Gardar drifts of the Iceland Basin (Fig. 12b) (Wold 1994). In contrast to the NW British margin drifts, the relatively simple, elongate geometry of these giant drifts is probably due to the largely unimpeded current flow within the morphologically less complex Iceland Basin (Faugères *et al.* 1993). However, the northern end of the Hatton Drift, on the northern flank of Hatton Bank, is noticeably more complex than along most of its length, probably because of its proximity to the channel between Rockall Plateau and Lousy Bank, linking the Rockall Trough and the Iceland Basin. In this area, the internal configuration of the Hatton Drift is more akin to the succession on the NW British margin, with complex sediment waves and erosional scours including an irregular, eroded, sea-bed surface (Fig. 13). A more flat-lying, sheeted seismic expression occurs on the upper part of Hatton Bank, which terminates in a small, elongate, mound with an associated moat. This complexity may be partly related to westward outflow from the Rockall Trough into the Iceland Basin interacting with northeastward flow along the Hatton Bank, and partly a function of specific topographic irregularities, such as the remnants (now buried) of early Palaeogene volcanic edifices.

(3) A significant shelf-margin input developed on the Hebridean margin with sediment pathways focused on the Barra and Sula Sgeir fans. From the mid-Miocene to early mid-Pleistocene, the fans were probably fed by fluvial systems draining from the Scottish landmass. However, significant climatic deterioration since late Pliocene time culminated in widespread glaciation of the Hebrides Shelf in the mid-Pleistocene, and grounded ice sheets reached the shelf-edge on at least two occasions during the last 0.5 Ma, depositing vast amounts of glacially reworked sediment directly onto the Hebrides Slope (Fig. 8b) (Stoker *et al.* 1994; Stoker 1995). Bottom-current activity prevailed on the Hebrides Slope throughout the whole period of shelf-margin input, albeit at reduced strength during the glacial stages. Sediment waves in the northeastern part of the Rockall Trough were active into the early Holocene (Howe 1996), and slope-currents persist at the present day (Howe & Humphery 1995).

Summary

There are several important points to emphasize about bottom-current activity and sediment-drift development on the continental margin off NW Britain.

(1) The onset of bottom-current activity on the continental margin occurred in the early late Eocene, following a major phase of subsidence in the Rockall Trough and Hatton–Rockall Basin.

(2) In the late Eocene to mid-Miocene interval, a vigorous, anticlockwise, southerly derived bottom-current regime may have prevailed, and was associated with significant lateral migration of deep-water sediment by upslope accretion onto both flanks of the Rockall Trough, and predominantly onto the eastern flank of the Hatton–Rockall Basin. The bulk of the Feni Ridge accumulated during this interval.

(3) Sedimentation patterns in the mid-Miocene to Holocene interval were modified as a result of the submergence of the Greenland–Scotland Ridge, linked to mid-Miocene tectonism, which led to the onset of deep-water exchange between the North Atlantic and

Arctic oceans, and the establishment of the modern oceanographic regime. On the NW British margin, drift accumulation persisted in the axial and eastern parts of the Rockall Trough, whereas a largely erosional regime prevailed on its western margin; the overall locus of drift accumulation shifted westwards onto the Rockall Plateau. A clastic wedge has built out along the Hebridean margin since the mid-Miocene, and downslope mass-flow deposits are locally interbedded with the sediment drifts.

(4) The variable character of the sediment-drift succession on the continental margin off NW Britain is most probably a function of the interaction of the bottom currents with the complex morphology of the margin. The semi-enclosed basins, narrow channels, and specific obstructions, such as seamounts, will tend to vary the intensity of bottom currents across the margin, which, in turn, will influence the degree of erosion, transport and deposition by these currents. This pattern of sedimentation contrasts markedly with the relatively simple, elongate, giant drifts developed in the open oceanic-basin system west of the continental margin, where relatively few topographic barriers hinder the deep circulation.

I would like to thank D. Evans, D. A. Ardus, D. A. V. Stow and J.-C. Faugères for their critical review of the paper; E. J. Gillespie for drafting the diagrams; and the following oil companies who, together with the BGS, make up the 'Rockall' continental margin consortium and without whose support this work could not have been undertaken: Agip, Amerada Hess, Arco, British Gas, BP, Conoco, Elf, Enterprise, Esso, Mobil, Phillips, and Statoil. The paper is published with the permission of the Director of the British Geological Survey (NERC).

References

ANDERTON, R., BRIDGES, P. H., LEEDER, M. R. & SELLWOOD, B. N. 1979. *A Dynamic Stratigraphy of the British Isles: a Study in Crustal Evolution*. George Allen & Unwin, London.

BERGGREN, W. A. & SCHNITKER, D. 1983. Cenozoic marine environments in the North Atlantic and Norwegian–Greenland Sea. *In*: BOTT, M. H. P., SAXOV, S., TALWANI, M. & THIEDE, J. (eds) *Structure and Development of the Greenland–Scotland Ridge: New Methods and Concepts*. Plenum, New York, 495–548.

BLANC, P.-L., RABUSSIER, D., VERGNAUD-GRAZZINI, C. & DUPLESSY, J.-C. 1980. North Atlantic Deep Water formed by the later middle Miocene. *Nature*, **283**, 553–555.

BOLDREEL, L. O. & ANDERSEN, M. S. 1993. Late Paleocene to Miocene compression in the Faeroe–Rockall area. *In*: PARKER, J. R. (ed.) *Petroleum Geology of Northwest Europe:*

Proceedings of the 4th Conference. Geological Society, London, 1025–1034.

—— & —— 1995. The relationship between the distribution of Tertiary sediments, tectonic processes and deep-water circulation around the Faeroe Islands. *In*: SCRUTTON, R. A., STOKER, M. S., SHIMMIELD, G. B. & TUDHOPE, A. W. (eds) *The Tectonics, Sedimentation and Palaeoceanography of the North Atlantic Region*. Geological Society, London, Special Publications, **90**, 145–158.

BOOTH, D. A. & ELLETT, D. J. 1983. The Scottish continental slope current. *Continental Shelf Research*, **2**, 127–146.

BULL, J. M. & MASSON, D. G. 1996. The southern margin of the Rockall Plateau: stratigraphy, Tertiary volcanism and plate tectonic evolution. *Journal of the Geological Society, London*, **153**, 601–612.

CLOETINGH, S., GRADSTEIN, F. M., KOOI, H., GRANT, A. C. & KAMINSKI, M. 1990. Plate reorganisation: a cause of rapid late Neogene subsidence and sedimentation around the North Atlantic. *Journal of the Geological Society, London*, **147**, 495–506.

DAVIES, T. A. & LAUGHTON, A. S. 1972. Sedimentary processes in the North Atlantic. *In*: LAUGHTON, A. S., BERGGREN, W. A. *et al.* (eds) *Initial Reports of the Deep Sea Drilling Project*, **12**. US Government Printing Office, Washington, DC, 905–934.

DICKSON, R. & KIDD, R. B. 1986. Deep circulation in the southern Rockall Trough – the oceanographic setting of site 610. *In*: RUDDIMAN, W. F., KIDD, R. B. *et al.* (eds) *Initial Reports of the Deep Sea Drilling Project*, **94**. US Government Printing Office, Washington, DC, 1061–1074.

EGERTON, P. D. 1997. Seismic characterization of Palaeogene depositional sequences: northeastern Rockall Trough. This volume.

ELDHOLM, O. 1990. Paleogene North Atlantic magmatic–tectonic events: environmental implications. *Memorie della Societa Geologica Italiana*, **44**, 13–28.

ELLETT, D. J. & ROBERTS, D. G. 1973. The overflow of Norwegian Sea Deep Water across the Wyville-Thomson Ridge. *Deep-Sea Research*, **20**, 819–835.

ENGLAND, R. W. & HOBBS, R. W. 1997. The structure of the Rockall Trough imaged by deep seismic reflection profiling. *Journal of the Geological Society, London*, **154**, 497–502.

FAUGÈRES, J. C., MEZERAIS, M. L. & STOW, D. A. V. 1993. Contourite drift types and their distribution in the North and South Atlantic basins. *Sedimentary Geology*, **82**, 189–203.

HAMILTON, E. L. 1985. Sound velocity as a function of depth in marine sediments. *Journal of the Acoustical Society of America*, **78**, 1348–1355.

HARLAND, W. B., ARMSTRONG, R. L., COX, A. V., CRAIG, L. E., SMITH, A. G. & SMITH, D. G. 1990. *A Geologic Time Scale 1989*. Cambridge University Press, Cambridge.

HARVEY, J. G. 1982. Theta–S relationships and water masses in the eastern North Atlantic. *Deep-Sea Research*, **29**, 1021–1033.

HEINRICH, R., WOLF, T. C. W., BOHRMANN, G. & THIEDE, J. 1989. Cenozoic paleoclimatic and paleoceanographic changes in the northern hemisphere revealed by variability of coarse-fraction composition in sediments from the Vøring Plateau – ODP leg 104 drill sites. *In*: ELDHOLM, O., THIEDE, J. *et al.* (eds) *Proceedings of the Ocean Drilling Program*, **104**. Ocean Drilling Program, College Station, TX, 75–188.

HILL, A. E. & MITCHELSON-JACOB, E. G. 1993. Observations of a poleward-flowing saline core on the continental slope, West of Scotland. *Deep-Sea Research*, **40**, 1521–1527.

HINZ, K., ELDHOLM, O., BLOCK, M. & SKOGSEID, J. 1993. Evolution of North Atlantic volcanic continental margins. *In*: PARKER, J. R. (ed.) *Petroleum Geology of Northwest Europe: Proceedings of the 4th Conference*. Geological Society, London, 901–913.

HOWE, J. A. 1995. Sedimentary processes and variations in slope-current activity during the last Glacial–Interglacial episode on the Hebrides Slope, northern Rockall Trough, North Atlantic Ocean. *Sedimentary Geology*, **96**, 201–230.

—— 1996. Turbidite and contourite sediment waves in the northern Rockall Trough, North Atlantic Ocean. *Sedimentology*, **43**, 219–234.

—— & HUMPHERY, J. D. 1995. Photographic evidence for slope-current activity, Hebrides Slope, NE Atlantic Ocean. *Scottish Journal of Geology*, **30**, 107–115.

——, STOKER, M. S. & STOW, D. A. V. 1994. Late Cenozoic sediment drift complex, northeast Rockall Trough, North Atlantic. *Paleoceanography*, **9**, 989–999.

JANSEN, E., RAYMO, M. & BLUM, P. 1995. North Atlantic Arctic Gateways II. *JOIDES Journal*, **21**, 37–43.

JONES, E. J. W., PERRY, R. G. & WILD, J. L. 1986. Geology of the Hebridean margin of the Rockall Trough. *Proceedings of the Royal Society of Edinburgh, Section B*, **88**, 27–51.

KENYON, N. H. 1986. Evidence from bedforms for a strong poleward current along the upper continental slope of NW Europe. *Marine Geology*, 72, 187–198.

Kidd, R. B. & Hill, P. R. 1986. Sedimentation on mid-ocean sediment drifts. *In*: SUMMERHAYES, C. P. & SHACKLETON, N. J. (eds) *North Atlantic Palaeoceanography*. Geological Society, London, Special Publications, **21**, 87–102.

KNOTT, S. D., BURCHELL, M. T., JOLLEY, E. W. & FRASER, A. J. 1993. Mesozoic to Cenozoic plate reconstructions of the North Atlantic and hydrocarbon plays of the Atlantic margins. *In*: PARKER, J. R. (ed.) *Petroleum Geology of Northwest Europe: Proceedings of the 4th Conference*. Geological Society, London, 953–974.

LAUGHTON, A. S., BERGGREN, W. A. *et al.* (eds) 1972. *Initial Reports of the Deep Sea Drilling Project*, 12. US Government Printing Office, Washington, DC.

MARTINI, E. 1971. Standard Tertiary and Quaternary calcareous nannoplankton zonation. *In*: FARINACCI,

A. (ed.) *Proceedings of the Second Planktonic Conference, Roma*, 2. Edizioni Tecnoscienza, Roma.

MASSON, D. G. & KIDD, R. B. 1986. Revised Tertiary seismic stratigraphy of the southern Rockall Trough. *In*: RUDDIMAN, W. F., KIDD, R. B., *et al.* (eds) *Initial Reports of the Deep Sea Drilling Project*, **94**. US Government Printing Office, Washington, DC, 1117–1126.

MCCAVE, I. N., LONSDALE, P. F., HOLLISTER, C. D. & GARDNER, W. D. 1980. Sediment transport over the Hatton and Gardar contourite drifts. *Journal of Sedimentary Petrology*, **50**, 1049–1062.

MILLER, K. G. & TUCHOLKE, B. E. 1983. Development of Cenozoic abyssal circulation south of the Greenland–Scotland Ridge. *In*: BOTT, M. H. P., SAXOV, S., TALWANI, M. & THIEDE, J. (eds) *Structure and Development of the Greenland–Scotland Ridge*. Plenum, New York, 549–589.

—— & WRIGHT, J. D. 1989. Subsidence history of Greenland–Scotland Ridge revisited: implications of uplift on deepwater circulation history. *Abstract, 28th IGC Congress*, **2**, 438.

MOUNTAIN, G. S. & TUCHOLKE, B. E. 1985. Mesozoic and Cenozoic geology of the U.S. Atlantic continental slope and rise. *In*: POAG, C. W. (ed.) *Geological Evolution of the United States Atlantic Margin*. Van Nostrand Reinhold, New York, 293–341.

MUSGROVE, F. W. & MITCHENER, B. 1996. Analysis of the pre-Tertiary rifting history of the Rockall Trough. *Petroleum Geoscience*, **2**, 353–360.

NUNNS, A. G. 1983. Plate tectonic evolution of the Greenland–Scotland Ridge and surrounding areas. *In*: BOTT, M. H. P., SAXOV, S., TALWANI, M. & THIEDE, J. (eds) *Structure and Development of the Greenland–Scotland Ridge*. Plenum, New York, 11–30.

ROBERTS, D. G. 1972. Slumping on the east margin of Rockall Bank. *Marine Geology*, **13**, 225–237.

—— 1975. Marine geology of the Rockall Plateau and Trough. *Philosophical Transactions of the Royal Society of London, Series A*, **278**, 447–509.

——, GINZBURG, A., NUNN, K. & MCQUILLIN, R. 1988. The structure of the Rockall Trough from seismic refraction and wide-angle reflection measurements. *Nature*, **332**, 632–635.

——, MORTON, A. C. & BACKMAN, J. 1984. Late Paleocene–Eocene volcanic events in the northern North Atlantic Ocean. *In*: ROBERTS, D. G., SCHNITKER, D. *et al.* (eds) *Initial Reports of the Deep Sea Drilling Project*, **81**. US Government Printing Office, Washington, DC, 913–923.

SCHNITKER, D. 1980. Global palaeoceanography and its deep water linkage to the Antarctic glaciation. *Earth-Science Reviews*, **16**, 1–20.

SCRUTTON, R. A. & STOW, D. A. V. 1984. Seismic evidence for early Tertiary bottom-current controlled deposition in the Charlie Gibbs Fracture Zone. *Marine Geology*, **56**, 325–334.

SHANNON, P. M., JACOB, A. W. B., MAKRIS, J., O'REILLY, B., HAUSER, F. & VOGT, U. 1994. Basin evolution in the Rockall region, North Atlantic. *First Break*, **12**, 515–522.

SHOR, A. W. & POORE, R. Z. 1979. Bottom currents and ice-rafting in the North Atlantic. *In*: LUYENDYK, B. P., CANN, J. R., *et al.* (eds) *Initial Reports of the Deep Sea Drilling Project*, **49**. US Government Printing Office, Washington, DC, 859–872.

STOKER, M. S. 1995. The influence of glacigenic sedimentation on slope-apron development on the continental margin off Northwest Britain. *In*: SCRUTTON, R. A., STOKER, M. S., SHIMMIELD, G. B. & TUDHOPE, A. W. (eds) *The Tectonics, Sedimentation and Paleoceanography of the North Atlantic Region*. Geological Society, London, Special Publications, **90**, 159–177.

—— 1997. Mid- to late Cenozoic sedimentation on the continental margin off NW Britain. *Journal of the Geological Society, London*, **154**, 509–515.

——, AKHURST, M. C., HOWE, J. A. & STOW, D. A. V. 1998. Sediment drifts and contourites on the continental margin off north-west Britain. *Sedimentary Geology*, in press.

——, HARLAND, R., MORTON, A. C. & GRAHAM, D. K. 1989. Late Quaternary stratigraphy of the northern Rockall Trough and Faeroe–Shetland Channel, northeast Atlantic Ocean. *Journal of Quaternary Science*, **4**, 211–222.

——, HITCHEN, K. & GRAHAM, C. C. 1993. *United Kingdom offshore regional report: the geology of the Hebrides and West Shetland shelves and adjacent deep-water areas*. HMSO, London, for the British Geological Survey.

——, LESLIE, A. B., SCOTT, W. D., BRIDEN, J. C., HINE, N. M., HARLAND, R., WILKINSON, I. P., EVANS, D. & ARDUS, D. A. 1994. A record of late Cenozoic stratigraphy, sedimentation and climate change from the Hebrides Slope, NE Atlantic Ocean. *Journal of the Geological Society, London*, **151**, 235–249.

STOW, D. A. V. 1982. Bottom currents and contourites in the North Atlantic. *Bulletin de l'Institut de Geologie du Bassin d'Aquitaine*, **31**, 151–166.

—— 1986. Deep clastic seas. *In*: READING, H. G. (ed.) *Sedimentary Environments and Facies*, 2nd edn. Blackwell Scientific, Oxford, 399–444.

—— & HOLBROOK, J. A. 1984. Hatton Drift contourites, northeast Atlantic. *In*: ROBERTS, D. G., SCHNITKER, D. *et al.* (eds) *Initial Reports of the Deep Sea Drilling Project*, **81**. US Government Printing Office, Washington, DC, 695–699.

VANNESTE, K., HENRIET, J.-P., POSEWANG, J. & THEILEN, F. 1995. Seismic stratigraphy of the Bill Bailey and Lousy Bank area: implications for subsidence history. *In*: SCRUTTON, R. A., STOKER, M. S., SHIMMIELD, G. B. & TUDHOPE, A. W. (eds) *The Tectonics, Sedimentation and Palaeoceanography of the North Atlantic Region*. Geological Society, London, Special Publications, **90**, 125–139.

WHITE, R. S. 1992. Crustal structure and magmatism of North Atlantic continental margins. *Journal of the Geological Society, London*, **149**, 841–854.

WOLD, C. N. 1994. Cenozoic sediment accumulation on drifts in the northern North Atlantic. *Palaeoceanography*, **9**, 917–941.

—— 1995. Palaeobathymetric reconstruction on a gridded database: the northern North Atlantic and southern Greenland–Iceland–Norwegian sea. *In*: SCRUTTON, R. A., STOKER, M. S., SHIMMIELD, G. B. & TUDHOPE, A. W. (eds) *The Tectonics, Sedimentation and Palaeoceanography of the North Atlantic Region*. Geological Society, London, Special Publications, **90**, 271–302.

Cyclic sedimentation on the Faeroe Drift 53–10 ka BP related to climatic variations

TINE L. RASMUSSEN[1], ERIK THOMSEN[2] & TJEERD C. E. VAN WEERING[3]

[1]*Institute of Geology, University of Copenhagen, Øster Voldgade 10, DK-1350 Copenhagen, Denmark*

[2]*Department of Earth Sciences, University of Aarhus, DK-8000 Aarhus C, Denmark*

[3]*Netherlands Institute for Sea Research (NIOZ), PO BOX 59, 1790 AB Den Burg, Texel, The Netherlands*

Abstract: More than 15 sedimentary cycles dated to the last glacial period (53–10 ka BP) have been recognized in a contourite deposit on the Faeroe Drift in the southern part of the Norwegian Sea. Each cycle consists of a silty, basaltic lower part and a clayey, acidic (siliceous) upper part. The sedimentary cycles can be accurately correlated with the Dansgaard–Oeschger temperature cycles in the Greenland ice cores, and it appears that the cyclic sedimentation was controlled by climatic and palaeoceanographic changes. The basaltic layers were deposited during warm interstadial periods in a current regime that resembles the modern circulation system in the North Atlantic region. Deep bottom-water created by thermohaline convection in the Norwegian–Greenland Seas flowed along the Iceland–Faeroe Ridge into the Atlantic Ocean, with the Faeroe–Shetland Channel as the main gateway. The source of the basaltic sediments was the volcanic rocks and detritus on the Iceland–Faeroe Ridge and on the shelf of eastern Iceland. The fine-grained acidic (siliceous) layers were deposited during intervening cold periods, in which convection took place in the subpolar North Atlantic Ocean and the circulation was reversed in the Faeroe–Shetland Channel. The acidic deposits were carried into the Norwegian Sea from the Hebrides and West Shetland shelves and the northeastern Atlantic Ocean.

Deep water is generated by thermohaline convection of surface waters in the Norwegian–Greenland Seas. The water mass (Norwegian Sea Deep Water) flows southward passing the Denmark Strait, the Iceland–Faeroe Ridge and the Faeroe–Shetland Channel. The deep-water outflow has created several contourite drifts in the North Atlantic Ocean south of the Faeroe Islands (Jones *et al.* 1970; McCave *et al.* 1980; Stow & Holbrook 1984; Faugères *et al.* 1993). The Faeroe Drift, located north of the Faeroe Islands (van Weering *et al.* 1993, 1994, this volume) (Fig. 1) is generated by the eastward flow of the Norwegian Sea Deep Water along the north flank of the Iceland–Faeroe Ridge (Nielsen *et al.* this volume), and is >100 m thick.

Drift deposits are controlled by deep-water currents and offer a good opportunity to study the relationship between sedimentary processes and palaeoceanographic changes. Furthermore, they are commonly characterized by a high rate of sedimentation and provide a high stratigraphical resolution (e.g. Keigwin & Jones 1989; van Weering & de Rijk 1991; Dowling & McCave 1993; Manighetti & McCave 1995*a*,*b*; McCave *et al.* 1995).

The present paper presents results from an 11 m long piston core, ENAM93-21, taken on the Faeroe Drift at a water depth of 1020 m (Fig. 1). The core covers isotope stages 1–3 and represents about 50,000 a with an average sedimentation rate of 25 cm/1000 a (Rasmussen *et al.* 1996*a*). The core is taken to the east of the northern end of the Faeroe–Shetland Channel, one of the main gateways between the Norwegian–Greenland Seas and the North Atlantic Ocean, in an area which during glacial periods was affected by numerous changes in bottom-water currents (Rasmussen *et al.* 1996*a*,*b*). The palaeobiological, sedimentological, and isotopic records of the core can be correlated in detail with the climatic fluctuations recorded in the Greenland ice cores, and the study is especially concerned with the relationship between the climatic and palaeoceanographic shifts in the Norwegian–Greenland Seas and the sedimentary history of the contourite. The focus is, in particular, on changes in the origin and the rate of sedimentation of the terrigenous components.

Material and methods

The core was sampled at 5 cm intervals taking fixed volume samples of 6 cm^3. A total of 219 samples were analysed. The samples were

RASMUSSEN, T. L., THOMSEN, E. & VAN WEERING, TJ. C. E. 1998. Cyclic sedimentation on the Faeroe Drift 53–10 ka BP related to climatic variations. *In*: STOKER, M. S., EVANS, D. & CRAMP, A. (eds) *Geological Processes on Continental Margins: Sedimentation, Mass-Wasting and Stability*. Geological Society, London, Special Publications, **129**, 255–267.

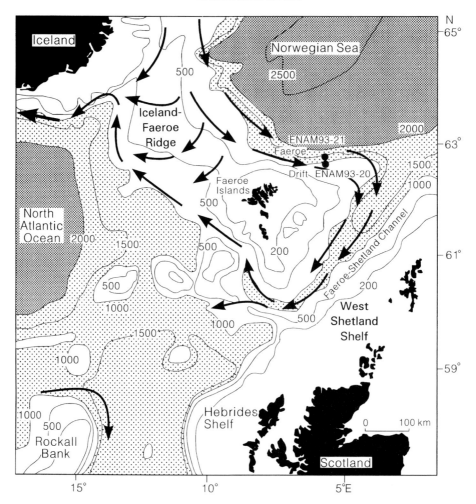

Fig. 1. Map of the Iceland–Faeroe Ridge and Faeroe–Shetland Channel showing present-day near-bottom currents in the area. The location of the Faeroe Drift and the positions of cores ENAM93-20 and ENAM93-21 are indicated. Bathymetric contours are given in metres. Current directions based on Meincke (1983).

weighed both wet and dry to calculate the wet and dry bulk densities. The samples were washed through 63 μm and 100 μm sieves as a measure of the grain-size distribution of the coarse fraction.

Coarse-fraction components

The >100 μm fractions were subsequently split by dry sieving through a 150 μm sieve. The number of grains >150 μm was counted as a measure of the content of ice-rafted detritus (IRD) and the composition determined microscopically (about 150–250 grains were counted in each sample). The 63–150 μm and >150 μm fractions were weighed and the proportion was calculated relative to the total dry weight of the samples.

Magnetic susceptibility, isotope composition, clay mineralogy, and calcium carbonate content

Magnetic susceptibility was measured at 5 cm intervals using a Barthington magnetometer before opening the core. The percentage of bulk calcium carbonate per total dry weight of sediment was measured at 10 cm intervals by titration. The composition of the planktonic and benthonic oxygen isotopes was obtained by analyses of the planktonic foraminifer *Neogloboquadrina pachyderma* (sinistral) and the benthonic foraminifer *Nonion zaandamae*. Bulk sample mineralogy and clay mineralogy were analysed in representative samples on the basis of X-ray diffraction (XRD).

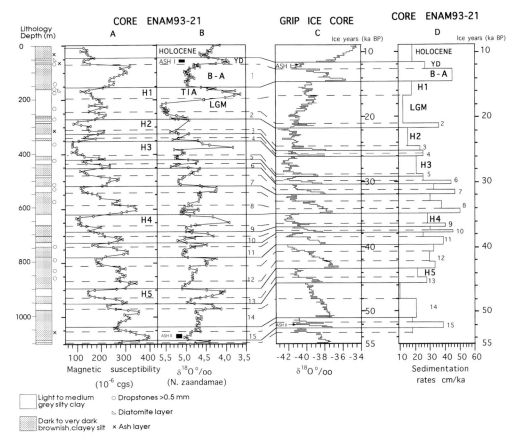

Fig. 2. Lithology and sedimentation rates of ENAM93-21 and correlation of ENAM93-21 with GRIP ice core. The correlation is based on (**a**) the magnetic susceptibility, (**b**) benthonic $\delta^{18}O$ records of ENAM93-21 core, and (**c**) the $\delta^{18}O$ record of GRIP ice core. The $\delta^{18}O$ record of the ice core is plotted against the ice core time scale (ice years, ka BP). Continuous horizontal lines mark boundaries between Dansgaard–Oeschger cycles (numbered 1–15). Dashed lines mark boundaries, placed at first significant cooling episode, between warm interstadials and cold stadials. The positions of ash I and ash II are shown (positions in ice core are from Grønvold *et al.* 1995). (**d**) Sedimentation rates are calculated separately for each interstadial and stadial interval. YD, Younger Dryas; B–A, Bølling-Allerød; H1–H5, Heinrich events 1–5; TIA, Termination IA; LGM, Last Glacial Maximum.

Stratigraphy and age control

The ENAM93-21 core has previously been investigated for palaeoceanographic purposes and shown to contain a detailed record of variations in magnetic susceptibility, oxygen isotopes, IRD, and foraminiferal faunas (Rasmussen *et al.* 1996*a,b,c*). These variations follow a characteristic pattern that is very similar to the cycles of $\delta^{18}O$ variations observed in the Greenland ice cores (see Johnsen *et al.* 1992; Dansgaard *et al.* 1993; Grootes *et al.* 1993), and the ENAM93-21 core and the ice cores can be correlated with great precision. Figure 2 shows the correlation between the magnetic susceptibility curve and benthonic $\delta^{18}O$ curve of the ENAM93-21 core

and the $\delta^{18}O$ curve of the ice core of the Greenland Ice-core Project (GRIP).

The $\delta^{18}O$ cycles in the Greenland ice cores are called Dansgaard–Oeschger cycles (Broecker & Denton 1989). Each of these cycles can be subdivided into a lower part with high $\delta^{18}O$ values and an upper part with low $\delta^{18}O$ values (Johnsen *et al.* 1992). The close correlation between the GRIP ice core and ENAM93-21 allows us to transfer the datings of the Dansgaard–Oeschger cycles and their internal subdivision from the ice core to the ENAM93-21 core. The interval from the base of the Holocene to 52 ka BP contains 15 Dansgaard–Oeschger cycles (Johnsen *et al.* 1992), and a total of 38 levels in the ENAM93-21 core have been dated by this method. Some

Table 1. *AMS* [14]*C dated levels in the ENAM93-21 core performed on* Neogloboquadrina pachyderma *(sinistral)*

Sample depth (cm)	Conventional ^{14}C age (ka)	Res. corrected ^{14}C age (ka) (Res. age: 400)	Lab. no.
30.0 (ash)	9.1*		
40.5	9.24 ± 80	8.84 ± 80	AAR-2559
57.5	10.3[†]		
67.5	11.34 ± 100	10.64 ± 100[‡]	AAR-2560
100.0	12.14 ± 100	11.74 ± 100	AAR-2561
140.75	12.82 ± 110	12.42 ± 110	AAR-2562
152.5	13.35 ± 120	12.95 ± 120[§]	AAR-2568
160.0	13.64 ± 160	13.24 ± 160	AAR-1642
175.0	13.65 ± 190	13.25 ± 190	AAR-2563
215.0	15.90 ± 170	15.50 ± 170	AAR-1643
245.0	18.65 ± 200	18.25 ± 200	AAR-1644
280.0	22.42 ± 210	22.02 ± 210	AAR-2564
310.0	23.39 ± 310	22.99 ± 310	AAR-1645
317.5	23.25 ± 190	22.85 ± 190	AAR-2565
530.0	30.70 ± 380	30.30 ± 380	AAR-2566

[*] Saksunavatn Ash (Mangerud *et al.* 1986).
[†] Vedde Ash (Mangerud *et al.* 1984; Bard *et al.* 1994).
[‡] Corrected for a reservoir age of 700 a (Bard *et al.* 1994).
[§] Age obtained from benthonic foraminifera *Cornuspira foliacea*.

particularly cold stadials correlate with the Heinrich events (Rasmussen *et al.* 1996c). Five Heinrich events are generally recognized in the investigated time period (Bond *et al.* 1992).

A number of samples from the upper part of the ENAM93- 21 core were dated by [14]C (Table 1 and Fig. 3). The ^{14}C dates are, generally, in good agreement with the dates transferred from the ice core, although a date of 8840 BP for a

sample taken 10 cm below the Saksunarvatn Ash is too young. The ash, according to Mangerud *et al.* (1986), is 9100 [14]C a in age.

The age–depth relationships were used to calculate the sedimentation rates and the bulk accumulation rates. The bulk accumulation rates were calculated on the basis of the sedimentation rates as $g/cm^2/ka$. The methods described by Ehrmann & Thiede (1985) and Wolf & Thiede (1991) were followed with a density correction factor for seawater of 1.025 (g/cm^3). The equations of Ehrmann & Thiede (1985) were used to calculate the accumulation rates of fine sands (63–150 μm). The accumulation rate of IRD was calculated as number of grains >150 $μm/cm^2/ka$.

Results

Sedimentary variations in the ENAM93-21 core

The lower parts of Dansgaard–Oeschger cycles in the Greenland ice cores represent relatively warm interstadial climates, the upper parts relatively cold, stadial climate (Johnsen *et al.* 1992; Dansgaard *et al.* 1993).

The recognition of interstadial and stadial intervals forms the most distinct subdivision of the core. Therefore, sedimentation rates have been calculated separately for each interval (Table 2). In large cycles 8 and 12, intervals of

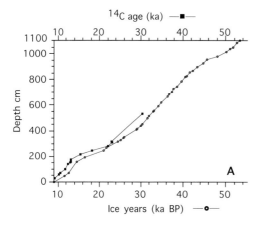

Fig. 3. [14]C ages and ice year ages plotted against depth for core ENAM93-21. Ice-year ages are based on correlation with the GRIP ice core. The two curves are nearly parallel, indicating a good correspondence between the two dating methods.

Table 2. *Sedimentation rates and ages of dated levels in the ENAM93-21 core*

Stratigraphical interval		Bottom (cm)	Thickness (cm)	Age of bottom (ice core years, ka)	Duration (ice core years, ka)	Rate of Sedimentation (cm/ka) Stadial	Interstadial
E. Holocene		46.3	46.3	11.6	2.6		17.8
Y. Dryas	S	72.5	26.2	12.6	1.0	26.3	
Bøl.–Aller.	I	157.5	85.0	14.5	1.9		44.8
TIA (Heinrich 1)	S	192.5	35.0	16.5	2.0	17.5	
LGM		245.0	52.5	20.9	4.4	11.9	
D–O cycle 2	I	270.0	25.0	21.6	0.7		35.7
Heinrich 2	S	313.8	43.8	24.5	2.9	15.1	
D–O cycle 3	I	327.5	13.7	25.1	0.6		22.9
D–O cycle 4	S	342.5	15.0	25.7	0.6	25.0	
	I	352.5	10.0	26.1	0.4		25.0
Heinrich 3	S	407.5	55.0	28.8	2.7	20.4	
D–O cycle 5	I	417.5	10.0	29.2	0.4		25.0
D–O cycle 6	S	435.0	17.5	29.9	0.7	25.0	
	I	452.5	17.5	30.3	0.4		43.8
D–O cycle 7	S	485.0	32.5	31.3	1.0	32.5	
	I	517.5	37.5	32.0	0.7		46.4
D–O cycle 8	S	547.5	30.0	33.0	1.0	30.0	
	I	622.5	75.0	34.8	1.8		43.3
Heinrich 4	S	667.5	45.0	36.4	1.6	28.1	
D–O cycle 9	I	687.5	20.0	36.9	0.5		40.0
D–O cycle 10	S	702.5	15.0	37.4	0.5	30.0	
	I	725.0	22.5	37.9	0.5		45.0
D–O cycle 11	S	742.5	17.5	38.6	0.7	25.0	
	I	785.0	42.5	39.7	1.1		38.6
D–O cycle 12	S	817.5	32.5	40.7	1.0	32.5	
	I	897.5	80.0	43.3	2.6		30.8
Heinrich 5	S	927.5	30.0	44.7	1.4	21.4	
D–O cycle 13	I	952.5	25.0	45.6	0.9		27.8
D–O cycle 14	S	977.5	25.0	48.1	2.5	10.0	
	I	1037.5	60.0	51.0	2.9		20.7
D–O cycle 15	S	1050.0	12.5	51.7	0.7	17.9	
	I	1085.0	35.0	52.6	0.9		38.9
D–O cycle 16	S	1097.5	12.5	53.3	0.7	17.9	

The dated levels are mainly boundaries between Dansgaard–Oeschger cycles, between stadials (S) and interstadials (I), or between Heinrich events and interstadials. The ages are obtained by correlation with the GRIP ice core (Johnsen *et al.* 1989, 1992). The version with a $\delta^{18}O$ average with a resolution of 100 a (Johnson *et al.* 1997) was used. Sedimentation rates were calculated separately for each warm and cold interval.

Table 3. *Subdivision of the large Dansgaard–Oeschger cycles 8 and 12 into three subintervals, a warm interstadial interval (I), a transitional cooling interval (T) and a cold stadial interval (S) (for further explanation, see Table 2 caption)*

Stratigraphical interval		Bottom of interval (cm)	Thickness (cm)	Age of bottom (ice core years, ka)	Duration (ice core years, ka)	Rate of sedimentation (cm/ka) Stadial	Transitional intervals	Interstadial
	S	547.5	30.0	33.0	1.0	30.0		
D–O cycle 8	T	592.5	45.0	34.2	1.2		37.5	
	I	622.5	30.0	34.8	1.0			50.0
	S	817.5	32.5	40.7	1.0	32.5		
D–O cycle 12	T	867.5	25.0	42.4	1.0		29.4	
	I	897.5	30.0	43.3	0.9			33.3

Table 4. *Mineralogy of bulk samples and clay fractions (<2 μm) of representative samples from the interstadial, stadial and transitional cooling intervals*

Stratigraphic interval	Bulk sample mineralogy (%)						Clay minerals in clay fraction (<2 μm) (%)		
	Quartz	K-feldspar	Plagioclase	Calcite	Dolomite	Clay	Smectite	Illite	Kaolinite/ chlorite
Interstadial intervals									
Interstadial 6 (451.0 cm)	30	11	19	7	4	25	57	28	15
Interstadial 8 (600 cm)	32	8	18	10	4	23	59	25	16
Interstadial 12 (890 cm)	35	12	18	8	3	21	74	12	14
Transitional interval									
D–O cycle 8 (542.5 cm)	37	9	17	9	3	19	39	39	22
Stadial intervals									
Heinrich 3 (370.0 cm)	38	8	11	7	3	31	21	58	21
Heinrich 4 (660.0 cm)	43	5	13	7	3	27	–	68	32
Heinrich 5 (910.0 cm)	41	6	12	4	3	33	–	73	27

gradual cooling between the interstadial and the stadial intervals could be distinguished (Table 3). These transitional intervals are regarded as a part of the interstadial intervals in the two-fold subdivision in Table 2.

Interstadial intervals. The deposits of the interstadial intervals consist of clayey silt with a relatively large proportion of coarse-fraction sediment (10–20% by weight in the 63–150 μm fraction). The colour is mostly dark brown (Fig. 2). The coarse fraction consists predominantly of grains of fresh vesicular volcanic glass and quartz. The relatively low (9–16%) calcium carbonate content is made up mainly of well-preserved tests of foraminifera (Rasmussen *et al.*

1996*c*). The magnetic susceptibility is high (Fig. 2). XRD analyses of three samples from interstadial intervals show a content of quartz between 30% and 35%, and of feldspars between 26% and 30%. The clay content varies from 21% to 25% (Table 4); smectite makes up between 57% and 74% of the clay minerals.

The interstadial intervals are characterized by relatively high rates of sedimentation ranging from 20.7 cm/ka in Dangaard–Oeschger cycle 14, to 46.4 cm/ka in Dansgaard–Oeschger cycle 7 (Fig. 2 and Table 2).

Stadial intervals. The deposits of the stadial intervals consist of silty clay. The sediment is finer grained than that of the interstadials,

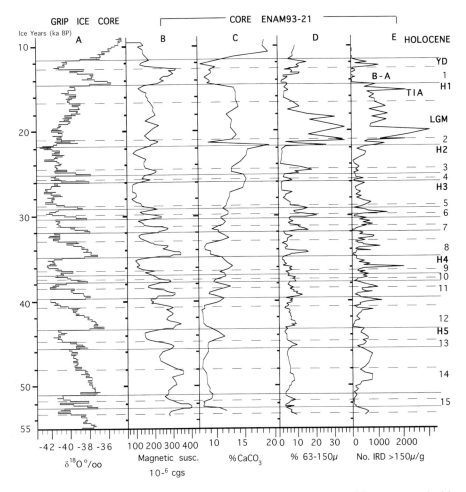

Fig. 4. Comparison of δ^{18}O of GRIP ice core (**a**) with (**b**) magnetic susceptibility, (**c**) percentage of calcium carbonate, (**d**) percentage of coarse fraction (63–150 μm) and (**e**) concentration of IRD (number of grains >150 μm per gram) in core ENAM93-21. The data are plotted v. the GRIP ice core time scale. YD, Younger Dryas; B–A, Bølling–Allerød; H1–H5, Heinrich events 1–5; TIA, Termination IA; LGM, Last Glacial Maximum.

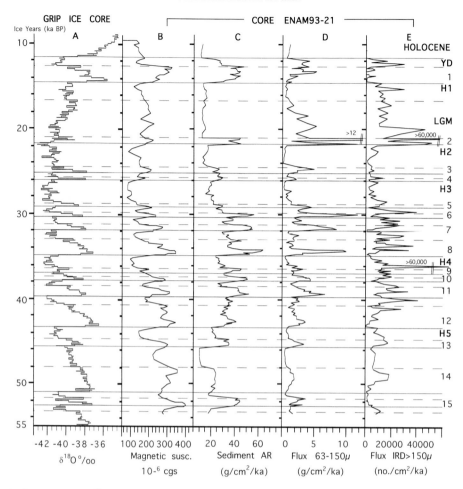

Fig. 5. Comparison of $\delta^{18}O$ record of GRIP ice core (**a**) with (**b**) magnetic susceptibility, (**c**) bulk accumulation rate, (**d**) flux of coarse fraction (63–150 μm), and (**e**) flux of IRD (number of grains >150 μm per gram) for core ENAM93-21. The data are plotted v. the GRIP ice core time scale. YD, Younger Dryas; B–A, Bølling–Allerød; H1–H5, Heinrich events 1–5; TIA, Termination IA; LGM, Last Glacial Maximum.

although the dropstone and IRD contents are higher. Between 2 and 10% by weight of the sediment falls in the 63–150 μm fraction as compared with 10–20% in the interstadials (Fig. 4). The colour is generally light grey (Fig. 2). Calcium carbonate, composed of small fragments of foraminifera and calcareous IRD particles, makes up 11–20% of the sediment in most stadials. The magnetic susceptibility is relatively low. XRD analyses of samples from Heinrich events 3, 4, and 5 show a content of quartz between 38% and 43%, and of feldspars between 18% and 19%. The clay content varies from 27% to 33% (Table 4). Illite makes up from 58% to 73% of the clay minerals. Smectite, which is the most important clay mineral in the interstadials, was not found in the samples from

Heinrich events 4 and 5 and makes up only about 21% in the sample from Heinrich event 3. Heinrich events 2 and 3 differ from the other stadial intervals in having a lower content of dropstones and IRD, and in a poorer preservation of the foraminifera.

The stadial intervals are characterized by relatively low rates of sedimentation ranging from 10.0 cm/ka in stadial 14 to 32.5 cm/ka in stadial 7 (Fig. 2 and Table 2).

Transitional cooling intervals. The transitional cooling intervals have only been studied in detail in Dansgaard–Oeschger cycles 8 and 12. The sedimentary and palaeobiological parameters of the transitional intervals are strongly fluctuating (Rasmussen *et al.* 1996c). Overall, however, the values seem to be intermediate between those found in the stadial and interstadial intervals (Fig. 2 and Tables 3 and 4). It should be noted that the preservation of foraminifera is very poor. The sedimentation rate in the transitional intervals was apparently relatively low (Table 3).

Rates of sedimentation and accumulation

The average sedimentation rate in the entire ENAM93-21 core is 25.2 cm/ka. The rates range from 20.7 cm/ka to 46.4 cm/ka in the interstadial intervals and from 10.0 cm/ka to 32.5 cm/ka in the stadial intervals. The mean of the sedimentation rates in the interstadials is 35.2 cm/ka as compared with 22.7 cm/ka in the stadials. The difference between the interstadials and the stadials is highly significant ($P < 0.001$) according to the non-parametric Wilcoxon test for paired samples. A similar difference between the stadial and the interstadial intervals is present in the ENAM93-20 core taken at a water depth of 850 m about 13 km south of the ENAM93-21 site. The difference appears to be a general feature of the Faeroe Drift.

The bulk accumulation rates closely parallel the sedimentation rates (Fig. 5). The rates were high during interstadial intervals and low during stadial and cooling intervals. The accumulation rates of the coarse fraction (63–150 µm) followed the same pattern, whereas the accumulation rate of IRD generally was highest during the stadials and cooling intervals (Fig. 5). Heinrich event 3 is an exception to this general pattern, as it shows a very low accumulation rate of IRD.

Discussion

Palaeoceanography

The distributions of planktonic and benthonic foraminifera and oxygen isotopes in the ENAM93-21 core indicate that during the last glacial period (53–10 ka BP) the circulation system in the southeastern Norwegian Sea changed repeatedly between an interstadial and a stadial mode (Rasmussen et al. 1996b,c) (Table 5).

The benthonic faunas of interstadials 3–15 in particular are very similar to the modern faunas in the Norwegian Sea (Rasmussen et al. 1996b,c); *Nonion zaandamae* is the dominant species (Table 5). The concentration of benthonic species is comparatively high and the ratio of planktonic to benthonic foraminifera is about one. In the planktonic fauna, *Neogloboquadrina*

pachyderma (dextral) is relatively common. The faunas suggest that the interstadial circulation system resembled the modern system in which warm saline Atlantic Surface Water flows from the Atlantic Ocean via the Faeroe–Shetland Channel into the Norwegian Sea, where it cools and sinks during the winter. The deep water generated flows south into the North Atlantic Ocean through the Denmark Strait, across the Iceland–Faeroe Ridge and through the Faeroe–Shetland Channel, which forms the main gateway (Dooley & Meincke 1981; Hopkins 1991).

The stadial intervals are characterized by an increase in the relative abundance of the polar species *N. pachyderma* (sinistral) and an increase in the amount of IRD, suggesting colder surface-water conditions as compared with the interstadials (Rasmussen et al. 1996b,c) (see also Table 5). The benthonic fauna is very different from that of the interstadials; the dominating species is *Cassidulina teretis* and there are several species with a relationship to Atlantic Intermediate Water, comprising *Sigmoilopsis schlumbergeri*, *Eggerella bradyi*, and *Epistominella decorata/Alabaminella weddellensis*. *Nonion zaandamae* is rare and the concentration of benthonic foraminifera is low. The stadial periods are further characterized by a strong depletion of ^{18}O in both the bottom and surface waters (Rasmussen et al. 1996b).

The large difference between the stadial and the interstadial intervals in the values of $\delta^{18}O$ and in the foraminifera points to great changes in the deep-water circulation in the Greenland–Norwegian Seas, with a strongly reduced thermohaline convection during the stadial periods (Rasmussen et al. 1996b,c). Probably convection shifted from the Greenland–Norwegian Seas to the subpolar North Atlantic Ocean south of the Scotland–Greenland Ridge. As a consequence, during the stadial intervals the bottom water flowed north through the Faeroe–Shetland Channel (Rasmussen et al. 1996a,b). The stadial benthonic faunas in the ENAM93-21 core are impoverished, indicating poor living conditions and probably slow currents over the site.

The transitional cooling intervals were periods with strongly fluctuating climates, and both the position and intensity of the convection probably shifted constantly (Rasmussen et al. 1996b,c).

Sedimentation

The palaeoenvironmental and sedimentological parameters in the ENAM93-21 core are closely

Table 5. *Summary of sedimentological, micropalaeontological and palaeoceanographical interpretations regarding the Dansgaard–Oeschger cycles between 21.6 and 52.6 ka BP*

Parameters	Interstadial intervals	Stadial intervals (excl. H2 and H3)	Stadial intervals (H2 and H3)	Transitional cooling intervals	Holocene
Sedimentological parameters					
Lithology	Dark brown sandy silt	Light grey, silty clay, dropstones	Olive grey silty clay	Grey brown sandy clay	Dark brown clayey silt
$CaCO_3$ content (%)	9–16	11–17	11–17	<8	15–20
Rate of sedimentation (cm/ka)	35.2	22.5	17.8	Low	Low
Coarse fraction (63–150 µm) %	10–35	1–6	1–6	2–10	5–10
IRD	Absent	Abundant	Common	Few	Absent
Biological parameters					
Abundance of foraminifera	Medium plankt., medium benth.	Very high plankt., very low benth.	Very low plankt., very low benth.	Very low plankt., very low benth.	High plankt., high benth.
Preservation of foraminifera	Good	Very good	Variable dissolution	Dissolution	Good
Plankt./benth. ratio	1	3–10	1.5–16	0.3–0.7	1
Subpolar plankt. foram. (%)	10–35	0–10	2–8	2–10	30–45
Dominant benth. species	*Nonion zaandamae*	*Cassidulina teretis*	*Cassidulina teretis*	Decrease *N. zaand.*, increase *C. teretis*	*Nonion zaandamae*
Palaeoceanographical interpretation					
Climate	Relatively warm	Very cold	Very cold	Fluctuating rather cold	Warm
Deep-water generation	'Normal' thermohaline convection	Convection south of Iceland–Scotl. Ridge	Convection south of Iceland–Scotl. Ridge	Shifting thermohaline convection	'Normal' thermohaline convection
Surface water	Alt. to marginal Atl. Surface Water	Polar Surface Water	Polar Surface Water	Marginal Atl. to Polar Surf. Water	Atlantic Surface Water
Deep currents	Relatively strong	Very weak	Very weak	Weak	Strong
Sea ice	Nearly as at present	High iceberg rafting	Extensive ice cover	Increasing ice	As at present

The micropalaeontological data and interpretations are from Rasmussen *et al.* (1996*b,c*).

coupled and can be correlated with the Dansgaard–Oeschger temperature cycles in the Greenland ice core. The pattern of sedimentation follows the rapid shifts between warm interstadial periods and cold stadial periods (see summary of data in Table 5).

The composition of the interstadial sediments is easily explained in terms of the interstadial circulation pattern and the occurrence of volcanic deposits in the North Atlantic region. Volcanic rocks are widespread north and west of the ENAM93-21 site. The Faeroe Islands are built mainly of basaltic lavas, and enormous amounts of sand-sized volcanic ash and glass, mixed with smectitic clays, are deposited on the shelf around Iceland and on the Iceland–Faeroe Ridge (Eisma & van der Gaast 1983). The current system of the interstadial periods was, as described above, undoubtedly similar to the modern system in the Norwegian–Greenland Sea (Fig. 1). It is suggested that the volcanogenic deposits in the interstadials on the Faeroe Drift came from Iceland and from the Iceland–Faeroe Ridge.

The deposits of the stadial intervals are finer grained than those of the interstadials. They contain few volcanic particles, and the clay minerals are composed mainly of illite. The stadial deposits are further characterized by a significant amount of dropstones and IRD and by a lower rate of deposition. These differences suggest a different source of sediments. It is proposed that the main source of the stadial deposits was in the northeastern Atlantic Ocean and, in particular, in the area along the Hebrides and West Shetland shelves. The sediments here are characterized by high concentrations of illite and quartz (Eisma & van der Gaast 1983) and they are essentially very similar to the stadial deposits in the ENAM93-21 core. A source in the northeastern Atlantic Ocean would also be in agreement with the circulation system inferred for the stadial periods. During the stadials convection took place in the North Atlantic Ocean south of the Scotland–Greenland Ridge, and the current system was reversed in the Faeroe–Shetland Channel. The bottom water flowed north, and the surface water flowed south.

It should also be noted that the stadial sediments in the ENAM93-21 core compare well with glacial sediments from the Feni Drift. These sediments consist of fine-grained mud with scattered dropstones and a calcium carbonate content between 10 and 30% (van Weering & de Rijk 1991), and were deposited during periods with low or no current activity.

One of the most conspicuous differences between the stadial and interstadial periods in the ENAM93-21 core is the consistent difference in rates of sedimentation and accumulation (Figs 2 and 5). The interstadial rates were generally twice those of the neighbouring stadials. A change in the rate of deposition is often difficult to explain as it may be caused by many factors. In this particular case it seems that current velocity must have been important. The currents of the interstadial periods were strong, very much like the present currents in the area, and were capable of transporting the relatively coarse material found on the Iceland–Faeroe Ridge and around Iceland. The currents of the stadial periods were much weaker, and were apparently only able to carry material in the clay fraction. The ENAM93-21 site is, furthermore, slightly offset from the main flow through the Faeroe–Shetland Channel, and the area probably received only a small fraction of the material carried into the Norwegian Sea.

Magnetic susceptibility and composition of sediments

The magnetic susceptibility curve in the ENAM93-21 core closely follows the sedimentological log, reflecting changes in the composition of the sediment. The magnetic susceptibility values, therefore, indirectly reflect climatic and palaeoceanographic changes.

The magnetic susceptibility readings are low in the stadial intervals including the Heinrich events, and high in the interstadials. A similar pattern is present in deep-sea core ODP-644 taken northeast of the ENAM93-21 core on the Vøring Plateau (see Fronval et al. 1995), and the pattern is apparently representative of the southern part of the Norwegian Sea. The record of the Norwegian Sea contrasts with the North Atlantic record, in which the highest values are found in the Heinrich layers (Grousset et al. 1993; Maslin et al. 1995; Robinson et al. 1995).

The magnetic susceptibility in the ENAM93-21 core is probably mainly controlled by the concentration of basaltic particles, which is highest in the interstadials and lowest in the stadials. The high magnetic susceptibility values recorded in the Heinrich events in the North Atlantic Ocean, are, in contrast, probably mainly caused by magnetic, iron-rich IRD particles brought out from the North American continent with icebergs from the Laurentide ice sheet (Grousset et al. 1993; Maslin et al. 1995; Robinson et al. 1995).

The different magnetic susceptibility records north and south of the Iceland–Faeroe Ridge may, therefore, be related chiefly to differences in the sources of the sediments and transport processes. These differences may then be enhanced or diminished by variations in the carbonate content related to variations in the production rate of planktonic foraminifera. In the North Atlantic Ocean production was low during the cold Heinrich events and high in the warm periods (Bond *et al.* 1992; Bond & Lotti 1995; Oppo & Lehman 1995; Robinson *et al.* 1995). In the southeastern Norwegian Sea production was generally highest in the stadials.

Summary and conclusions

The deposits from the last glacial period on the Faeroe Drift consist of alternating layers of acidic clays, rich in quartz, and basaltic, sandy silt, rich in volcanic glass. In core ENAM93-21 recovered from the Faeroe Drift at a water depth of 1020 m, the thickness of the glacial deposits reaches almost 11 m and more than 15 sedimentary cycles have been observed. The large mineralogical difference between the basaltic and the acidic layers suggests that they have different origins.

The sequence in the ENAM93-21 core can be accurately correlated with the Dansgaard–Oeschger temperature cycles in the Greenland ice cores. The correlation shows that the sandy, basaltic layers were deposited in periods with a relatively warm interstadial climate. The circulation pattern of the interstadial periods was similar to that of the modern system, and the sandy and silty volcanic sediments were carried to the Faeroe Drift from the volcanic-rich area between the Faeroe Islands and Iceland by bottom currents along the Iceland–Faeroe Ridge.

The fine-grained acidic layers were deposited in cold stadial periods. Mineralogically, the stadial deposits resemble the sediments on the Shetland–Scotland shelf and nearby drifts in the northeastern Atlantic Ocean. Such an origin is in agreement with the stadial current system. During the stadials, convection took place south of the Iceland–Faeroe Ridge, and the current system was reversed in the Faeroe–Shetland Channel as compared with that of the interstadials. Thus, the Faeroe Drift presents an excellent example of climatic and oceanographic control of deep-sea sedimentation.

This study forms part of Project ENAM (European North Atlantic Margin: Sediment pathways, processes and fluxes) supported by EEC Grant MAST II CT93-0064. We thank S. Johnsen for the use of the GRIP $\delta^{18}O$ data and L. Labeyrie for the isotope data of the ENAM93-21 core. The accelerator mass spectrometry datings were performed by the AMS Laboratory, Institute of Physics and Astronomy, University of Aarhus. The XRD analyses were performed at the Department of Earth Sciences, University of Aarhus, under the supervision of O. B. Nielsen and at the University of Copenhagen under the supervision of T. B. Zunic. A. Kuijpers is thanked for helpful discussions. S. B. Andersen, H. Friis and O. B. Nielsen read an early draft of the manuscript.

References

BARD, E., ARNOLD, M., MANGERUD, J. *et al.* 1994. The North Atlantic atmosphere-sea surface ^{14}C gradient during the Younger Dryas climatic event. *Earth and Planetary Science Letters*, **126**, 275–287.

BOND, G. & LOTTI, L. 1995. Iceberg discharges into the North Atlantic on millennial time scales during the last glaciation. *Science*, **267**, 1005–1010.

——, HEINRICH, H., BROECKER, W. S. *et al.* 1992. Evidence for massive discharges of icebergs into the North Atlantic Ocean during the last glacial period. *Nature*, **360**, 245–249.

BROECKER, W. S. & DENTON, G. H. 1989. The role of ocean–atmosphere reorganizations in glacial cycles. *Geochimica et Cosmochimica Acta*, **53**, 2465–2501.

DANSGAARD, W., JOHNSEN, S. J., CLAUSEN, H. B. *et al.* 1993. Evidence for general instability of past climate from a 250-kyr ice-core record. *Nature*, **364**, 218–220.

DOOLEY, H. D. & MEINCKE, J. 1981. Circulation and water masses in the Faeroese Channels during Overflow '73. *Deutsche Hydrographische Zeitung*, **34**, 41–55.

DOWLING, L. M. & McCAVE, I. N. 1993. Sedimentation on the Feni Drift and late glacial bottom water production in the northern Rockall Trough. *Sedimentary Geology*, **82**, 79–87.

EHRMANN, W. U. & THIEDE, J. 1985. History of Mesozoic and Cenozoic sediments fluxes to the North Atlantic Ocean. *Contributions to Sedimentology*, **15**, 109.

EISMA, D. & VAN DER GAAST, S. J. 1983. Terrigenous Late Quaternary sediment components north and south of the Scotland–Greenland Ridge and the Norwegian Sea. *In*: BOTT, M., SAXOV, S., TALWANI, M. & THIEDE, J. (eds) *Structure and Development of the Greenland–Scotland Ridge, New Methods and Concepts*. Plenum, New York, 601–636.

FAUGÈRES, J., MÈZERAIS, M. L. & STOW, D. A. V. 1993. Contourite drift types and their distribution in the North and South Atlantic basins. *Sedimentary Geology*, **82**, 189–203.

FRONVAL, T., JANSEN, E., BLOEMENDAL, J. & JOHNSEN, S. J. 1995. Oceanic evidence for coherent fluctuations in Fennoscandian and Laurentide ice sheets on millennium timescales. *Nature*, **374**, 443–446.

GRØNVOLD, K., OSKARSON, N., JOHNSEN, S. J., CLAUSEN, H. B., HAMMER, U., BOND, G. & BARD, E. 1995. Ash layers from Iceland in the Greenland GRIP ice core correlated with oceanic and land sediments. *Earth and Planetary Science Letters*, **135**, 149–155.

GROOTES, P. M., STUIVER, M., WHITE, J. W. C., JOHNSEN, S. J. & JOUZEL, J. 1993. Comparison of oxygen isotope records from the GISP2 and GRIP Greenland ice cores. *Nature*, **366**, 552–554.

GROUSSET, F. E., LABEYRIE, L., SINKO, J. A. *et al.* 1993. Patterns of ice-rafted detritus in the glacial North Atlantic (40–55°N). *Paleoceanography*, **8**, 175–192.

HOPKINS, J. S. 1991. The GIN Sea – a synthesis of its physical oceanography and literature review 1972–1985. *Earth-Science Reviews*, **30**, 175–318.

JOHNSEN, S. J., CLAUSEN, H. B., DANSGAARD, W. *et al.* 1992. Irregular glacial interstadials recorded in a new Greenland ice core. *Nature*, **359**, 311–313.

——, ——, —— *et al.* 1997. The δ¹⁸O record along the GRIP deep ice core and the problem of possible Eemian climatic instability. *Journal of Geophysical Research*, in press.

——, DANSGAARD, W. & WHITE, J. W. C. 1989. The origin of Arctic precipitation under present and glacial conditions. *Tellus*, **41B**, 452–468.

JONES, E. J. W., EWING, M., EWING, J. I. & EITTREIM, S. L. 1970. Influences of Norwegian Sea overflow water on sedimentation in the northern North Atlantic and Labrador Sea. *Journal of Geophysical Research*, **75**, 1655–1680.

KEIGWIN, L. D. & JONES, G. A. 1989. Glacial–Holocene stratigraphy, chronology, and paleoceanographic observations on some North Atlantic sediment drifts. *Deep-Sea Research*, **36**, 845–867.

MANGERUD, J., FURNES, H. & JOHANSEN, S. 1986. A 9000-year old ash bed on the Faeroe Islands. *Quaternary Research*, **26**, 262–265.

——, LIE, S. E., FURNES, H., KRISTIANSEN, I. L. & LØMO, L. 1984. A Younger Dryas ash bed in Western Norway, and its possible correlations with tephra in cores from the Norwegian Sea and the North Atlantic. *Quaternary Research*, **21**, 85–104.

MANIGHETTI, B. & McCAVE, I. N. 1995a. Depositional fluxes, palaeoproductivity, and ice rafting in the NE Atlantic over the past 30 ka. *Paleoceanography*, **10**, 579–592.

—— & —— 1995b. Late glacial and Holocene palaeocurrents around Rockall Bank, NE Atlantic Ocean. *Paleoceanography*, **10**, 611–626.

MASLIN, M. A., SHACKLETON, N. J. & PFLAUMANN, U. 1995. Surface water temperature, salinity, and density changes in the northeast Atlantic during the last 45,000 years: Heinrich events, deep water formation, and climatic rebounds. *Paleoceanography*, **10**, 527–544.

McCAVE, I. N., LONSDALE, P. F., HOLLISTER, C. D. & GARDNER, W. D. 1980. Sediment transport over the Hatton and Gardar contourite drifts. *Journal of Sedimentary Geology*, **50**, 1049–1062.

——, MANIGHETTI, B. & BEVERIDGE, N. A. S. 1995. Circulation in the glacial North Atlantic inferred from grain-size measurements. *Nature*, **374**, 149–152.

MEINCKE, J. 1983. The modern current regime across the Greenland–Scotland Ridge. *In*: BOTT, M.,

SAXOV, S., TALWANI, M. & THIEDE, J. (eds) *Structure and Development of the Greenland–Scotland Ridge, New Methods and Concepts*. Plenum, New York, 637–650.

NIELSEN, T., VAN WEERING, T. C. E. & ANDERSEN, M. S. 1997. Cenozoic changes in the sedimentary regime on the northeastern Faeroes margin. This volume.

OPPO, D. W. & LEHMAN, S. J. 1995. Suborbital timescale variability of North Atlantic Deep Water during the past 200,000 years. *Paleoceanography*, **10**, 901–910.

RASMUSSEN, T. L., THOMSEN, E., LABEYRIE, L. & VAN WEERING, T. C. E. 1996b. Circulation changes in the Faeroe–Shetland Channel correlating with cold events during the last glacial period (58–10 ka). *Geology*, **24**, 937–940.

——, VAN WEERING, T. C. E. & LABEYRIE, L. 1996a. High resolution stratigraphy of the Faeroe–Shetland Channel and its relation to North Atlantic paleoceanography: the last 87 ka. *Marine Geology*, **131**, 75–88.

——, ——, —— & —— 1996c. Rapid changes in surface and deep water conditions at the Faeroe Margin during the last 58,000 years. *Paleoceanography*, **11**, 757–771.

ROBINSON, S. G., MASLIN, M. A. & McCAVE, I. N. 1995. Magnetic susceptibility variations in Upper Pleistocene deep-sea sediments of the NE Atlantic: implications for ice rafting and paleocirculation at the last glacial maximum. *Palaeoceanography*, **10**, 221–250.

STOW, D. A. V. & HOLBROOK, J. A. 1984. North Atlantic contourites: an overview. *In*: STOW, D. A. V. & PIPER, D. J. W. (eds) *Fine-grained Sediments: Deep-water Processes and Facies*. Geological Society, London, Special Publications, **4**, 245–256.

VAN WEERING, T. C. E. & DE RIJK, S. 1991. Sedimentation and climate-induced sediments on Feni Ridge, Northeast Atlantic Ocean. *Marine Geology*, **101**, 49–69.

——, ANDERSEN, M. S., KUIJPERS, A. & SHIPBOARD SCIENTISTS 1993. *Project ENAM, European North Atlantic Margin: Sediments Pathways, Processes and Fluxes*. Shipboard Report Cruise R.V. *Pelagia*, 2–29 August 1993.

——, NIELSEN, T., KENYON, N. H., AKENTIEVA, K. & KUIJPERS, A. H. 1998. Large submarine slides at the NE Faeroe continental margin. *This volume*.

——, VAN BENNEKOM, J. A., KUIJPERS, A. & SHIPBOARD SCIENTISTS 1994. *Project ENAM, European North Atlantic Margin: Norwegian Sea and Faeroe–Shetland Margins*. Shipboard Report Cruise R.V. *Pelagia*, 17 October–4 November 1994.

WOLF, T. C. W. & THIEDE, J. 1991. History of terrigenous sedimentation during the past 10 m.y. in the North Atlantic (ODP legs 104 and 105 and DSDP leg 81). *Marine Geology*, **101**, 83–102.

Late Quaternary stratigraphy and palaeoceanographic change in the northern Rockall Trough, North Atlantic Ocean

J. A. HOWE[1]*, R. HARLAND[2], N. M. HINE[3] & W. E. N. AUSTIN[4]

[1]*British Geological Survey, Murchison House, West Mains Road, Edinburgh EH9 3LA, UK, and Department of Geology, Southampton University, Southampton, SO9 5NH, UK*
*Present address: British Antarctic Survey, High Cross, Madingley Road, Cambridge CB3 0ET, UK. e-mail: jaho@pcmail.nerc-bas.ac.uk
[2]*Centre for Palynological Studies, University of Sheffield, Mappin Street, Sheffield S1 3JD, and DinoData Services, 50 Long Acre, Bingham, Nottingham NG13 8AH, UK*
[3]*Industrial Palynology Unit, Centre for Palynological Studies, University of Sheffield, Mappin Street, Sheffield S1 3JD, UK*
[4]*Environmental Research Centre, Department of Geography, University of Durham, South Road, Durham DH1 3LE, UK*

Abstract: A series of six shallow gravity cores, taken from a variety of sedimentary settings in the northern Rockall Trough, have been analysed using microfossil and sedimentological techniques. Cores from sediment waves on the Barra Fan are interpreted as being sequences of hemipelagites, turbidites and hemiturbidites. Northeastern Rockall Trough cores, from slope apron, escarpment and sediment drift areas are interpreted as hemipelagites, with glaciomarine deposits interbedded with and overlain by muddy–silty and sandy contourites. The dinoflagellate cyst, planktonic foraminifera and nannofossil biostratigraphy reveals a four-fold deglaciation record, with a single long core from the Barra Fan seemingly containing all four divisions of the Late Glacial, Allerød–Bölling, Younger Dryas and Holocene intervals. The sedimentary record suggests that deglaciation in the North Atlantic Ocean was not a simple linear process but an irregular, non-linear series of rapid events characterized by sudden sea-surface temperature changes, and fluctuating bottom-current activity.

There has been a great deal of interest in the palaeoceanography of the North Atlantic Ocean, especially with respect to the development of deep-water circulation and its response to climate change. Previous studies such as CLIMAP (Climate Long Range Investigation Mapping and Predictions) and the continuing results of the DSDP (Deep Sea Drilling Project) and the ODP (Ocean Drilling Program) have contributed a wealth of new data (Kellogg 1976; McIntyre *et al.* 1976; Ruddiman & McIntyre 1976, 1981). Present work on the oceanic response to climate change is given added impetus by forecasts of global warming. The last glacial–deglaciation event in particular has recently been the subject of several papers from the mid- to high-latitude areas of the northeastern Atlantic Ocean (Bond *et al.* 1992; Lehman & Keigwin 1992; Austin & Kroon 1996).

This study concentrates on the eastern margin of the Rockall Trough and the Hebrides Slope, off northwest Scotland, north of 56°N. Previous work in the Rockall Trough has concentrated on both the microflora (Harland 1988, 1989, 1994; Hine 1990) and the sediments from cores on the slope, and in deeper water from the Feni Ridge sediment drift on the western side of the trough (Keigwin & Jones 1989; Stoker *et al.* 1989, 1994; Dowling & McCave 1993). A four-fold division of the Late Pleistocene–Holocene interval has previously been applied to cores from the Hebrides (Graham *et al.* 1990; Peacock *et al.* 1992), the Faeroe–Shetland Channel and the northeastern Rockall Trough (Stoker *et al.* 1989). This study uses the same stratigraphic subdivision based on the original core (59/-08 34CS) used by Stoker *et al.* (1989). This stratigraphic correlation is further supported by evidence from ash horizons, notably the Vedde Ash, which has recently been dated from marine sediments of Younger Dryas age on the Hebrides Shelf (Austin *et al.* 1995).

This preliminary study focuses on the oceanographic response to climate change during the last glacial–deglaciation as recorded in six short gravity cores recovered by the British Geological Survey (BGS) from the Hebrides Slope. Microfossil and sedimentological data are used to reconstruct the processes active across the area over the last 14 ka. The cores were taken

HOWE, J. A., HARLAND, R., HINE, N. M. & AUSTIN, W. E. N. 1998. Late Quaternary stratigraphy and palaeoceanographic change in the northern Rockall Trough, North Atlantic Ocean. *In*: STOKER, M. S., EVANS, D. & CRAMP, A. (eds) *Geological Processes on Continental Margins: Sedimentation, Mass-Wasting and Stability*. Geological Society, London, Special Publications, **129**, 269–286.

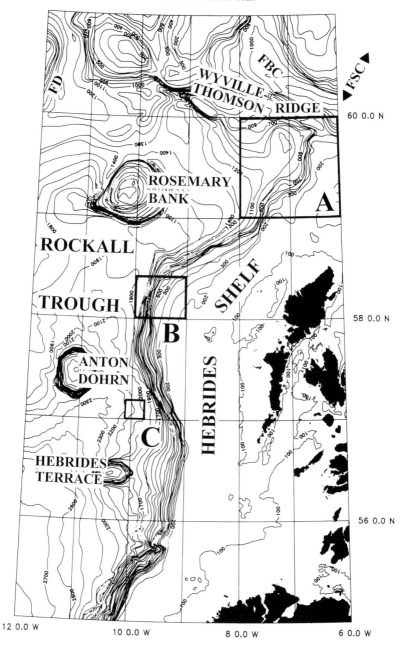

Fig. 1. Regional bathymetric setting of the northern Rockall Trough with the location of the study areas: A, northeastern Rockall Trough (Fig. 2); B, Geikie Escarpment (Fig. 3); C, northern Barra Fan (Fig. 4). FSC, Faeroe–Shetland Channel; FBC, Faeroe–Bank Channel; FD, Feni Drift.

from four different settings: (1) a slope apron, built up from clastic glaciomarine deposition into a discrete sheet-like body that mantles the northern Hebrides Slope (Stoker *et al.* 1994); (2) a sediment drift and moated area produced by preferential deposition and non-deposition of sediments under bottom-current flow, adjacent to the Wyville-Thomson Ridge in the north (Howe *et al.* 1994); (3) the steeper slope of the Geikie Escarpment (Stoker *et al.* 1994); (4) sediment waves at the distal edge of the Barra Fan (Howe 1996).

Regional setting

The Rockall Trough is a northeast-trending basin that separates the Hebrides Shelf to the east from the Rockall Plateau to the west (Fig. 1). At its northeastern limit, the trough is separated from the Faeroe–Shetland Channel by the Wyville-Thomson Ridge. To the southwest the trough deepens to 4000 m, opening out to the Porcupine Abyssal Plain (Roberts *et al.* 1979). The width of the trough at the 1000 m isobath is relatively uniform, between 200–250 km. The Feni Ridge, a major depositional drift, flanks the western margin of the trough, whereas fans and slope aprons locally encroach onto the basin floor along its eastern margins (Stoker *et al.*

1994). Punctuating the centre of the trough are the isolated bathymetric highs of the Rosemary Bank, and the Anton Dohrn and Hebrides Terrace seamounts.

The present bathymetric configuration of the trough has probably been in existence since the mid-Tertiary sub-era. During the Neogene period (including the Pleistocene epoch), clastic sedimentation was dominant along the eastern margin of the trough, depositing the fans and slope aprons which mantle the Hebrides Slope (Stoker *et al.* 1993).

The asymmetry of the Neogene sedimentation suggests that it was controlled, not by subsidence and tectonics, but by sediment availability linked to the oceanographic

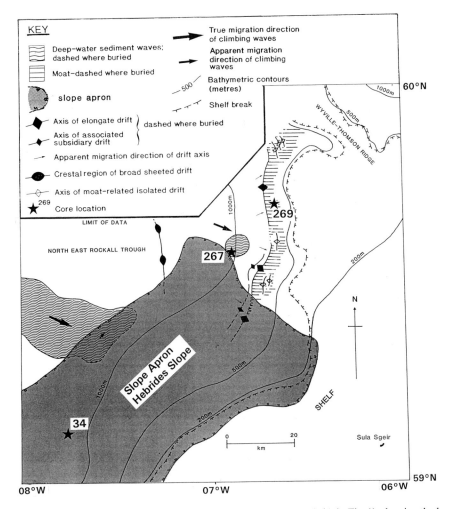

Fig. 2. Principal morphological features of the northeastern Rockall Trough (A in Fig. 1), showing the location of the three cores: 59/-08 34CS on the lower slope apron, 59/-07 267CS from the small sediment waves on the flank of an elongate sediment drift and 59/-07 269CS from the moated area at the base of the Hebrides Slope.

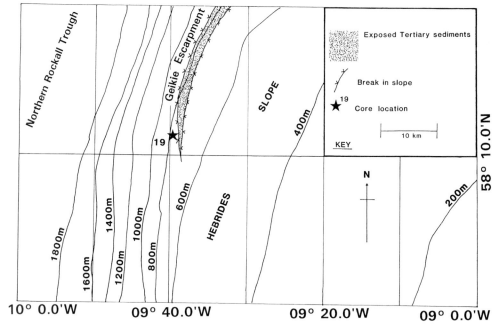

Fig. 3. Principal morphological features of the Geikie Escarpment (B in Fig. 1), showing the location of core 58/-10 19CS on the mid-Hebrides Slope beneath the escarpment.

development of the northeast Atlantic Ocean. Bottom-current activity has been significant in this area since late Eocene–early Oligocene time

(Miller & Tucholke 1983). During its inception, bottom-current flow was very strong, but it stabilized and decreased during mid-Miocene

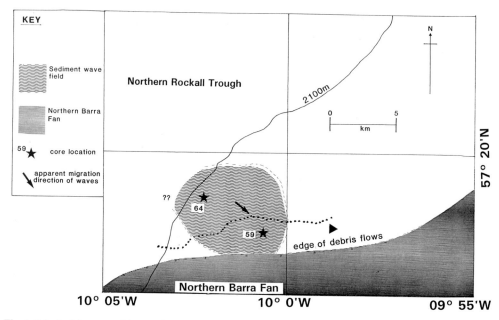

Fig. 4. Principal features of the northern Barra Fan (C in Fig. 1), showing the location of cores 57/-11 59CS and 57/-11 64CS in the small sediment wave field developed at the distal northern edge of the main debris flows of the fan.

times (Stow & Holbrook 1984). In the northeast region of the trough adjacent to the Wyville-Thomson Ridge the vigorous circulation produced a phase of cut and fill, with subsequent development of a Neogene sediment drift complex comprising elongate, isolated and moat-related drifts and sediment waves, with an erosive moat developed at the base of the Hebrides Slope (Howe *et al.* 1994). On the Hebrides Slope south of the drift complex, Neogene clastic sedimentation has produced an apron deposit, developed as a result of mass-wasting and glaciomarine processes (Fig. 2). Farther south along the Hebrides margin, the Geikie Escarpment (Fig. 3) is probably the result of erosion by strong mid-Oligocene current activity, enhanced by the exposure and steepness of the slope (Stoker *et al.* 1994). Downslope sedimentation dominates in the area of the Barra Fan, constructed by debris flows, turbidites and glaciomarine sedimentation throughout the Neogene period (Fig. 4).

Oceanographic setting

The physical oceanography of the Rockall Trough is complex, involving several disparate water masses (Fig. 5). North Atlantic surface water flows northeast as the North Atlantic Current (NAC), cooling and sinking because of increased density in the Norwegian Sea. This returns as Norwegian Sea Deep Water (NSDW), flowing south through the Faeroe–Shetland Channel and over the western end of the Wyville-Thomson Ridge (Dickson & Kidd 1986). On entering the trough, NSDW partially mixes with Labrador Sea Water and Antarctic Bottom Water to form North Atlantic Deep Water (NADW) termed the Deep Northern Boundary Current (DNBC), as it circulates within the trough (McCartney 1992). NADW flows as an anticyclonic gyre within the southern region of the trough (Stow & Holbrook 1984; Dickson & Kidd 1986). In addition, the Hebrides Slope is affected by a northward-

flowing slope current to depths of 1000 m (Booth & Ellett 1983; Huthnance 1986; Kenyon 1986; D. J. Ellett, pers. comm. 1992). The origin of the slope current is uncertain although it may have originated from the continuation of the NADW gyre circulating in the southern part of the trough, flowing north along the Hebrides Slope. However, at higher levels, on the mid–upper slope, the slope current may be of North East Atlantic Water (NEAW) origin (Harvey 1982). Ultimately, the slope current may become detached from the slope as the trough shallows in the north, to flow westward around the base of the Wyville-Thomson Ridge, where it becomes entrained by the overflowing NSDW.

Current-meter evidence supports the existence of a slope current with flow orientations to the NNE, following the bathymetric contours. Maximum current velocities, recorded on the slope are: 19 cm s^{-1} in 784 m of water on the Barra Fan; increasing to 25 cm s^{-1} in 457 m of water on the Geikie Escarpment; and up to 30 cm s^{-1} from 518 m water depth on the slope apron south, and adjacent to the Wyville-Thomson Ridge in the northeastern part of the trough (Huthnance 1986; Howe & Humphery 1995).

Materials and methods

Details of the six gravity cores presented in this study are given in Table 1.

The archive sections of each core were X-radiographed in half-core sections, using a Scanray AC120L at settings of 75 kV and 5 mA for between 2 and 2.5 min. Ice-rafted debris (IRD) counts were made from the X-radiograph negatives, counting clasts >2 mm in size in each 2 cm interval. Particle size analysis was conducted on 50–100 g samples. Sand and coarse gravel percentages were obtained by sieving, and the <4.0 phi fraction (silt sized and less) was analysed using a Micromeritics Sedigraph 5000ET. Carbonate content was analysed by a 'Karbonate-Bombe' following the method of Muller & Gastner (1971).

Table 1. *Summary of the information on core location, length (recovery), water depth and position*

Area	Core	Length (m)	Water depth (m)	Position
Northeastern	267	1.8	1022	59°35.0′N, 06°55.0′W
Rockall Trough	269	1.2	870	59°42.6′N, 07°46.6′W
(A)	34	1.9	1191	59°07.1′N, 07°46.6′W
Geikie				
Escarpment (B)	19	2.1	747	58°11.03′N, 09°39.70′W
Northern Barra	59	3.6	2089	57°00.85′N, 10°00.60′W
Fan (C)	64	2.4	2097	57°01.37′N, 10°02.21′W

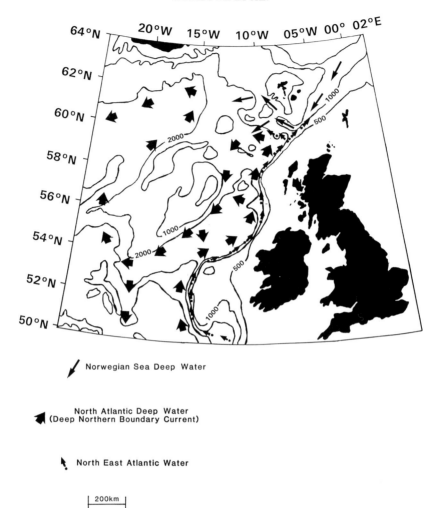

Fig. 5. Present-day deep-water circulation in the Rockall Trough. The main influences on the study areas are North East Atlantic Water (NEAW) on the mid–lower slope, and North Atlantic Deep Water (NADW) in the basin and on the lower slope (adapted from Stow & Holbrook 1984; Dickson & Kidd 1986; McCartney 1992).

Samples were taken every 0.10 m for calcareous nanofossil analysis, except for cores 57/-11 59CS and 58/-10 19CS where only the top 0.5 m was examined. Slides were prepared for light microscope study with a minimum of 25 fields of view or 300 calcareous nanofossils counted from each slide, following the method of Backman & Shackleton (1983). Dinoflagellate cyst analysis was conducted on cores 59/-08 34CS (previously described in Stoker *et al.* 1989) 58/-10 19CS, 57/-11 59CS and 57/-11 64CS. Samples were taken every 0.10 m and processed for dinoflagellate cysts using normal palynological techniques involving acid digestion with hydrochloric and hydrofluoric acids. Oxidation reagents were avoided so as not to lose any of the fragile

peridiniacean cysts (Dale 1976). To calculate the number of cysts/gram of sediment the technique of Harland (1989) was employed.

Core sedimentology

The northeastern Rockall Trough cores and the Geikie Escarpment cores have been divided into four lithofacies; the Barra Fan cores contain three additional lithofacies. Brief summary descriptions of each of the lithofacies are given below.

Northeastern Rockall Trough and Geikie Escarpment

From this area, core 59/-07 267CS is included from the sediment drift, core 59/-07 269CS from the moat at the base of the slope, core 59/-08 34CS from the slope apron and core 58/-10 19CS from the escarpment (Fig. 6).

Lithofacies 1. This lithofacies occurs at the top of all the cores, and also at 0.7–0.8 m in core 59/-07 269CS, and is greyish yellow muddy sand up to

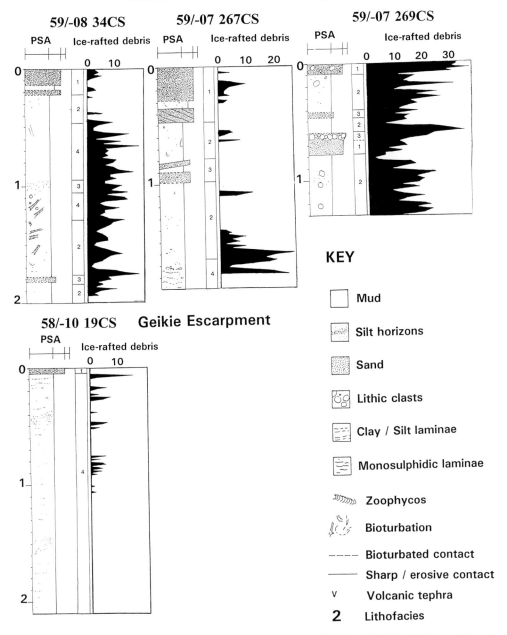

North-Eastern Rockall Trough

KEY

☐ Mud

▨ Silt horizons

▨ Sand

▨ Lithic clasts

▨ Clay / Silt laminae

▨ Monosulphidic laminae

)))) Zoophycos

Bioturbation

– – – – Bioturbated contact

——— Sharp / erosive contact

v Volcanic tephra

2 Lithofacies

Fig. 6. Graphic logs of cores 59/-08 34CS (slope apron), 59/-07 267CS (drift) and 59/-07 269CS (moat) from the northeastern Rockall Trough, and core 58/-10 19CS from the Geikie Escarpment showing core lithology, lithofacies divisions and ice-rafted debris component (clasts per 2 cm slice). Core 59/-08 34CS as described by Stoker *et al.* (1989) and cores 59/-07 267CS and 59/-07 269CS as described by Howe *et al.* (1994). PSA, Particle size analysis.

0.28 m thick. Moderately well-sorted fine- to medium-grained sand forms 30–80%, with 10–45% mud. Gravel clasts are common. Bioturbation is intense, with most primary structures destroyed, although core 59/-07 267CS preserves some cross-lamination. Contact with the lower lithofacies varies from sharp and erosive (cores 59/-07 269CS and 58/-10 19CS) to irregular as a result of bioturbation (cores 59/-08 34CS and 59/-07 267CS).

Lithofacies 2. This is up to 0.55 m thick and consists of olive–brown sandy muds, with up to 40% fine-grained sand and up to 70% mud. This lithofacies is absent from core 58/-10 19CS. Lithic clasts are present throughout and bioturbation is moderate to intense with commonly preserved

Zoophycos, Trichichnus and *Chondrites* burrow types. Planar lamination of the silts is visible with <0.05 m thick inverse to normally graded laminated units. Lower contacts are commonly gradational and bioturbated, although some are sharp and erosive.

Lithofacies 3. This lithofacies is a <0.2 m thick unit of poorly sorted yellow–brown sands and muds. Only the trio of most northerly cores contain this lithofacies. Gravel is commonly present, especially in core 59/-07 269CS. The unit is totally structureless and lacks any indication of bioturbation, perhaps because of the coarse, disorganized nature of the lithofacies. Contacts with continuous lithofacies above and below are gradational.

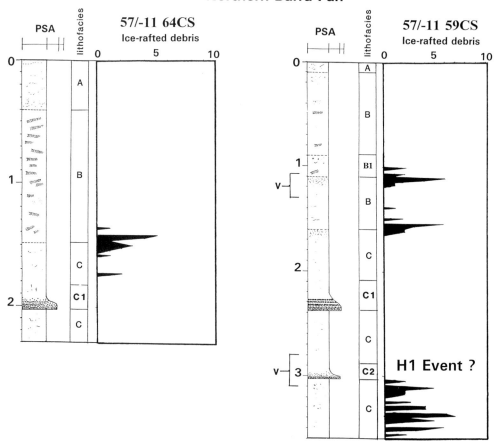

Fig. 7. Graphic logs of cores 57/-11 59CS (edge of the wave field) and 57/-11 64CS (wave crest) from the northern Barra Fan sediment wave field. Core lithology, lithofacies divisions and ice-rafted component shown (for key to symbols, see Fig 6).

Lithofacies 4. This lithofacies is best developed in core 58/-10 19CS, where it is up to 2.00 m thick. Core 59/-07 269CS, from the moated area of the drift complex, lacks this lithofacies. It consists of dark olive brown homogeneous mud with <5% sand. Monosulphidic lamination is common throughout, and present in the more northerly cores are matrix-supported lithic clasts; the sediment is intensely bioturbated throughout. The upper contact with lithofacies 1 is sharp and erosive.

Northern Barra Fan

The two cores from this deeper-water, sediment wave area (Fig. 7) are included for comparison with the more northerly cores. Core 57/-11 59CS is upslope from the best-developed waves, core 57/-11 64CS is from a near wave-crest region.

Lithofacies A. This lithofacies is a 0.1–0.4 m thick unit of poorly sorted greyish sandy mud. Bioturbation is intense throughout, although some faint planar lamination is visible in X-radiographs. The lower contact is intensely bioturbated.

Lithofacies B. This lithofacies ranges from 1.10 to 1.50 m thick, comprising a poorly sorted, dark olive, homogeneous mud. Intense bioturbation is present throughout, with well-developed *Zoophycos* traces. X-Radiographs reveal a totally homogenized unit with some *Mycelia* development. The upper and lower contacts are intensely bioturbated. One interval in core 57/-11 59CS contains a sub-lithofacies, B1.

Lithofacies B1. This (<0.2 m) lithofacies occurs within lithofacies B and consists of poorly sorted, bioturbated dark olive muds. B1 is absent from core 57/-11 64CS, and out of seven other cores taken from the wave-field only three display the B1 lithofacies. The cores containing lithofacies B1 occur upslope of the waves' development, with the unit thinning in a downslope direction (Howe 1996). Contacts with the surrounding lithofacies are totally bioturbated.

Lithofacies C. Lithofacies C is represented by up to 2.0 m of dark olive sandy–silty poorly sorted muds in core 57/-11 59CS. The sequence has been homogenized by intense bioturbation, although it lacks the well-developed *Zoophycos* traces seen in lithofacies B. This lithofacies has been further subdivided into lithofacies C1 and C2.

Lithofacies C1 and C2. Occurring within lithofacies C, at 2.00–2.30 m depth in cores 57/-11 59CS and 57/-11 64CS is a sequence of medium-grained well-sorted sands fining-up to silts and muds. The lower contact is sharp and erosive but the upper contact is more indistinct as the sequence grades into lithofacies C. Bioturbation is only found in the upper sections. Lithofacies C2 is identical in character although found lower down at 3.00–3.08 m in core 57/-11 59CS.

Interpretation of the lithofacies

Northeastern Rockall Trough and Geikie Escarpment. Lithofacies 1, containing a moderately sorted sand fraction with rare cross-bedding and intense bioturbation, is comparable with a sandy contourite (Stow & Holbrook 1984). Where it occurs at the sediment surface it can be interpreted as the coarser component of a coarsening-upward (negative) sequence, indicating high current velocities winnowing away the fines (Gonthier *et al.* 1984). In contrast, the silty lamination of lithofacies 2 with recognizable coarsening and fining sequences, associated with bioturbated fine-grained sediments, are features consistent with muddy contourites (Stow & Lovell 1979; Stow & Piper 1984). The coarsening- and fining-upward sequences may therefore be interpreted as resulting from fluctuations in the current flow. Source material for both the sandy and muddy contourites may have originated from a combination of hemipelagic and glaciomarine sedimentation. Direct glaciomarine sedimentation, through overturning and unloading of sediment-laden icebergs, may have also triggered thin debris flows and given rise to lithofacies 3. Similar deposits have been described by Molnia (1983) and Eyles *et al.* (1985) as iceberg rain-out and unloading. Lithofacies 4, the bioturbated, homogeneous muds, can be attributed to hemipelagic sedimentation producing a predominantly fine-grained deposit as a result of the slow accumulation of sediment (Stow & Piper 1984). The monosulphidic lamination seen in core 58/-19 19CS is possibly a secondary diagenetic feature.

Northern Barra Fan. The poorly sorted, bioturbated sandy muds of lithofacies A, B and C are all interpreted as predominantly hemipelagites. However, the faint lamination visible in lithofacies A could be indicative of a slight current influence, possibly the lateral transfer of a muddy cloud (Stow & Piper 1984). Wetzel (1984) described the *Zoophycos* ichnofacies, well developed in lithofacies B, as indicative of slow, stable accumulation of fine-grained sediment. The poorly sorted bioturbated muds of lithofacies B1 suggest a hemipelagite, although the localized occurrence and downslope thinning are more indicative of distal low-concentration turbidites. Stow & Wetzel (1990) described a similar deposit from the distal portion of the Bengal Fan as being a hemiturbidite, resulting from settling of a dilute, distal turbidity current. Lithofacies C1 and C2 are thin sandy–silty turbidites on the evidence of a lower sharp, erosive base and fining-upward silty muds with upper bioturbation (Stow & Piper 1984; Stow 1985).

Ice-rafted debris

Ice-rafted debris (IRD) analysis reveals gross differences in the volume of clasts being transported between the three study areas (Figs 6 and 7). The more northerly cores, from the northeastern Rockall Trough, contain highest percentage of IRD, with core 59/-7 269CS, from the moat around the base of the Hebrides Slope, containing the largest amount with up to 38 clasts (>2 mm) per 2 cm slice. Core 59/-07 267CS from the sediment drift also contains large amounts of clastic material, with up to 24 clasts per 2 cm slice at >1.4 m core depth. Core 59/-08 34CS from the slope apron contained large amounts of material

although individual peaks were smaller, the largest consisting of 18 clasts. A decrease in the amount of IRD in the southerly study areas can be seen in core 58/-10 19CS from the escarpment, which contained up to 16 clasts per 2 cm slice near the top of the core, with smaller peaks down throughout. Cores 57/-11 59CS and 57/-11 64CS from the northern Barra Fan are in marked contrast to the trio of northerly cores, with a much reduced IRD signature. Peaks of 4–7 clasts per 2 cm slice are evident at intervals in both the cores notably at 1.00–1.6 m and from 3.00 m downwards in 57/-11 59CS, and at 1.4–1.8 m in 57/-11 64CS. Compositionally, all the IRD was consistent with a Scottish mainland, Outer Isles source with clasts typically of Lewisian and Moinian high-grade metamorphic terranes with some Torridonian sandstone and Tertiary volcanic rocks.

Biostratigraphy

Nannofossil data

Nannofossil analysis was conducted on all the cores except core 57/-11 64CS. The nannofossil assemblages can be assigned to three biostratigraphic units (Fig. 8), based on

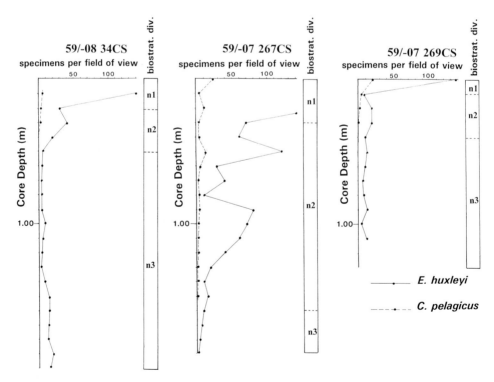

Fig. 8. Downcore variations in elements of the nannoflora from cores 59/-08 34CS, 59/-07 267CS and 59/-07 269CS.

variations in the nannofossil assemblages (n1–n3), described below in ascending stratigraphic order.

(n3) A unit of impoverished nannofossil assemblages, containing abundant reworked Tertiary and Cretaceous forms. This may be interpreted as a cold-water glaciomarine environment, with relative proximity to shelf-wide ice-sheets suppressing surface water productivity.

(n2) A unit characterized by an increase in the total number of nannofossils, with some *Coccolithus pelagicus* observed. This suggests an amelioration in climate with the retreat of ice-sheets and an increase in surface water productivity.

(n1) A unit characterized by rich nannofossil assemblages dominated by *Emiliania huxleyi* and common *Coccolithus pelagicus*. This suggests a retreat of the ice-sheets, with high levels of productivity. The predominance of *E.huxleyi* allows an assignment to the eponymous zone of Gartner (1977), which in the northeast Atlantic at this latitude has a base at <65 ka BP. The *C. pelagicus* bloom has been dated at 0–7.5 ka BP by Gard (1988, 1989) and 0–10 ka BP (Baumann & Matthiessen 1992), suggesting a Holocene assignment for this unit.

The three northerly cores (59/-07 269CS, 59/-07 267CS and 59/-08 34CS) display all three units (see Fig. 8). Only the uppermost 0.5 m of cores 58/-10 19CS and 57/-11 59CS were analysed for their calcareous nannofossils. The 0.5 m of core 58/-10 19CS consisted entirely of unit n3 (impoverished assemblages and abundant reworking). The 0.5 m of core 57/-11 59CS was indicative of unit n1 (rich assemblages of nannofossils dominated by *E. huxleyi*); however, the bloom of *C. pelagicus* was poorly developed and this was attributed to the removal of the core top possibly by bottom currents.

Dinoflagellate data

Dinoflagellate cyst analysis was conducted on cores 57/-11 59CS, 57/-11 64CS (Fig. 9), 58/-10 19CS and 59/-08 34CS (described by Stoker *et al.*

1989). Four clearly distinguished biostratigraphic units (d1–d4) based on the cyst assemblages were identified.

(d1) The oldest unit (only in core 57/-11 59CS) is characterized by low cyst recovery of <500 cysts/g, low diversity and the presence of low numbers of *Operculodinium centrocarpum* (Deflandre and Cookson) Wall, *Bitectatodinium tepikiense* Wilson, round brown *Protoperidinium* cysts and some *Spiniferites* species. *B. tepikiense* and the round brown *Protoperidinium* species are the most prominent members of the assemblages.

(d2) The next unit contains a richer, higher diversity cyst assemblage of >500–3500 cysts/g of sediment, containing high numbers of *O. centrocarpum* towards the top of the unit; *B. tepikiense*, more noticeable towards the base; and round brown *Protoperidinium* cysts together with small amounts of *Spiniferites* spp., *Protoperidinium pentagonum* (Gran) Balech and *Impagidinium* species. None the less, the unit tends to be dominated by *B. tepikiense* (up to 2000 cysts/g) and round brown *Protoperidinium* cysts (up to 1000 cysts/g).

(d3) In the third unit there is a return to the poor cyst assemblages first observed in unit d1, with low diversity, low cyst numbers of <300 cysts/g, with small amounts of *O. centrocarpum* and *B. tepikiense* and round brown *Protoperidinium* cysts.

(d4) The youngest unit contains a rich, diverse assemblage with high numbers of cysts (>500 cysts/g to >10,000 cysts/g). High numbers of *O. centrocarpum* and *Nematosphaeropsis labyrinthus* (Ostenfeld) Reid occur, with moderate amounts of *Protoperidinium* cysts, including *P. pentagonum*, *Spiniferites* spp. and *Impagidinium* spp.

Dinoflagellate and nannofossil chronostratigraphy

Using both the detailed calcareous nannofossil data and the dinoflagellate cyst evidence the cores can be tentatively placed within the

Table 2. *Summary of the biostratigraphic divisions of the cores, and relative stratigraphic units*

Nannofossil unit	Dinoflagellate unit	Foraminiferal unit (core 59 only)	Stratigraphy
n1	d4	p4	Holocene
n2			
n3	d3	p3	Younger Dryas
	d2	p2	Allerød–Bölling
	d1	p1	Late Glacial

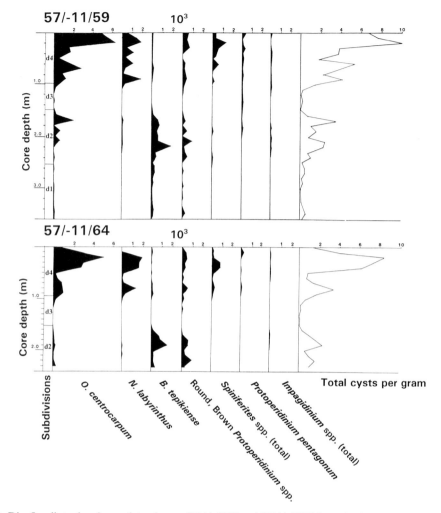

Fig. 9. Dinoflagellate abundance plots of cores 57/-11 59CS and 57/-11 64CS from the Northern Barra Fan.

established late Quaternary chronostrati-graphic framework (see Table 2) with reference to Gard (1988, 1989), Graham *et al.* (1990), Hine (1990) and Baumann & Matthiessen (1992). A conservative approach precludes dis-cussion of other possibilities until further evi-dence is available, but the presence of the Vedde Ash in core 57/-11 59CS supports our dating. A similar dinoflagellate cyst bio-stratigraphy has been described from the north-east Atlantic Ocean by Harland (1994), and from the Norwegian Sea by Baumann & Matthiessen (1992).

Dinoflagellate unit d1, included within nanno-fossil unit n3, is considered to be of Late Weich-selian age, >13 ka BP, with impoverished floras typical of cold, arctic conditions.

Dinoflagellate unit d2, still within nannofossil

unit n3, may be related to the Allerød–Bölling Interstadial at 13,000–11,000 a BP. The dino-flagellate assemblages rich in *B. tepikiense* may be indicative of lower than normal salinities (Wall *et al.* 1977; Dale 1985), suggesting a melt-water input into the system. However, the pres-ence of cysts of the heterotrophic dinoflagellate *Protoperidinium* may indicate the proximity of sea ice (Dale 1985).

Both the nannofossil unit n3 and the dino-flagellate unit d3 display low diversity, poor assemblages. This is interpreted as a return to cold, arctic conditions of the Younger Dryas stadial of 11–10 ka BP.

Nannofossil units n2 and n1 and the dino-flagellate unit d4 can be assigned to the early Holocene (Hine 1993; Harland 1994), and, therefore must be younger than 10 ka BP.

Planktonic foraminifera data, core 57/-11 59CS

Planktonic foraminifera were examined from the sieve fraction >125 μm from core 57/-11 59CS. The assemblages can be loosely defined as having an arctic (i.e. containing predominantly sinistrally coiled *Neogloboquadrina pachyderma* (Ehrenberg), (sinistral), and a North Atlantic Current (NAC) aspect. The NAC assemblages are highly variable throughout the core and may warrant further quantitative examination. The most common NAC species are: *Globigerina bulloides*, d'Orbigny, *Turborotalia quienqueloba* (Natland), *Neogloboquadrina pachyderma* (dextral), *Globorotalia inflata* d'Orbigny, *Globigerinita glutinata* (Egger) and *Globorotalia scitula* (Brady).

The assemblage boundaries (p1–p4) within the core are generally sharp and well defined at the following core depths:

(p1) Extends from the base of core to 3.00 m. Arctic assemblage with >90% *N. pachyderma* (sin.). Major decline in specimen abundance from 3.38 to 3.00 m. The transition to unit 2 is not well defined and *N. pachyderma* (sin.) specimens extend upcore to 2.78 m.

(p2) 3.00 to 1.70 m. NAC assemblages with increased species diversity and specimen abundance throughout. Specimens of *N. pachyderma* (sin.) are rare, apart from a brief occurrence at about 2.25–2.30 m, where they are associated with the C1 turbidite.

(p3) 1.70–0.97 m. Arctic assemblage >75% *N pachyderma* (sin.). Occasional specimens of *N. pachyderma* (dex.) and *G. bulloides*.

(p4) 0.97 m to the top of the core. NAC assemblages with a marked increase in abundances above 0.41 m and higher species diversity.

Interpretations

The foraminiferal assemblages indicate major oceanographic change at the site of core 57/-11 59CS (see Table 2). Preliminary analyses indicate the presence of clear, winged shards characteristic of rhyolitic Vedde Ash as well as brown basaltic shards characteristic of the additional consitituents of North Atlantic Ash Zone 1 (Kvamme *et al.* 1989). These occur at a core depth of 1.10–1.30 m, suggesting that unit p3 accumulated during the Younger Dryas cold phase (*c.* 11–10 ka BP). This suggests that the unit p3–p4 boundary dates to about 10 ka BP, whereas the unit p2–p3 boundary dates to 11 ka BP. Tephra from about 2.80 m have a geochemical signature which distinguishes them from the

Younger Dryas tephra (J. Hunt, pers. comm. 1994). We note that older tephra are recorded in the Greenland–Norwegian Sea and that they date to *c.* 14 ka BP (H. Haflidason, pers. comm. 1994). If the latter belong to the same volcanic event and this age is correct, then the unit p1–p2 boundary dates to about 14 ka BP.

Employing the stratigraphic framework which the tephra constrain, comparisons can be made with published planktonic isotope stratigraphies from two other North Atlantic cores. These are cores NA87-22 from the Rockall Plateau at 55°30′N, 14°42′W (Duplessy *et al.* 1992) and V23-81 from the Feni Drift at 54°2′N, 16°8′W (Jansen & Veum 1990). The V23-81 planktonic oxygen isotope stratigraphy is measured entirely upon *N. pachyderma* (sin.). In view of the significant changes in the abundance of this species, associated with the reorganization of surface water masses in the northeast Atlantic Ocean, we will make comparisons only with units p1 and p3 of core 57/-11 59CS.

Discussion

Although, in this study, a division of the Late Glacial is unsupported by accurate radiometric dating, we feel confident that by combining the described biostratigraphic chronology a four-fold subdivision of the Late Glacial period can be constructed. Each subdivision is described below in general palaeoenvironmental terms.

Cores 58/-10 19CS and 57/-11 59CS from the Geikie Escarpment and northern Barra Fan display the oldest unit (1) which is of Late Glacial age (>14 ka BP) (Fig. 10). On the Barra Fan the predominant process appears to be the slow accumulation of fine-grained sediment with hemipelagite deposits. A sandy–silty turbidite recovered from this interval in the lower part of core 57/-11 59CS may indicate that there was some downslope input of clastic material possibly associated with Late Glacial lowered sea levels. Normark *et al.* (1993) have described the lowstands associated with the Late Glacial as giving rise to increased downslope clastic input on a prograding fan system. The turbidites may have been triggered by debris flows upslope on the fan (Howe 1996). On the Geikie Escarpment, slow sedimentation by hemipelagic processes appeared to dominate the Late Glacial, with the limited influence of sea-ice and proximal shelf ice.

The Allerød–Bölling Interstadial (13–11 ka BP), unit 2, is represented in both cores from the Barra Fan and core 59/-08 34CS from the northeastern Trough slope apron. On the Barra Fan

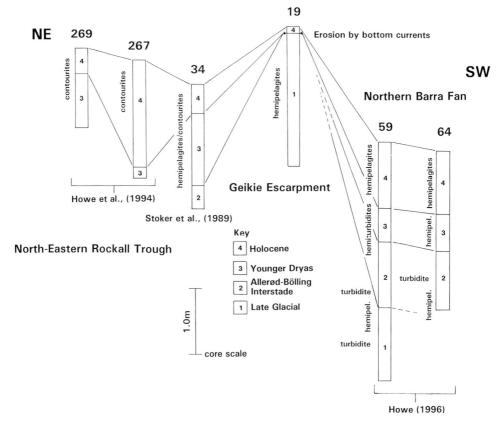

Fig. 10. Overall stratigraphy and principal sedimentary units of the northern Rockall Trough, from the three study areas. Stratigraphic divisions of cores 59/-07 269CS and 59/-7 267CS based on nannofossils, core 59/-08 34CS from dinoflagellate analysis, and cores 58/-10 19CS, 57/-11 59CS and 57/-11 64CS using dinoflagellates, nannofossils and planktonic foraminifera (57/-11 59CS only).

hemipelagic processes dominated, with turbidites in both cores indicating clastic input into a quiet, deep-water area. Core 59/-08 34CS from the slope apron penetrated interstadial sediments which comprise muddy contourites and sandy ice-rafted horizons. Climate warming into the interstadial may have reactivated bottom currents with perhaps a reduced sea-ice cover in the Norwegian Sea allowing the sinking and circulation of NADW (Lehman & Keigwin 1992; Koc *et al.* 1993). Perhaps an associated flux of meltwater from the retreating ice could drive bottom-current activity by increasing density differences and therefore thermohaline circulation (Howe 1995).

The return to cold, glacial conditions of the interpreted Younger Dryas Stadial (11–10 ka BP), unit 3, is marked in most of the cores by both an increase in IRD and a decrease in bottom-current activity. On the Barra Fan hemipelagic sedimentation was predominant on

the wave field, although in the vicinity of core 57/-11 59CS some distal turbidite activity is inferred from the deposition of hemiturbidites. In the northeastern Rockall Trough there was slow, quiet hemipelagic sedimentation with occasional influxes of poorly sorted ice-rafted glaciomarine material. Bottom currents appear to have been suppressed, with both the slope apron and the drift cores displaying little or no current influence.

The end of glacial conditions and the beginning of the youngest unit (4), the Holocene (<10 ka BP), produced a dramatic change in sedimentation in the northeastern Trough and Geikie Escarpment areas (Fig. 11). An increase in bottom-current activity resulted in the deposition of muddy contourites overlain by sandy contourites as the currents increased in strength and winnowed away the fines. In the area of the Geikie Escarpment contour current activity eroded the underlying muddy contourites and

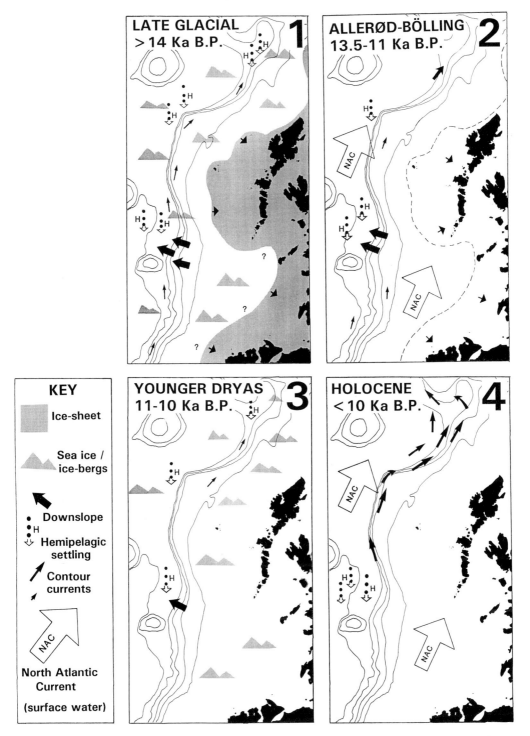

Fig. 11. Palaeoceanographic maps of the northern Rockall Trough during the last deglaciation based on the four-fold subdivision of Stoker *et al.* (1989) and this paper, showing the dominant processes during each interval. (Adapted from a number of sources, notably: (Lowe & Walker 1987; Selby 1989; Stoker *et al.* 1993).

any record of the preceding deposits. This may, in part, be due to the concentrating effects of steeper slope angles focusing and strengthening current activity. On the northern Barra Fan, away from the effects of the slope current, deposition was controlled by hemipelagic processes with little or no current influence. As with the slight increase in current activity during the Allerød–Bölling, the retreat of sea ice from the Norwegian Sea, associated with meltwater fluxes and increases in sea-surface temperatures may all contribute to increasing current strengths (Duplessy *et al.* 1992; Veum *et al.* 1992; Faugères & Stow 1993).

Conclusions

The cores from the northeastern Rockall Trough display a four-fold biostratigraphic division of a deglaciation record constrained for the moment by tephrochronology and linked to lithology. This allows an interpretation of the changing oceanographic conditions in the northeastern Atlantic Ocean. Late Glacial processes were influenced by the proximal position of ice sheets on the Hebrides Shelf, with a subsequent suppression of bottom-current activity. The onset of deglaciation led to a retreat of ice, and an increase of bottom-current activity and turbidites in the Barra Fan region. A single core from the Barra Fan sediment wave field contains a four-fold deglaciation division, with a highly detailed Holocene record (Harland & Howe 1995).

The dinoflagellate record indicates the presence of meltwater pulses in the Allerød–Bölling Interstadial with fluctuating amounts of *B. tepikiense*; however, a meltwater pulse typified by this species is absent from the Holocene record of the examined cores. The nannofossil bloom of *C. pelagicus*, indicating the late Holocene, is absent from the northern Barra Fan core 57/-11 59CS. This may be a localized effect of sea-surface productivity or the removal of any evidence of a bloom by bottom-current activity.

The authors are grateful to M. Stoker at the British Geological Survey, D. A. V. Stow at the Southampton Oceanography Centre, and D. Kroon at Edinburgh University for helpful suggestions, and to C. Pudsey of the British Antarctic Survey for reading an early draft of the manuscript. D. Long at the British Geological Survey identified and sampled the Barra Fan sediment waves, and J. Hunt conducted ash shard analysis. The paper was greatly improved by the critical reviews of C. Evans and I. Harding. N. Hine wishes to thank Conoco (UK) Ltd for their support. W. E. N. Austin acknowledges the receipt of NERC grant G.S.T/02/723. The first author acknowledges the receipt of Southampton University contract studentship No. 220. This paper is published with the permission of the Directors, British Geological Survey, British Antarctic Survey (NERC).

References

AUSTIN, W. E. N. & KROON, D. 1996. Late Glacial sedimentology, foraminifera and stable isotope stratigraphy of the Hebridean Continental Shelf, northwest Scotland. *In*: ANDREWS, J. T., AUSTIN, W. E. N., BERGSTEN, H. & JENNINGS, A. E. (eds) *Late Quaternary Palaeoceanography of the North Atlantic Margins.* Geological Society, London, Special Publications, **111**, 187–213.

——, BARD, E., HUNT, J. B., KROON, D. & PEACOCK, J. D. 1995. The ¹⁴C age of the Icelandic Vedde Ash: implications for Younger Dryas marine reservoir age corrections. *Radiocarbon*, **37**, 53–62.

BACKMAN, J. & SHACKLETON, N. J. 1983. Quantitative biochronology of Pliocene and Early Pleistocene calcareous nannofossils from the Atlantic, Indian and Pacific Oceans. *Marine Micropalaeontology*, **8**, 141–170.

BAUMANN, K. H. & MATHIESSEN, J. 1992. Variations in surface water mass conditions in the Norwegian Sea: evidence from Holocene coccolith and dinoflagellate cyst assemblages. *Marine Micropalaeontology*, **20**, 129–146.

BOND, G., HEINRICH, H., BROECKER, W., LABEYRIE, L., MCMANUS, J., ANDREWS, J., HUON, S., JANTSCHIK, R., CLANSEN, S., SIMET, C., TEDESCO, K., KLAS, M., BONANI, G. & IVY, S. 1992. Evidence for massive discharges of icebergs into the North Atlantic Ocean during the last glacial period. *Nature*, **360**, 245–249.

BOOTH, D. A. & ELLETT, D. J. 1983. The Scottish continental slope current. *Continental Shelf Research*, **12**, 127–146.

DALE, B. 1976. Cyst formation, sedimentation, and preservation: factors affecting dinoflagellate assemblages in Recent sediments from Trondheimsfjord, Norway. *Review of Palaeobotany and Palynology*, **22**, 39–60.

—— 1985. Dinoflagellate cyst analysis of Upper Quaternary sediments in core GIK 15530-4 from the Skagerrak. *Norsk Geologisk Tidsskrift*, **65**, 97–102.

DICKSON, R. & KIDD, R. B. 1986. Circulation in the Southern Rockall Trough, oceanographic setting of Site 610. *In: Initial Reports of the Deep-Sea Drilling Project*, **XCIV**. US Government Printing Office, Washington, DC, 1061–1073.

DOWLING, L. M. & MCCAVE, I. N. 1993. Sedimentation on the Feni drift and Late Glacial bottom water production in the Northern Rockall Trough. *Sedimentary Geology*, **82**, 79–89.

DUPLESSEY, J. C., LABEYRIE, L., ARNOLD, M., PATERNE, M., DPRAT, J. & VAN WEERING, T. C. E. 1992. Changes in surface salinity of the North Atlantic Ocean during the last deglaciation. *Nature*, **358**, 485–487.

EYLES, E. H., EYLES, N. & MIALL, A. D. 1985. Models of glaciomarine sedimentation and their application to the interpretation of ancient glacial sequences. *Palaeogeography, Palaeoclimatology and Palaeoecology*, **51**, 15–84.

FAUGÈRES, J.-C. & STOW, D. A. V. 1993. Bottom current controlled sedimentation: a synthesis of the contourite problem. *Sedimentary Geology*, **82**, 287–299.

GARD, G. 1988. Late Quaternary calcareous nannofossil biozonation, chronology and palaeo-oceanography in areas north of the Faeroe–Iceland ridge. *Quaternary Science Review*, **7**, 65–78.

—— 1989. Variations in coccolith assemblages during the last glacial cycle in the mid–high latitude Atlantic and Indian Oceans. *In*: CRUX J. A. & VAN HECK, S. E. (eds) *Nannofossils and their Applications*. Ellis Horwood, Chichester, 108–121.

GARTNER, S. 1977. Calcareous nannofossil biostratigraphy and revised zonation of the Pleistocene. *Marine Micropaleontology*, **2**, 1–25.

GONTHIER, E. G., FAUGÈRES, J.-C. & STOW, D. A. V. 1984. Contourite facies of the Faro drift, Gulf of Cadiz. *In*: STOW, D. A. V. & PIPER, D. J. W. (eds) *Fine-grained Sediments: Deep-water Processes and Facies*. Geological Society, London, Special Publications, **15**, 275–291.

GRAHAM, D. K., HARLAND, R., GREGORY, D. M., LONG, D. & MORTON, A.C. 1990. The biostratigraphy and chronostratigraphy of BGS Borehole 78/4, North Minch. *Scottish Journal of Geology*, **26**, 65–75.

HARLAND, R. 1988. Dinoflagellates, their cysts and Quaternary stratigraphy. *New Phytologist*, **108**, 111–120.

—— 1989. A dinoflagellate cyst record for the last 0.7 Ma from the Rockall Plateau, northeast Atlantic Ocean. *Journal of the Geological Society, London*, **146**, 945–951.

—— 1994. Dinoflagellate cysts from the Glacial/Postglacial transition in the NE Atlantic Ocean. *Palaeontology*, **37**(2), 263–283.

—— & HOWE, J. A. 1995. Dinoflagellate cysts and the Holocene oceanography of the north-eastern Atlantic Ocean. *Holocene*, **5**(2), 220–228.

HARVEY, J. G. 1982. Theta-S relationships and water masses in the North Atlantic. *Deep-Sea Research*, **29**, 1021–1033.

HINE, N. M. 1990. *Late Cenozoic calcareous nannofossils from the Northeast Atlantic*. PhD thesis, University of East Anglia, Norwich, UK.

—— 1993. *Calcareous nannofossil analysis of three gravity cores, 34, 267 and 269, Northern Rockall Trough, Sula Sgeir sheet*. British Geological Survey Technical Report, Biostratigraphy and Sedimentology Group **WH93/76 R 1-6**.

HOWE, J. A. 1995. Sedimentary processes and variations in slope-current activity during the last glacial–interglacial episode on the Hebrides Slope, Northern Rockall Trough, North Atlantic. *Sedimentary Geology*, **96**, 201–230.

—— 1996. Turbidite and contourite sediment waves from the Northern Rockall Trough, North Atlantic Ocean. *Sedimentology*, **43**, 219–235.

—— & HUMPHERY, J. D. 1995. Photographic evidence for slope-current activity on the Hebrides slope, North-east Atlantic Ocean. *Scottish Journal of Geology*, **30**, 107–115.

——, STOKER, M. S. & STOW, D. A. V. 1994. A Late Cenozoic sediment drift complex, northern Rockall Trough, North Atlantic Ocean. *Paleoceanography*, **9**(6), 989–999.

HUTHNANCE, J. M. 1986. Rockall slope current and shelf edge processes. *Proceedings of the Royal Society of Edinburgh, Section B*, **88**, 83–101.

JANSEN, E. & VEUM, T. 1990. Evidence for two-step deglaciation and its impact on North Atlantic deep-water circulation. *Nature*, **343**, 612–616.

KEIGWIN, L. D. & JONES, G. A. 1989. Glacial–Holocene stratigraphy, chronology, and paleoceanographic observations on some North Atlantic sediment drifts. *Deep-Sea Research*, **36**(6), 845–867.

KELLOGG, T. B. 1976. Late Quaternary climatic changes: evidence from deep-sea cores of Norwegian and Greenland seas. *Geological Society of America Memoir*, **145**, 77–107.

KENYON, N. H. 1986. Evidence from bedforms of a strong poleward current along the upper continental slope, Northwest Europe. *Marine Geology*, **72**, 187–198.

KOC, N., JANSEN, E. & HAFLIDASEN, H. 1993. Paleoceanographic reconstructions of surface ocean conditions in the Greenland, Iceland and Norwegian Seas through the last 14 ka based on diatoms. *Quaternary Science Review*, **12**, 115–140.

KVAMME, T., MANGERUD, J., FURNES, H. & RUDDIMAN, W. F. 1989. Geochemistry of Pleistocene ash zones in cores from the North Atlantic. *Norsk Geologisk Tidsskrift*, **69**, 251–272.

LEHMAN, S. J. & KEIGWIN, L. D. 1992. Sudden changes in North Atlantic circulation during the last deglaciation. *Nature*, **356**, 757–762.

LOWE, J. J. & WALKER, M. J. C. 1987. *Reconstructing Quaternary Environments*. Longman, London, 389 pp.

MCCARTNEY, M. S. 1992. Recirculating components to the deep boundary current of the northern North Atlantic. *Progress in Oceanography*, **29**, 283–383.

MCINTYRE, A., KIPP, N. G., BE, A.W. H., CROWLEY, T., KELLOG, T., GARDNER, J. V., PRELL, W. & RUDDIMAN, W. F. 1976. Glacial North Atlantic 18000 years ago: a CLIMAP reconstruction. *Geological Society of America Memoir*, **145**, 43–75.

MILLER, K. G. & TUCHOLKE, B. E. 1983. Development of Cenozoic abyssal circulation south of the Greenland–Scotland ridge. *In*: BOTT, M. H. P., SAXOV, S., TALWANI, M. & THEIDE, J. (eds) *Structure and Development of the Greenland–Scotland Ridge*. Plenum, New York, 549–589.

MOLNIA, B. F. 1983. Distal glaciomarine sedimentation: abundance, composition and distribution of North Atlantic Ocean Pleistocene ice-rafted sediment. *In*: MOLNIA, B. F. (ed.) *Glacial-marine Sedimentation*. Plenum, New York, 593–625.

MULLER, G. & GASTNER, M. 1971. The 'Karbonate Bombe' a simple device for the determination of carbonate content in sediments, soils and other materials. *Mineralogie Monatshefte*, **10**, 446–469.

NORMARK, W. R., POSAMENTIER, H. & MUTTI, E. 1993.

Turbidite systems: state of the art and future directions. *Reviews in Geophysics*, **31**, 91–116.

PEACOCK, J. D., AUSTIN, W. E. N., SELBY, I,. HARLAND, R., WILKINSON, I. P. & GRAHAM, D. K. 1992. Late Devensian and Holocene palaeoenvironmental changes on the Scottish Continental Shelf west of the Outer Hebrides. *Journal of Quaternary Science*, **7**, 145–161.

ROBERTS, D. G., HUNTER, P. M. & LAUGHTON, A. S. 1979. Bathymetry of the northeast Atlantic: continental margin around the British Isles. *Deep-Sea Research*, **26A**, 417–428.

RUDDIMAN, W. F. & McINTYRE, A. 1976. Northeast Atlantic Paleoclimatic changes over the past 600,000 years. *Geological Society of America Memoir* **145**, 111–146.

—— & —— 1981. The north Atlantic Ocean during the last deglaciation. *Palaeogeography, Palaeoclimatology and Palaeoecology*, **35**, 145–214.

SELBY, I. 1989. *The Quaternary geology of the Hebridean continental margin.* PhD thesis, University of Nottingham.

STOKER, M. S., HARLAND, R., MORTON, A. C. & GRAHAM, D. K. 1989. Late Quaternary stratigraphy of the northern Rockall Trough and Faeroe–Shetland Channel, northeast Atlantic Ocean. *Journal of Quaternary Science*, **4**, 211–222.

——, HITCHEN, K. & GRAHAM, C. C. 1993. *The geology of the Hebrides and West Shetland shelves, and adjacent deep-water areas.* British Geological Survey UK Regional Offshore Report. HMSO, London, 149 pp.

——, LESLIE, A. B., SCOTT, W. D., BRIDEN, J.C. HINE, N. M., HARLAND, R., WILKINSON, I. P., EVANS, D. & ARDUS, D. A. 1994. A record of late Cenozoic stratigraphy, sedimentation and climate change from the Hebrides Slope, NE Atlantic Ocean. *Journal of the Geological Society, London*, **151**, 235–249.

STOW, D. A. V. 1985. Deep-sea clastics: where are we and where are we going? *In*: BRENCHLY, P. J. & WILLIAMS, B. J. P. (eds) *Sedimentology: Recent developments and Applied Aspects.* Geological Society, London, Special Publications, **18**, 67–93.

—— & HOLBROOK, J. A. 1984. North Atlantic contourites: an overview. *In*: STOW, D. A. V. & PIPER, D. J. (eds). *Fine-grained Sediments: Deep-water Processes and Facies.* Geological Society, London, Special Publications, **15**, 245–256.

—— & LOVELL, J. P. B. 1979. Contourites: their recognition in modern and ancient sediments. *Earth Science Reviews*, **14**, 251–291.

—— & PIPER, D. J. W. 1984. Deep-water fine grained sediments: facies models. *In*: STOW, D. A. V. & PIPER, D. J. W. (eds) *Fine-grained Sediments: Deep-water Processes and Facies.* Geological Society, London, Special Publications, **15**, 611–645.

—— & WETZEL, A. 1990. Hemiturbidites: a new kind of deep water sediment. *In: Proceedings of the Deep Sea Drilling Project, Scientific Results*, **116**. US Government Printing Office, Washington, DC, 25–34.

VEUM, T., JANSEN, E., ARNOLD, M., BEYER, I. & DUPPLESSEY, J.-C. 1992. Water mass exchange between the North Atlantic and the Norwegian Sea during the past 28,000 years. *Nature*, **356**, 783–884.

WALL, D., DALE, B., LOHMANN, G. P. & SMITH, W. K. 1977. The environmental and climatic distribution of dinoflagellate cysts in modern marine sediments from regions in the North and South Atlantic Oceans and adjacent seas. *Marine Micropalaeontology*, **2**, 121–200.

WETZEL, A. 1984. Bioturbation in deep-sea fine grained sediments: influence of sediment texture, turbidite frequency and rates of environmental change. *In*: STOW, D. A. V. & PIPER, D. J. W. (eds) *Fine-grained Sediments: Deep-water Processes and Facies.* Geological Society, London, Special Publications, **15**, 595–608.

Upper slope sand deposits: the example of Campos Basin, a latest Pleistocene–Holocene record of the interaction between alongslope and downslope currents

A. R. VIANA[1,2] & J.-C. FAUGÈRES[1]

[1]Université Bordeaux I, DGO, URA 197 CNRS, Talence, 33405, France, e-mail:
viana@geocean.u-bordeaux.fr; faugeres@geocean.u-bordeaux.fr;

[2]PETROBRAS–Petróleo Brasileiro S.A., E&P/GEREX-BC/GEOMAR, Macaé, RJ,
27.913-350, Brazil

Abstract: Upper slope sand deposits constitute large accumulations of well-sorted very fine to coarse sand that are of great economic interest. A depositional model is here proposed based on hydrographic, physiographic and sedimentological characteristics of the modern Campos Basin margin, SE Brazil. Data comprise long-term bottom-current measurements, long and short cores, high-resolution seismic lines and side-scan sonar records. Results indicate that offshelf bed-load transfer of sand is driven by shelf bottom currents forced by combined oceanographic and meteorological factors, among them the 'sea-floor polishing effect', the rotatory sweeping of the sea bed by mesoscale eddies. The main conditions for development of upper slope sand accumulations are: (1) convex outer shelf–slope morphology; (2) sandy sediment available at the shelf edge; (3) net offshelf transport of shelf sands induced by shelf-edge bottom currents; (4) the presence of a relatively strong slope boundary current. Transfer to the slope occurs as successive, short-distance-flowing sand fluxes. The latter occur mainly through shelf-edge incisions and lead to elongated and narrow sand lobes on the upper slope. The convex morphology of the upper margin controls the Brazil Current (BC) flow and two main zones can be identified: a BC 'funnelling zone' where the current is accelerated, and a BC 'expansion zone' where the current has its speed decelerated. Reworking by the southward-flowing Brazil Current develops gravel lag deposits, erosional scours, and interchannel sand waves on the upper slope (between 200 and 350 m water depth). Downslope, the reduction of slope current intensity enhances the development of flat sand sheets, controlled by the northward flow of the Brazil Counter-Current. The upper Quaternary vertical facies succession is there characterized by a coarsening-to fining-up asymmetric sequence. Upper Pleistocene sand–mud intercalations are overlain gradually or with a sharp or erosive contact by lower Holocene coarse to fine-grained clean sands that grade to upper Holocene fine-grained muddy sands toward the top.

It has long been considered that outer shelf and slope areas of continental margins constitute sediment transit zones, with little significant accumulation. The depositional mud-line that delineates sand and mud accumulation zones, and reflects the environmental energy, is usually placed around the 300 m isobath (Stanley & Wear 1978; Stanley et al. 1983, Vogt & Tucholke 1986). Sand deposits on the outer shelf are generally related to relict sediments supplied to the outer shelf during lowstands of sea level, and/or reworked by nearshore processes during early stages of marine transgressions (Emery 1968; Swift et al. 1971; Trumbull 1972; Walker 1984; Swift & Niedoroda 1985). The transfer of shelf sands to deep waters is primarily related to transport by turbidity currents (Heezen & Ewing 1952; Mutti 1985, 1992; Mutti & Normark 1987; Vail 1987; Vail et al. 1991), or by other gravity processes (McCave 1972; Middleton &

Hampton 1973, 1976; Cacchione & Southard 1974; Doyle et al. 1979; Lowe 1979, 1982; Nardin et al. 1979; Stanley et al. 1983; Gorsline et al. 1984). These long-distance sediment transport processes are usually responsible for deposits found on the continental rise and abyssal plains. The occurrence of sand accumulations on the shelf edge available for downslope transport implies bed-load transport over this area, with a strong near-bottom water circulation. The sand accumulation on the upper slope has generally been thought to be temporary or non-significant. Gullies and submarine canyons are thought to be the major sediment conduits feeding deep environments.

Some oceanographic studies have shown the high energy and complexity of the circulation pattern of the shelf edge (Csanady 1973; Walsh et al. 1988; Pingree & LeCann 1989; Huthnance 1992, 1995; Houghton et al. 1994; Nitrouer &

Viana, A. R. & Faugères, J.-C. 1998. Upper slope sand deposits: the example of Campos Basin, a latest   287
Pleistocene–Holocene record of the interaction between alongslope and downslope currents. In:
Stoker, M. S., Evans, D. & Cramp, A. (eds) Geological Processes on Continental Margins: Sedimentation,
Mass-Wasting and Stability. Geological Society, London, Special Publications, **129**, 287–316.

Wright 1994; Shaw *et al.* 1994). However, few works have dealt with the quantification of off-shelf sand transfer. A number of investigations have measured bottom-current speeds in the shelf-edge zone, but for short periods only (Hill & Bowen 1983; Karl *et al.* 1983; McGrail & Carnes 1983; Pietrafiesa 1983). All these studies have suggested a tendency towards offshore transport of shelf-edge sediments forced by hydrodynamic processes whose relative import-ance depends on the setting. More recently, results from long-term observations have cor-roborated those ideas, illustrating the complex-ity of outer shelf–slope bottom circulation and its influence on sediment transport (Viana *et al.* 1998*a*).

The sediment transfer from the continental shelf to the deep ocean is controlled by geo-logical and oceanographic factors which change with relative sea-level oscillations (Karl *et al.* 1983). Geological factors comprise margin physiography, sediment supply and grain-size. Oceanographic factors comprise all those mechanisms that induce seawater circulation (shelf–ocean fronts interaction, meteorologi-cally induced currents, tides, onshelf pen-etration of slope currents, internal and surface waves). The observation of strong slope cur-rents, both on western and eastern ocean margins (Huthnance 1992, 1995; Chang & Richards 1994) sweeping upper slope and outer shelf areas provides a potentially important mechanism of sand dispersal. Some workers have shown that outer shelf areas may be invaded by slope boundary currents with the consequent development of bedforms (Hunt *et al.* 1977; Flemming, 1978, 1980, 1981; Stanley *et al.* 1981; Chen *et al.* 1992; Ramsay 1994; Viana *et al.* 1998*a*). Shanmugam *et al.* (1993*a,b*) pro-posed the term bottom-current-reworked sand deposits for the predominant alongslope reworking of pre-existent sediments indepen-dently of their bathymetric emplacement. Those workers presented facies and grain-size characteristics that differ from those suggested by Stow & Lovell (1979) for the sandy con-tourites. The model of Shanmugam *et al.* (1993*a,b*) amplifies the original concept of con-tourites as proposed by Heezen & Hollister (1971). Viana *et al.* (1998*b*) proposed a bathy-metric distinction for bottom-current-con-trolled sand deposits. Their classification is based on modern examples, and consists of deep-water, mid-depth water and outer shelf–upper slope current-controlled sand deposits. Those workers described outer shelf–upper slope sand deposits as being the result of alongslope reworking of outer shelf

sediments, driven to the upper slope by the shelf-edge oceanographic forces and trapped by local morphology.

In this paper we present the analysis of upper slope sand deposits from the Campos Basin off SE Brazil (Fig. 1). We consider local physio-graphic, sediment facies and hydrodynamic characteristics, and the impact of relative sea-level fluctuations in the development of those sand bodies that extend as deep as 700 m water depth. Investigation was based on the interpre-tation of long-term bottom-current data coupled with sediment cores, high-resolution seismic sec-tions and side-scan sonar images. Additional data include sea-surface temperature (SST), satellite images, CTD (conductivity, tempera-ture and density) and conventional current measurements. These data allowed the develop-ment of a conceptual model for the deposition of the Campos Basin upper slope sand deposits and their preservation through geologic time. This model stresses the role played by shelf and slope currents.

Campos Basin setting

The main controls on the modern sedimentation in the Campos Basin include physiographic characteristics, hydrologic pattern and sediment availability.

Physiography

The Campos Basin is located on the southeast-ern Brazilian continental margin, between 21°S and 24.5°S (Fig. 1), and may be characterized as a convex passive margin. It comprises a 100 km wide continental shelf, with a depth between 0 and 120 m, and a 45 km wide continental slope extending down to 2000 m. A 240 km wide province, the São Paulo Plateau, is located between the continental slope and the continen-tal rise (2000–3500 m). The São Paulo Plateau is morphologically controlled by halokinesis of the Aptian salt and evaporitic deposits (Castro 1992; Miller *et al.* 1996).

The convexity, or seaward projection of the margin, is controlled by deep structures (Aptian salt-seated listric faults, Carminatti & Scarton 1991) and marked by the similarity between the modern coastline and the shelf break. Large sub-marine canyons are developed with a radial divergent trend from this margin projection: the Almirante Camara and Itapemirim to the north, and São Tomé to the south. To the north, the continental shelf narrows to 50 km wide and has a NW–SE trend. Southwards it changes to a NE–SW trend and is more than 100 km wide.

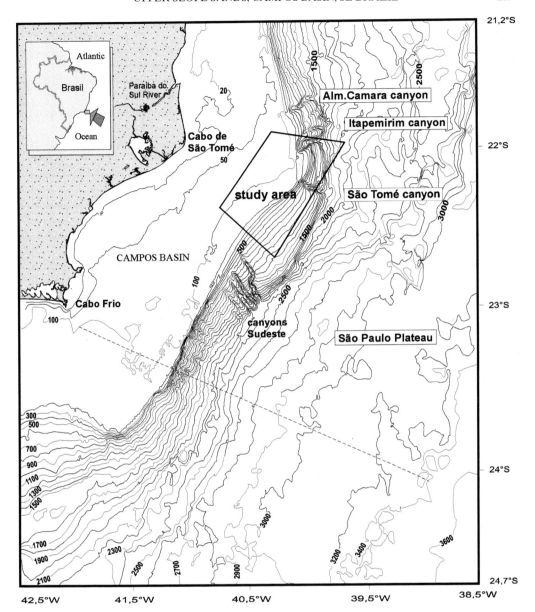

Fig. 1. Southeastern Brazilian margin. Location of the study area and main physiographic features are displayed. Dotted line indicates the southern limit of Campos Basin. The northern limit is about 50 km north of the presented chart. Isobaths are in metres.

The shelf edge is marked by carbonate banks several kilometres long which form local topographic highs. The shelf break is marked by a conspicuous scarp between the 120 and 220 m isobaths locally reaching up to 14° dip. Systems of gullies are observed locally incising the shelf edge. Mass-movement scars and buried canyon heads smooth the scarp in the southernmost part of the study area.

The northern upper slope (Fig. 2) is separated from the southern slope by the Sao Tomé submarine canyon (STC) and constitutes a flat erosional terrace, the Albacora Terrace, 10 km wide, that extends from the base of the shelf break scarp to the 450 m isobath. The terrace is marked by bottom-current-related erosional ridges elongated parallel or slightly oblique to the isobaths, and trending south towards the São

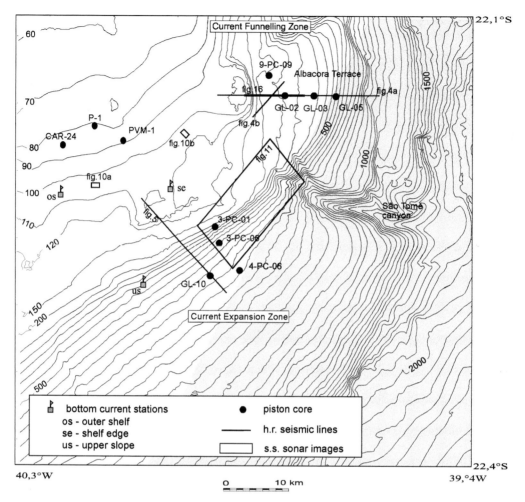

Fig. 2. Study area showing data distribution and location of figures referred to in the text.

Tomé canyon head. These ridges (Figs 3 and 4) are tens of metres high, a few kilometres long and hundreds of metres wide. Shallow channels are developed at their foot. The truncation of seismic reflectors against the sea floor is a prominent feature (Fig. 4b).

The southern upper slope (Fig. 5) is marked by a narrow terrace developed at the foot of the shelf-edge scarp, about the 200 m isobath. The shelf break is gentle at the head of São Tomé canyon and in the southernmost portion of the study area. The shelf-edge scarp is very steep immediately south of Sao Tomé canyon, reaching locally 10° dip. The upper slope shows a generally concave profile from 250 m to 550 m water depth, on the upper–middle slope boundary. To the south of the study area, the heads of large mass movements penetrate onto the shelf edge.

Hydrography

The hydrographic characteristics of the southeastern Brazilian margin have been studied by many researchers (Signorini 1978; Evans *et al.* 1983; Evans & Signorini 1985; Garfield 1990; Stech & Lorenzzetti 1992; Castro & Lee 1995). The shallow circulation may be separated into shelf currents and the Brazil Current (BC). Shelf currents are the result of combined meteorological and tidal forcing upon the shelf waters. The observations made by Campos & Miller (1995) suggest a northeastward propagation of shelf waters between the BC and the coast. The BC is a southward-flowing western boundary current (Fig. 3), driven by the wind-controlled South Atlantic Gyre. The inner margin of the BC roughly coincides with the shelf break. The BC has a northward countercurrent component, the

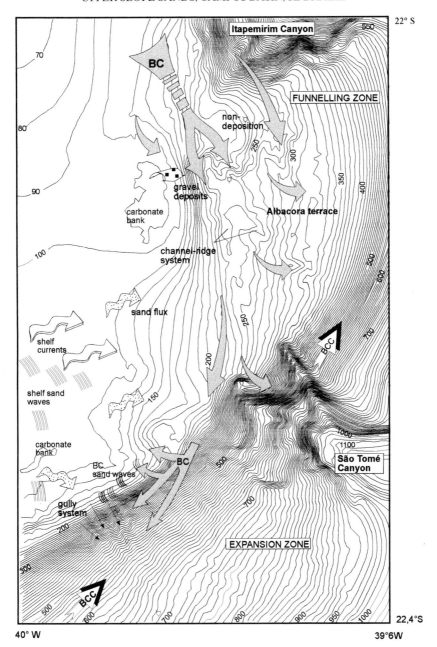

Fig. 3. High-resolution bathymetric map (contour interval 10 m) showing schematic representation of the main physiographic, geological and oceanographic features (see text for details).

Brazil Counter-Current (BCC) that flows below 350 and 400 m isobaths and constitutes a water mass known as South Atlantic Central Water (SACW). The forcing mechanisms which drive the outer shelf and upper slope bottom currents were recently analysed by Viana *et al.* (1995, 1998*a*) and Lima *et al.* (1998). The most

energetic phenomena observed are low-frequency oscillations in the form of eddies and meanders associated with the onshelf penetrations of the BC. They are responsible for more than 50% of the recorded energy of the bottom currents. Strong cross-shelf currents were identified south of São Tomé canyon. They have peak

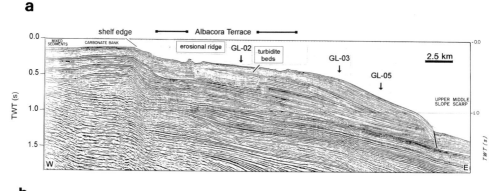

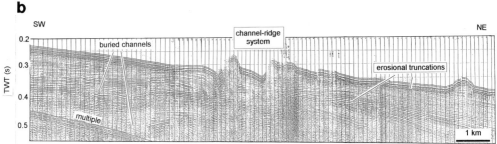

Fig. 4. Northern slope. (**a**) High-resolution multi-channel seismic profile (air-gun) showing the distribution of long cores (boreholes GL-02, -03, -05) collected along this profile, and major geological features (modified from Viana *et al.* 1997*a*). (**b**) Single-channel sparker line illustrating the erosional channel–ridge system shaped from the sub-bottom up to the sea floor, and the multiple episodes of erosional truncation of seismic reflectors. Location of both sections is shown in Fig. 2. TWT, Two-way travel time.

velocities of 50 cm/s downslope, both at the shelf edge and at the upper slope. Average speeds were around 20 cm/s, with higher speeds occurring for about 30% of the long-term experiment (Lima *et al.* 1998). Current displacement plots (Fig. 6), carried out from July to December 1992, point out the northward trend of outer shelf bottom currents, marked by long period landward and slopeward shifts. Bottom currents at the shelf edge and at 400 m water depth present a downslope trend of propagation.

Analyses of SST satellite images indicate a

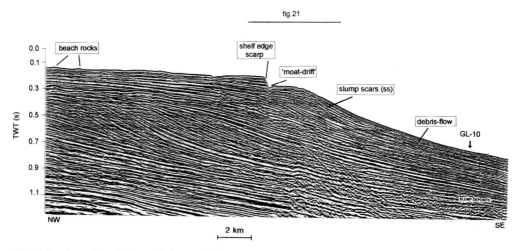

Fig. 5. Southern slope: high-resolution multi-channel seismic profile (air-gun) showing the major geological features and the location of the long core GL-10 collected along this profile. Location of section is shown in Fig. 2. The continuous line above the profile indicates the section shown in Fig. 21.

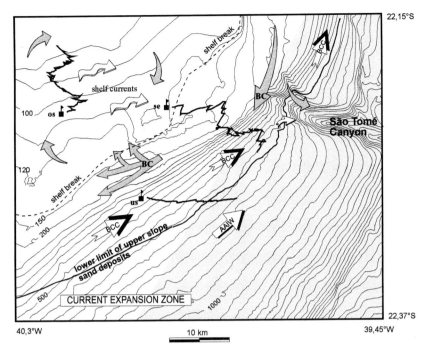

Fig. 6. Bottom-current displacement plot, indicating the hypothetical trajectory of the bottom currents at each recording station, from July 1992 to December 1992. Outer shelf bottom-currents plot displays coastward and slopeward shifts in their northward propagation, with seasonal predominance of one direction over the other. Shelf-edge bottom-currents plot shows an offshelf trend (towards the east), disturbed by southwestward shifts, probably related to the onshelf penetration of the Brazil Current. Upper slope bottom currents (400 m isobath) show a net downslope trend, occasionally disturbed by minor southwestward shifts, probably related to oceanic eddy passage (see text for details). Arrows represent the near-to-the-bottom expression of the currents discussed in the text. BC, Brazil Current; BCC, Brazil Counter-Current; AAIW, Antarctic Intermediate Water; os, se, and us are outer shelf, shelf-edge and upper slope bottom-current stations represented in Fig. 2.

morphologic control on the characteristics of the BC (Garfield 1990; Viana *et al.* 1998*a*). The convexity of the shelf edge induces the narrowing and acceleration of the BC north of Sao Tomé canyon. In this area, the BC shifts coastward and sometimes penetrates onto the shelf (Fig. 7). This zone, corresponding to the Albacora Terrace, plays the role of a 'funnelling zone'. Near-bottom speeds of BC measured downstream of the 'funnelling zone', at the 210 m isobath, reach more than 70 cm/s.

To the south of the São Tomé canyon, the BC widens and shifts seaward in response to the change in the trend of the margin; this area is called the 'expansion zone'. Upper slope bottom currents may reach more than 40 cm/s at 400 m water depth. Internal waves were recorded near the bottom on the upper slope and have been related by Lima *et al.* (1998) to changes in the upper slope topography. Thermal contrast of more than 5°C at the sea surface was observed between the cold waters that flow over the shelf and the warm waters of

the Brazil Current during the winter (Fig. 7). Such a contrast in temperature may induce the downwelling of the shelf waters. The energy of such effects may be enhanced when flow is inside the gullies' walls, propelling the downslope transfer of the shelf-edge sediments. Additional oceanographic data and numerical modelling will allow the future evaluation of the sedimentological impact of the downwelling process.

The morphology of the margin induces additional hydrologic perturbations in the BC flux. Along the 'expansion zone', the BC meanders and develops oceanic eddies that penetrate onto the shelf, shearing shelf waters in a cyclonic (clockwise) sense (Fig. 7). The eddy moves back to the slope immediately south of the canyon head, increasing locally the sediment export towards the slope. Shelfward penetrations of western boundary currents were numerically modelled by Chang & Richards (1994), who reproduced the oceanic features observed in the Campos Basin.

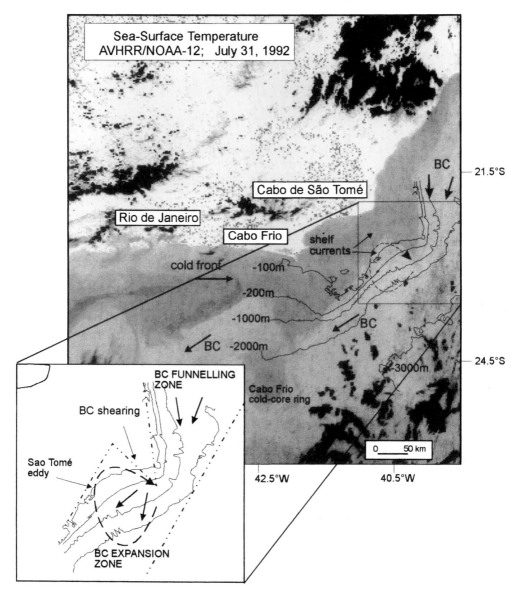

Fig. 7. Sea-surface temperature (SST) satellite image of the southeastern Brazilian margin circulation. Warm waters have pale shades, and cold waters are dark. The arrival of a southerly cold front, related to the northward winter penetration of the Falkland Current, is clearly indicated. Warm waters of the Brazil Current (BC) are shifted offshore downstream of the Cabo de São Tomé, shearing and developing the São Tomé eddy. The inset at the left schematically displays the area of this eddy. NOAA satellite image processed and provided by C. L. Silva, Jr (INPE/BR).

The impact of the eddies on bottom-current activity can be assessed by combining the SST image of 31 July 1992 (Fig. 7) and data from the bottom-current displacement plot (Fig. 8). The image shows the development of the Sao Tomé eddy, a clockwise gyre, inducing a transversal shearing of the BC (Figs 7 and 8a). The eddy begins its development over the upper slope, shifting the normal downslope circulation towards an along-isobath generation seems to follow an onshelf penetration of the BC. Such a perturbation of the flux of the BC is recorded at the shelf break station as a southward shift of the bottom current (Fig. 8b). The eddy begins its development over the upper slope, shifting the normal downslope circulation towards an along-isobath

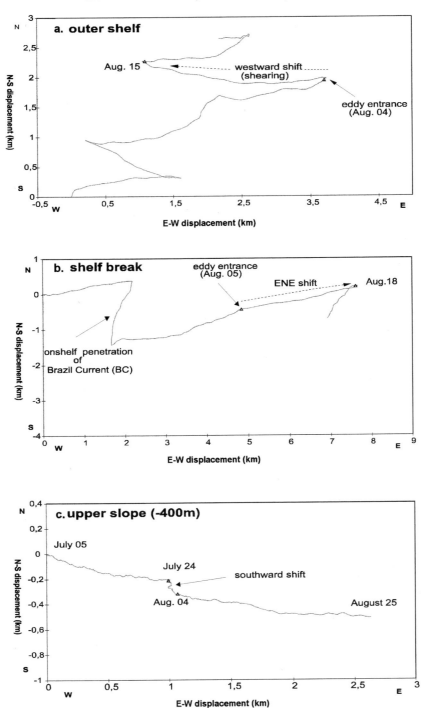

Fig. 8. The influence of the São Tomé eddy on the bottom currents, represented by the shift observed in the plot of bottom-current displacement. (See text for details.)

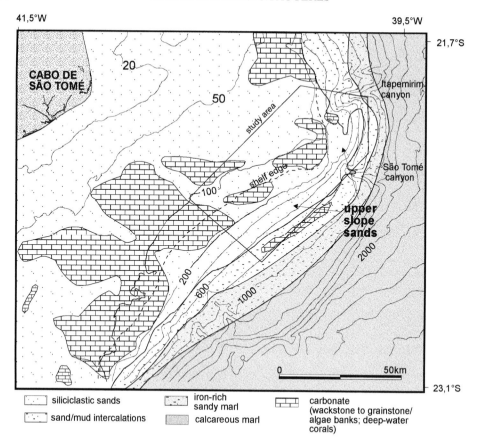

Fig. 9. Surficial facies distribution in the region of the study area (modified from Viana *et al.* 1998*a*). Dashed line indicates shelf-edge position. Bold line outlines the studied area.

trend (Fig. 8c). Its migration towards the shelf shears the shelf waters, thereby inducing a clockwise gyre. The BC warm waters penetrate onto the shelf to the south (Fig. 8a), and, to the north, the cold shelf waters are exported towards the slope, trapped in the core of the ring (Fig. 8b). The shelf waters are then driven southward along the slope by the BC slope boundary flux. These records strongly reinforce the idea that the eddies are one important factor controlling bottom currents on the outer shelf–slope boundary. They act as giant rotating brushes as they travel along the margin, sweeping sediments from the shelf to the slope. This effect has been called 'sea-floor polishing' by Viana *et al.* (1995, 1998*a*).

Sedimentology

Several workers have already dealt with shelf sedimentation on the southeastern Brazilian margin (Schaller 1973; Rocha *et al.* 1975;

Kowsmann & Costa 1979; Palma 1979; Della Piazza *et al.* 1983). In the Campos Basin area, samples from three geotechnical boreholes drilled at the outer continental shelf, in the centre of the study area (CAR-24, P-1 and PVM-1; Fig. 2), are here examined, to characterize outer shelf vertical and lateral facies composition, and grain-size distribution. More than 300 bottom samples collected during several surveys were also integrated into the study to map the distribution of superficial bottom sediments (Fig. 9).

Slope sediments were investigated from a large set of short Kullenberg and long hydraulic piston cores (each core more than 100 m long). The sediments were analysed with special attention given to grain-size, sedimentary structures and colour (Caddah *et al.* 1998). In this study, long cores GL-02, GL-03 and GL-05, and Kullenberg core 9-PC-09 (Fig. 2) were used to characterize the northern slope sediments. Kullenberg cores 3-PC-01 and 3-PC-06, and long

Table 1. *Core distribution, depth of recovery (water depth) and cored length*

	Water depth (m)	Cored section (m)
Northern slope		
9-PC-09	202	5.15
GL-02	275	145.50
GL-03	345	146.95
GL-05	630	149.05
Southern slope		
3-PC-01	245	2.40
3-PC-06	425	7.60
GL-10	620	75.60
4-PC-06	686	5.75

core GL-10 (Fig. 2) illustrate the southern area facies. Table 1 presents the bathymetric distribution of these cores.

Sediment ages and rates were obtained from ^{14}C dating and palaeoclimatic biozonation of planktonic foraminifera (biozones Z, Y, and X of Ericson & Wollin 1968 and the 40 ka BP biostratigraphical datum *Pulleniatina obliquiloculata*; Vicalvi 1994, 1995). Some analogue side-scan sonar images recorded for drilling sites illustrate the sediment and bedform types occurring on the shelf (Fig. 10). On the shelf break and continental slope, an SMS960 E.G.G. side-scan sonar survey was undertaken. Analogue data were scanned and sonar image mosaics were produced on a personal computer (Fig. 11). Seismic records comprise 1 kJ sparker sections and high-resolution multichannel sections acquired with the parameters described by Guimaraes *et al.* (1991).

Shelf sediments. Superficial sediments on the outer shelf are composed of siliciclastic sands, carbonate accumulations and mixed sediments (Fig. 9). Siliciclastic sands extend throughout the outer shelf area, developing large fields of sand waves. Fields of dunes and megaripples are tens of kilometres long and kilometres to tens of kilometres wide, being developed from the outer shelf to the shelf break (Figs 10 and 11). They are well preserved in the central and southern portion of the study area. To the north, the shelf edge is dominated by a carbonate bank whose positive relief prevents the migration of the dunes. The dunes are 0.5–1.5 m high on average and 5–15 m in wavelength. Fossil beach rocks, corresponding to periodic stabilization of the shoreline during the last sea-level rise, are observed at the 110 m, 90 m and 75 m isobaths (Fig. 10). Between the 85 m and 130 m isobaths, sand waves have rectilinear crests (2D geometry in the sense of Harms *et al.* 1982) and show a NE

transport direction (Fig. 10) towards the shelf break. This direction corresponds to the resultant trend of the bottom hydrodynamics of the outer shelf as previously discussed in the section 'Hydrography'. Offshore, between the 130 m and 200 m isobaths, sets of curvilinear-crested sand waves (3D geometry), are observed migrating towards the SW. They are 2–5 m high, tens of metres in wavelength and suggest a sediment transport controlled by the Brazil Current passage (Fig. 11). Locally, at the shelf edge in the 'expansion zone', sand waves are irregular with no clear trend of migration. These bedforms could be the result of the interaction between the forcing mechanisms (shelf currents and BC) flowing in opposite directions (respectively NE and SW).

The siliciclastic sands are yellow to greenish, with an average grain-size between 0.2 and 0.7 mm. They are rounded to sub-angular, comprising quartz, some feldspars, often impregnated with Fe-oxide, and <20% bioclasts (encrusting and articulate red algae, benthic foraminifera, bryozoa, pelecypods, agglutinated foraminifera). The mud content is usually <5%. Glauconite and heavy minerals (zircon, tourmaline) locally constitute more than 5% of the sediment composition. The sand accumulations are coarser (0.4 mm to 1 cm, grain diameter) and thicker (a few metres to >10 m thick) towards the shelf edge.

Carbonate occurs in banks some 10–15 km long and 5–10 km wide, lying 10 to 20 km apart from one another (Fig. 9). They are located at the shelf edge, between the 100 m and 120 m isobaths. They display a positive relief of 1–5 m, and are slightly elongated parallel to the isobaths. Carbonate banks are bioherms mainly comprising red algae with rare corals. They constitute coarsening–shallowing-upward sequences constituted of wackestone at the base and grainstone to bioherms at the top. Wackestone displays marine cement; grainstone and bioherms usually display grain dissolution and fresh water cement at the top of the accumulations and are in contact with continental sediments, which indicates subaerial exposure (Spadini *et al.* 1992). Red algae nodules dispersed in muddy carbonate sediments were collected in the northern area at the shelf edge, in 90 m water depth, and dated by C14. The ages obtained are up to 4.7 ka BP, indicating that carbonate growth has locally survived the Holocene drowning of the continental shelf (R. O. Kowsmann, pers. comm.).

Mixed siliciclastic and carbonate sands to bioclastic sands are observed in the areas adjacent to the carbonate banks. This suggests an origin from the reworking of the carbonate banks by

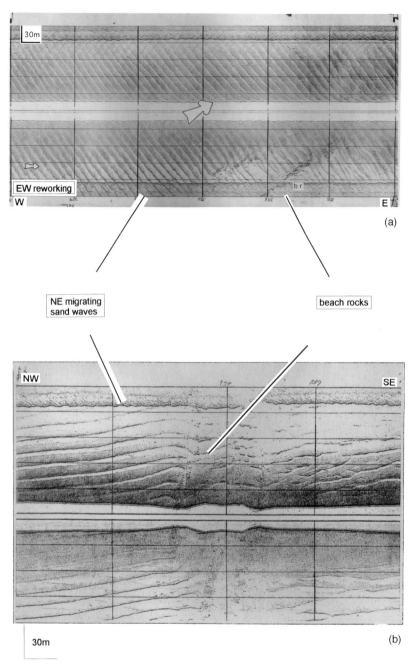

Fig. 10. Side-scan sonar images of outer shelf–shelf-edge sand waves 15 km apart from each other. Both (**a**) and (**b**) present bedforms migrating to the northeast in the 'expansion zone'. The -110 m beach rocks observed in both images disturb the sand waves' migration, inducing dragging and diffraction of the bedforms.

physical processes that could have occurred from the early sea-level rise to the modern highstand (e.g. wave action, longshore currents and shelf currents). In addition, fine-grained carbonate sediments occur in current protected zones. These sediments are characterized by an olive–grey marl, with some silty to very fine sand intercalated (up to 20%), with bioclasts

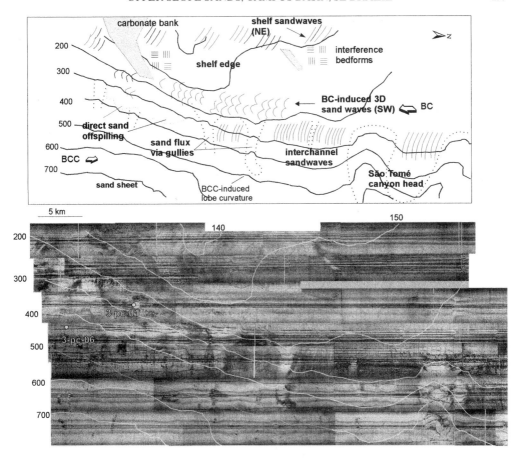

Fig. 11. Side-scan sonar mosaic and schematic interpretation (see text) of the shelf-edge–upper slope environment south of the São Tomé canyon, in the 'expansion zone'. Opposite trends of sand waves are observed related to the opposite sense of the shelf currents and the Brazil Current circulation. Interference bedforms are developed in the contact zone of both currents. The BC flux induces 3D bedforms between the 200 and 300 m isobaths, 2D short-wavelength sand waves in the intergullies–interchannel area, and direct spilling of sand onto the slope in non-channelled zones. The Brazil Counter-Current (BCC) bends the sand lobes along its path and develops a wide sand sheet. Location of area of mosaic is shown in Fig. 2.

comprising red algae fragments, bryozoa, and some gastropods and corals. Planktonic foraminifera are common.

Recurrent episodes of siliciclastic and carbonate sedimentation are observed in the samples from geotechnical boreholes (Fig. 12). The uppermost siliciclastic sands are a few metres to 15 m thick. Similar siliciclastic sand beds are found further down-section. They alternate with beds of a metre to a few metres thickness composed of mixed bioclastic sands (up to 70% quartz) and calcareous deposits (wackestone and packstone–grainstone beds and/or bioherms). Rare silty–clayey sand beds with plant debris indicating shallow water deposits are present. The vertical facies variation (Fig. 12) shows that: (1) about 50% of shelf sediments

are sand grade (siliciclastic and bioclastic) and constitute a major source of sand supply to deeper waters; (2) carbonates occur either as isolated beds composed of a unique type of facies or as a vertical succession of different types of carbonate facies that are coarser and cleaner upward, capped sometimes by a bioherm bed, in a shallowing-upward sequence.

The contact between the uppermost siliciclastic sands and the underlying carbonates corresponds to a major discontinuity, probably the Pleistocene–Holocene boundary. This discontinuity seems to result from the initiation of the last sea-level rise, with the shoreline retreat landward and subsequent drowning of the continental shelf corresponding to that observed by Kowsmann & Costa (1979). This interpretation

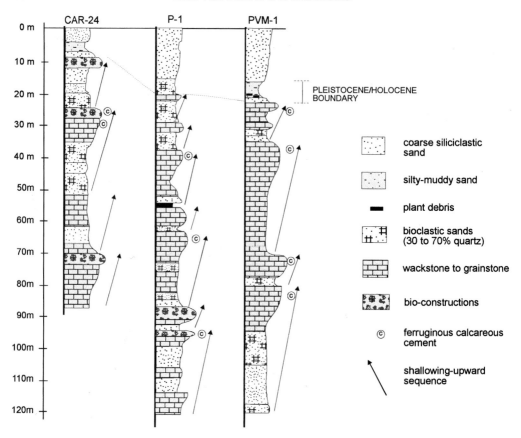

Fig. 12. Schematic lithological logs based on sedimentological analysis of samples from geotechnical boreholes CAR-24, P-1 and PVM-1 (modified from Caddah 1990). Location of boreholes is shown in Fig. 2.

is supported by the fresh water cementation and dissolution of the top of the carbonate bed, and by the presence of plant debris-rich silty clay at the base of the siliciclastic bed.

To summarize, the outer-shelf environment seems to have worked as a good source of sand for offshelf reworking from the early sea-level rise to highstand. Siliciclastic sand deposition seems to occur from transgressive to highstand periods and is related to different sedimentation mechanisms (retreating-shore dynamics to outer-shelf bottom-current reworking). Carbonate accumulation apparently occurs during relative sea-level falls, before subaerial exposure of the shelf. Some samples from geotechnical boreholes were analysed by Spadini *et al.* (1992) and confirmed the association of shallowing-up trends with the sea-level variations. This does not imply a direct relationship between changes in the processes controlling sedimentation and water depth. These processes change responding to the palaeoceanographic changes (intensity of bottom and surface currents, turbidity,

etc.) related to sea-level oscillations and local physiography. The occurrence and thickness of bioclastic sandy deposits mainly depend on the reworking intensity of carbonate banks and the rate of sand-wave migration on these banks. The high sand mobility and the irregular topography of carbonate banks, as seen in the modern environment, prevent a reliable correlation of sedimentary episodes between the cores.

Slope sediments. The upper slope superficial sediments generally comprise shelf-derived sands extending to the 750 m isobath in the north and to the 600 m isobath in the south. Sand accumulations are distributed along the isobaths and become finer downslope. Three main types of slope deposits are evident on the sea floor: clean sands on the uppermost slope (down to the 400 m isobath), muddy sands, between the 400 m and 600–700 m isobaths, and sand–mud intercalations further downslope.

The vertical pattern of upper slope sand deposits displays a coarsening to fining-up trend

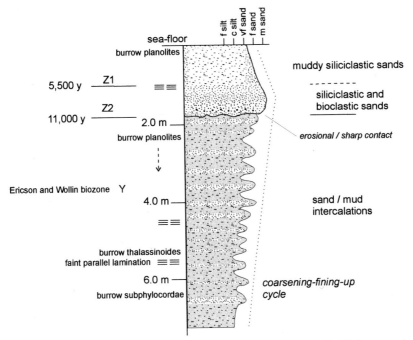

Fig. 13. Schematic log based on cores collected in the upper slope area representing Holocene and upper Pleistocene section about the 400 m isobath.

schematically represented in Fig. 13. At the base, the deposits are composed of sand–mud intercalations, tens of metres thick. Upwards, they pass gradually or abruptly–erosionally to clean sands a few metres thick. The clean sand layer becomes muddier to the top of the section (Fig. 13). Such a sequence is mainly observed between the 300 and 600 m isobaths, and is also generally coarser-grained upslope and finer-grained downslope.

The sand–mud intercalations comprise centimetric sandy layers interbedded with centimetric to decimetric silty–mud layers (Fig. 14). The sandy layers are composed of very fine- to fine-grained muddy to clean sands. Rare traction structures (ripples) are preserved, being generally disturbed by intense bioturbation (Fig. 13). The layers usually present a gradual passage from the muddy to the sandy fraction (inverse grading). Sharp contacts are often observed at the top of the sand layers.

The clean sands are mainly siliciclastic with a variable bioclastic content (5–65%). The sands are greenish grey to yellowish grey, very fine- to coarse-grained (0.07–2.0 mm), and locally gravelly. These sediments are well sorted where predominantly siliciclastic, have <15% clay content, with glauconite and mica, and are slightly to moderately bioturbated. The main bioclastic components included are echinoids, benthic and planktonic foraminifera, molluscs, pteropods and algae fragments. These sands were sampled from the shelf-edge scarp to the 400 m isobath (Fig. 2: GL-02 and GL-03; 3-PC-01, 3-PC-06 and 4-PC-06). Locally they develop accumulations >5 m thick (3-PC-01).

The muddy sands are mixed siliciclastic–bioclastic in composition, generally very fine-grained, with 15–50% mud fraction (Fig. 13). They have a high content of bioclasts and are usually bioturbated, with burrows filled with finer material. They occur in units 1–15 m thick. Under the core of the BC, in water depths between 200 and 350 m, the high energy of the bottom currents prevents the deposition of fine-grained sediments, and locally leads to the development of gravel-lag, shelf-derived deposits (in the 'funnelling zone').

Biostratigraphic analysis (Vicalvi 1994, 1995) indicates that the changeover from sand–mud intercalations to massive sands corresponds to the Pleistocene–Holocene boundary. The Holocene deposits are composed of clean and coarse sands at the base, and muddy sands at the top. Vicalvi (1994) associated the transition from clean to muddy sands with climatic changes occurring in the mid-Holocene. In the Pleistocene section plant debris is abundant in sediments older than 26 ka BP and is better preserved in the cores collected in deeper water in the northern area (Caddah *et al.* 1998).

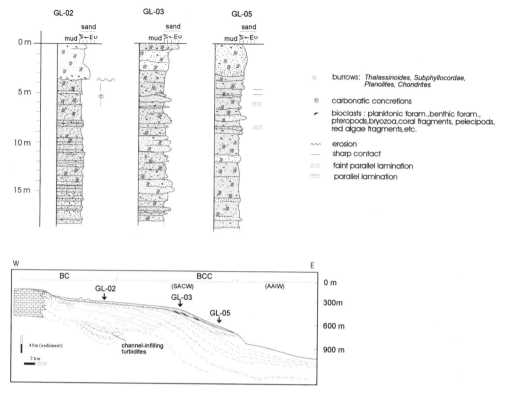

Fig. 14. Logs from the upper portion of the long cores collected in the 'funnelling zone'. Location of cores is shown in Fig. 2. Inset section shows location of cores along a dip section from the shelf edge to the middle slope in the northern area. BC, BCC (SACW and AAIW) are projections of the zone of action of these currents over the sea bed.

The middle slope is covered by sand–mud intercalations with a hemipelagic Fe-rich sandy mud drape at the very top. The lower slope is draped by a pelagic marl (Caddah *et al.* 1998; Viana *et al.* 1998a).

On the northern slope, the upper slope terrace is covered by Holocene sand extending >10 km down to the 700 m isobath (Fig. 14). In addition to the three facies previously described, two other types of facies are observed: gravel deposits at the foot of the shelf-edge scarp, and channel-filling turbidites, on the uppermost slope. Bioclastic gravel and pebbly sand are found on the sea floor between the 200 m and 250 m isobaths, below the inner border of the BC, where they are developed into lag-deposits. In some cores, bioclastic gravelly deposits (13 ka BP rhodolite nodules) are found at the erosive base of Holocene sands, overlying late Pleistocene (older than 26 ka BP) sand–mud intercalations, as observed in the core 9-PC-09 (Fig. 15). In cores, along the whole northern upper slope, sand/mud intercalations are truncated at the top and are all older than 26 ka (Caddah *et al.* 1995).

Thin to thick turbiditic layers are found in borehole GL-02 between 87 and 92 m below the sea bed. Seismic profiles (Fig. 4a) indicate that these layers are restricted to wide channels (hundreds of metres to a few kilometres wide) that have been excavated from oblique to parallel to the isobaths. Turbiditic layers and beds are several centimetres to a few metres thick, and comprise fine- to medium-grained sands. They display erosional bases, normal grading and mud clasts, and are associated with intraformational conglomerates (Caddah *et al.* 1995). In addition, high-resolution seismic profiles obtained at the shelf–slope boundary (Fig. 16) indicate the presence of a modern system of vertically stacked cut-and-fill features, tens to hundreds of metres wide, and up to 30 m deep (Fig. 4a). They display the same reflection pattern as the turbiditic layers. This implies modern accumulation of turbiditic deposits infilling erosional features, at the base of the shelf-edge scarp.

In the northernmost portion of the study area, at the entrance of the 'funnelling zone', immediately south of the Itapemirim canyon head, a

9-pc-09 (-220m)

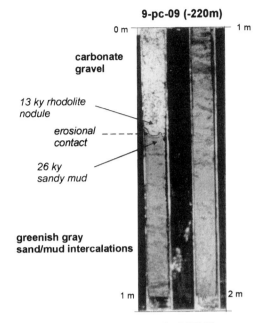

carbonate gravel

13 ky rhodolite nodule

erosional contact

26 ky sandy mud

greenish gray sand/mud intercalations

Fig. 15. Interpreted photo from the 9-PC-09 core collected on the uppermost slope of the 'funnelling zone'. (Note the erosional contact between modern carbonate coarse-grained sediments and upper Pleistocene sand–mud intercalations.) Hiatus represents about 13 ka and is related to the restart of the strong Brazil Current circulation. Location of core is shown in Fig. 2.

zone of non-deposition is developed where Pleistocene sand–mud intercalation outcrops (Fig. 3). This is linked to the extreme energy of the bottom currents in this area, which prevent local sedimentation.

On the southern upper slope, south of the São Tomé canyon, the deposits are finer grained than in the north, and no gravelly sediments are found. The general characteristics of the upper slope sands described for the northern slope are also present in this area. The transition between the Pleistocene sand–mud intercalations and the coarser Holocene sands is gradual. Sand deposits develop a wedge-shaped body, several metres thick at the base of the shelf edge scarp, which becomes thin and extends >5 km downslope (Fig. 17). Typical turbiditic features such as massive to normal grading, erosional bases, and top-truncated beds are only observed downslope of the gully mouths, between the 600 m and 700 m isobaths (cores 4-PC-06 and GL-10, Fig. 17). They may represent isolated events of turbulent sand flows which travel for relatively longer distances than the other more common processes operating the transfer of shelf sands to the upper slope (sand cascading, grain-flow, quasi-steady turbidity currents; see the Discussion). The observed turbidites may also derive from the instability of previously deposited upper slope sands. Both flows occurred within the interval of accumulation of the sand–mud intercalations which were deposited during the latest Pleistocene, between

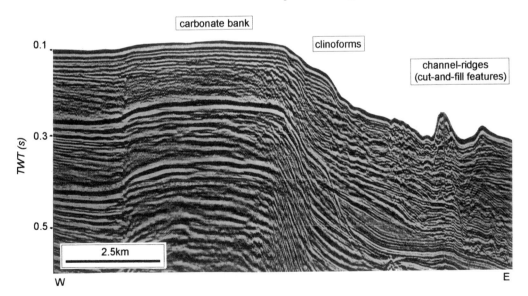

Fig. 16. High-resolution multi-channel seismic profile of the shelf-edge–uppermost slope region of the 'funnelling zone'. The positive relief related to the carbonate banks is highlighted. The shelf-edge scarp-attached clinoforms are related to shelf-edge sweeping by the internal border of the Brazil Current. The channel ridge system is also related to the passage of the BC. The narrow channels are infilled by high-amplitude reflectors correlated with gravity deposits developed by turbidity currents. First and second seismic multiples disturb the section of the shelf. Location of section is shown in Fig. 2.

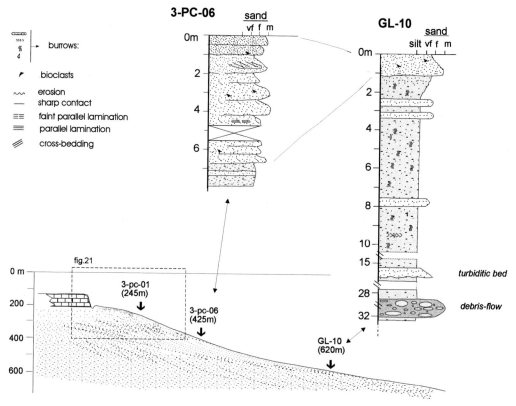

Fig. 17. Schematic logs from cores collected in the southern slope 'expansion zone'. Only the upper part of GL-10 is illustrated. Rectangle indicates the region analysed in Fig. 21. Location of cores is shown in Fig. 2.

40 ka and 18 ka BP. Locally, debris-flow-deposited intraformational paraconglomerates are found (e.g. at 28 m sub-bottom in borehole GL-10; Kowsmann *et al.* 1993). Outcropping Pleistocene sandy muds are found in direct connection with the steepest zones of shelf-edge scarp, where modern sedimentation was prevented by the steepness of the escarpment or by removal by traction currents (along- or down-slope).

The presence of finer-grained sediments, the development of bedforms at the foot of the shelf-edge escarpment, an absence of a widespread erosive surface at the base of the Holocene, and the absence of large erosional features, indicate a relative decay of the BC intensity in the 'expansion zone'. Deposition controlled by the BC is the predominant process in this area, whereas in the northern 'funnelling zone', erosion and deposition controlled by the BC are both active. Additionally, gravity processes such as submarine grain-flows and turbidity currents, play an important role in supplying shelf sediments to the upper slope.

The general characteristics of the upper slope sand deposits, synthesized in Fig. 13, are: (1) different degrees of burrowing and bioturbation, from almost absent in the coarsest facies, to abundant in the finest; burrows are mainly *Planolites*, and also *Thalassinoides*, *Chondrites*, *Subphylocorda* and *Helmintopsis* in finer sediments (Figs 13–15 and 17); (2) absence or rare presence of primary structures, typically masked by bioturbation, or disturbed by coring processes (Fig. 18); (3) the presence of abundant bioclasts with shelf-derived forms (Fig. 14); (4) sand–mud intercalations developing centimetric to decimetric sequences, showing a general inverse grading (mud at the base and sand at the top), a gradual transition between the two facies and usual sharp contact at the top of the sands; (5) clean and muddy sands that develop massive beds with internal erosional surfaces (Fig. 18). These characteristics resemble some of those indicated by Stow & Lovell (1979), Gonthier *et al.* (1984) and Shanmugam *et al.* (1993*a,b*) for alongslope current-controlled accumulations or contourite deposits. Evidence of turbiditic

3-PC-01 (245m)

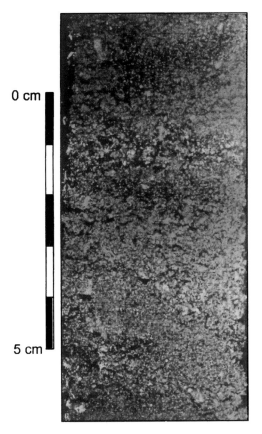

0 cm

5 cm

Fig. 18. Photo detail from the Kullenberg piston core 3-PC-01, collected at 250 m isobath. The core is composed entirely of clean, medium- to coarse-grained, siliciclastic to bioclastic, massive sand. Location of core is shown in Figs 2 and 17.

events are rare, and always interbedded with the finer-grained facies.

Slope sedimentation rates. The 'funnelling' and 'expansion' zones have each behaved differently regarding sedimentation rates from the latest Pleistocene to the Holocene. Sedimentation rates have been determined from several piston cores (Vicalvi 1994, 1995), and averaged for 500 m water depth.

The sedimentation rate during the Holocene (0–11 ka) is lower than during the late Pleistocene. During the Holocene, the northern upper slope has accumulated four times more sediment than the southern upper slope (respectively 17 cm/ka and 4.5 cm/ka). Conversely, during the latest Pleistocene (11 ka to 40 ka BP),

the southern slope has accumulated twice as much sediment as the northern slope, respectively 74 cm/ka for the south and 38 cm/ka for the north.

A modern high rate of shelf-derived sand seems to be continuously transferred through gullies. An annual supply of >1 m³/m was suggested by Viana *et al.* (1998*a*) after submarine video inspections carried out at the mouth of a gully, in the southern upper slope region. Nevertheless, all cores collected in this gullied region record a relatively slow sediment accumulation rate, <20 cm/ka. This low rate suggests that the sediments bypass the gullies, being continuously reworked and retransported along the isobaths by the upper slope currents (BC and BCC).

Offshelf sediment transfer. Side-scan sonar images obtained at the shelf–slope boundary in the southern area, immediately south of Sao Tomé canyon heads, reveal the introduction of shelf sands onto the upper slope through passageways such as gullies, incised valleys and canyon heads. A gentle slope between the shelf and the slope is developed in the canyon head region (Figs 3 and 11), and favours the persistent sediment supply to the upper slope by the hydrodynamic forces. Southward-migrating 3D bedforms are developed across the canyon head, extending >5 km, between the 150 m and 300 m isobaths. They merge with the northward-trending sand waves coming from the shelf edge, developing complex interference bedforms (Fig. 11). Echosounding records collected simultaneously with the sonar surveys suggest that the gullies are only partially filled by sediments, confirming their role as a zone of sediment bypassing. In the inter-gullies area, between 200 m and 300 m isobaths, 2D linear sand waves are observed. They are 0.5–2 m high, and 10–20 m in wavelength, propagating southward alongslope (Fig. 11). Downslope, elongated fan-like deposits are attached to the mouth of the gullies, at about the 400 m isobath. These elongated sediment lobes develop short (hundreds of metres to a few kilometres long), and narrow (<1 km wide) accumulations (Fig. 11). They are curved slightly northward, suggesting the action of the BCC that drives South Atlantic Central Waters (SACW) to the north. These data demonstrate the interaction between downslope and alongslope processes occurring in the shelf–slope sediment transfer zone.

Bed-load transport on the outer shelf and upper slope. Bed-load transport calculations were carried out to quantify the sediment flux along the study area. Calculations were made based

Table 2. *Bed-load transport calculations for outer shelf, shelf-edge and upper slope bottom-current stations (see Fig. 2 for locations) developed for 80 µm and 200 µm sized sediments (see comments in the text)*

	80 µm		200 µm	
	Longitudinal	Transverse	Longitudinal	Transverse
Outer shelf				
East/north	2.3×10^{-2} m³/m.s	1.01×10^{-1} m³/m.s	1.1×10^{-5} m³/m.s	1.3×10^{-3} m³/m.s
West/south	-3.2×10^{-3} m³/m.s	-8.1×10^{-2} m³/m.s	-6.2×10^{-6} m³/m.s	-8.7×10^{-4} m³/m.s
Annual	6.3×10^{5} m³/m.a	6.5×10^{5} m³/m.a	139 m³/m.a	1.5×10^{4} m³/m.a
Shelf edge				
East/north	3×10^{-2} m³/m.s	1.3×10^{-1} m³/m.s	9.1×10^{-5} m³/m.s	1.3×10^{-3} m³/m.s
West/south	-9.7×10^{-2} m³/m.s	-2.1×10^{-2} m³/m.s	-1.6×10^{-3} m³/m.s	-7.9×10^{-6} m³/m.s
Annual	-2.14×10^{6} m³/m.a	3.31×10^{6} m³/m.a	-4.8×10^{4} m³/m.a	3.9×10^{4} m³/m.a
Upper slope				
East/north	1.4×10^{-2} m³/m.s	1.2×10^{-1} m³/m.s	8.9×10^{-6} m³/m.s	1.3×10^{-3} m³/m.s
West/south	-4.1×10^{-3} m³/m.s	-1.9×10^{-3} m³/m.s	-2×10^{-6} m³/m.s	-3.9×10^{-6} m³/m.s
Annual	3.1×10^{5} m³/m.a	3.9×10^{6} m³/m.a	217 m³/m.a	4.2×10^{4} m³/m.a

on the work of Soulsby (1994), who regrouped the different approaches already developed by various researchers. The threshold Shields parameter (θ_{cr}) and threshold bed shear stress (τ_{cr}) were obtained for 200 µm and 80 µm diameter particles. The results are presented in Table 2 and confirm the offshelf exportation from both shelf stations and the downslope trend of the sediment transport across the 400 m isobath (upper slope). Cross-isobath sediment transport prevails in the study area. It increases from the outer shelf to the upper slope. Along-isobath transport is also important for finer sediments (80 µm) and is reinforced at the shelf edge where even 200 µm sediments are longitudinally transported. The increase of longitudinal transport at the shelf edge highlights the importance of this sector for the upper slope sediment supply and redistribution.

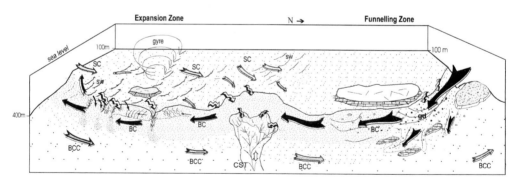

Fig. 19. Block diagram representing the view of the modern environment of the 'funnelling zone and 'expansion zone' (see Fig. 2 for location). Shelf currents (SC) induce the sand migration as sand waves (sw) towards the shelf edge, to the south. Shelf sediments are spilled over the slope (twisted arrows) by gravity processes. The western boundary Brazil Current (BC, black arrow) penetrates the slope from the north, with its inner border sweeping the shelf edge and transporting shelf sediments (twisted arrows) to the upper slope (clinoforms of Fig. 16). The BC is also responsible for the erosion of the sea bed (channel–ridge system and non-deposition zones), and the deposition of gravel and coarse-grained sediments (gd). Downstream the BC penetrates the 'expansion zone', where it decelerates and expands. The Brazil Counter Current (BCC, grey arrow) induces a weak alongslope reworking of the finer sediments. In the 'expansion zone', the passage of a gyre increases the shelf-current ability to transport sediment. Sand is exported towards the upper slope (twisted arrows), confined inside canyons and gully walls, or non-confined and avalanching directly over the uppermost slope. The decelerated BC develops only minor erosional features and sand waves. The double arrow inside the São Tomé canyon (STC) indicates the up- and down-canyon confined currents which are not considered by this paper. (See text for details.)

Discussion: the depositional model for upper slope sand deposits of the Campos Basin

From the example of the Campos Basin margin described in this study, the critical morphological, hydrological and sedimentological parameters for the development of upper slope sedimentary bodies are outlined. They consist essentially of a seaward projection of the shelf margin, a strong superficial slope boundary current and sand availability on the outer shelf. The distinctions between the 'funnelling' and 'expansion' zones correspond to the different ways in which each zone responds to the interplay between ocean currents and sedimentation processes, which are synthetically represented in the block diagram of Fig. 19.

Funnelling zone

North of São Tomé canyon, in the 'funnelling zone', where the BC is accelerated, alongslope processes dominate and are strongly influenced by sea-level variations and climate changes. The distribution of upper slope sand deposits and erosive features is a function of the relative position of the BC core.

During the lowering of sea level (episode a, Fig. 20), latest Pleistocene sediments were deposited during a period when river mouths had migrated towards the shelf edge. River-derived coarse sediments were redistributed along the coast by longshore currents. River-derived suspended sediments were introduced onto the now narrow continental shelf. Thus, suspended plumes easily reached the upper slope areas, and were then redistributed southward by the superficial BC circulation. Sand–mud intercalations rich in plant debris were deposited in this period and correspond to the truncated reflectors observed in the seismic line of Fig. 4b. High sedimentation rates are observed related to the sediment-rich slope water. Such southward transport of suspended material induced the higher sedimentation rate to the south of São Tomé canyon (78 cm/ka) even though the main supply sources were located tens of kilometres to the north (34 cm/ka).

There is evidence of a rather low superficial circulation during this period: the presence of fine sediments on the uppermost slope, the reduced quantity of sandy sediments which occur in a continuum with muddy sediments, and the absence of widespread erosional scours. The absence of a large quantity of sand over the upper slope in this interval may be explained by the lack of an efficient winnowing system at this time. However, it may also be due to the capture of the main river courses by submarine canyon heads that backstep during sea-level lowstands (Wescott 1993). The sea level was somewhere below the 75 m isobath from 70 ka to 10 ka BP (SPECMAP relative sea-level stand data). During that time, the main river, Paraiba do Sul, could have been connected directly to the head of Almirante Camara canyon (Kowsmann & Costa 1979). In such a scenario, most of the coarse-grained material should be transported directly to the deep sea through the canyons. The high sedimentation rate may have favoured sea-floor instability. Locally, some mass movements may have evolved into turbidity currents infilling channels, as observed in the borehole GL-02 (Fig. 4a).

During the maximum lowstand (episode b, Fig. 20), from 18 ka to 13 ka BP, the BC core shifted to its deepest position, between the 350 m and 450 m isobaths, where it influenced the shaping of the Albacora Terrace, between 250 and 550 m water depth. The terrace was formed by the removal of a large quantity of sediments; geotechnical data indicate that up to 30 m of sediment column were removed (Viana et al. 1994, 1998a). Excavation of the erosional ridges on the terrace, elongated parallel to the BC direction, is demonstrated in Figs 3, 4, 16 and 20. The erosional features seem to correspond elsewhere to the hiatus observed in some cores (Figs 14 and 15). This hiatus is emphasized by the abrupt contact between 26 ka fine-grained sediments and 13 ka coarse sediments. It implies a strong increase in the energy of the environment at that time as a result of a drastic increase in the BC circulation. Thus, the BC acceleration began during the Last Glacial Maximum (18 ka), and continued up to the middle Holocene. The major part of the sediment transported by the BC should be trapped inside the São Tomé canyon. Only a minor part would by-pass the canyon heads to be deposited on the southern slope.

During the present-day highstand (episode c, Fig. 20) the strong BC core has shifted landward accompanying the last sea-level rise. Consequently, it developed a diachronous erosional surface, younger towards the shelf, along which fine-grained upper Pleistocene sediments (sand–mud intercalations) crop out or are overlain by Holocene coarse-grained or gravelly sediments. On the modern highstand, the BC penetrates and crosscuts the shelf edge in the northernmost part of the area, at the entrance to the 'funnelling zone' (Figs 3 and 19). Downstream, the BC re-enters onto the slope with

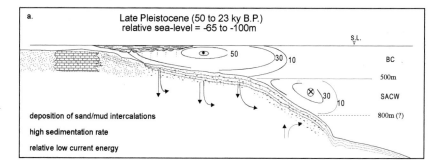

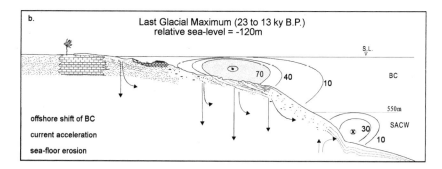

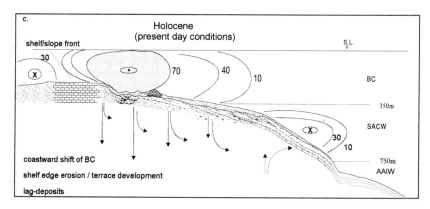

Fig. 20. Diagram illustrating the evolution of late Quaternary circulation in the 'funnelling zone' in three episodes (a, b, and c) corresponding to different sea-level stands and Brazil Current position. Numbers indicate the hypothetical current speeds in cm/s. X indicates flow direction into the page, filled circle indicates flow direction towards the reader. Arrows indicate the direction of bed-load transport. A longshore current, installed at the shelf-edge–upper slope passage, is suggested during the Last Glacial Maximum. (See text for details.) Horizontal Scale is 1 cm to 3 km

increased speeds, supplying the upper slope with coarse-grained and gravelly sediments swept from the shelf (Fig. 19). Small clinoforms prograding out from the shelf edge are developed (Fig. 16). No fine-grained sedimentation occurs on the uppermost slope, between the 200 m and 300 m isobaths. The termination of the fining-upward sedimentation cycle is observed in sediments deposited downslope from the 400 m isobath. This has been related to a small deceleration of the BC linked to the climatic change of mid-Holocene age (see 'Slope sediments' section). Such a deceleration is enhanced by the maximum landward excursion of the sea level occurring at that time. This shift constrained the BC core against the shelf-edge scarp (Fig. 20), with a muddy sand accumulation consequent upon weakening in the longitudinal transport downslope at the 350 m isobath. The sand transported by the BC that previously bypassed the area (episode b, and early stage of episode c) and supplied the São Tomé canyon head is now settled in the uppermost slope and/or transferred downslope. Minor turbiditic events occurred at the base of the shelf-edge scarp, infilling the erosional ridges excavated by the BC passage, as suggested by the observation of seismic lines (Fig. 16).

Expansion area

In the southern area, the expansion of the BC flux induces a deceleration in the current speeds. The narrower and steeper upper slope, and the altered BC behaviour (with the generation of shelf-penetrating eddies) induces a different depositional response to the circulation pattern.

The modern processes involved in the sediment transfer from the shelf edge to the upper slope are: (1) sediment preparation at the outer shelf, with a temporary accumulation (migrating sand waves) at the shelf edge, induced by the shelf currents and eddy effect; (2) sand remobilization along the uppermost slope by the BC, with the development of sand waves; (3) downslope transfer under gravity processes; (4) alongslope reworking by the BCC. This sequence of processes is illustrated in Fig. 19.

Sediment preparation. Sandy sediments are driven towards the shelf break (150 m water depth) under bed-load transport resulting from shelf-current action (the combined action of surface waves, tides, storm and trade winds, eddies and other less expressive phenomena). Oceanic eddies penetrating the continental shelf seem to sweep the fine sediments away in a winnowing process like a brushing machine. This 'sea-floor polishing effect' results in outer shelf sediments made up almost exclusively of medium- to coarse-grained sands. The sum of the various oceanographic and meteorological forces impinging on the sea bed results in tens of thousands of cubic metres of fine- to medium-grained sand crossing one linear metre per year, at the shelf edge (Table 2). Slopeward-migrating sand waves are the geological representation of these processes (Fig. 10).

Sand remobilization along uppermost slope. The inner border of the BC sweeps the shelf edge and the uppermost slope where a narrow terrace occurs. The BC has its energy reduced but still maintains strong velocities (>30 cm/s). It is responsible for a southward along-isobath reworking of the shelf sediments, which occurs

in an opposite direction to that induced by the shelf currents. Some irregular bedforms are developed as the result of the interference between northward-flowing shelf currents and the part of the BC that penetrates onto the shelf. Sand waves with curvilinear crests (3D) are developed in the narrow terrace area (Fig. 11). Linear-crested sand waves (2D) are developed in the steeper intergullies area further downslope between the 200 m and 300 m isobaths (Fig. 11), suggesting a more regular bottom-current flow in this area.

Gravity processes. A downslope transfer of shelf sediments occurs when the migrating sands encounter the shelf break. The São Tomé canyon and slope gully heads capture the great majority of the sediments. However, sonar records (Fig. 11) suggest that sand may also be spilled directly over the upper slope (Fig. 19).

The numerous gullies that dissect the shelf edge act as coarse sediment conduits to the slope, as shown by very high, local sedimentation rates and the development of sedimentary lobes attached to the gully mouths. The main gravity process that seems to take part in this shelf sand spillover is a persistent downslope sand flux triggered by the shelf-edge bottom currents. This sand flux seems to comprise processes occurring both as the submarine grain-flows described by Lowe (1979, 1982), Nardin *et al.* (1979) and Stow (1994), and the high-density quasi-steady turbidity currents proposed by Kneller & Branney (1995). Submarine grain-flows are supported by grain dispersive pressure, and steady uniform grain-flows can only be maintained on steep slopes and are generally <5 cm thick (Lowe 1979, 1982; Nardin *et al.* 1979). On lower-angle slopes, the frictional strength of colliding grains exceeds the applied tangential component of gravity, freezing the flow (Lowe 1979) and providing short-term transport. The development of such a mechanism can be enhanced by the steep slope of the shelf-edge scarp, and by the availability of well-sorted granular sediments.

Stow (1994) considered that the steep slopes required for grain-flow transport make them similar to small-scale sand avalanches occurring in the heads of submarine canyons, as observed by Shepard & Dill (1966). To be sustained, this kind of process must flow under laminar conditions. An increase in turbulence results in the evolution of the flow to a high-density turbidity current, which may travel for longer distances before deposition. Such a process is locally witnessed by the presence of a turbiditic layer further downslope from the gully mouths, near

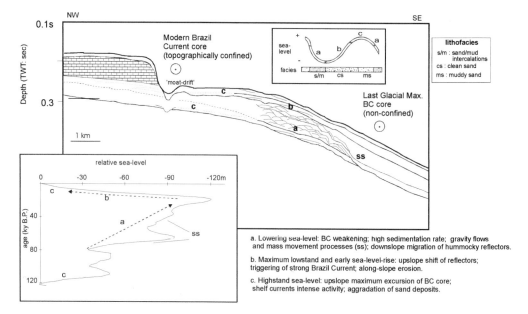

Fig. 21. Scheme of late Quaternary sedimentation on the upper slope of the 'expansion zone' (see Figs 2 and 17 for location). Three episodes (see comments in text) are recognized by lithological and seismic facies and geometric characteristics, and compared with different sea-level stands (see text for details).

the 700 m isobath (core 4-PC-06). The low quantity of muddy sediments on the Campos outer shelf, as a result of strong bottom hydrodynamics, inhibits the development of long-distance-flowing low-density turbidity currents.

To accumulate thick deposits under grain-flow-like processes, successive and continuous grain flows need to occur (Nardin *et al.* 1979). Kneller & Branney (1995) considered high-density, quasi-steady depletive currents, inducing grain-supporting mechanisms other than fluid turbulence, as the process responsible for massive sand accumulations. As a result of this mechanism, deposits will not present erosive sharp base discontinuities, and may be thicker than the original current. Loosely packed to uncompacted accumulations commonly result from such a process (Allen 1972; Kneller & Branley 1995). Sediments collected in the upper slope of the 'expansion zone' display a >5 m thick succession of loosely packed, centimetric layered massive sand (3-PC-01) corresponding to the characteristics proposed for those mechanisms (successive submarine grain-flow-like deposits).

Alongslope reworking by the BCC. The north-ward-flowing BCC impinges on the upper slope sea floor in the 'expansion zone' around the 400 m isobath. Its effect upon the sediments is observed in two distinct ways: (1) the shift, along the current direction, of the sediment lobes

attached to the gully mouths (Figs 11 and 19), and (2) the deposition of a flat sand sheet that extends down to the 650 m or 700 m isobaths (Figs 11 and 19). This sand sheet is composed of massive, moderately to highly bioturbated, fine- to very-fine-grained sand layers. Downslope, these layers become thin and grade to sand–mud intercalations. Bioturbation seems to disturb the preservation of traction structures related to such a high-energy environment.

In the 'expansion zone', the impact of sea-level variations on the upper slope sediment distribution is less marked than in the 'funnelling zone'. However, the BC acceleration for the Last Glacial Maximum is recorded in the deposits. If there is no erosional contact between the upper-most Pleistocene sand–mud intercalations and the sandy Holocene deposits, the transition is fairly sharp, with an abrupt coarsening of the sediment grain-size at the base of the Holocene deposits. Furthermore, the grain-size becomes finer within the upper Holocene section. This vertical facies development is very similar to that observed on the northern slope, and records the same variation of the BC circulation: (1) low during the late Pleistocene; (2) strong increase during the Last Glacial Maximum; (3) slight lowering since the mid-Holocene.

The same three major depositional episodes (a, b and c) observed on the northern slope are represented on the southern slope. They are well recorded in high-resolution seismic lines and piston cores, which reflect the impact of the slope currents and gravity processes on the general sedimentation pattern. The episodes, which are contained in a fourth-order sequence of about 100 ka (Vail *et al.* 1991), are vertically stacked in the upper slope area. They are schematically represented in terms of their seismic characteristics in Fig. 21.

Episode a (Fig. 21) represents sedimentation when the sea level falls. The sea level fall is accompanied by decreasing BC intensity, as observed on the northern slope. A high sedimentation rate is related to the supply of the northern derived suspended sediments (see Fig. 20, episode a), and to the direct supply of sands from the now narrow shelf. The reduced intensity of the BC energy induces a predominance in the downslope transport by gravity flows (cores GL-10 and 4-PC-06). The BC core shifts downslope towards a steeper zone. This shift is seismically characterized by downslope-migrating seismic reflectors occurring as downlapping hummocky clinoforms, which grade downslope to wavy reflectors. The development of local mass movements is witnessed by slump scars (ss in Fig. 21) in the seismic sections (Fig. 5), and by debris-flow deposits, as observed in the GL-10 borehole (Fig. 17). Debris-flow deposits in GL-10 show ages corresponding to this period (between 40 and 85 ka, Kowsmann *et al.* 1993). Sediments are essentially characterized by sand–mud intercalations which develop deposits several metres thick. In a seismic-stratigraphic classification, this episode should correspond to a slope-perched forced-regression parasequence (Posamentier *et al.* 1992).

Episode b (Fig. 21) corresponds to sedimentation from the lowest sea-level stand to the onset of sea-level rise. The episode is marked by the restart of the strong BC flux and by the landward shift of its core. The alongslope sediment transport by the BC prevails. With the sea-level rise and restriction of suspended sediment supply, a backstepping of the hummocky clinoforms is observed. The upslope shift of the BC core is accompanied by the development of a lithologic unconformity and, locally, by an erosional surface. This episode develops thin accumulations on the upper slope. The thickness of the related deposits is a function of the rate of the sea-level rise, which controls the position of the BC core and the rate of sediment supply.

Episode c (Fig. 21) corresponds to the sediments deposited during sea-level highstands. These sediments are characterized by shelf sands, exported to the slope by the shelf currents. They develop an aggradational accumulation,

seismically characterized as a partially transparent facies, with discontinuous wavy reflectors. They constitute relatively thin (few metres thick) coarse- to fine-grained sand accumulations, which grade downslope to a downlapping drape of fine-grained sediments (Fig. 5). Erosion of the internal border of the uppermost slope is observed, probably caused by the landward shift of the BC core. This episode corresponds to the modern sedimentation pattern. The rapid development of a carbonate bank at the shelf edge, at the latest Pleistocene, has confined the internal border of the BC. Such a confinement has increased the local competence of this current, shaping a moat-drift system at the foot of the shelf-edge scarp (Figs 5 and 21).

The analysis of facies distribution, depositional geometry and physiographic features indicates that the development of upper slope sand bodies is the result of: (1) offshelf-trending bottom currents, (2) linear multiple source supply, with direct shelf sediment offspilling, (3) short downslope gravity transport, and (4) alongslope sediment reworking and redistribution by slope boundary currents. The presence of coarser facies and large erosional features on the northern slope indicates the higher energy of that area and corroborates the hydrographic observations. The homogeneous sands deposited as an elongate wedge, extending >50 km along the isobaths, support the idea of the installation of a multiple linear source supply at the shelf edge.

Similarities with the model proposed by Shanmugam *et al.* (1993*a,b*) are found in the 'expansion zone'. The interchannel area of their model (deep turbidite channels) could correspond to the intergullies area in the Campos model. Both are swept by alongslope bottom currents induced by a strong superficial circulation (Loop Current v. Brazil Current), developing bedforms migrating transversally to the gully axes.

Industry applications

The conditions for the development of such sedimentary bodies are not rare on other continental margins (Eastern USA, Gulf of Mexico, Japan, SE Africa, Western UK; see some analogues in Viana *et al.* 1998*b*). The large amount of sand accumulated on the Campos Basin upper slope suggests that those deposits may constitute excellent hydrocarbon reservoirs if preserved in the geological record. The preservation of upper slope sand deposits in the geological record is demonstrated by the analysis of Campos Basin upper Quaternary seismic data. Similar accumulations are also observed in ancient analogues such as those preserved in the Mackenzie Mountains, Canada (Sheepbed Fm, Windermere Supergroup, Neoproterozoic; see Dalrymple & Narbonne 1996), Quebec, Canada (Tourelle Fm, Lower Ordovician; see Hiscott & Middleton 1979), Delaware Basin, West Texas, USA (Brushy Canyon Member, Permian; see Mutti *et al.* 1992), Vocontian Basin (SE France, Middle Cretaceous; see Rubino & Delamette, 1985; Rubino *et al.* 1995), and Hokkaido, Japan (Tobetsu, Morai, Bannosawa and Astuta Fms, Neogene; see Klein *et al.* 1979) among others. The alongslope upper slope sand deposits, whose coarsest facies are mainly deposited during highstand of sea level, are not exhaustively taken into account in sequence stratigraphic models (Vail 1987; Posamentier & Vail 1988; Vail *et al.* 1991). It is essential to integrate palaeo-circulation studies with sequence-stratigraphic interpretations, so as to couple the punctual slope-transverse supply, mainly active during the sea-level lowstands, sustained by Vail's theory, with the linear multiple source supply, presenting alongslope reworking, as proposed in our model, and observed on various continental margins. The occurrence of longitudinal sand bodies in upper bathyal environments indicates that deep-water economic targets must not be expected to occur exclusively in a palaeo-isobath transverse direction (turbidite channels and basin floor fans). Investigation must also include along-isobath directions, in search of current-elongated bodies.

Summary and conclusions

Upper slope sand deposits were developed on the modern Campos Basin margin because of particular characteristics of margin morphology and superficial hydrodynamics. The main conditions for development of upper slope sand accumulations are: (1) convex outer shelf–slope morphology; (2) sandy sediment available at the shelf-edge; (3) net offshelf transport of shelf sands induced by shelf-edge bottom currents; (4) the presence of a relatively strong slope boundary current. The interaction between the shelf–slope morphology and the hydrological factors are key processes leading to the deposition of upper slope sands. Two key physiographic sectors are involved:

(1) The seaward projection of the shelf edge and the region immediately to the north. The morphology of this zone induces a funnelling effect on the southward circulation of the superficial BC. In this 'funnelling zone' the BC is accelerated to very high speeds and sweeps the shelf edge and the uppermost slope.

(2) South of the seaward projection of the shelf. In this sector, the BC expands in response to the change of the shelf–slope morphology, which shows a landward inflexion. In this 'expansion zone', the BC meanders and generates eddies whose activity interferes with the shelf currents.

In the 'funnelling zone', the modern sedimentation reflects very high energy environments. As a consequence, the shelf edge is swept clean of sediment and the BC introduces coarse sediments onto the upper slope. Gravelly to coarse-grained sand deposits and erosional features are frequent. Downslope, sediments become fine grained (very fine sand at the transition between upper and middle slope, around 600 m water depth).

In the 'expansion zone', outer shelf environments are dominated by shelf currents which induce the ENE migration of sand waves towards the shelf edge. Oceanic eddies constitute an important mechanism in the sand transport across the outer shelf. Downwelling of shelf waters seems to increase the downslope transfer of shelf sands, which are prolonged in the form of gravity sand fluxes. Sand fluxes behave as successive and continuous grain-flows, and, more locally, as high-density turbidite flows. Sandy sediments flow downslope, mainly through canyons and gullies, and also as a direct spillover. In contrast to the northern area, where erosive features predominate, the uppermost slope is characterized by bedforms migrating along the trend of the BC southward flow. They develop curvilinear-crested (3D) sand waves in the head of the gullies and rectilinear-crested (2D) sand waves in the intergullies area. Downslope, the sand is redistributed as sand sheets by the BCC.

The vertical passage to underlying fine-grained sediments is erosive in the 'funnelling zone'. In the 'expansion zone', the vertical passage to the underlying fine-grained facies is sharp or gradual. The general vertical facies succession observed in both zones defines a coarsening to fining upward sequence.

The general geometry of the upper slope sand deposits is a thin wedge-shaped accumulation, tens of metres thick in the proximal zone, thinning downslope. The wedge is 5–10 km wide downslope and tens of kilometres along the isobaths. The internal seismic pattern of the coarse-grained sandy layers is transparent to discontinuous and hummocky. Fine-grained sand to sand–mud intercalations develop hummocky clinoforms grading to wavy reflections and then to a parallel pattern in the steepest southern zone, and a parallel pattern in the flattened northern zone.

The model here proposed suggests that

important sand accumulations occur at the upper slope during highstands of sea level. The highstand activity is due to the presence of shelf currents, which sweep off the shelf large volumes of sand, and to strong superficial slope-boundary currents, that rework the sand on the slope. These currents were inhibited during the sea-level fall, and the shelf currents no longer existed at the maximum lowstand, when the shelf areas were subaerially exposed.

Coarse-grained, well-sorted sand deposits covering relatively large areas make the upper slope sand deposits an attractive play to be investigated for potential hydrocarbon reservoirs. The main difficulty in the application of this model to hydrocarbon reservoir analogues seems to be the vertical and lateral development of sealing facies. The Campos Basin example suggests that the hydrodynamic changes which accompany the climatic and sea-level fluctuations may be severe enough to induce the needed facies alternation. A more systematic study of modern environments must be coupled with research on ancient rock outcrops to evaluate the geological record of different expressions of the controlling mechanisms of our model.

The authors thank D. A. V. Stow and J. A. Howe for constructive comments for the improvement of the manuscript. We gratefully acknowledge the discussions with R. O. Kowsmann, D. A. V. Stow, J. G. Rizzo, and E. Gonthier concerning the depositional model. Review of the manuscript by M. Stoker helped to clarify the presentation of some ideas. J. A. Lima and O. O. Moeller closely collaborated in the oceanographic data analysis and interpretation. L. F. Caddah and the team of the PETROBRAS sedimentology laboratory were fundamental in the core analysis and facies interpretation. C. L. da Silva from INPE (Brazilian Spatial Agency) provided the AVHRR satellite image. J. Hermida provided invaluable advice in image processing and figure development. A. Gonçalves made the last orthographic revision of the manuscript. The authors thank PETROBRAS for granting permission to publish this paper. Financial support was provided by PETROBRAS and CNRS-URA197 (Centre National de Recherche Scientifique).

References

ALLEN, J. R. L. 1972. Intensity of deposition from avalanches and the loose packing of avalanche deposits. *Sedimentology*, **18**, 105–111.

CACCHIONE, D. A. & SOUTHARD, J. B. 1974. Incipient sediment moving by shoaling internal waves. *Journal of Geophysical Research*, **70**, 2237–2242.

CADDAH, L. F. G. 1990. *Análise sedimentar do Mb. Grussaí/Fm. Enbosê a partir de poços geotécnicos*. PETROBRAS/DESUD Internal Report DINTER/SELA B-007/90.

——, CASTRO, R. D., RANGEL, M. D. & LANGENDONCK, M. F. V. 1995. *Facies sedimentares do talude quaternário do campo de Albacora com base nos furos geológicos BU91-GL02/2A, BU91-GL03, BU91-GL05 e BU91-GL07*. Macaé, PETROBRAS/DESUD Internal Report SIEX 002/95, 58 p.

——, KOWSMANN, R. O. & VIANA, A. R. 1998. Sedimentary facies of the Campos continental slope. *In*: STOW, D. A. & FAUGÈRES, J. C. (eds) *Bottom-current, Turbiditic-current and Other Deep-marine Processes. Sedimentary Geology Special Volume*, in press.

CAMPOS, E. J. & MILLER, J. L. 1995. Hydrography of the South Brazil Bight as observed during Project COROAS. *EOS Transactions*, American Geophysical Union Fall Meeting, **76**(46), F324.

CARMINATTI, M. & SCARTON, J. C. 1991. Sequence stratigraphy of the Oligocene turbidite complex of the Campos Basin, offshore Brazil: an overview. *In*: WEIMER, P. & LINK, M. H. (eds) *Seismic Facies and Sedimentary Processes of Submarine Fans Turbidite Systems*. Frontiers in Sedimentary Geology, **18**. Springer, New York, 241–246.

CASTRO, D. D. 1992. *Morfologia da margem continental sudeste-sul Brasiliera e estratigrafia sismica do sopé continental*. MSc thesis, Universidade Federal de Rio de Janeiro.

CASTRO, B. M. & LEE, T. N. 1995. Wind-forced sea level variability on the southeast Brazilian shelf. *Journal of Geophysical Research*, **100**(C8), 16045–16056.

CHANG, K. I. & RICHARDS, K. J. 1994. The shelfward penetration of western boundary currents. *EOS Transactions*, Western Pacific Geophysics Meeting, 21 June, 45.

CHEN, M.-P., LO, S.-C. & LIN, K.-L. 1992. Composition and texture of surface sediment indicating the depositional environments off Northeast Taiwan. *Terrestrial, Atmospheric and Oceanic Sciences*, **3**(3), 395–418.

CSANADY, G. T. 1973. Wind-induced baroclinic motions at the edge of the continental shelf. *Journal of Physical Oceanography*, **3**, 274–279.

DALRYMPLE, R. W. & NARBONNE, G. M. 1996. Continental slope sedimentation in the Sheepbed Formation (Neoproterozoic, Windermere Supergroup), Mackenzie Mountains, N.W.T. *Canadian Journal of Earth Sciences*, **33**, 848–862.

DELLA PIAZZA, H., CASTRO, J. R. J. & SILVA, H. P. 1983. Discordância Pré-Holocênica e sua importância no estudo de fundações na Bacia de Campos. *Boletim Técnico da PETROBRAS*, **26**(2), 91–114.

DOYLE, L. J., PILKEY, O. H. & WOO, C. C. 1979. Sedimentation on the Eastern United States continental slope. *In*: DOYLE, L. J. & PILKEY, O. H. (eds) *Geology of Continental Slopes*. Society of Economic Paleontologists and Mineralogists, Special Publication, **27**, 119–129.

EMERY, K. O. 1968. Relict sediments on continental shelves of the world. *Bulletin, American Association of Petroleum Geologist*, **52**, 445–464.

ERICSON, D. B. & WOLLIN, G. 1968. Pleistocene climates and chronology in deep-sea sediments. *Science*, **162**(3859), 1227–1234.

EVANS, D. L. & SIGNORINI, S. S. 1985. Vertical structure of Brazil Current. *Nature*, **315**(6014), 48–50.

——, —— & MIRANDA, L. B. 1983. A note on the transport of the Brazil Current. *Journal of Physical Oceanography*, **13**(9), 1732–1738.

FLEMMING, B. W. 1978. Underwater sand dunes along the southeast African continental margin – observations and implications. *Marine Geology*, **26**, 177–198.

—— 1980. Sand transport and bedform patterns on the Continental Shelf between Durban and Port Elizabeth (southeast Africa continental margin). *Sedimentary Geology*, **26**, 179–205.

—— 1981. Factors controlling shelf sediment dispersal along the southeast African continental margin. *Marine Geology*, **42**, 259–277.

GARFIELD, III, N. 1990. *The Brazil Current at subtropical latitudes*. PhD thesis, University of Rhode Island.

GONTHIER, E., FAUGÈRES, J-C. & STOW, D. A. V. 1984. Contourite facies of the Faro Drift, Gulf of Cadiz. *In*: STOW, D. A. V. & PIPER, D. J. W. (eds) *Fine-grained Sediments: Deep-water Processes and Facies*. Geological Society, London, Special Publications, **15**, 245–256.

GORSLINE, D., KOLPACK, R. L., KARL, H. A., DRAKE, D. E., FLEISCHER, P., THORNTON, S. E., SCHWALBACH, J. R. & SAVRDA, C. E. 1984. Studies of fine-grained sediment transport processes and products in the California Continental Borderland. *In*: Stow, D. A. V. & Piper, D. J. W (eds) *Fine-grained Sediments: Deep-water Processes and Facies*. Geological Society, London, Special Publications, **15**, 395–415.

GUIMARAES, M. A. G., OLIVEIRA, C. H. B. & LENGLER, R. 1991. Technique applied to obtain high resolution. *In*: *2° Congresso Internacional de Geofisica, Sociedade Brasileira de Geofísica, Salvador, Annals*, 699–704.

HARMS, J. C., SOUTHARD, J. B. & WALKER, R. G. 1982. *Structures and Sequences in clastic rocks*. SEPM Short Course No. 9, Tulsa.

HEEZEN, B. C. & EWING, M. 1952. Turbidity currents and submarine slumps and the 1929 Grand Banks earthquake. *American Journal of Science*, **250**, 849–873.

—— & HOLLISTER, C. D. 1971. *The Face of the Deep*. Oxford University Press, Oxford, 659 pp.

HILL, P. R. & BOWEN, A. J. 1983. Modern sediment dynamics at the shelf-slope boundary off Nova Scotia. *In*: STANLEY, D. J. & MOORE, G. T. (eds) *The Shelfbreak: Critical Interface on Continental Margins*. Society of Economic Paleontologists and Mineralogists Special Publication **33**, 265–276.

HISCOTT, R. N. & MIDDLETON, G. V. 1979. Depositional mechanics of thick-bedded sandstones at the base of a submarine slope, Tourelle Formation (Lower Ordovician) Quebec, Canada. *In*: DOYLE, L. J. & PILKEY, O. H. (eds) *Geology of Continental Slopes*, Society of Economic Paleontologists and Mineralogists Special Publication, **27**, 307–326.

HOUGHTON, R. W., FLAGG, C. N. & PIETRAFFESA, L. J. 1994. Shelf-slope frontal structure, motion and eddy heat flux in the southern Middle Atlantic Bight. *Deep-Sea Research, II*, **41**, 273–306.

HUNT, R. E., SWIFT, D. J. P. & PALMER, H. 1977. Constructional shelf topography, Diamond Shoals, North Carolina. *Geological Society of America Bulletin*, 88, 299–311.

HUTHNANCE, J. M. 1992. Extensive slope currents and the ocean-shelf boundary. *Progress in Oceanography*, **29**, 161–192.

—— 1995. Circulation, exchange and water masses at the ocean margin: the role of physical processes at the shelf edge. *Progress in Oceanography*, **35**, 353–431.

KARL, H. A., CARLSON, P. R. & CACCHIONE, D. A. 1983. Factors that influence sediment transport at the shelfbreak. *In*: STANLEY, D. J. & MOORE, G. T. (eds) *The Shelfbreak: Critical Interface on Continental Margins*. Society of Economic Paleontologists and Mineralogists Special Publication, **33**, 219–231.

KLEIN, G. V., OKADA, H. & MITSUI, K. 1979. Slope sediments in small basins associated with a Neocene active margin, western Hokkaido Island. *In*: DOYLE, L. J. & PILKEY, O. H. (eds) *Geology of Continental Slopes*, Society of Economic Paleontologists and Mineralogists Special Publication, **27**, 359–374.

KNELLER, B. C. & BRANNEY, M. J. 1995. Sustained high-density turbidity currents and the deposition of thick massive beds. *Sedimentology*, **42**(4), 607–616.

KOWSMANN, R. O. & COSTA, M. P. A. 1979. *Sedimentação Quaternária da Margem Continental Brasileira e das Areas Adjacentes*. Projeto REMAC, PETRO-BRAS, Rio de Janeiro, Vol. 8, 55 pp.

——, VIANA, A. R., CADDAH, L. F. G., COSTA, A. M. & AMARAL, C. S. 1993. *Integraçao de dados geologicos com perfis geotecnicos de resistência (Su) no talude de Marlim, Bacia de Campos*. PETRO-BRAS, Rio de Janeiro, Internal Report Cenpes/Sintep **650-14.965**, 39 pp.

LIMA, J. A. M., VIANA, A. R. & MIDDLETON, J. 1998. Bottom boundary currents on the Southeast Brazilia Shelf and Slope. *Continental Shelf Research*, in press.

LOWE, D. R. 1979. Sediment gravity flows: their classification and some problems of application to natural flows and deposits. *In*: DOYLE, L. J. & PILKEY, O. H. (eds) *Geology of Continental Slopes*, Society of Economic Paleontologists and Mineralogists Special Publication, **27**, 75–82.

—— 1982. Sediment gravity flows: II. Depositional models with special reference to the deposits of high-density turbidity currents. *Journal of Sedimentary Petrology*, **52**, 279–297.

McCAVE, I. N. 1972. Transport and escape of fine-grained sediments from shelf areas. *In*: SWIFT, D. J. P., DUANE, D. B. & PILKEY, O. H. (eds) *Shelf Sediment Transport*. Dowden, Hutchinson and Ross, Stroudsbourg, PA, 225–248.

McGRAIL, D. W. & CARNES, M. 1983. Shelf edge dynamics and the nepheloid layer in the northwestern Gulf of Mexico. *In*: STANLEY, D. J. & MOORE, G. T. (eds) *The Shelfbreak: Critical Interface on Continental Margins*. Society of Economic Paleontologists and Mineralogists Special Publication, **33**, 251–264.

MIDDLETON, G. V. & HAMPTON, M. A. 1973. Sediment gravity flows: mechanics of flow and deposition. *In: Turbidites and Deep-water Sedimentation.* Society of Economic Paleontologists and Mineralogists, Pacific Section Short Course Lecture Notes, 1–18.

—— & HAMPTON, M. A. 1976. Subaqueous sediment transport and deposition by sediment gravity flows. *In:* STANLEY, D. J. & SWIFT, D. J. P. (eds) *Marine Sediment Transport and Environmental Management.* John Wiley, New York, 197–218.

MILLER, D. J., KOWSMANN, R. O. & RIZZO, J. G. 1996. Feições fisiográficas da Bacia de Campos à luz de novos dados batimétricos. *In: 34 Congresso Brasileirode Geologia, Annals,* Vol. 3, 397–398.

MUTTI, E. 1985. Turbidite systems and their relation to depositional sequences. *In:* ZUFFA, G. G. (ed.) *Provenance of Arenites. NATO-ASI Series,* Riedel, Dordrecht, 65–93.

—— 1992. Ancient deep-water siliciclastic systems: what do we really know? *British Sedimentological Research Group Workshop on Deep-water Clastics: Dynamics of Modern and Ancient Systems, Oxford Programme and Abstracts,* 2 pp.

—— & NORMARK, W. 1987. Comparing examples of modern and ancient turbidite systems: problems and concepts. *In:* LEGGETT, J. K. & ZUFFA, G. G. (eds) *Marine Clastic Sedimentology: Concepts and Case Studies.* Graham & Trotman, London, 1–38.

——, DAVOLI, G., SEGADELLI, S., CAVALLI, C., CARMINATTI, M., STECCHI, S., MORA, S. & ANDREOZZI, M. 1992. *Turbidite sandstones.* AGIP/Instituto di Geologia–Università di Parma, Parma, 236 pp.

NARDIN, T. R., HEIN, F. J., GORSLINE, D. S. & EDWARDS, B. D. 1979. A review of mass-movements processes, sediment and acoustic characteristics, and contrasts in slope and base-of-slope system versus canyon–fan–basin floor systems. *In:* DOYLE, L. J. & PILKEY, O. H. (eds) *Geology of Continental Slopes.* Society of Economic Paleontologists and Mineralogists Special Publication, **27**, 61–74.

NITROUER, C. A. & WRIGHT, L. D. 1994. Transport of particles across continental shelves. *Reviews of Geophysics,* **32**, 85–113.

PALMA, J. J. C. 1979. Minerais pesados. *In:* AMARAL, C. A. B. (ed.) *Recursos minerais da margem continental brasileira e das áreas oceânicas adjacentes.* Projeto REMAC, PETROBRAS, Rio de Janeiro, Vol. 10.

PIETRAFESA, L. J. 1983. Shelfbreak circulation, fronts and physical oceanography: east and west coast perspectives. *In:* STANLEY, D. J. & MOORE, G. T. (eds) *The Shelfbreak: Critical Interface on Continental Margins.* Society of Economic Paleontologists and Mineralogists Special Publication, **33**, 233–250.

PINGREE, R. D. & LECANN, B. 1989. Celtic and Armorican slope and shelf residual currents. *Progress in Oceanography,* **23**, 303–338.

POSAMENTIER, H. W. & VAIL, P. R. 1988. Eustatic control on clastic deposition II: Sequence and systems tracts models. *In:* WILGUS, C. K., POSAMENTIER, H. W., VAN WAGONER, J., ROSS, C. A. & KENDALL, G. S. T. C. (eds) *Sea-level Changes: an Integrated Approach.* Society of Economic Paleontologists and Mineralogists Special Publication, **42**, 125–155.

——, ALLEN, G. P., JAMES, D. P. & TESSON, M. 1992. Forced regressions in a sequence stratigraphic framework: concepts, examples, and exploration significance. *Bulletin, American Association of Petroleum Geologists,* **76**(11), 1687–1709.

RAMSAY, P. J. 1994. Marine geology of the Sodwana Bay shelf, southeast Africa. *Marine Geology,* **120**, 225–247.

ROCHA, J. M., MILLIMAN, J. D., SANTANA, C. I. & VICALVI, M. A. 1975. Southern Brazil. *Contributions to Sedimentology. Upper Continental margin off Brazil.* Stuttgart, **4**, 117–150.

RUBINO, J. L. & DELAMETTE, M. 1985. The Albian Shelf of South East France: an example of clastic sand distribution dominated by oceanic currents. *In: 6th European Regional Meeting of Sedimentology, Lleida. Book of Abstracts,* 399–402.

——, IMBERT, P. & PARIZE, O. 1995. Shelf to basin facies relationship on a ramp setting: the Albian sandstones of the Vocotian Basin (South-East France). *In: IAS 16th Regional Meeting of Sedimentology, Books of Abstracts.* Publication ASF, Paris, **22**, 129.

SCHALLER, H. 1973. Estratigrafia da Bacia de Campos. *In: Congresso Brasileiro de Geologia, 27, Aracaju, Annals.* Sociedade Brasileira de Geologia, **3**, 247–258.

SHANMUGAM, G., SPALDING, T. D. & ROFHEART, D. H. 1993a. Traction structures in deep-marine, bottom-current-reworked sands in Pliocene and Pleistocene, Gulf of Mexico. *Geology,* **21**, 929–932.

——, —— & —— 1993b. Processes, sedimentology and reservoir quality of deep-marine bottom-current reworked sands (sandy contourites): an example from the Gulf of Mexico. *Bulletin, American Association of Petroleum Geologists,* **77**(7), 1241–1259.

SHAW, P. T., PIETRAFESA, L. J., FLAGG, C. N., HOUGHTON, R. W. & SU, K.-H. 1994. Low-frequency oscillations on the outer shelf in the southern Mid-Atlantic Bight. *Deep-Sea Research, II,* **41**, 253–271.

SHEPARD, F. & DILL, R. F. 1966. *Submarine Canyons and Other Sea-valleys.* Rand McNally, Chicago, IL.

SIGNORINI, S. R. 1978. On the circulation and the volume transport of the Brazil Current between Cape of São Tomé and Guanabara Bay. *Deep-Sea Research,* **25**(6), 481–490.

SOULSBY, R. L. 1994. *Manual of Marine Sands.* HR Wallingford, UK. Report **SR 351**.

SPADINI, A. R., MONTEIRO, M. C., KOWSMANN, R. O. & VIANA, A. R. 1992. *Análise geológica dos furos GT-23, GT-24 e GT-25 da área da Plataforma de Albacora-1.* PETROBRAS/CENPES/DIGER, Internal Report **057/92**, 23 pp.

STANLEY, D. J. & WEAR, C. M. 1978. The 'mud-line': an erosion–deposition boundary on the upper continental slope. *Marine Geology,* **28**, M19–M29.

——, ADDY, S. K. & BEHERENS, E. W. 1983. The

mudline: variability of its position relative to the shelfbreak. *In*: STANLEY, D. J. & MOORE, G. T. (eds) 1983. *The shelfbreak: Critical Interface on Continental Margins*. Society of Economic Paleontologists and Mineralogists Special Publication, **33**, 279–298.

——, SHENG, H., LAMBERT, D. N., RONA, P. A., MCGRAIL, D. W. & JENKYNS, J. S. 1981. Current-influenced depositional provinces, continental margin off Cape Hatteras, identified by petrologic method. *Marine Geology*, **40**, 215–235.

STECH, J. L. & LORENZZETTI, J. A. 1992. The response of the South Brazil Bight to the passage of wintertime cold fronts. *Journal of Geophysical Research*, **97**(C6), 9507–9520.

STOW, D. A. V. 1994. Deep-sea processes of sediment transport and deposition. *In*: PYE, K. (ed.) *Sediment Transport and Depositional Processes*. Blackwell Scientific, Oxford, 257–291.

—— & LOVELL, J. P. B. 1979. Contourites: their recognition in modern and ancient sediments. *Earth-Science Review*, **14**, 251–291.

SWIFT, D. J. P. & NIEDORODA, A. 1985. Fluid and sediment dynamics on continental shelves. *In*: TILLMAN, R. W., SWIFT, D. J. P. & WALKER, R. G. (eds) *Shelf sands and Sandstone Reservoirs*. Society of Economic Paleontologists and Mineralogists Short Course, **13**, 47–133.

——, STANLEY, D. J. & CURRAY, J. R. 1971. Relict sediments on continental shelves: a reconsideration. *Journal of Geology*, **79**, 322–346.

TRUMBULL, J. V. A. 1972. Atlantic continental shelf and slope of the United States – sand-sized fraction of bottom sediments, New Jersey to Nova Scotia. *US Geological Survey Professional Paper*, **529K**, 45 pp.

VAIL, P. R. 1987. Seismic stratigraphy interpretation using sequence stratigraphy. Part 1: seismic stratigraphy interpretation procedure. *In*: BALLY, A. W. (ed.) *Atlas of Seismic Stratigraphy*, American Association of Petroleum Geologists, Studies in Geology, **27**, 1–10.

——, AUDEMARD, F., BOWMAN, S. A., EISNER, P. N. & PEREZ-CRUZ, C. 1991. The stratigraphic signature of tectonics, eustasy and sedimentology – an overview. *In*: EINSELE, G., RICKEN, W. & SEILACHER, A. (eds) *Cycles and Events in Stratigraphy*. Springer, Berlin, 617–659.

VIANA, A. R., FAUGÈRES, J. C., KOWSMANN, R. O., LIMA, J. A. M., CADDAH, L. F. G. & RIZZO, J. G. 1998a.

Sand sedimentation on the continental slopes: an overview from the modern Campos Basin data, SE Brazilian continental margin. *In*: STOW, D. A. & FAUGÈRES, J. C. (eds) *Bottom-current, Turbiditic-current and other Deep-marine Processes. Sedimentary Geology*, Special Volume, in press.

——, —— & LIMA, J. A. 1995. The role of outer shelf currents in feeding deep-water systems. *In*: *16th IAS Regional Meeting of Sedimentology, Aix-les-Bains, Book of Abstracts*. Publication ASF, Paris, **22**, 151.

——, ——. & STOW, D. A. V. 1998b. Bottom current controlled sand deposits – a review from modern shallow and deep water environments. *In*: STOW, D. A. & FAUGÈRES, J. C. (eds) *Bottom-current, Turbiditic-current and other Deep-marine Processes. Sedimentary Geology*, Special Volume, in press.

——, KOWSMANN, R. O. & CADDAH, L. F. G. 1994. Architecture and oceanographic controls on the sedimentation of Campos Basin continental slope. *In*: *14th International Sedimentological Congress, Recife, Brazil, August 1994, Abstracts*. International Association of Sedimentologists, D20–21.

VICALVI, M. A. 1994. *Resultados dos estudos micropaleontologicos da seção Quaternaria dao talude da Bacia de Campos*. PETROBRAS, Rio de Janeiro, Relatorio Interno CENPES/DIVEX/SEBIP.

——, 1995. *Zoneamento bioestratigráfico e paleoclimático dos sedimentos do Quaternário superior do talude da Bacia de Campos, RJ, Brasil*. PETROBRAS, Rio de Janeiro, Comunicação Técnica CENPES/DIVEX/SEBIP.

VOGT, P. R. & TUCHOLKE, B. E. (eds) 1986. *The Geology of North America, Vol. M, The Western North Atlantic Region*. Geological Society of America Special Publication.

WALKER, R. G. 1984. *Facies Models*. Geoscience Canada, Reprint Series **1**, 317 pp.

WALSH, J. J., BISCAYE, P. E. & CSANADY, G. T. 1988. The 1983-84 Shelf Edge Exchange Processes (SEEP)-I experiment: hypothesis and highlights. *Continental Shelf Research*, **8**, 435–456.

WESCOTT, W. A. 1993. Geomorphic thresholds and complex response of fluvial systems – some implications for sequence stratigraphy. *Bulletin, American Association of Petroleum Geologists*, **77**(7), 1208–1218.

Hemipelagites: processes, facies and model

DORRIK A. V. STOW[1] & ALI R. TABREZ[2]

[1]*Department of Geology, University of Southampton, Southampton Oceanography Centre, European Way, Southampton SO14 3ZH, UK*

[2]*National Institute of Oceanography, Karachi, Pakistan*

Abstract: Detailed sedimentological studies have been carried out on Pleistocene to Recent sediments from the Makran margin and on Miocene to Recent sediments from the Oman margin of the NW Indian Ocean. In both areas, hemipelagites make up *c.* 50% of the succession, being closely interbedded with turbidites (Makran) and pelagites (Oman). Their sedimentary features (structure, texture, fabric and composition), bedding style and rates of sedimentation are reported and compared with previous detailed studies of both modern and ancient hemipelagites. A composite facies model is presented. Hemipelagites are fine-grained sediments typically occurring in deep-water settings. They generally comprise an admixture of >10% biogenic pelagic material and >10% terrigenous or volcanigenic material, in which >40% of the terrigenous (volcanigenic) fraction is silt size or greater (i.e. >4 μm). They are deposited by a combination of vertical settling and slow lateral advection. Hemipelagites are mostly thoroughly bioturbated with ichnofossils including *Zoophycos*, *Planolites*, *Chondrites*, *Phycosiphon* and others. They are generally devoid of primary sedimentary structures, except for an organic-rich facies deposited in anoxic conditions which has a distinctive fissile lamination. The submicroscopic fabric is closely packed with subparallel alignment of clays. The grain size is strongly influenced by the composition, although in general the deposits are fine grained (mean size 5–35 μm), and poorly sorted, in some cases with bimodal, trimodal or polymodal grain-size distributions. The composition includes a biogenic fraction that is calcareous and/or siliceous, and a terrigenous fraction dependent on the nature of the supply pathway, i.e. river plumes, aeolian dust, glacigenic input or volcaniclastic fallout. Chemogenic components (phosphorites, glauconite, ferromanganese nodules) are common in some settings, whereas high organic carbon contents characterize others. Cyclic bedding, typical of climatic forcing, is common although bed boundaries are generally very gradational. Sedimentation rates typically vary from <5 cm/ka to >20 cm/ka and greatly influence development of the different characteristics outlined above. Extremely high rates (>100 cm/ka) occur in zones of hemipelagic sediment focusing, such as slope canyon systems and sediment pathways.

Hemipelagites are fine-grained sediments comprising mixtures of biogenic and terrigenous material that have been deposited by a combination of vertical (pelagic) settling and slow lateral advection. They include muddy oozes, calcareous and siliceous muds or marls, as well as organic-rich, volcaniclastic-rich and glacigenic-rich muds, and are very widespread in most deep-sea environments, especially those close to continental margins. An estimated 20% of the present-day sea floor is composed of hemipelagites.

Although hemipelagites are widely recognized, the only general facies model to have been proposed so far is that of Stow (1982, 1985*a,b*, 1986). This is a very rudimentary model that does not fully represent the variety of hemipelagic facies that exist, nor the complex interplay of processes involved in their sedimentation.

The aim of this paper, therefore, is threefold: (1) to report on very detailed studies carried out on two continental margin systems in the NW Indian Ocean in which contrasting styles of hemipelagic sedimentation are important; (2) to review briefly a number of previous studies on modern and ancient hemipelagites; (3) to present a composite facies model for hemipelagites.

Methods

The detailed studies of hemipelagites from the Makran and Oman margins were carried out by one of us (A. R. Tabrez) as part of a PhD thesis at the University of Southampton. Cores were all described, photographed and X-radiographed, and then subsampled for laboratory analysis. Standard techniques were used throughout (e.g. Tucker 1989) and have been reported more fully by Tabrez (1995).

Grain-size analysis of over 350 samples was carried out using a Malvern 2600 Particle Analyser, with a focal length of 100 mm. Bulk

STOW, D. A. V. & TABREZ, A. R. 1998. Hemipelagites: processes, facies and model. *In*: STOKER, M. S., EVANS, D. & CRAMP, A. (eds) *Geological Processes on Continental Margins: Sedimentation, Mass-Wasting and Stability.* Geological Society, London, Special Publications, **129**, 317–337.

sediment analysis of dried and powdered samples, followed by clay mineral analysis of the <2 μm fraction, was carried out on 175 samples by X-ray diffraction. A Philips 1730 X-ray Diffractometer was used with Cu radiation and Ni filtering, at 35 mA, 40 kV and 1.7°/min scanning speed. Further compositional studies utilized conventional petrographic techniques on smear slides, grain mounts and washed residues. Sediment fabric was examined and characterized on a limited number of samples (<20) using a JEOL 6400 Scanning Electron Microscope. An additional 300 samples were analysed for both inorganic carbonate and organic carbon–hydrogen–nitrogen using a Carlo Elba EA 1108 Elemental Analyser.

Makran margin hemipelagites

General setting

The Makran continental margin in the Gulf of Oman (Fig. 1) forms the seaward extremity of an accretionary sediment prism which extends several hundred kilometres inland (Minshull & White 1989). The Makran accretionary complex marks a zone of convergence between oceanic lithosphere of the Arabian plate and continental lithosphere of the Eurasian plate, which has resulted in an irregular stepped slope

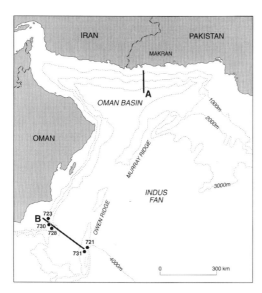

Fig. 1. General setting and bathymetry of the NW Indian Ocean showing line of transect across the Makran margin (line A; see Fig. 2), along which cores MKM-468 – MKM-472 were taken, and transect across the Oman margin (line B; see Fig. 5), along which ODP Leg 117 sites were located.

morphology of basins and part basins separated by intervening structural highs or ridges that represent the seaward edge of a series of imbricate thrust slices (Fig. 2).

Between the uplifted ridges, the basins are filled with sediment, the amount of infill generally increasing towards the coast. The sediment in the basins is derived partly from reworking of the material of the ridges, and partly from surrounding land masses of Arabia and the Makran. The complex morphology and active nature of the margin has resulted in a series of channel-like pathways connecting the basins rather than any distinct erosive channels. Both low-density turbidity currents and slow hemipelagic advection are believed to follow these downslope pathways, although hemipelagic dispersion is widespread over the intervening highs as well.

Five piston cores, MKM-468, MKM-469, MKM-470, MKM-471 and MKM-472, were recovered in water depths of between 1325 and 3274 m. They form a single N–S transect along the line of the seismic section described by White (1982) (Figs 1 and 2). They range in length from 5.5 to 14 m and provide a complete late Quaternary–Holocene succession back to about 26 ka (core 469). Tabrez (1995) was able to correlate and date the cores on the basis of lithostratigraphy, oxygen isotope analysis and magnetostratigraphy.

Sediment facies and structures

The principal facies present are closely interbedded fine-grained turbidites and hemipelagites, each making up between 40% and 60% of the cores examined. True pelagites are present only at the deepest site and make up 10–15% of Core 472. Hemipelagite units can be distinguished from the intervening turbidites on the basis of texture, composition, sedimentary structures and colour. The hemipelagites consist of a foraminiferal and nannofossil biogenic fraction, and terrigenous silty clays of similar composition to the turbidites. They are apparently fully homogenized and mottled by bioturbation, although distinct burrow traces are not very common. Sediment colour varies from olive grey to pale grey. The upper parts of hemipelagic units are typically rich in foraminifera and nannofossils, with indistinct bioturbation. The lower bedding contacts with underlying turbidites are gradational whereas the base of overlying turbidites is sharp. No lamination or other primary sedimentary structures are evident. Bed thickness ranges from 5 cm to 50 cm.

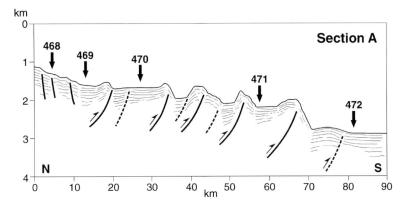

Fig. 2. Line drawing interpretations of seismic profile from White (1982) across the Makran margin, showing locations of core sites MKM-468–MKM-472. Depths are recalculated to below sea level. Location of section A is shown in Fig. 1.

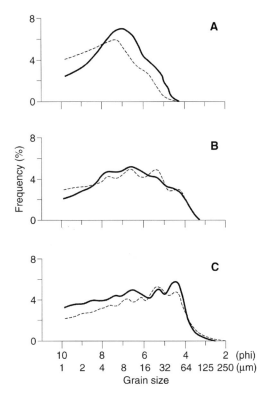

Fig. 3. Selected grain-size distribution smoothed frequency curves from hemipelagites in cores MKM-469 and MKM-472. The three main types of grain-size distribution are illustrated: (**a**) very fine silt–clay median size, unimodal; (**b**) fine silt median, irregular or slightly bimodal; (**c**) fine to medium silt median, irregular to polymodal.

Grain-size characteristics

The median grain size of all hemipelagite samples analysed varies between about 5 μm and 15 μm, and for most of these samples lies within the narrower range of 5–10 μm. The mean grain-size values tend to be slightly higher, typically 7–16 μm, and range up to 27 μm. Sorting is generally very poor. Most samples show slight to moderate positive skewness.

The variation in shape of the grain-size distribution curves is illustrated for cores MKM-469 and MKM-472 (Fig 3a and b). Three types are evident: (1) sediments with a relatively fine median grain size (i.e. 5–8 μm), which present a unimodal distribution with a peak between 5 and 8 μm; (2) sediments with coarser median grain size (8–15 μm) with either bimodal or less regular distributions; (3) sediments with a median grain size up to about 20 μm and a polymodal distribution in which individual peaks are more or less distinct from the background curve at values of about 5, 10, 30 and 40 μm. In all three cases, the apparent 'peak' at 0.5 μm is due to summation of all the finer grain sizes as the lower limit of grain-size measurement was 0.5 μm.

The vertical variation of grain-size properties, for hemipelagites analysed throughout each of the five cored intervals, is extremely interesting. This is most clearly illustrated by median grain-size variation shown in Fig. 4. Three features are noteworthy. First, there is a zone of minimum grain size between 2 and 3 m depth in cores MKM-468, -469 and -470, and between 7 and 10 m in cores MKM-471 and -472. The median grain size shows a general decrease up to this level from the base of the cores and then an increase upwards from this minimum zone.

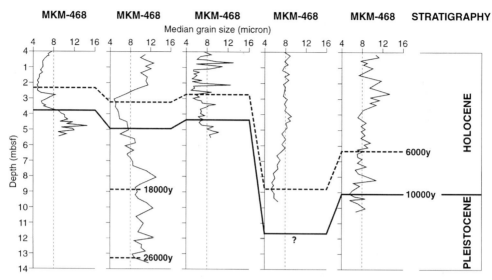

Fig. 4. Plots of median grain size v. depth for all Makran margin cores. Depths in metres below sea floor (mbsf). The grain-size low apparent in most cores is shown correlated with a dashed line and has been dated at around 6000 a BP. The Holocene–Pleistocene boundary was determined principally from oxygen isotope analysis of foraminifera from cores MKM-469 and MKM-472, as were the other ages indicated on the longer section of MKM-469. (See Tabrez (1995) for full details.)

Second, there is a small-scale oscillation or cyclicity of median grain size evident on a vertical scale of 0.2–1.0 m. Third, this oscillation is particularly marked, with variation in median grain size of 4–6 μm, only at certain horizons in each of the cores. The uppermost 2.5 m of section in core MKM-468 does not show this cyclicity.

If the grain-size oscillation is smoothed, then the average values for median or mean grain size might be expected to show some consistent variation with distance from shore. The innermost (i.e. shoreward) sites, MKM-468 and -469, both show smoothed average values between 5 and 11 μm; the two slope basin sites, MKM-470 and -471, show average values of 6–8 μm; and

the abyssal plain site, MKM-472, values of 7–10 μm. No clearcut pattern is evident.

Composition

The composition of both hemipelagites and turbidites, based on smear-slide analysis of over 100 samples, is shown in Table 1. The main components of hemipelagites typically include 60–80% terrigenous material (quartz, feldspars and clays being dominant) and 20–40% biogenic material (mainly nannofossils). The main terrigenous components of the interbedded turbidites are very similar to those of the hemipelagites, but the biogenic fraction is notably less. Non-specific calcite, which could be

Table 1. *Composition of Makran margin sediments; per cent estimates from smear slides*

	Hemipelagites	Turbidites
Quartz	15–40	30–40
Feldspar	1–10	10–15
Non–specific calcite	5–25	15–20
Dolomite	1–3	<1
Opaques	<1	<1
Terrigenous silt and clay	20–30	20–30
Nannofossils	20–50	1–5
Foraminifera	<5	<2
Other biogenic material	<5	<1

derived from land or from fragmented biogenic material, is a significant component of both facies. Generally, there is very uniform composition between sites, but there are marked vertical changes in hemipelagite composition coincident with grain-size change. There is a decrease in terrigenous and increase in nannofossil content coincident with finer grain size.

The clay mineral assemblage identified by X-ray diffraction (XRD) analysis includes illite, kaolinite, chlorite and a small amount of smectite. There is no apparent change in this mineralogy either within or between cores, and there is only very slight variation in the relative amounts of these clays recorded between samples, although this shows no consistent pattern and is well within the bounds of analytical error. A sample from a mud volcano present in the Makran coastal region was also analysed and revealed exactly the same clay mineralogy.

Organic carbon contents in all facies are low, ranging from 0.06 to 1.33%. The higher values are found in hemipelagites from core MKM-468. Other sites of mid-slope basins and the abyssal plain show lower values. The $CaCO_3$ content is in the range of 15–20%, with lower values from the slope region and higher values in hemipelagites of cores 468 and 472.

Oman margin hemipelagites

General setting

The Oman margin is a complex, stepped divergent margin directly off the narrow Oman shelf (Fig. 5). The Owen Ridge originally formed as a component of the Oman slope-apron system but is now separated from it by the narrow oceanic Oman Basin. Superficially, the morphology of this margin is similar to that of the Makran margin and comprises a series of slope basins and partial basins separated by structural highs (Figs 1 and 5). These latter are remnant highs from the rift break-up of the margin rather than thrust slices, and tectonic activity has remained very low during at least the Neogene (White & Louden 1982; Minishull *et al.* 1992).

The Oman outer shelf and upper slope is currently a region affected by seasonal (monsoon-induced) upwelling and associated high primary productivity. An expanded oxygen-minimum zone within the water column has periodically been well developed and has impinged on the sea floor in the upper-slope region (Fig. 5), thereby affecting the activity of macrobenthos and the preservation of organic carbon in the sediments.

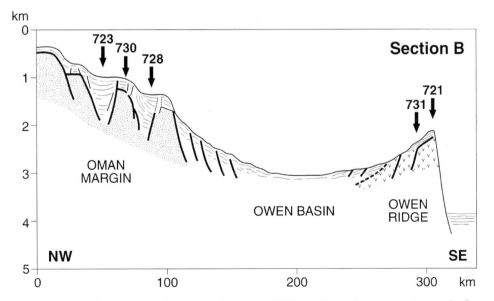

Fig. 5. Line drawing interpretation of seismic profiles from ODP Leg 117 showing a transect across the Oman margin and Owen Ridge. Locations of sites used for this study are shown. Location of section B is shown in Fig. 1.

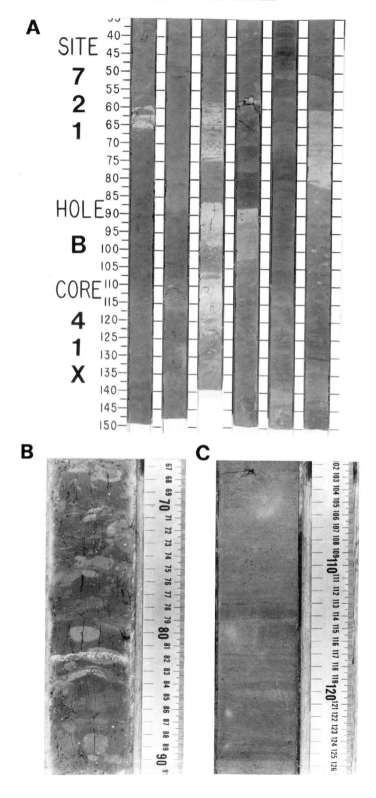

This study has utilized existing cores from Ocean Drilling Program (ODP) Leg 117, drilled along a cross-margin slope transect similar to that across the Makran margin (Prell *et al.* 1989). Five sites were selected for detailed study, including three sites (723, 728 and 730) from the Oman margin, and two sites (721 and 731) from the Owen Ridge (Figs 1 and 5). Selection was based on site distribution, good core recovery and on the occurrence of a range of different facies. Core sections from each of these sites have been redescribed by one of the authors (A. R. Tabrez) at the ODP core repository, College Station, Texas, and *c.* 500 samples were taken for subsequent analysis. Smear slides were also made where the required amount of sample was not available. The age of the succession has been determined by detailed micropalaeontological studies as early Miocene to Recent (Prell *et al.* 1989).

Sediment facies and structures

Hemipelagites and pelagites are the dominant facies on this margin, generally occurring as alternating layers of light, carbonate-rich intervals and darker, clay- and organic-rich intervals (Fig. 6). The lithological units described by Prell *et al.* (1989) are calcareous ooze and clayey silts, which were deposited under the influence of monsoonal-induced upwelling and the related oxygen minimum zone. Each of the selected sites shows a particular variation in lithology that reflects the different settings as well as oceanographic conditions. The upwelling conditions have led to a rather more varied suite of hemipelagite facies and structures than observed on the Makran margin. These fall into three main facies groups: (1) laminated oozes–muddy oozes with faint to well-defined parallel lamination (millimetre to centimetre scale) and minor, small-scale bioturbation (Fig. 6a); (2) bioturbated oozes–muddy oozes with moderate to intense bioturbation; distinct burrow types including *Chondrites*, *Planolites*, *Zoophycos* and *Trichichnus* can be recognized together with a well-developed three-tier ichnofossil distribution in some cases (Fig. 6c); (3) chaotic oozes–muddy oozes in which the original laminated or bioturbated structure has been modified by slumping (Fig. 6b).

There is a complete gradation from intensely bioturbated to well-laminated facies as well as from pure calcareous ooze (pelagite) to muddy ooze (hemipelagite). The laminated facies are less widespread than the bioturbated facies, and are best interpreted as the result of partial to complete suppression of a burrowing macrobenthos under low oxygen or anoxic bottom waters and a stratified water column. Both facies groups may be resedimented downslope via slumps and slides or turbidity currents.

Grain-size characteristics

Grain-size analysis of selected samples from different facies of the upper basin, lower basin and inter-basinal sites has been carried out. Each sample has been analysed twice to investigate different grain-size components; first, bulk samples were analysed to determine the overall grain-size distribution; second, the same samples were analysed after the removal of carbonate. In general, median values for bulk samples range between 5 and 31 μm, and mean values of the same samples are in the range of 13–47 μm. Sorting is moderate to poor. Removal of carbonate has different effects on different samples in that the median or mean size may be significantly decreased, remain about the same, or be increased. This clearly depends on the composition of the original sample: where nannofossils are the dominant carbonate component then dissolution tends to increase the mean size, whereas where foraminifers and/or coarse detrital carbonate are significant then the mean size may decrease on dissolution.

The different facies groups tend to show characteristic size distribution patterns as follows (Fig. 7):

(1) Unimodal distribution with long fine tail: very fine-grained input of nannofossils and fine clay, the latter probably from aeolian–fluvial suspension; typical of the laminated facies group.

(2) Unimodal or bimodal distribution or no distinct mode with broad spread of sizes, poorly sorted: input of nannofossils together with terrigenous material of probable aeolian and fluvial derivation; typical of the bioturbated facies group.

(3) Polymodal distribution, more or less distinctly peaked: input of nannofossils, mixed biogenic and terrigenous fraction with some foraminifera, probably reflecting greater influence of direct fluvial or reworked shelf input at lowered sea level; typical of slump facies group.

Fig. 6. Photographs of Oman margin cores showing typical hemipelagite facies. (**a**) Owen Ridge Site 721B, Core 41X, parts of Sections 1–6, depth scale in centimetres. (**b**) Detail of bioturbated ooze from Margin Site 728A, Core 29X, Section 1, scale in centimetres. (**c**) Detail of laminated ooze from Margin Site 723B, Core 4H, Section 2, scale in centimetres.

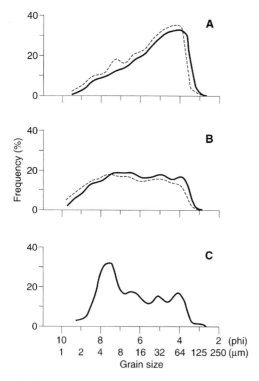

Fig. 7. Selected grain-size distribution smoothed frequency curves from the Oman margin hemipelagites. (**a**) Unimodal distribution typical of laminated facies; (**b**) irregular very poorly sorted distribution typical of bioturbated facies; (**c**) polymodal to irregular distribution typical of some bioturbated facies and of slump facies.

Composition

The principal components of the hemipelagic facies are shown in Table 2. There is a mixed terrigenous and biogenic composition, together with a relatively large broken calcite fraction (up to 20%), which may be fragmented biogenic material or terrigenous debris, or a mixture of both. Separation of the clay fraction and XRD

Table 2. *Composition of Oman margin hemipelagic sediments; per cent estimates from smear slides*

Quartz	10–15
Feldspar	2–5
Non–specific calcite	5–20
Dolomite	2–5
Opaques	<1
Silt–clay	20–30
Nannofossils	20–50
Foraminifera	<5–25
Diatoms	0–20

analysis shows a clay composition dominated by illite and chlorite, with variable amounts of palygorskite, smectite and mixed layer species.

Organic carbon contents (TOC; total organic carbon) vary widely both between sites on the margin and as a function of sedimentary facies. The highest TOC concentrations (*c.* 7%) occur in sporadic intervals at Site 723B, whereas the average organic carbon content for the laminated facies is 2–3% and for the bioturbated facies is <1%. The Owen Ridge sediments are moderately rich in organic matter, with organic carbon concentrations averaging 0.9–1.2% in the pelagic–hemipelagic sediments. The underlying turbidites typically contain <0.2% organic carbon.

$CaCO_3$ concentrations range from 36 to 94%. Comparatively higher percentages are noted in the distal slope basin. Most of the bioturbated facies show >70% $CaCO_3$, whereas the laminated and organic carbon rich facies show <70% $CaCO_3$.

Microfabric

Thirty samples were selected for microfabric studies using scanning electron microscopy (SEM) from representative facies at each core site. Some of the important features of sediment fabric are illustrated in Fig. 8. However, lack of resolution at these very fine grain sizes, even using high-power SEM backscatter images, precluded the obtaining of definitive results. Preliminary transmission electron microscopic studies were more promising, although this work is still in progress.

The following general observations and interpretations can be made:

(1) Randomly oriented particles in the laminated facies suggest deposition, in part, as flocs and at relatively high rates of sedimentation. Some intervals within the laminated facies show better orientation, suggesting periods of lower rates of sedimentation.

(2) Relatively more preferred orientation of fabric in the bioturbated facies suggests fewer flocs and lower rates of sedimentation. Areas with more random particle arrangements are interpreted to result from bioturbation.

(3) The preferred orientation observed in some slump horizons may result from particle reorientation by within-sediment slippage during downslope slide movement.

(4) In general, studies of the clay microfabric of these sediments are compatible with pelagic–hemipelagic settling of organic aggregates, clay flocs and silt particles (Bennett *et al.* 1991; O'Brien *et al.* 1980; Reynolds & Gorsline 1991).

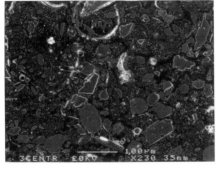

Fig. 8. Microfabric characteristics of Oman margin hemipelagites. Backscatter electron micrographs. Scale bars represent 10 or 100 μm as shown. (**a**) Minor bioturbation, weak parallel fabric; (**b**) parallel lamination and weak bioturbation, random fabric; (**c**) thorough bioturbation, random fabric.

Discussion

Although there are many studies of both modern and ancient deep-sea successions that refer to hemipelagites, often as the 'background' sediment within which the 'more interesting' sediments occur, there are relatively few detailed accounts of hemipelagic facies. Several of these are summarized below to illustrate the range that exists in both marginal and open ocean settings and everywhere from high latitudes to the equator. Some are dominated by terrigenous input whereas others, especially those associated with upwelling systems, have abundant biogenic material together with preserved organic matter.

Modern examples

Nova Scotia Continental Slope. (Stanley *et al.* 1972; Hill 1984)

Setting:	Passive margin, water depth 1000–5000 m, late Quaternary –Recent.
Structure:	Bioturbated hemipelagic mud.
Texture:	Poorly sorted coarse sandy mud and sandy–silty mud.
Composition:	Sand fraction: quartz, glauconite and minor heavy minerals; dominant clays, minor biogenics material.
Sedimentation rate:	Holocene (5 cm/ka), late glacial period (20 cm/ka).
Other features:	Interbedded with fine-grained turbidites and influenced by bottom curents between 4 and 5 km depth.

Canadian Polar Continental Margin. (Hein *et al.* 1990)

Setting:	Passive margin, water depth: 140–283 m, Tertiary–Recent.
Structure:	Bioturbated and burrowed hemipelagic mud.
Texture:	Pebbly sandy mud–clayey mud, poorly sorted.
Composition:	Gravel–sand fraction: quartz, feldspar, augite, clinopyroxene, rock fragments, basalt, granite, etc; dominant clays, minor biogenics material.
Other features:	Eight facies are recognized on this margin, of which only one is hemipelagic and the rest are generally of turbiditic origin. The hemipelagic facies contains ice-rafted meltout and fallout deposits.

Western Mediterranean Sea. (Rupke 1975)

Setting:	Marginal ocean basin, water depth 2588–2719 m, late Quaternary.
Structure:	Bioturbated–burrowed hemipelagic mud.
Texture:	Average 8% sand, very poorly sorted mud.
Composition:	Sand fraction: foraminifera and

pteropods; $CaCO_3$: 35–40%; clays abundant.

Sedimentation rate: 23 cm/ka.

Other features: This abyssal plain basin is bordered by narrow shelves and slopes. Hemipelagic mud occurs at the surface of a thin sediment fill, which is estimated to be 300–400 m thick and to comprise mainly turbidites.

Levantine Sea–Nile Cone. (Stanley & Maldonado 1979)

Setting: Marginal ocean basin, water depth >500 m, late Quaternary.

Structure: Bioturbated hemipelagic mud.

Texture: Sand fraction 5–10%, poorly sorted mud.

Composition: Sand fraction: pteropods, planktonic foraminifera; $CaCO_3$ 5–45%; clays abundant, organic carbon rich in parts.

Sedimentation rate: 2.9–62.7 cm/ka.

Other features: Cyclic nature of sediment sections (turbidite–hemipelagite cycles).

Mozambique Basin, Southwest Indian Ocean. (Kolla *et al.* 1980)

Setting: Abyssal plain, water depth 2000–5000 m, Quaternary–Holocene.

Structure: Structureless.

Texture: Fine sand–silty clay.

Composition: Mn nodules; sand fraction: foraminifera and diatoms; $CaCO_3$ 10–85%, clays 10–80%.

Sedimentation rate: <1 cm/ka during Quaternary, 1–2 cm/ka during Holocene.

Other features: Hemipelagites influenced by bottom curents in part.

Sierra Leone Rise. (Dean *et al.* 1981)

Setting: Passive margin, water depth 3200–3700 m, Tertiary.

Structure: Structureless.

Textures: Poorly sorted.

Composition: Chalk; marl; $CaCO_3$ *c.* 80%; clays minor.

Sedimentation rate: 5 cm/ka.

Other features: Chalk, marl and limestone show 20–60 cm thick cyclic alternations of clay-rich and clay-poor beds.

California Borderland Basins. (Hein 1985)

Setting: Slope-aprons of active margin strike-slip basins, water depth 180–1800 m, Pleistocene–Recent.

Structure: Bioturbated–burrowed; laminated where organic carbon rich.

Texture: Poorly sorted.

Composition: Clay minerals include smectite (>50%), vermiculite, illite, chlorite and kaolinite; organic C 0.5–5.3%; $CaCO_3$ 0.5–13%.

Sedimentation rate: Range from >20 cm/ka for inner basins to <8 cm/ka for outer basins.

Other features: Some of region under the influence of upwelling system; hemipelagites interbedded with various resedimented facies.

Okushiri Ridge, Northeast Japan Arc. (Tokuyama *et al.* 1992)

Setting: Back-arc basin, water depth 3500–3700 m, Miocene–Recent.

Structure: Structureless.

Texture: Poorly sorted.

Composition: Sand fraction: quartz, feldspar, volcanic grains, clays dominant, biogenic material common.

Sedimentation rate: 'High' rate of sedimentation.

Other features: Alternation of siliceous hemipelagites and turbidites; thick basin fill.

Japan Sea. (Tamaki *et al.* 1992)

Setting: Back-arc basin, water depth 3500–4000 m, Miocene Pliocene–Recent.

Structure: Bioturbated.

Texture: Poorly sorted.

Composition: Sand fraction: quartz, feldspar, glass, pyrite, inorganic calcite, biotite; organic carbon 0.02–7.45; $CaCO_3$ 1–74% (generally low); clays dominant.

Sedimentation rate: 4.9–5.6 cm/ka during Miocene; 7.7 cm/ka during Pliocene; 7.4 cm/ka during Quaternary.

Other features: Light–dark (organic-poor–organic-rich) cyclicity; interbedded turbidites in parts; thick basin fill.

Walvis Ridge, SE Atlantic. (Hay *et al.* 1984; Stow 1987)

Setting: Crest of aseismic ridge, ocean margin; water depth 365 m, Miocene–Recent.

Structure: Thoroughly bioturbated and burrowed.

Texture: Silty clay grade, with biogenic sand fraction (<15%).

Composition: Sand fraction; diatoms and foraminifera, minor terrigenous debris; principal components, calcareous and siliceous biogenic material; proportions of clays vary through time and cyclically; organic carbon 0.5–8%.

Sedimentation rate: 4–8 cm/ka.

Other features: This region lies within the influence of the Benguela upwelling system off Namibia, which results in high primary productivity and high rates of sedimentation rich in biogenic silica and organic carbon. There is a clear cyclic bedding (0.4–1.0 m thick beds) of organic-rich–organic-poor hemipelagite, but bioturbation is thorough and no lamination is preserved.

Peru Margin. (Kemp 1990; Brodie & Kemp 1994)

Setting: Forearc basin, water depth 500–1000 m, late Quaternary.

Structure: Slightly bioturbated, well laminated.

Texture: Poorly sorted.

Composition: Sand fraction: quartz, feldspar, foraminifera, diatoms; organic carbon 3–12% (site 680) or 1–5% (Site 686); clays abundant.

Sedimentation rate: 6.7 cm/ka (Site 680), and 18.25 cm/ka (Site 686).

Other features: The waters of the Peruvian shelf are the site of one of the world's most persistent wind-driven coastal upwelling systems. The sustained high flux of organic matter beneath this highly productive region rapidly depletes oxygen and creates anoxic conditions both within the sediment and in the water column, resulting in extensive preservation of laminated sediments in the forearc basins of the shelf and upper slopes (Suess *et al.* 1988).

Weddell Sea, Abyssal Plain. (Pudsey *et al.* 1988; Pudsey 1992)

Setting: Abyssal plain–distal rise, water depth >450 m, Pleistocene–Recent.

Structure: Bioturbated, mottled and burrowed hemipelagic mud.

Texture: Mostly very fine-grained, poorly sorted, silty mud. Ice-rafted sand and pebbles locally.

Composition: Dominant terrigenous fraction: quartz, feldspar and clays; minor siliceous biogenic material; locally ice-rafted rock fragments.

Sedimentation rate: Generally low: <0.5 to >1.6 cm/ka.

Other features: Interbedded with fine-grained turbidites and ash-fall layers and influenced by bottom currents along northern margin of basin.

NW Hebridean Continental Margin. (Stoker 1990; Howe 1994; Howe *et al.* 1994)

Setting: Passive margin, slope and rise, water depth 500–1500 m, Pleistocene–Recent.

Structure: Bioturbated, mottled and burrowed hemipelagic mud.

Texture: Poorly sorted silty mud, locally with ice-rafted sand and pebbles.

Composition: Mixed terrigenous (60–90%) and biogenic (10–40%) fractions; quartz, clays and carbonate biogenic material dominant; ice-rafted rock fragments.

Sedimentation rate: Difficult to separate from other facies, probably <10 cm/ka.

Other features: Interbedded with turbidite and contourite facies.

West Iberian Margin. (Young 1995)

Setting: Passive margin, Lisbon and Setubal canyon fill, water depth 200–2500 m, Holocene.

Structure: Bioturbated and burrowed hemipelagic mud.

Texture: Poorly sorted mud, clay–silty-clay grade.

Composition: Mixed terrigenous (65–90%) and biogenic (10–35%) fractions; quartz, clays and calcareous biogenic material dominant.

Sedimentation rate: Extremely high: 58–838 cm/ka.

Other features: High sedimentation rates may be explained by sediment focusing in slope canyons. Area of moderate upwelling and organic carbon contents up to 2%.

Ancient examples

Still fewer detailed studies of ancient hemipelagites are reported in the literature. Hesse (1975) provided a careful account from the Upper Cretaceous flysch sequence in the eastern Alps, and showed how to distinguish between interbedded calcareous turbidites and carbonate-free hemipelagic mud deposited below the CCD. These hemipelagites are very different from those described from the Cretaceous Tertiary Scaglia Rossa formation of the central Apennines, in which the calciturbidites and pelagites are equally carbonate rich and difficult to distinguish, whereas the hemipelagites are more typical marls (Stow et al. 1984).

Limestone–marl cyclic sedimentation is commonly reported from ancient successions (see papers in Einsele et al. 1991; De Boer & Smith 1994). In some cases, the marls are best interpreted as hemipelagites, whereas in others they are more properly pelagites. In both instances, however, the original characteristics and composition have been extensively modified by subsequent diagenesis, although the primary control on cyclicity was climatic.

Those successions without significant cyclic bedding that have been interpreted as hemipelagic slope deposits include part of the Cretaceous Humps Island formation in the Antarctic (Pirrie 1989), and part of the Pissouri Marlstone member of the Nicosia Formation in Cyprus (Stow et al. 1994). In the latter case, the hemipelagite marlstones become interbedded upwards with volcaniclastic turbidites, the succession showing many features in common with the Mio-Pliocene Misaki Formation of south central Japan (Stow et al. 1997).

Hemipelagite definition and distinction from related facies

Various workers have proposed specific definitions in terms of composition and grain size.

Stow & Piper (1984) defined hemipelagites as those sediments with >10% biogenic and >10% terrigenous material, with over 40% of the terrigenous component being silt sized (4–63 μm). The American Geophysical Institute (AGI) definition states that hemipelagic deposits in the deep sea are sediments in which more than 25% of the fraction coarser than 5 μm is of terrigenous, volcanigenic, and/or neritic origin. Such deposits usually accumulate near the continental margin and its adjacent abyssal plain, so that continentally derived sediment is more abundant than in eupelagic deposits, and the sediment has undergone lateral transport. Berger (1974) proposed a very similar definition but also noted that the median grain size of the terrigenous component is >5 μm. The more recent review of deep-sea sediments and environments by Stow et al. (1996) does not give rigorous compositional or textural boundaries for hemipelagites, referring to them simply as 'fine-grained sediments with both biogenic and terrigenous components' and emphasizing the combination of vertical settling and lateral (hemipelagic) advection for their accumulation.

Clearly there is a general consensus on what the term means. However, as hemipelagites have been recovered from many different parts of the world's oceans, from the poles to the tropics and from the upper slope to abyssal plain, they display a wide range of characteristics. It is equally clear, therefore, that over-restrictive limits placed on their structural, textural and compositional attributes are meaningless. We therefore propose the following definition: 'Hemipelagites are fine-grained sediments typically occurring in marginal deep-water settings. They comprise an admixture of biogenic pelagic material (generally >10%) and terrigenous or volcanigenic material (>10%), in which a significant proportion (over c. 40%) of the terrigenous (volcanigenic) fraction is of silt size or greater (i.e. >4 μm), and the overall grain-size distribution is poorly sorted. They are deposited by a combination of vertical settling and slow lateral advection.'

They have certain similarities with other deep-water sediments with which they are commonly found in close association, including: (1) pelagites, both biogenic oozes and red clays; (2) fine-grained turbidites and other mud-rich mass-flow deposits; (3) muddy contourites; (4) hemi-turbidites. There is a complete gradation between pelagites and hemipelagites (Stow et al. 1996). True pelagic oozes have a high percentage of biogenic material, typically >70%, and a terrigenous component that is dominantly clay sized, without a significant silt or sand fraction. Abyssal red clays have low biogenic content (<30%) and a high clay content (>70%), also without a significant silt or sand fraction. Accumulation rates for pelagites are mostly <1 cm/ka or ≪1 cm/ka, whereas hemipelagic rates are >1 cm/ka and typically between 5 and 15 cm/ka.

Fine-grained turbidites are more readily distinguished from hemipelagites. They are mostly well bedded, clearly structured and/or graded, and moderately to well sorted. Where interbedded, the two facies generally have different compositions, and show a more or less gradational and bioturbated contact passing upwards from turbidite into hemipelagite. Certain other rapidly deposited mass-flow facies such as muddy debrites, disorganized fine-grained turbidites, fine-grained bioclastic turbidites and muddy megabeds (homogenites or unifites) (Stanley 1981; Stow 1985b), can show superficial similarities to hemipelagites. However, they are structureless rather than bioturbated and mostly show well-defined beds, even if these may be tens of metres thick.

At the other end of the spectrum, hemiturbidites are fine-grained sediments with partly turbiditic and partly hemipelagic characteristics (Stow & Wetzel 1990). They have indistinct bedding, poor to moderate sorting, very subtle grading, distinctive continuous bioturbation and a composition equivalent to that of the interbedded distal turbidites. They have only recently been recognized as the deposits that result from an essentially stationary suspension cloud formed by reversing buoyancy during the terminal stage of turbidity current flow (Sparks *et al.* 1993). Hemiturbidites are truly transitional to hemipelagites and, in many cases, cannot be readily distinguished.

Muddy contourites can also display very similar features to those of hemipelagites and we must recognize that the boundary between these facies is partly gradational. Both types of deposits are typically bioturbated, fine grained, poorly sorted and of mixed biogenic–terrigenous composition. Very subtle differences in sorting, grading and primary sedimentary structures may be observable on close inspection at the scale of the core or outcrop, whereas in many cases differentiation must rely on larger-scale factors such as depositional setting and geometry, known bottom-current patterns, hiatuses and sedimentation rates.

A summary of the principal differences between these various deep-water facies is given in Table 3.

Facies model

Based on our detailed studies of the Makran and Oman margin hemipelagites, coupled with an extensive review of previous work by various workers, we have established below the main features of a facies model for hemipelagites (Fig. 9). These include sedimentary characteristics (structure, fabric, texture and composition), the nature of bedding and cyclicity, the distribution and rates of sedimentation, and the processes of deposition.

Structures. Hemipelagites deposited in open-water oxygenated conditions are completely devoid of primary sedimentary structures, but are generally highly bioturbated. Ichnofacies are those for quiet, oxygenated, deep water, and typically include *Zoophycos*, *Planolites*, *Chondrites*, *Phycosiphon* and others. Trace fossil zonation (or tiering) is most evident in more rapidly deposited hemipelagites (Wetzel 1984), especially where they are interbedded with turbidites or other resedimented facies. Complete bioturbational mottling is more common in slowly accumulated deposits.

Where bottom waters are low in oxygen or completely anoxic, then weak to well-developed parallel lamination is preserved, with low to absent bioturbation (Kemp 1990; Brodie & Kemp 1994). The lamination is most commonly a 'fissile-lamination' (Stow 1987; Stow & Atkin 1987) in which the laminae have a subparallel wavy anastomosing pattern. Lamination tends to be better developed at relatively high rates of sedimentation, but shows no evidence of current influence. In passing from oxic to anoxic conditions the ichnofossils decrease in size, abundance and variety from a highly mixed assemblage through to no bioturbation (Stow 1987; Wignall 1994).

Fabric. The submicroscopic fabric of hemipelagites requires more detailed study. However, some results from this study together with those reported by O'Brien (1980) and Shepard & Rutledge (1991) suggest that open-water hemipelagites have a relatively close-packed fabric with subparallel alignment of clays, nannofossils and other platey components. Larger grains and bioturbation cause disturbance of this fabric. Under anoxic conditions, well-laminated hemipelagites show a still tighter, uniform, subparallel fabric (Kemp 1990).

Textures (Fig. 9). Grain size of hemipelagites is strongly influenced by the sediment composition

Table 3. *Diagnostic criteria for the recognition and distinction of fine-grained turbidites, muddy contourites, hemipelagites and pelagites; modified and expanded from Stow (1985b)*

	Turbidites (fine grained, thin bedded)	Turbidites–Unifites (fined grained, very thick bedded)	Hemiturbidites (Stow & Wetzel 1990)	Contourites (fine grained, depositional)	Hemipelagites	Pelagites (biogenic ooze)	Pelagites (abyssal red clay)
Bedding	Usually well defined, continuous, thin bedded, regular	Usually well defined, regular, extensive; beds >1 m, and may exceed 25 m thick	Poorly defined bedding	Poorly defined and irregular, may be absent; irregular, variation of very thin to very thick beds	Poorly defined or moderate, regular, but may be absent; beds when present may be even bedded and medium thick	Poorly defined to absent	Poorly defined to absent
Structures	*Lamination* lenticular and parallel, regular or indistinct; micro-cross-lamination, low-amplitude climbing ripples, fading ripples and convolute lamination common	*Structureless* or with very indistinct parallel lamination near base of bed	*Primary structures* absent, but may cap fine-grained laminated turbidite	*Lamination* in parts only, mainly irregular, wavy, indistinct or lenticular, cross-lamination only rarely present in silt or fine-sand layers; irregular mottling very common	*No primary structures,* but may develop fine fissile lamination where deposited under anoxic conditions	*No primary structures*	*No primary structure*
	Contacts usually sharp at bases and sharp or gradational at top of laminae; micro-scours, loading and injection structures	*Contacts* sharp at bases and sharp or gradational at top; little or no scouring evident	*Contacts* gradational and bioturbated	*Contacts* can be sharp or gradational at tops and bases of layers and laminae; often gradations between the two along same contact; often irregular, sometimes erosive	*Contacts* between beds always gradational and bioturbated	*Contacts* gradational and bioturbated	*Contacts* gradational and bioturbated
	Bioturbation episodic, concentrated near tops of beds, often small scale, sometimes absent, rarely destroys all primary structures	*Bioturbation* in upper part of beds	*Bioturbation* continuous throughout bed; may be distinctive monospecific ichnofacies	*Bioturbation* continuous and intensive, throughout sequence; several tiers of burrows, types vary according to contourite facies; can markedly alter or destroy primary structures	*Bioturbation* continuous and intensive throughout sequence; several tiers of burrows, uniform ichnofacies; may become homogenized	*Bioturbation* continuous and intensive as for hemipelagites	*Bioturbation* continuous; may be less evident than in oozes and hemipelagites; mottling and homogenization common
Textures	*Grain size* from fine to sand to clay grade	*Grain size* typically from silt to clay	*Grain size* very fine silt and clay	*Grain size* from sand to clay grade	*Grain size* from sand to clay grade. with >40% of terrigenous fraction being silt sized; locally coarser	*Grain size* from sand to clay grade	*Grain size,* clay and fine silt

Distribution and moderate to good sorting indicate current deposition; silt and mud laminae usually well separated; silts often positively skewed (fine tail)	*Sorting* moderate to poor	*Sorting* moderate to poor	*Sorting* usually poor to moderate, but *distribution* does indicate current deposition; silt and mud often irregularly mottled; silts often have low positive or negative skew (i.e. both coarse and fine tails)	*Sorting* usually poor to moderate, but *distribution* does indicate current deposition; silt and mud often have low positive or negative skew (i.e. both coarse and fine tails)	*Sorting* poor to moderate, no current indications	*Sorting*, poor to moderate
Grading positive, often in regular graded–laminated units	*Grading* absent or very slight positive grading	*Grading* generally absent	*Grading* irregular, both positive and negative sequences	*No true grading*, but irregular grain-size fluctuation may be present	*No grading*	*No grading*
Fabric / *Grain alignment* (silts) parallel to downslope currents	*Fabric* not studied; presumed similar to thin-bedded turbidites	*Fabric* not studied	*Grain alignment* (silts) may be parallel to alongslope currents; more often disturbed by bioturbation	*No grain alignment*	*No grain alignment*	*No grain alignment*
Mud fabric may show large particle clusters (flocs) with random orientation			*Mud fabric* may show small particle clusters, with horizontal orientation where not bioturbated	*Mud fabric* may show small particle clusters and isolated particles; bed parallel	*Fabric* with small clusters and isolated particles; bed parallel	*Fabric with small clusters and isolated particles, bed parallel*
Magnetic fabric (?) parallel to downslope currents			*Magnetic fabric* (?) parallel to alongslope currents	*No magnetic fabric*	*No magnetic fabric*	*Random magnetic fabric*
Composition / *Allochthonous* elements introduced into an area, so that turbidite composition often differs markedly from that of interbedded sediments	*Allochthonous* composition	Mainly *allochthonous* composition as for associated turbidites, plus admixture of hemipelagic input	*Uniform* composition at scale of drift or margin deposit; part may be far-travelled, but most derived locally from pelagic and turbiditic input and bottom-current resuspension	*Uniform* composition, local and far-travelled in surface currents, winds; varied inputs	*Uniform* composition derived from primary productivity in surface waters	*Uniform* composition
Nature can be terrigenous, biogenic, volcanigenic, or mixed, often containing shallow-water elements; may show compositional grading	Terrigenous, biogenic, volcanigenic or mixed; may show compositional grading		*Nature* usually a mixture of terrigenous and biogenic; can be >80% one or other, can also include volcanigenic debris; reworked biogenic material common, often as broken and iron-stained debris; Fe–Mn rich in parts	*Nature* usually a mixture of terrigenous and biogenic (dominantly pelagic) elements, can include volcanigenic debris or glacio-marine input	*Nature*, calcareous, siliceous or mixed; with rare volcanigenic, terrigenous and cosmogenic input; Fe–Mn nodules and crusts locally	*Terrigenous* and *volcaniclastic* clays and fine silts; wind-blown dust; cosmogenic input, Fe–microtektites, Fe–Mn nodules and crusts

continued

Table 3. *Continued*

	Turbidites (fine grained, thin bedded)	Turbidites–Unifites (fined grained, very thick bedded)	Hemiturbidites (Stow & Wtzel 1990)	Contourites (fine grained, depositional)	Hemipelagites	Pelagites (biogenic ooze)	Pelagites (abyssal red clay)
Distribution	Vertical sequence often a regular succession of positively graded beds, or graded–laminated units (2–20 cm thick); these can form part of thicker coarsening- or fining-upward sequences	Typically occur as isolated megabeds	Typically occur within distal turbidite basin-plain succession	*Vertical sequence* often an irregular succession of positively and/or negatively graded intervals (10–100 cm thick); larger-scale sequences not yet clearly defined	*Vertical sequence* absent or as regular cycles of more and less biogenic-rich composition	*Vertical sequence* absent or as regular cycles of more and less biogenic-rich composition	No *vertical sequence*
Sedimentation rates	*Horizontal trends* of sedimentary features (e.g. bed thickness, grain size, composition) along turbidity current pathways, i.e. downslope trends	*Horizontal trends* generally absent or very subtle	*Horizontal trends* absent	*Horizontal trends* of sedimentary features (e.g. grain size, composition) along bottom-current pathways (i.e. alongslope trends parallel to the margin or drift)	*Horizontal trends* not present or weakly developed over large area	*Horizontal trends* not present or weakly developed over large area	No *Horizontal trends*
	Current evidence (ripples, flute-casts, fabric) also shows downslope trends			*Current evidence* (ripples, fabric) where preserved, also shows alongslope trends	*No bottom-current evidence*	*No bottom-current evidence*	*No bottom-current evidence*
	Episodic turbidite sedimentation, background sedimentation continuous, hiatuses uncommon except when associated with coarser-grained turbidites	*Episodic*, typically less frequent than thin-bedded turbidites	*Episodic* events but with very long settling periods (e.g. 0.5–1 a)	*Semi-continuous* sedimentation, with irregularly spaced, often prolonged hiatuses when bottom currents particularly strong	*Continuous* sedimentation, no hiatus	*Continuous* sedimentation, but may be very reduced in places	*Continuous* sedimentation
	Rate very variable, <10 to 1000 cm/ka			*Rates* variable, low to moderate, <2 to 15 cm/ka	*Rates* relatively constant, commonly low, <10 cm/ka; may vary with carbonate cycles; locally may be moderate to very high (>100 cm/ka)	*Rates* very low, typically <1 cm/ka; rarely up to 5 or 10 cm/ka	*Rates* extremely low, ≪1 cm/ka; may be <0.1 cm/ka

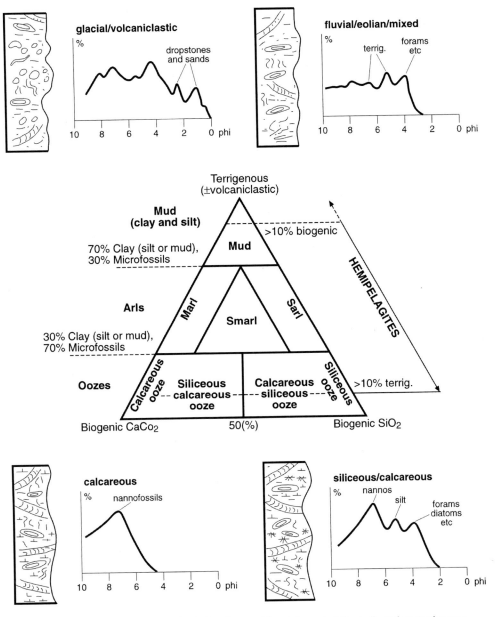

Fig. 9. Composite model for hemipelagite facies. Compositional characteristics indicated on terrigenous mud–calcareous ooze–siliceous ooze triangular plot. Most of the ooze, marl and mud fields can be hemipelagites. Typical grain-size characteristics shown for hemipelagites of different compositions and principal sediment sources. Schematic sediment logs illustrate minor grain-size variation related to components, intense bioturbation, and terrigenous or biogenic components.

and hence source. In general, however, they are fine grained with a mean size of 5–35 μm, poorly sorted with a broad spread of grain sizes, in some cases bimodal, trimodal or polymodal in distribution. Skewness values are slightly to markedly negative; kurtosis shows no standard pattern, although platykurtic distributions are more typical. The fine silt-sized mode (5–7 μm) is probably most common, and is the result of a dominant nannofossil contribution. Smaller peaks or modes at coarse silt sizes are dependent on specific terrigenous inputs of fluvial, aeolian, volcanic or glacial origin and therefore vary with the tectonic or oceanographic setting. Coarser grain sizes (sand and even gravel) may be introduced by floating ice or volcaniclastic activity. Other biogenic pelagic components (e.g. foraminifers, radiolarians, diatoms) can give minor grain-size peaks at coarse silt and fine sand grades.

Composition (Fig. 9). The composition of hemipelagites is very variable within the broad range allowed by definition, i.e. normally >10% biogenic and >10% terrigenous. The biogenic component is dominatd by open-ocean planktonic microfossils, either calcareous (nannofossils and foraminifers) or siliceous (radiolarians and diatoms). Minor amounts of other biogenic material (e.g. sponge spicules, silicoflagellates, dinoflagellates, benthic foraminifers, etc.) may also be present. We take >90% biogenic fraction as the absolute upper limit for hemipelagites, above which the sediment is a pelagite. Biogenic contourites and turbidites can also show >90% biogenic material. Many pelagic oozes in fact have >70% biogenic material, but the terrigenous fraction is dominantly clay rather than silt size.

The terrigenous fraction is dependent on the nature of the supply pathway and hence on the tectonic and geographic setting. Surface plumes of major rivers will carry clays and silts far into the ocean basins, the mineralogy being dependent on source area. Aeolian transport of silt and dust from deserts and arid lands can include chemogenic particles (e.g. palygorskite) as well as terrigenous grains (e.g. quartz, feldspar and clays). Ice-rafting at high latitudes is a major supplier of both fine and coarse terrigenous fraction, much of which may be compositionally immature (e.g. lithic grains). Along some active margins (island-arc settings) and around volcanic seamounts, chains or plateaux, a volcaniclastic component can dominate the terrigenous fraction. Hemipelagites that accumulate close to a dominant terrigenous source may have >90% terrrigenous material as, for example, in high-latitude glacio-marine hemipelagites, where there is negligible biogenic productivity because of sea ice.

Chemogenic components form authigenically in some slowly deposited hemipelagites, including phosphorites, glauconite and ferromanganese nodules or crusts. Iron monosulphides as precursors to pyrite are very common, especially in organic-rich hemipelagites. High total organic carbon content is characteristic of laminated hemipelagites deposited under anoxic conditions.

Cyclicity and bedding. Cyclic alteration of composition, colour, grain size or structures within hemipelagic sequences and cyclic interbedding with associated facies are both common. The periodicity of such cycles is typically on a Milankovitch time-scale (i.e. approximately 19–23, 41 or 100 ka cycles being most common). Where hemipelagites are interbedded with turbidites, then the turbidity current frequency will dominate the periodicity. Hemiturbidites, contourites and pelagites are other associated facies that occur in cyclic interbeds with hemipelagites.

Where cyclic changes are very marked then bedding may be distinct. More typically, however, the changes are subtle and gradual so that bedding is indistinct with bioturbated and gradational bed boundaries. Bed thickness varies from a few centimetres to a few metres, with 10–100 cm being most common.

Occurrence and rates of sedimentation. Hemipelagites occur in all tectonic and oceanographic settings, at all latitudes, and are most common within about 300 km of the coastline surrounding both continental landmasses and oceanic islands. Provided that there is adequate supply of silt-size terrigenous material, then hemipelagites may also occur far from land in open-ocean settings, but this is not normally the case. Water depths range from that of the shelf-break (typically 100–200 m) to oceanic depths in excess of 4–5 km. Outer shelf hemipelagites are here considered as a different facies that have a shallow-water signature and are affected by shelf processes. Clearly there will be a transition, however, between outer shelf and upper slope hemipelagites.

Rates of hemipelagic sedimentation typically range from <5 cm/ka (i.e. relatively slow) to >20 cm/ka (i.e. relatively rapid). In many cases, the sedimentation rate varies consistently with changes in the sedimentary characteristics outlined above. Hemipelagites formed in areas of high sedimentation rate, for example, tend to have a coarser mean grain size, a higher terrigenous/biogenic ratio and a better developed tiering structure of burrows than hemipelagites that have accumulated more slowly. Regions of upwelling and high biogenic productivity commonly show moderately high rates of sedimentation but with increased siliceous biogenic and organic carbon contents. Exceptionally high rates of sedimentation (e.g. >100 cm/ka) are associated with zones of hemipelagic sediment focusing, such as shelf-to-slope canyon systems.

Depositional processes (Fig. 10). There is a range of depositional processes responsible for deep-water fine-grained sedimentation, including turbidity currents and associated mass-flow events, bottom currents, hemipelagic advection and pelagic settling (e.g. Stow *et al.* 1996). Conceptually, these different processes are relatively well defined although the nature of hemipelagic sedimentation is particularly complex.

Turbidity currents and associated mass-flow processes are distinctly episodic, short duration, and infrequent events. They are bottom-hugging flows that mainly travel down-gradient on steeper slopes. Even the reverse buoyancy and upward mixing that produce hemiturbidites are episodic events, though their duration may be from 3 to 12 months (Stow & Wetzel 1990). Bottom currents also flow over the sea floor, but generally parallel to contours, and are semi-permanent (i.e. continuous through a measurable episode of geological time). They show significant variation in nature, intensity and location that contributes to their distinctive vertical sequences and other primary features. Pelagic sedimentation is the result of slow vertical settling under the influence of gravity in the absence of either turbidity currents or bottom currents.

Between vertical slow settling and low-density turbidity currents is a range of overlapping mechanisms that can be called hemipelagic processes (Fig. 10). The sedimentary materials involved are an admixture of terrigenous and primary biogenic components. The origin and deposition of primary biogenic material is governed by the same controls (productivity, dissolution and masking) as for pelagic sedimentation. The terrigenous input is from a combination of aeolian, fluvial, glacial, volcanic or coastal–shelf mechanisms, all of which serve to introduce suspended plumes and dispersed particles to the sea surface. In addition, the finest portions of low-concentration turbidity currents are in some cases stripped off at density discontinuities within the water column. Large eddies can become detached from bottom currents and further contribute their very dilute suspensions to the hemipelagic fallout. Internal tides and waves, interacting with the sea floor at the shelfbreak, within canyons (clear-water canyon currents) or on the upper slope, stir up fine-grained surficial sediments and mix them upwards into the overlying waters.

Some suspensions are dispersed across the sea surface by tides and other currents, whereas those with sufficient excess density will sink and move downslope as very dilute slow-moving density flows (or turbid layer flows). Where these encounter a density discontinuity they detach from the bed and flow out into the water column as an interflow. Such flows decelerate and material settles vertically to regroup at a deeper density interface and flow further basinwards before settling and regrouping once more. This process is known as suspension cascading. Less distinct and very dilute suspensions form thick mid- and bottom-water nepheloid layers

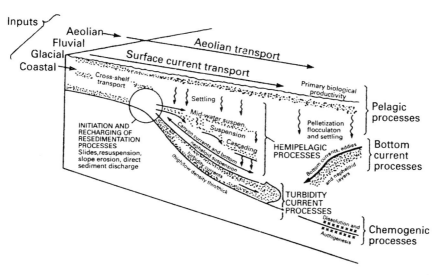

Fig. 10. Composite model for the deposition of fine-grained sediments in deep-water environments (modified from Stow 1985; Stow *et al.* 1996). The group of hemipelagic processes shown in the central part of the water column involve part vertical settling and part slow lateral advection. This distinguishes hemipelagic sedimentation from the other processes illustrated.

that may move along- and downslope by slow lateral advection. This combination of vertical settling and lateral advection has been documented by various workers (McCave 1972; Drake *et al.* 1978; Stow 1985b, 1994; Stow *et al.* 1996) and constitutes hemipelagic deposition.

Much of the work for this paper was carried out by A.R.T. during tenure of a Pakistan Government Scholarship at the University of Southampton. A.R.T. would also like to thank the National Institute of Oceanography (Pakistan) for encouraging his secondment, the Chief Scientist, officers and crew on the R.V. *Tyro*, G. van der Linden, J. Stel and G. Postma of Utrecht University, Netherlands, and the curators at the Ocean Drilling Program in College Station, Texas. We both thank the technical and secretarial staff at the Department of Geology, Southampton, as well as C. Pudsey and M. Akhurst for superb reviews of an earlier version of this paper.

References

BENNETT, R. H. *et al.* (eds) 1991 *Microstructures of Fine-grained Sediments, from Mud to Shale.* Springer-Verlag, New York.

BERGER, W. H. 1974. Deep-sea sedimentation. *In*: BURK, C. A. & DRAKE, C. L. (eds) *The Geology of Continental Margins.* Springer-Verlag, New York, 213–241.

BRODIE, I. & KEMP, A. E. S. 1994. Variation in biogenic and detrital fluxes and formation of laminae in late Quaternary sediments from the Peruvian coastal upwelling zone. *Marine Geology*, **116**, 385–398.

DEAN, W. E., GARDNER, J. V. & CEPEK, P. 1981. Tertiary carbonate-dissolution cycles on the Sierra Leone Rise, eastern equatorial Atlantic Ocean. *Marine Geology*, **1–2**, 81–101.

DE BOER, P. L. & SMITH, D. G. (eds) 1994. *Orbital Forcing and Cyclic Sequences.* Special Publications of the International Association of Sedimentologists, **19**, 559 pp.

DRAKE, D. R., HATCHER, P. G. & KELLER, G. H. 1978. Suspended particulate matter and mud deposition in Upper Hudson submarine canyon. *In*: STANLEY, D. J. & KELLING, G. (eds) *Sedimentation in Submarine Canyons, Fans and Trenches.* Dowden, Hutchison & Ross, Stroudsburg, PA, 33–41.

EINSELE, G., RICKEN, W. & SEILACHER, A. (eds) 1991. *Cycles and Events in Stratigraphy.* Springer-Verlag, Berlin, 955 pp.

HAY, W. W., SIBUET, J-C., *et al.* 1984. *Initial Reports of the Deep Sea Drilling Project Leg 75.* US Government Printing Office, Washington, DC, 1303 pp.

HEIN, F. J. 1985. Fine-grained slope and basin deposits, California continental borderland: facies, depositional mechanism and geotechnical properties. *Marine Geology*, **67**, 237–262.

——, VAN WAGONER, N. A. & MUDIE, P. J. 1990. Sedimentary facies and processes of deposition: Ice Island cores, Axel Heiberg Shelf, Canadian Polar Continental Margin. *Marine Geology*, **93**, 243–265.

HESSE, R. 1975. Turbiditic and non-turbiditic mudstone of Cretaceous flysch sections of the Western Alps and other basins. *Sedimentology*, **22**, 387–416.

HILL, P. R. 1984. Sedimentary facies of the Nova Scotia upper and middle continental slope, offshore eastern Canada. *Sedimentology*, **31**, 293–309.

HOWE, J. A. 1994. *Bottom currents, contourites and related sedimentation in the northern Rockall Trough, North Atlantic Ocean.* PhD Thesis, University of Southampton.

——, STOKER, M. S. & STOW, D. A. V. 1994. Late Cenozoic sediment drift complex, Northwest Rockall Trough, North Atlantic. *Paleoceanography*, **9**, 989–999.

KEMP, A. E. S. 1990. Sedimentary fabrics and variation in lamination style in Peru continental margin upwelling sediments. *In*: *Proceedings of the Ocean Drilling Program, Scientific Results* **112**. Ocean Drilling Program, College Station, TX, 43–58.

KOLLA, V., EITTREIM, S., SULLIVAN, L., KOSTECKI, J. A. & BURCKLE, L. H. 1980. Current-controlled abyssal microtopography and sedimentation in Mozambique basin, Southwest Indian Ocean. *Marine Geology*, **34**, 171–206.

McCAVE, I. N. 1972. Transport and escape of fine-grained sediment from shelf areas. *In*: SWIFT, D. J. P., DUANE, P. B. & PILKEY, O. H. (eds) *Shelf Sediment Transport: Process and Pattern.* Dowden, Hutchison & Ross, Stroudsburg, PA, 225–248.

MINSHULL, T. A. & WHITE, R. S. 1989. Sediment compaction and fluid migration in the Makran accretionary prism. *Journal of Geophysical Research*, **94**(B6), 7387–7402.

——, ——, BARTON, P. J. & COLLIER, J. S. 1992. Deformation of plate boundaries around the Gulf of Oman. *Marine Geology*, **104**, 265–277.

O'BRIEN, N. R. 1987. The effect of bioturbation on the fabric of shale. *Journal of Sedimentary Petrology*, **57**, 449–455.

——, NAKAZAWA, K. & TOKUHASHI, S. 1980. Use of clay fabric to distinguish turbiditic and hemipelagic siltstones and silt. *Sedimentology*, **27**, 47–61.

PIRRIE, D. 1989. *Sedimentology of the Marambio Group, Larsen Basin, Antarctica.* PhD Thesis, University of Nottingham.

PRELL, W. L., NIITSUMA, N. *et al.* 1989. *Proceedings of the Ocean Drilling Program, Initial Reports Leg 117.* Ocean Drilling Program, College Station, TX.

PUDSEY, C. J. 1992. Late Quaternary changes in Antarctic Bottom Water velocity inferred from sediment grain size in the northern Weddell Sea. *Marine Geology*, **107**, 9–33.

——, BARKER, P. F. & HAMILTON, N. 1988. Weddell Sea abyssal sediments: a record of Antarctic Bottom Water flow. *Marine Geology*, **81**, 289–314.

REYNOLDS, S. & GORSLINE, D. S. 1991. Silt microfabric of detrital, deep-sea mud(stone)s (California Continental Borderland) as shown by backscattered electron microscopy. *In*: BENNETT, R. H. *et al.* (eds) *Microstructures of Fine-grained Sediments, from Mud to Shale.* Springer-Verlag, New York, 203–212.

RUPKE, N. A. 1975. Deposition of fine grained sediments in the abyssal environment of the

Algero-Balearic Basin, West Mediterranean Sea. *Sedimentology*, **22**, 95–109.

SHEPHARD, L. E. & RUTLEDGE, A. K. 1991. Clay fabric of fine grained turbidite sequences from the southern Nares Abyssal plain. *In*: BENNETT, R. H. *et al.* (eds), *Microstructures of Fine-grained Sediments, from Mud to Shale.* Springer-Verlag, New York, 61–72.

SPARKS, R. S. J., BONNECAZE, R. T. *et al.* 1993. Sediment-laden gravity currents with reversing buoyancy. *Earth and Planetary Science Letters*, **114**, 243–257.

STANLEY, D. J. 1981. Unifites: structureless muds of gravity-flow origin in Mediterranean basins. *Geo-Marine Letters*, **1**, 77–83.

—— & MALDONADO, A. 1979. Levantine Sea – Nile Cone lithostratigraphic evolution: quantitative analyses and correlation with paleoclimate and enstatic oscillations in the late Quaternary. *Sedimentary Geology*, **23**, 37–65.

——, SWIFT, D. J. P., SILVERBERG, N., JAMES, N. P. & SUTTON, R. C. 1972. *Late Quaternary progradation and sand spillover on the outer continental margin off Nova Scotia, Southeast Canada. Smithsonian Contributions to the Earth Sciences* **8**.

STOKER, M. S. 1990. Glacially influenced sedimentation on the Hebridean slope, NW U.K. continental margin. *In*: DOWDESWELL, J. A. & SCOURSE, J. D. (eds) *Glacimarine Environments: Processes and Sediments.* Geological Society, London, Special Publications, **53**, 349–362.

STOW, D. A. V. 1982. Bottom currents and contourites in the North Atlantic. *Bulletin de l'Institut de Géologie du Bassin d'Aquitaine*, **31**, 151–166.

—— 1985a. Deep-sea clastics: where are we and where are we going? *In*: BRENCHLEY, P. J. & WILLIAMS, B. P. J. (eds) *Sedimentology: Recent Developments and Applied Aspects* Geological Society, London, Special Publications, **18**, 67–93.

—— 1985b. Fine-grained sediments in deep water: an overview of processes and facies models. *Geo-Marine Letters*, **5**, 17–23.

—— 1986. Deep clastic seas. *In*: READING, H. G. (ed.) *Sedimentary Environments and Facies.* Blackwell Scientific Publications, Oxford, 399–444.

—— 1987. South Atlantic organic-rich sediments: facies, processes and environments of deposition. *In*: BROOKS, J. & FLEET, A. J. (eds) *Marine Petroleum Source Rocks.* Geological Society, London, Special Publications, **26**, 287–299.

—— 1994. Deep sea processes of sediment transport and deposition. *In*: PYE, K. (ed.) *Sediment Transport and Depositional Processes.* Blackwell Scientific Publications, Oxford, 257–291.

—— & ATKIN, B. 1987. Sediment facies and geochemistry of Upper Jurassic mudrocks in the central North Sea area. *In*: BROOKS, J. & GLENNIE, K. W. (eds) *Petroleum Geology of NW Europe.* Graham & Trotman, London, 797–808.

—— & PIPER, D. J. W. 1984. Deep-water fine-grained sediments: facies models. *In*: STOW, D. A. V. & PIPER, D. J. W (eds) *Fine-grained Sediments: Deep-water Processes and Facies.* Geological Society, London, Special Publications, **15**, 611–646.

—— & WETZEL, A. 1990. Hemiturbidite: a new type of deep-water sediment. *In*: COCHRAN, J. R., STOW, D. A. V. *et al.* (eds) *Proceedings of the Ocean Drilling Program, Scientific Results*, **116**. Ocean Drilling Program, College Station, TX, 25–34.

——, BRAAKENBURG, N. E. & XENOPHONTOS, C. 1994. The Pissouri fan delta complex, SW Cyprus. *Sedimentary Geology*, **98**, 245–262.

——, READING, H. G. & COLLINSON, J. C. 1996. Deep seas. *In*: READING, H. G. (ed.) *Sedimentary Environments and Facies.* Blackwell Scientific Publications, Oxford, 395–453.

——, TAIRA, A., OGAWA, Y., SOH, W., TANIGUCHI, H. & PICKERING, K. T. 1997. Volcaniclastic sediments, process interaction and depositional setting of the Miocene-Pliocene Miura Group, SE Japan. *Sedimentary Geology*, (in press).

——, WEZEL, F. C., SAVELLI, D., RAINEY, S. C. R. & ANGELL, G. 1984. Depositional model for calcilutites: Scaglia Rossa Limestones, Umbro-Marchean Apennines. *In*: STOW, D. A. V. & PIPER, D. J. W. (eds) *Fine-grained Sediments: Deep-water Processes and Facies.* Geological Society, London, Special Publications, **15**, 223–241.

SUESS, E., VON HUENE, R. *et al.* (eds) 1988. *Proceedings of the Ocean Drilling Program, Initial Report*, **112**. Ocean Drilling Program, College Station, TX, 1015 pp.

TABREZ, A. R. 1995. *Slope sedimentation around the northwest Indian Ocean*. PhD Thesis, University of Southampton.

TAMAKI, K., SUYEHIRO, K. *et al.* (eds) 1992. *Proceedings of the Ocean Drilling Program, Scientific Results*, **127/128**, *Pt. 2*. Ocean Drilling Program, College Station, TX.

TOKUYAMA, H., KURAMOTO, S., SOH, W., MIYASHITA, S., BYRNE, T. & TANAKA, T. 1992. Initiation of ophiolite emplacement – a modern example from Okushiri Ridge, northeast Japan Arc. *Marine Geology*, **103**, 323–334.

TUCKER, M. 1989. *Techniques in Sedimentology.* Blackwell Scientific Publications, Oxford.

WETZEL, A. 1984. Bioturbation in deep-sea fine-grained sediments: influence of sediment texture, turbidite frequency and rate of environmental change. *In*: STOW, D. A. V. & PIPER, D. J. W. (eds) *Fine-grained Sediments: Deep-Water Processes and Facies.* Geological Society, London, Special Publications, **15**, 595–608.

WHITE, R. S. 1982. Deformation of the Makran accretionary sediment prism in the Gulf of Oman (north-west Indian Ocean). *In*: LEGGETT, J. K. (ed.) *Trench–Forearc Geology: Sedimentation and Tectonics on Modern and Ancient Active Plate Margins.* Geological Society, London, Special Publications, **10**, 357–372.

—— & LOUDEN, K. F. 1982. The Makran continental margin: structure of a thickly-sedimented convergent plate boundary. *American Association of Petroleum Geologists Bulletin*, **34**, 499–518.

WIGNALL, P. B. 1994. *Black Shales*, Clarendon Press, Oxford.

YOUNG, M. 1995. *The foraminiferal and sedimentological dynamics of a Portuguese submarine canyon system*. PhD Thesis, University of Southampton.

Late Glacial to Recent accumulation fluxes of sediments at the shelf edge and slope of NW Europe, 48–50°N

I. R. HALL & I. N. McCAVE

Department of Earth Sciences, University of Cambridge, Downing Street, Cambridge, CB2 3EQ, UK (e-mail: ih10006@esc.cam.ac.uk)

Abstract: Data were collected on magnetic susceptibility, volcanic tephra abundance, coarse silt mean grain-size, and total and fine inorganic carbon for a suite of cores collected from the NW European Margin between 48 and 50°N. Age models were constructed based on accelerator mass spectrometry (AMS) ^{14}C dating and stratigraphy based on δ^{18}O of foraminiferal carbonate together with correlatable instantaneous marker horizons of the Vedde Ash and the 'Heinrich' layers of ice-rafted detritus. Glacial vertical flux over the transect (2200–4500 m) on the open slope of Goban Spur yields an approximately 15 cm ka^{-1} deposition rate. This decreases in the Holocene to a present deposition rate of between 3 and 6 cm ka^{-1}. Deposition rates within the Porcupine Seabight are in close agreement with those seen at a similar depth on the open slope, whereas enhanced resuspension and sediment focusing leads to an approximate doubling in the depositional rate of hemipelagic sediment found near the base of King Arthur Canyon. Carbonate burial flux shows a maximum in the early part of isotope stage 2, during the deglaciation (17.5 ka to 12 ka) and in the late Holocene. This is seen at depths shallow enough to suggest it is not due to dissolution. It therefore could reflect enhanced productivity. Changes in current speed inferred from the sortable silt component of the sediment suggest slow currents during isotope stage 2 increasing irregularly through to a maximum just before the Younger Dryas, decreasing sharply in the early Holocene, then increasing again in the late Holocene. Rates of deposition at OM-2K follow this pattern fairly well for much of the time, suggesting lateral flux is important. However, during isotope stage 2 there is high flux of sediment with low current speed, suggesting supply from ice-rafting or directly downslope.

Coastal seas with enhanced productivity and a strong influence of continental inputs are an important source of dissolved and particulate matter exported to the open ocean. Understanding the processes and modelling shelf-sea and coastal responses to changes in climate, sea level and human activities is of the highest importance (Mantoura *et al.* 1991). Hence, it is necessary to quantify fluxes across the shelf–ocean margin, which is fundamental to evaluating budgets of carbon, nutrients and trace elements between the continents, coastal zone and ocean. Accordingly, overall objectives of the Ocean Margin EXchange (OMEX) project of the European Union have been to measure and to model exchange processes at contrasted shelf–ocean margins as a basis for the development of global models to predict the impact of environmental changes on the oceanic system, and more specifically on the coastal zone.

The first site chosen by the OMEX programme for study was Goban Spur, on the western edge of the Celtic Sea. It has a broad low-relief topography and is 'downstream' of the strong on- and off-shelf Celtic Sea currents. A shelf-break front showing a clear transition in

properties (salinity, temperature and transmission) was found during spring and autumn, with high values of new production at the shelf break as a result of vertical mixing of nitrate-rich deep water with surface water (Joint & Rees 1996). Microzooplankton and mesozooplankton are dominant; vertical transport of organic carbon is limited, as determined by sediment trap measurements and estimation of sedimentation rates and organic carbon burial, and estimations are complicated by lateral flux of different carbonaceous material resuspended from the slope (strong along-slope currents were found). Deposited organic matter was utilized rapidly by benthic organisms.

Contemporaneous mass flux studies provide a snapshot of a variable system. The long-term output sink for sediment, and organic and inorganic carbon is the sea bed of the margin and the nearby slope. It is important to know whether present conditions are in some way anomalous with respect to the recent history of the area. How do modern conditions sit in the spectrum of fluxes that have occurred in times of radical climatic change from the deglaciation through the Holocene Climatic Optimum of about 6000 a ago to the present? What is the dominant control

HALL, I. R. & McCAVE, I. N. 1998. Late Glacial to Recent accumulation of sediments at the shelf edge and slope of NW Europe, 48–50°N. *In*: STOKER, M. S., EVANS, D. & CRAMP, A. (eds) *Geological Processes on Continental Margins: Sedimentation, Mass-Wasting and Stability*. Geological Society, London, Special Publications, **129**, pp. 339–350.

on accumulation on this slope; along-slope advection or vertical flux and local downslope transport? We have attempted to determine this via analysis of core samples and dating them by stratigraphic methods. Here we report on the age models, accumulation fluxes and sedimentation mechanisms of sediments found within the OMEX core suite.

Sampling and methods

The OMEX coring was carried out on RRS *Charles Darwin* cruises 84 and 94 (Statham 1994, 1995), along a transect across Goban Spur from *c.* 11°W to 14°W, and from within the Porcupine Seabight at 50°N (the OMEX Region). Five Kasten cores were recovered using either a 3 m or 4 m long, 0.15 m square corer barrel (Zangger & McCave 1990). Core locations are shown in Fig.1 and sampling data in Table 1. In addition, a (15 m) giant piston core recovered from the base of King Arthur Canyon was kindly donated by C. Pudsey (British Antarctic Survey).

Kasten cores were logged descriptively before sub-sampling for X-ray slabs, sub-cores and syringe samples. X-ray slabs were imaged using a Faxitron 8050-010 X-radiography set. Water contents (WC) were determined on syringe

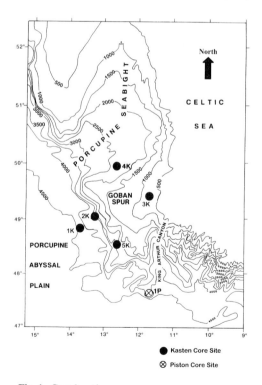

Fig. 1. Core locations.

samples taken at 4 cm intervals, and salt-corrected dry bulk density of sediment calculated assuming a particle density of 2.65 g cm^{-3}. Samples were wet sieved at 63 µm, dried and weighed to determine the percentage of coarse and fine fractions. Variation in magnetic susceptibility in deep-sea sediments reflects changes in lithology, such as the ratio of biogenic to lithogenic components, and therefore provides a rough guide to concentrations of terrigenous material (Robinson & McCave 1994). Bulk magnetic susceptibility (BMS) measurements also records variation in WC, with peaks in BMS coincident with WC troughs. Certain BMS and WC events are valuable for correlation between cores. Measurements of BMS were made every 2 cm downcore using a Bartington Instrument MS2 meter with a 'probe' type sensor held against sediment slabs. Magnetic susceptibility of the coarse and fine fraction was also measured on dried samples using a Bartington BS2B sensor with a 36 mm internal diameter cup.

Accelerator mass spectrometry (AMS) ^{14}C dating was carried out on monospecific samples of the planktonic foraminifera *Neoglobigerina pachyderma* (s) picked from the 250–300 µm fraction. Oxygen and carbon isotopic measurements were also made on two species of planktonic foram, *Globigerina bulloides* and *Neoglobigerina pachyderma*, in core OM-1K and a single benthic species, *Uvigerina* sp., in OM-2K. Determinations were made using a VG Sira Series II and Prism mass spectrometers (Shackleton & Hall 1989). The abundance of clear (rhyolitic) and brown (basaltic) volcanic ash shards, and lithic grains was determined microscopically on the coarse fraction (>63 µm), to determine the position and mixing of the Vedde Ash.

Total, organic and inorganic carbon was determined by instrumental CHN analysis (Carlo Erba EA1106 elemental analyser). Analyses were made on both the bulk and fine fraction, allowing the carbonate and non-carbonate content to be calculated as a percentage of the total sediment. Calibration was made using appropriate standards and analytical error was typically <2% of the determined value. Grain-size distributions of the fine (<63 µm) terrigenous and carbonate fractions were determined using a Micromeritics SediGraph 5000ET particle size analyser with a computer interface devised by Jones *et al.* (1988).

Results

Age model

The age model for the OMEX Kasten cores is based upon limited AMS ^{14}C dates from OM-2K

Table 1. *Kasten core sampling data*

Cruise	Core OM-	Date	Type	Core length (cm)	Latitude (N)	Longitude (W)	Water depth (m)
CD 84	1K	23/1/94	4 m Kasten	337	48° 58.60'	13° 39.83'	4494
CD 84	2K	24/1/94	4 m Kasten	254	49° 5.29'	13° 25.90'	3658
CD 84	3K	31/1/94	3 m Kasten	265	49° 22.70'	11° 11.62'	806
CD 94	4K	7/6/95	4 m Kasten	371	49° 59.25'	12° 31.06'	2280
CD 94	5K	12/6/95	4 m Kasten	391	48° 50.07	12° 39.95'	2333
BAS	1P	9/93	15 m Piston	1532	47° 58.80'	11° 48.50'	3652

and -4K, features in the oxygen isotopic profiles, together with distinct correlatable horizons such as the Vedde Ash and deposits of the latest two ice-rafting pulses known as 'Heinrich layers' (Heinrich 1988; Broecker *et al.* 1992).

The most prominent features of BMS and WC profiles in the continental margin setting of the OMEX study area are turbidites and pulses of ice-rafted debris, which both deliver a large quantity of lithic material of diverse mineralogy to the sea bed in a short time.

Turbidites commonly show up in X-radiographs as slightly X-ray opaque bands with sharp bases and sharp tops, and often displaying internal stratification patterns preserved (e.g. Fig. 2a). In contrast, Heinrich layers typically form dark, X-ray opaque bands, with sharp bases, gradational tops and abundant dropstones (e.g. Fig. 2b).

Analysis of X-radiographs in conjunction with BMS and water content profiles indicate the presence of Heinrich layer one (H1) and two (H2) in each of the OMEX Kasten and piston cores, except OM-3K, the shallowest core. In core OM-1K three large BMS peaks occur between H1 and present. X-Radiographs indicate that these peaks correspond to turbidites. No turbidites are evident in any other of the Kasten cores. In the case of OM-1K the turbidites have been spliced from the record in calculation of fluxes.

AMS radiocarbon dating on hand-picked foram shells from ice-rafted debris (IRD) peaks (Broecker *et al.* 1992) indicates that Heinrich layer 1 is dated at *c.* 15 ka and Heinrich layer 2 at *c.* 20 ka. Grousset *et al.* (1993) provided evidence that there is a distinct temporal

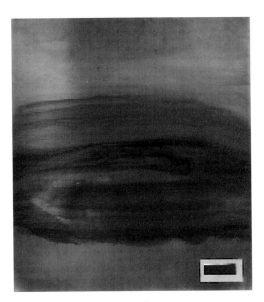

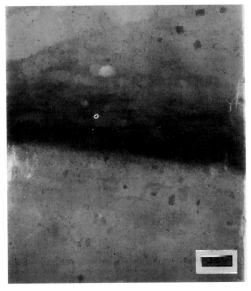

Fig. 2. Photographic prints of X-radiographs of core OMEX-1K. (**a**) Turbidite showing climbing ripple structure. (**b**) Heinrich layer 1 with preserved fine burrow structures beneath. Scale bar represents 2.0 cm.

correlation in these layers between the basins of the North Atlantic. Recent AMS ^{14}C dating of core material from the Biological Ocean Flux Study (BOFS) suite (60°N, 20°W) (Manighetti *et al.* 1995) placed calendar ages of 17,497 and 25,159 a on Heinrich layers 1 and 2, respectively. It is probable that deposition over the NE Atlantic during these ice-rafted events was essentially synchronous, thus the Heinrich layers are considered a reliable basis for true time correlation.

An instantaneous marker horizon, described by Ruddiman & Glover (1972) as North Atlantic ash zone 1, and correlated with the Vedde Ash in western Norway by Mangerud *et al.* (1984), is present in each of the OMEX cores. As a result of investigation on ash layers present in the central Greenland ice core GRIP, the ^{14}C age has recently been revised from the earlier value of 10,600 a (Mangerud *et al.* 1984) to *c.* 10,300 a (Grönveld *et al.* 1995), which corresponds to a true calibrated age of 11,980 a determined by counting ice layers. Figure 3 shows the abundance of clear ash shards >63 μm g[-1] total sediment with depth in core OM-2K. The peak ash concentration of *c.* 9000 shards g[-1] occurs at 52 cm and is similar to concentrations seen in the other cores. Brown as well as clear glass shards are present in the samples, although the clear

glass is used in preference to total glass. The shape of the ash abundance curve generally resembles the theoretical effect of bioturbation on an instantaneously deposited layer. In the ideal case, homogeneous bioturbation displaces such a signal to *x* cm below its original position, where *x* is the biological mixing depth (Guinasso & Schink 1975; Manighetti *et al.* 1995). The ash abundance curve suggests a homogeneous mixing depth of *c.* 3.5 cm. The base of the ash abundance curve is not as sharp as in the simple bioturbation model, with relatively high ash abundances occurring at least 4 cm below the peak. This suggests that the base of the signal may have been smeared out by heterogeneous mixing. Excess ^{210}Pb measurements in a box core from this station suggest a modern mixed layer depth of 3.3 cm (Hall & McCave 1996). Thus, it appears that homogeneous mixing has displaced the ash peak downwards by *c.* 4 cm, with the base smoothed by heterogeneous bioturbation for a further 2–4 cm. As a result, the 11,980 cal a BP horizon in OM-2K is placed at 48 cm, i.e. 4 cm above the glass shard peak. The relative position of the Vedde Ash marker horizon has been adjusted in a similar way in each of the OMEX cores.

AMS dates have been generated for a total of four samples from OM-2K and OM-4K (Table 2). They provide additional chronological tie points for these cores within the Holocene and between the Vedde Ash and H1.

Discussion

Sediment and mass accumulation rates

Sediment accumulation in the oceans is controlled by three main factors: the input of biogenic and terrigenous constituents; sediment redistribution under gravity and as a result of bottom-current activity, and the dissolution of soluble components.

Average sedimentation rates (cm ka[-1]) between age control intervals from H2 to the present in the OMEX core suite are presented in Table 2. Sedimentation rates over the depth transect on Goban Spur (cores OM-1K, -2K, -3K and -5K) generally decrease with depth. Average sedimentation rates between the Vedde Ash and the present decrease from 6.6 cm ka[-1] at *c.* 800 m (OM-3K) to 3.6 cm ka[-1] below depths of *c.* 3650 m, reflecting topographic variation and distance from the major sediment source. Values are comparable with those from the higher sedimentation sites in the open North Atlantic found during BOFS (Manighetti & McCave 1995*a*) and the rates compiled for the

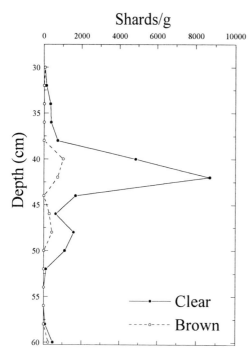

Fig. 3. Number of volcanic shards >63 μm g[-1] of total sediment with depth in core OM-2K.

Table 2. *Average sedimentation rates in cm ka⁻¹ in OMEX core suite*

Interval	Horizon (cm)	Age (cal. a BP)	Sediment accumulation rate (cm ka^{-1})
OM-3K: depth 806 m			
Present*–Vedde Ash	+2.5–76	0–11,980	6.6
OM-5K: depth 2333 m			
Present*–Vedde Ash	+2.5–64	0–11,980	5.6
Vedde Ash–H1	64–152	11,980–17,500	15.9
H1–H2	152–332	17,500–25,150	23.5
OM-2K: depth 3658 m			
Present–7584	+2.5–36.5	0-7584	5.1
7584–Vedde	36.5–48	7584–11,980	2.6
Vedde–14,846	48–73	11,980–14,846	8.7
14,846–H1	73–82	14,846–17,500	3.4
H1–LGM	82–112	21,000–21,150	8.6
LGM–H2	112–178	21,000–21,150	15.9
OM-1K: depth 4494 m			
Present*–Vedde Ash	+2.5–42	0–11,980	3.7
Vedde Ash–H1	42–98	11,980–17,500	10.1
H1–LGM	98–132	17,500–21,000	9.7
LGM–H2	132–194	21,000–25,150	14.9
OM-4K Porcupine Seabight: depth 2280 m			
Present–5123	+2.5–36.5	0–5123	7.6
5123–Vedde	36.5–76	5123–11,980	5.8
Vedde–14462	76–115.5	11,980–14,462	15.9
14462–H1	115.5–136	14,462–17,500	6.8
H1–H2	136–308	17,500–25,150	22.5
OM-1P King Arthur Canyon: depth 3652 m			
Present*†–Vedde Ash	+2.5–70	0–11,980	6.1
Vedde Ash†–H1	70–159	11,980–17,500	16.1
H1–H2	159–339	17,500–25,150	23.5

* Present is at +2.5 cm core top; estimated surface loss from cores.
† Excluding turbidites.

North Atlantic by Balsam & McCoy (1987). This is surprising, as higher sedimentation rates were expected given the nature of the continental margin setting.

Glacial–interglacial differences in the sedimentation rates are evident in all cores with stratigraphic control below the Vedde Ash. Average sedimentation rates between H2 and the Vedde Ash are typically 4–5 times greater than the Holocene average, with the shallower cores also having proportionally higher rates than the deeper ones. This primarily reflects the increased input flux from terrigenous sources (ice rafting and the more direct influence of fluvial discharge) during low sea-level stand of the last glacial; a point to which we will return. Sedimentation rates seen in OM-4K at 2280 m within the Porcupine Seabight are in close agreement with those of OM-5K at a similar depth (2333 m) on the open slope of Goban Spur.

In comparison, hemipelagic sediment accumulation rates at OM-1P in the bottom of King Arthur Canyon are consistently about double the rates seen at OM-2K on Goban Spur at a near identical depth. Geologically, canyon systems have been important conduits for the dispersal of shelf sediments to the deep ocean (Carson *et al.* 1986). During low sea-level stands within the Pleistocene, canyon-fed transport was dominated by gravity-controlled mechanisms (e.g. turbidites, debris flows and slumps) which transported large volumes of sediment from the upper slope and shelf to abyssal depths (Kelling & Stanley 1976; Nelson 1976; Nelson *et al.* 1978). However, since the Holocene transgression and retreat of shorelines away from the canyon heads, sediment supply has decreased and this has often led to the sedimentary regime altering from dominantly gravity flows to hemipelagic deposition (Nelson 1976; Nelson *et al.* 1978). This alteration is well recorded in core OM-1P (Fig. 4) with >40 individual gravity flow events in the magnetic susceptibility record between the base (*c.* 15 m; unknown age) and Heinrich layer

Magnetic Susceptibility (si)

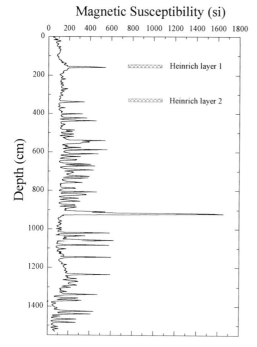

Fig. 4. Magnetic susceptibility with depth in core OM-1P.

1 (*c.* 1.5 m; 17,500 cal a BP), whereas within the Holocene there are no such mass-flow events present.

A number of studies (Drake & Gorsline 1973; Drake 1974; Baker 1976; Drake *et al.* 1978; Carpenter *et al.* 1982; Thornbjarnarson *et al.* 1986) have suggested that modern submarine canyons are areas where both suspended sediment concentrations and sediment accumulation rates are enhanced relative to those of the open slope. Certainly in this area the canyon received more material, both in the Holocene and late Pleistocene time. The modern high sedimentation rate is probably due to sediment focusing, i.e. resuspension of hemipelagic sediment from the sides of the canyon by internal waves and tides (Hotchkiss & Wunsch 1982; Hickey *et al.* 1986; Gardner 1989) and rain-out onto the canyon floor.

The mass accumulation rate (MAR), expressed as g cm^{-2} ka^{-1}, is a product of the sedimentation rate and the dry bulk density of the sediment, corrected for salt content. Sediment mass accumulation rates between the age control intervals from H2 to the present in the cores of the Goban Spur sediment trap transect (OM-1K, -2K, -3K and -5K) are presented in Fig. 5. Mass accumulation rates have decreased substantially between the last glacial and the

present and also generally show the depth-related succession as seen in the sedimentation rate.

Terrigenous and carbonate fluxes

Using the mass accumulation rates, the carbonate and non-carbonate contents of the coarse and fine fractions expressed as a percentage of the total sediment can be converted to flux. Total carbonate burial flux on the OMEX depth transect over Goban Spur are shown in Fig 5. The biogenic silica content of the OMEX cores is <1% and therefore we suggest that that the non-carbonate fraction is a good estimate of the 'terrigenous' flux. The total terrigenous flux therefore is represented by the difference between the mass accumulation curve and the carbonate accumulation curve as depicted in Fig. 5.

As the terrigenous sand fraction (>63 µm) is too coarse and dense to be influenced by moderate currents (less than *c.* 25 cm s^{-1}) the flux of non-carbonate sand (>63 µm) has been taken as a suitable indicator of the ice-rafted flux (e.g.

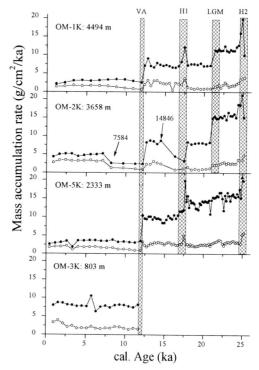

Fig. 5. Mass accumulation rates of total sediment (●) and total carbonate (○) in OMEX cores on Goban Spur transect over the past 25 ka. The difference between the curves represents the terrigenous flux.

Manighetti & McCave 1995*a*). Whereas this is the case in the open North Atlantic, on the continental margin setting reported in this study it probably reflects a combination of the IRD flux and the terrigenous flux supplied by the adjacent continent. Figure 6 shows the non-carbonate flux for core OM-2K. In the glacial, non-carbonate fluxes are generally high, between 1 and 1.5 g cm⁻² ka⁻¹, and show spikes associated with the 'Heinrich layers' recorded at *c.* 17.5 and 25.15 ka (Manighetti *et al.* 1995). Holocene values are low, around 0.3 g cm⁻² ka⁻¹. The distribution in time seen here is broadly similar to those reported for the BOFS cores (Manighetti & McCave 1995*a*), in particular from the southernmost station on the East Thulean Rise (BOFS-5K and -6K) only slightly to the north of the OMEX suite, but at *c.* 21°W. The magnitude of the OMEX fluxes are elevated slightly over the BOFS cores, especially at the Heinrich layers. These trends suggest that, during the glacial, Goban Spur and the East Thulean Rise responded similarly to ice-rafted material. The somewhat higher flux at the OMEX sites may reflect the more direct influence of icebergs from the north European ice sheet and of fluvial discharge issuing to the region at this time. This is probable given the estimates of the position of palaeo-coastlines and the presence of the 'Great

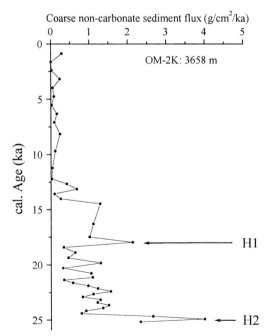

Fig. 6. Flux of coarse non-carbonate sediment (>63 µm), i.e. terrigenous sand, over the past 25 ka in OM-2K. H1 and H2 mark the position of Heinrich layers.

Channel River' discharging material across the present outer continental shelf during the sea-level minimum of the last glacial (Gibbard 1988; Lambeck 1995). The greatly reduced flux observed in the Holocene reflects both the lack of IRD and the retreat of the palaeo-coastline from the outer shelf, close to Goban Spur, towards its present position and the decrease of fluvial input to the channel by the flow of the Rhine and Thames into the present North Sea. The influence of material of British and Hercynian (Cornwall, Brittany) origin is suggested by Sr and Nd isotopic data from the nearby Armorican Seamount (le Noroit; 46.5°N, 12.5°W) to be greater than at sites further to the west (Revel *et al.* 1996).

Average late Holocene carbonate burial flux ranges between 1.6 and 2.6 g cm⁻² ka⁻¹ (16–26 g m⁻² a⁻¹) in cores on the open slope, showing generally decreasing fluxes with depth reflecting the dominance of the lower sediment accumulation rates at the deeper stations rather than the increasing carbonate percentage composition of the sediment. The highest burial flux is seen at the base of King Arthur Canyon, reflecting the topographic control of sediment accumulation at this station.

The average export carbonate flux recorded in the shallow sediment traps deployed over Goban Spur (OM-2K and -3K) between 1993 and 1995 was 9.7 g m⁻² a⁻¹ (A. Antia, IFM, Kiel, pers. comm.), which is considerably less than the Holocene average. As pointed out by Manighetti & McCave (1995*a*), there are some difficulties in comparing Holocene averages with short-term fluxes recorded in sediment traps. These arise because Holocene averages also include periods which may have been more favourable for productivity than the present, such as the Climatic Optimum *c.* 6000 BP. This is evident as an increase in the carbonate burial flux in some profiles. More recently, the Little Ice Age, which occurred between the fifteenth and nineteenth centuries AD, may have been marked by lower productivity and reduced carbonate accumulation, though no cores yet recovered have the requisite sedimentation rate to resolve this.

The coarse fraction consists primarily of foram shells, whereas the fine fraction is composed mainly of coccoliths and fragmented forams. In the shallow core (OM-3K; *c.* 800 m) aragonitic mollusc remains such as pteropod shells are present, making a contribution to the burial flux which is absent at greater depths. The burial flux of carbonate is dominated by the fine fraction, and even though total sediment accumulation rates vary two-fold, the magnitude

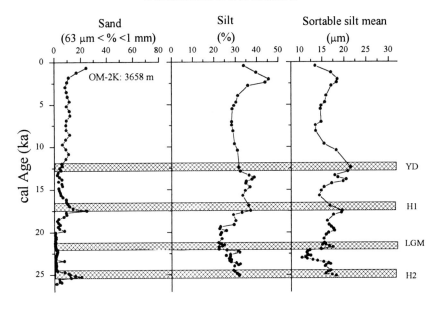

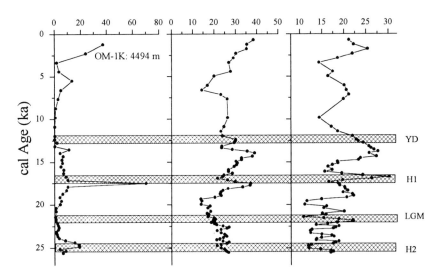

Fig. 7. Grain-size parameters for cores OM-2K and OM-1K over the last 25 ka.

of the coarse (>63 (m) carbonate flux changes very little between cores, with typically <2 % of the total carbonate flux being coarse.

The glacial period flux is in marked contrast to the late Holocene flux. Strong variability is evident, particularly in the deeper cores, with periods of enhanced burial occurring between *c.* 12 and 18 ka and periods of reduced burial between *c.* 18 and the Last Glacial Maximium (LGM). These may primarily be due to variations in surface productivity. The former may be a response to the Bölling–Allerød warming (e.g. OM-2K); alternatively, they may simply reflect dilution or periodic enhanced dissolution at the deeper depth. Evidence from other cores in the NE Atlantic, taken during the BOFS study, suggest that bottom waters may have been more corrosive during the glacial, up to at least 3500 m below present-day sea level (Manighetti 1993).

Peaks in the accumulation rates seen in each of the profiles, except the shallowest core (OM-3K), at 17.5 and 25.5 ka are unrelated to biogenic productivity. Instead, they reflect the influx of large quantities of ice-rafted debris including detrital limestone in Heinrich layers.

Grain-size and current flow speed

Temporal and spatial variation in grain-size may result from differences in input such as those associated with the presence of Heinrich layers or changes in the productivity of the overlying waters (Manighetti & McCave 1995a). McCave et al. (1995b) termed material in the 10–63 μm fraction as sortable silt because it is fine enough to be transported by currents of moderate strength, but is above the threshold of cohesive behaviour at the stresses required to move it and so is affected on a primary particle basis (see fig. 4 of McCave et al. 1995b). The mean particle size of this fraction reflects the prevailing current regime acting at the time of deposition; the mean particle size increases with stronger currents as the finer silts are winnowed away or prevented from being deposited.

In this margin setting, an important question is whether the mass flux is dominated by current-controlled lateral advection or by a vertical productivity–IRD-driven flux, and how the balance between these end-members changes between glacial and interglacial times?

Size analysis of the non-carbonate fraction displays features related to both ice-rafted sediment delivery and bottom-current sorting. Particle size parameters for cores OM-2K and OM-1K are presented in Fig. 7. In OM-2K the percentage of sand fluctuates around 10% during the Holocene, but in the glacial higher value spikes are present. The Heinrich layers at 25 ka and 17.5 ka are clearly marked by excursions of over 20%. This is not paralleled by either the percentage of silt or size of coarse silt (sortable silt mean), which have been argued to be current strength indicators (McCave et al. 1995a,b).

Figure 8 shows similar features in the sortable silt mean records of OM-1K and OM-2K, together with the record in sediments recovered from BOFS 8K from 4045 m depth in the South Rockall Gap (52°30.1′N, 22°04.2′W; Manighetti & McCave 1995b), which lies downstream on the advective flow path of Lower Deep Water (LDW) (McCartney 1992).

The relative current speed inferred appears to increase in velocity at OM-1K compared with that deduced from OM-2K, whereas both experience a slower current regime than BOFS-8K. These differences can be explained by topographic variation at the core sites; OM-2K implies the slowest velocity and is situated on a flat segment of the base of Goban Spur, whereas OM-1K was recovered from the steeper slope below and BOFS-8K is from the South Rockall

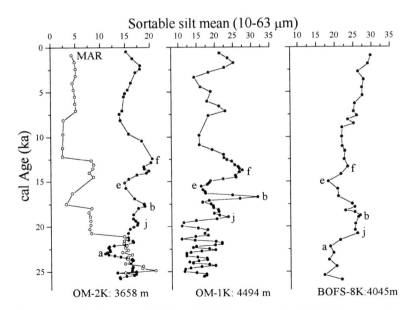

Fig. 8. Sortable silt mean (10–63 μm) in OM-2K and OM-1K together with BOFS-8K from the South Rockall Gap; data from Manighetti (1993). Mass accumulation rate (○) is shown for OM-2K, expressed as g cm^{-2} ka^{-1}.

Gap where LDW is funnelled and so current velocity increases.

Manighetti & McCave (1995*b*) reported a strong covariation between the sortable silt mean in cores influenced at present by Southern Source Water (SSW) and North Atlantic Deep water (NADW). They suggested that there was a strong sympathetic linkage between the SSW velocities and NADW production.

The major features of the record for OM-2K, at 3658 m, and OM-1K at 4495 m, which are both at present located in the northward-flowing southern source Lower Deep Water and probably were during the glacial, are similar to those found south of Rockall (landmarks a, b, e, f and j, Fig. 8): a smaller silt size (low speed–low NADW production) in isotope stage 2 and an abrupt increase (NADW production) starting around the LGM; a rise toward the end of the glacial (Termination 1A beginning *c.* 18 ka ago) followed by an abrupt decline during T1A ('switch off'); a rapid sharp pulse of increased current speed during the Bölling–Allerod (14–16 ka BP); another abrupt slow-down during the Younger Dryas; and, finally, a steady increase in speed to the present day. These current records are also similar to that found on the Blake–Bahama Outer Ridge in the NW Atlantic, most notably the low at the glacial maximum, rise to T1A and sharp fall during the termination (Haskell *et al.* 1991; McCave *et al.* 1995*a*). Such a consistent picture from different locations along the flow path of LDW gives added confidence in the method and suggests that the main circulation is being recorded, not some local phenomenon. The palaeoceanographic implications of such events, namely, increased strength of LDW in stage 2, have been discussed elsewhere (Manighetti & McCave 1995*b*; McCave *et al.* 1995).

The mass accumulation rate in OM-2K also tracks the sortable silt mean, between T1A and the present (Fig. 8). Such a relationship suggests that during this time the depositional environment is current dominated and the major flux of sediment is not a primary vertical flux but rather a secondary source transported through lateral advection to the area. As discussed above, south of Goban Spur is a canyon system which is an area of high turbidity and enhanced deposition. It is most likely that much of the sediment deposited on the open slope of Goban Spur is material which has escaped the confines of the canyon system and became entrained within the northerly flowing current of LDW. The canyon-fed supply of sediment to the open continental slope has been reported by Baker & Hickey (1986) and Gardner (1989).

Conclusions

On the open slope of Goban Spur sediment deposition and mass accumulation rates show strong glacial–interglacial variability. This is primarily related to the flux of ice-rafted debris, with high deposition and accumulation occurring during the LGM and reduced fluxes at the present. Deposition rates found in the Porcupine Seabight are of a similar magnitude to those seen on the open slope. In comparison, substantially elevated sedimentation rates are seen at the foot of King Arthur Canyon, because it underlies a region of high turbidity probably caused by enhanced resuspension and channelized flows within the canyon system. Carbonate burial is typically reduced around the period of the LGM and shows a maximum during the early part of isotope stage 2. This is probably a reflection of enhanced productivity, which may relate to the Bölling–Allerød warming.

Relative current speed inferred from the sortable silt mean in cores OM-2K and OM-1K shows considerable similarity to that in BOFS-8K situated in the same water mass (LDW) but further downstream along its flow path than Goban Spur. This provides additional confidence in the use of grain-size as a palaeo-current strength indicator, as suggested by McCave *et al.* (1995*a,b*). We suggest that that the deposition of sediment on Goban Spur as far back as T1A is dominated by a lateral advective flux, which is probably supplied from suspended particulate material escaping from the canyon systems to the south. Around the LGM there is evidence that the current speed of LDW dropped and that sediment was supplied predominantly via ice-rafting or direct downslope transport.

We are grateful to the following for their help in the laboratory: G. Foreman, C. Fogwill, M. Chapman, M. Hall and A. Gerrard, and S. Brown for her invaluable help at sea during CD94. P. Statham is warmly thanked. We also thank M. Dobson and A. Cramp for their reviews of this manuscript. This work was supported by the European Union MAST programme (Contract MAS2-CT93-0069 for OMEX).

References

BAKER, E. T. 1976. Distribution, composition, and transport of suspended particulate matter in the vicinity of Willapa submarine canyon, Washington. *Geological Society of America Bulletin*, **87**, 625–632.

—— & HICKEY, B. M. 1986. Contemporary sedimentation processes in and around an active West coast submarine canyon. *Marine Geology*, **71**, 15–35.

BALSAM, W. L. & MCCOY, F. W. 1987. Atlantic sediments: glacial/interglacial comparisons. *Paleoceanography*, **2**, 531–542.

BROECKER, W., BOND, G., KLAS, M., CLARK, E. & MCMANUS, J. 1992. Origin of the northern Atlantic Heinrich events. *Climate Dynamics*, **6**, 265–273.

CARPENTER, R., PETERSON, M. L. & BENNETT, J. T. 1982. Pb-210 derived sediment accumulation and mixing rates for the Washington continental slope. *Marine Geology*, **48**, 135–164.

CARSON, B., BAKER, E. T., HICKEY, B. M., NITTROUER, C. A., DEMASTER, D. J., THORBJARNARSON, K. W. & SNYDER, G. W. 1986. Modern sediment dispersal and accumulation in Quinault submarine canyon – a summary. *Marine Geology*, **71**, 1–14.

DRAKE, D. E. 1974. Distribution and transport of suspended particulate matter in submarine canyons off southern California. *In*: GIBBS, R. J. (ed.) *Suspended Solids in Water*. Plenum, New York, 133–153.

—— & GORSLINE, D. S. 1973. Distribution and transport of suspended particulate matter in Hueneme, Redondo, Newport, and La Jolla submarine canyons. *Geological Society of America Bulletin*, **84**, 3949–3968.

——, HATCHER, P. G. & KELLER, G. H. 1978. Suspended particulate matter and mud deposition in Upper Hudson submarine canyon. *In*: STANLEY, D. J. & KELLER, G. H. (eds) *Sedimentation in Submarine Canyons, Fans, and Trenches*. Dowden, Hutchinson and Ross, Stroudsburg, PA, 33–41.

GARDNER, W. D. 1989. Baltimore Canyon as a modern conduit of sediment to the deep sea. *Deep-Sea Research*, **36**, 323–358.

GIBBARD, P. L. 1988. The history of the great northwest European rivers during the past three million years. *Philosophical Transactions of the Royal Society of London, Series B*, **318**, 559–602.

GRÖNVELD, K., OSKARSSON, N., JOHNSEN, S. J., CLAUSEN, H. B., HAMMER, C. U., BOND, G. & BARD, E. 1995. Ash layers from Iceland in the Greenland GRIP ice core correlated with oceanic and land sediments. *Earth and Planetary Science Letters*, **135**, 149–155.

GROUSSET, F. E., LABEYRIE, L., SINKO, J. A., CREMER, M., BOND, G., DUPRAT, J., CORTIJO, E. & HUON, S. 1993. Patterns of ice rafted detritus in the glacial North Atlantic (40–55°N). *Paleoceanography*, **8**, 175–192.

GUINASSO, N. L. & SCHINK, D. R. 1975. Quantitative estimates of biological mixing rates in abyssal sediments. *Journal of Geophysical Research*, **80**, 3032–3043.

HALL, I. R. & MCCAVE, I. N. 1996. *Transport and accumulation fluxes of sediments at the shelf edge and slope*. Ocean Margin Exchange, Final Report **E1–E15**.

HASKELL, B. J., JOHNSON, T. C. & SHOWERS, W. J. 1991. Fluctuations in deep western north Atlantic circulation on the Blake Outer Ridge during the last deglaciation. *Paleoceanography*, **6**, 21–31.

HEINRICH, H. 1988. Origin and consequences of cyclic ice rafting in the northeast Atlantic Ocean during the past 130,000 years. *Quaternary Research*, **29**, 143–152.

HICKEY, B., BAKER, E. & KACHEL, N. 1986. Suspended particle movement in and around Quinault submarine canyon. *Marine Geology*, **71**, 35–83.

HOTCHKISS, F. S. & WUNSCH, C. 1982. Internal waves in Hudson Canyon with possible geological implications. *Deep-Sea Research*, **29**, 415–442.

JOINT, I. & REES, A. 1996. *Primary and new production*. Ocean Margin Exchange. Final Report **D-73**.

JONES, K. P. N., MCCAVE, I. N. & PATEL, P. D. 1988. A computer interfaced Sedigraph for modal size analysis of fine grained sediments. *Sedimentology*, **35**, 163–172.

KELLING, G. & STANLEY, D. J. 1976. Sedimentation in canyon, slope, and base of slope environments. *In*: STANLEY, D. J. & SWIFT, D. J. P. (eds) *Marine Sediment Transport and Environmental Management*. Wiley–Interscience, New York, 379–436.

LAMBECK, K. 1995. Late Devensian and Holocene shorelines of the British Isles and North Sea from models of glacio-hydro-isostatic rebound. *Journal of the Geological Society, London*, **152**, 437–448.

MANGERUD, J., LIE, S. E., FURNEES, H., KRISTIANSEN, I. L. & LOMO, L. 1984. A Younger Dryas ash bed in Western Norway, and its possible correlations with tephra in cores from the Norwegian Sea and the north Atlantic. *Quaternary Research*, **21**, 85–104.

MANIGHETTI, B. 1993. *The Glacial to Holocene sedimentary regime in the northeast Atlantic Ocean*. PhD thesis, University of Cambridge, 218 pp.

—— & MCCAVE, I. N. 1995a. Deposional fluxes paleoproductivity and ice-rafting in the northeast Atlantic over the past 30 ka. *Paleoceanography*, **10**, 657–667.

—— & —— 1995b. Late glacial and Holocene palaeocurrents around Rockall Bank, N.E. Atlantic. *Paleoceanography* **10**, 611–626.

——, ——, MASLIN, M. & SHACKLETON, N. J. 1995. Chronology for climate change: developing age models for the Biogeochemical Ocean Flux Study cores. *Paleoceanography*, **10**, 513–525.

MANTOURA, R. F. C., MARTIN, J.-M. & WOLLAST, R. (eds) 1991. *Ocean Margin Processes: on Global Change, Dahlem Workshop Report*, John Wiley, Chichester, 469 pp.

MCCARTNEY, M. S. 1992. Recirculating components of the deep boundary current of the northern North Atlantic. *Progress in Oceanography*, **29**, 283–383.

MCCAVE, I. N., MANIGHETTI, B. & BEVERIDGE, A. S. 1995a. Circulation in the glacial north Atlantic inferred from grain-size measurements. *Nature*, **374**, 149–152.

——, —— & ROBINSON, S. G. 1995b. Sortable silt and fine sediment size/composition slicing: parameter for palaeocurrent speed and palaeoceanography. *Paleoceanography*, **10**, 598–610.

NELSON, C. H. 1976. Late Pleistocene and Holocene depositional trends, processes, and history of Astoria deep-sea fan, northeast Pacific. *Marine Geology*, **20**, 129–174.

——, NORMARK, W. R., BOUMA, A. H. & CARLSON, P. R. 1978. Thin bedded turbidites in modern submarine canyons and fans. *In*: STANLEY, D.J. &

KELLING, G. R. (eds) *Sedimentation in Submarine Canyons, Fans, and Trenches*. Dowden, Hutchinson, and Ross, Stroudsburg, PA, 177–189.

REVEL, M., SINKO, J. A., GROUSSET, F. E. & BISCAYE, P. E. 1996. Sr and Nd isotopes as tracers of north Atlantic lithic particles–paleoclimatic implications. *Paleoceangraphy*, **11**, 95–113.

ROBINSON, S. G. & McCAVE, I. N. 1994. Orbital forcing of bottom-current enhanced sedimentation on Feni Drift, northeast Atlantic, during the mid-Pleistocene. *Paleoceanography*, **9**, 943–972.

RUDDIMAN, W. F. & GLOVER, L. K. 1972. Vertical mixing of ice-rafted volcanic ash in north Atlantic sediments. *Geological Society of America Bulletin*, **83**, 2817–2836.

SHACKLETON, N. J. & HALL, M. A. 1989. Stable isotope history of the Pleistocene at ODP site 677. *Proceedings of the Ocean Drilling Program Scientific Results*, **111**. Ocean Drilling Program, College Station, TX, 295–316.

STATHAM, P. J. 1994. *Ocean Margin Exchange Study*. Cruise Report. RRS *Charles Darwin* **84**, University of Southampton, Department of Oceanography, 36 pp.

—— 1995. *Ocean Margin Exchange Study*. Cruise Report. RRS *Charles Darwin* **95**, University of Southampton, Department of Oceanography, 52 pp.

THORNBJARNARSON, K. W., NITTROUER, C. A. & DEMASTER, D. J. 1986. Accumulation of modern sediment in Quinault submarine canyon. *Marine Geology*, **71**, 107–124.

ZANGGER, E. & McCAVE, I. N. 1990. A redesigned Kasten core barrel and sampling technique. *Marine Geology*, **94**, 165–171.

Index

accelerated mass spectroscopy (AMS), carbon
  dating, 258, 340–2
airgun profiles
  Barra Fan, 242
  Campos Basin, 292
  Feni Ridge, 234
  Rockall plateau, 245
    NW flank, 250
  Wyville–Thompson Ridge, 241
Alaska, Gulf of, 43–66
  Fairweather–Queen Charlotte Fault and transfer
    margin, 45–57
  location map, 44
  physiography
    tectonic setting and stratigraphy
      discussion, 59–64
      interpretation, 45–58
Albacora Terrace, 289
Aleution Trench, 45, 58
Alika landslide, tsunami, 154
Allerød–Bölling interstadial, meltwater pulses, 284
alongslope processes, 3–4
Alsek River and Trench, 58, 60
Amazon Fan, 133–44
  area, 13
  levee systems
    compaction, 133–44
    stability analysis, 141–2
  location maps, 133, 134
  soil mechanics: gravitational compaction, 135–7
Anatolian Rise, Cyprus–Eratosthenes seamount,
  sediments supply to Herodotus Basin, 19
Andean Uplift, 134
Antarctic
  Weddell Sea, abyssal plain, hemipelagites, 327
  Western Antarctic Peninsula, Central Bransfield
    Basin
    acoustic character, 206–7
    geological setting, 207–13
    geological significance, 213–15
Antarctic Bottom Water (AABW) flow, 236
Anton Dohrn Seamount, 70–2, 235, 270
Atlantic Intermediate Water, foraminifera, 263
Atlantic Ocean, Northeast, sediment-drift
  development, 229–54
Atlantic Ocean, Southeast, Walvis Ridge,
  hemipelagites, 327
Australia, Burdekin River, 159

Barra Fan, Hebrides Slope
  area, 81, 97
  debrites, 67–80
    fan geometry and regional stratigraphy, 68–70
    large-scale slide events, 73
    seabed physiography, 70–3
    slide transfer pathways, timing and volumes,
      73–6
  dinoflagellates, 279–80
  earthquakes, 78
  features, 272
  location map, 68

morphology and sedimentation, 81–104
  bathymetry, 83, 85–9
  canyons, 89–90
  debris-flow deposits and lobes, 90
  echo character, 90–4
  elongate bedforms, 90
  ice scour marks, 89
  methods, 83–5
  seabed photography, 94–101
  slides and slumps, 89
  sedimentology, 277–8
Biscay, Bay of, slope failure, 13
bottleneck slides, debris flows, 12
Bransfield Basin see Antarctic
Brazil
  Amazon Fan, 133–44
  Campos Basin, 4, 287–316
    location maps, 289, 290
Brazil Counter-Current, 291
Britain, NW, continental margin
  bathymetric setting, 230
  Cenozoic sediments, 229–54
  present bottom-water circulation, 235–6
  tectono–stratigraphic framework, 246
Burdekin River, Australia, 159

California Borderland Basins, hemipelagites, 326
Campos Basin, SE Brazil
  alongslope and downslope currents, 4, 287–316
  depositional model, 307–11
    industry applications, 311–13
  expansion zone, 304, 306, 309–11
  funnelling zone, 302–3, 306, 307–9
  hydrography, 290–6
  location maps, 289, 290
  physiography, 288–90
  Pleistocene–Holocene boundary, 301
  Sao Toméddy, 293, 295
  sea-surface temperature (SST) satellite image, 294
  sedimentology, 296–306
Canadian Polar Continental margin, hemipelagites,
  325
canyons, shelfbreak, 89
carbon dating, AMS, 258, 341–2
Cassidulina teretis, 263, 264
Charlie Gibbs Fracture Zone, 229
Chondrites, 304, 329
Coccolithus pelagicus, 279
continental margin, model, 2
continental rise and shelf
  sediment transit zone, 287
  seismic character, Southeast Greenland glaciated
    margin, 178–85
contourites
  diagnostic criteria for recognition and
    differentiation, 330–2
  drifts, Central Bransfield Basin, 213
  Eocene, Faeroe–Rockall Plateau, 5–17, 9–15
  Faeroe Drift, cyclic sedimentation, 255–67
crag-and-tail structures, current scour, 96
current-induced features, scour, 95–6

cyclic sedimentation, Faeroe Drift, 255–67
Cyprus–Eratosthenes seamount (Anatolian Rise),
    sediments supply to Herodotus Basin, 19

Dansgaard–Oeschger cycles, 6, 257–63
    correlation with ENAM cores, 266
Darcy's Law, overpressured sediments, 136
debris flow deposits, 90
    Amazon Fan, 134–5
    Hebrides Slope
        Barra Fan, 67–80
        Sula Sgeir Fan, 105–16
    lobes, 10–12
        bottleneck slides, 12–13
debrites
    defined, 67
    Herodotus Basin, 31
Deep Northern boundary current (DNBC), 273
dinoflagellates
    chronstratigraphy, 279–80
    cyst analysis, 279
Donegal Fan, Hebrides Slope, seabed physiography,
    70–3, 81
downslope proceses, 1–3

Edoras Bank, and Rockall Trough, comparisons, 227,
    233
Eggerella bradyi, 263
Emiliana huxleyi, 279
Eocene
    basalts, Faeroe–Rockall Plateau, 5
    contourites, 5–17
Epistomella decorata, 263
European North Atlantic Margin (ENAM)
    cores, correlation with Dansgaard–Oeschger
        cycles, 266
    sediment accumulation fluxes, 339–50
European Union, Ocean Margin Exchange (OMEX)
    project, 339–50
evaporites, Campos Basin, 288

Faeroe Drift, cyclic sedimentation, 255–67
    origins, 265
    stratigraphy and age control, 257–63
Faeroe–Shetland channel, 118
    geotechnical profile, 124–6
    location map, 256
    Rona Wedge, 119, 125
Faeroes continental margin
    Cenozoic changes, 167–71
    location maps, 6, 167
    seismic recording, 5–17
    seismic stratigraphy, 167–70
    submarine slides, 2, 5–17
    turbidites, extension into Norwegian Sea, 13–16
Fairweather–Queen Charlotte Fault and transfer
    margin, 45–51
    with localized transpression, 52–7
Feni Drift, and Faeroe Drift, cyclic sedimentation,
    265
Feni Ridge
    age, 233
    airgun profile, 234
    seismic expression, 236–8

foraminifera
    accelerated mass spectroscopy (AMS), 340–2
    Atlantic Intermediate Water, 263
    Globigerina ooze, void ratio, 136
    Herodotus Basin, Levantine Sea, 32–3
    NAC species, 281
    Neoglobigerina pachyderma, carbon dating, 258,
        340

Geikie Bulge and Escarpment, 105, 270, 272
    slope, 117
    stability analysis, 111–13
Geological Long-Rang Inclined Asdic (GLORIA)
    side-scan sonar, 24, 44–5
glaciers, sourcing of terrigenous sediments, 59–64
glacigenic debris flows, Sula Sgeir Fan, Hebrides
    Slope, 105–16
Goban Spur, OMEX programme, 339, 342, 344–5
gravitational compaction, soil mechanics, 135–7
Greenland, Southeast, glaciated margin, 173–203
    bathymetric maps, 176
    correlations and interpretations, 188–97
    location map, 174
    morphology, 174–5
    oceanographic currents of Irminger Basin, 175–8
    seismic character of continental rise, 185–8
    seismic character of shelf, 178–85
    seismic data, 174
Greenland–Scotland Ridge, 229

halokinesis, 288
Hatton–Rockall *see* Rockall Trough
Hebridean continental margin
    extensional basins, 217–19
    hemipelagites, 327
Hebridean Escarpment, 69
Hebrides Slope
    Barra Fan
        debrites, 67–80
        morphology and sedimentation, 81–104
    bottom-water circulation, 236
    geotechnical profile, 122–4
    gross depositional environment maps, Cenozoic,
        249
    Late Quaternary stratigraphy and change, 269–86
    Sula Sgeir Fan, 105–16
Hebrides Terrace Seamount, 87–9, 270
    earthquakes, 110
Hebrides–West Shetland margin, 117–32
    location map, 118
Hellenic Trench, sediment trapping, 24
Helmintopsis, 304
hemipelagites, 4, 317–38
    composition, 334
    definition and differentiation from related facies,
        328–34
    depositional processes, 334–6
    diagnostic criteria for recognition and
        differentiation, 330–2
    examples
        ancient, 328
        modern, 325–8
    model for deposition, 335
    occurrence and sedimentation rate, 334

hemiturbidites, diagnostic criteria for recognition and differentiation, 330–2
Herodotus Abyssal Plain, deformations, 24
Herodotus Basin, Levantine Sea, 19–41
    correlation and dating, 31–4
    geological setting and previous work, 21–4
    isopach map, 38
    location map, 20
    morphology and acoustic facies–physiography, 24
        map, 22
    sediment types, 24–31
    seismic profiles, 22–3
    turbidites, dimensions, 38
Holocene Climatic Optimum, 339
hyperpycnal events, 149–50
    rapid sediment transfer in geological record, 157–61
    and tsunami, 154, 157

Iberian (West) continental margin, hemipelagites, 327
Iceland–Faeroe Ridge, location map, 256
Indian Ocean
    Makran and Oman margins, 317–25
    Mozambique Basin, hemipelagites, 326
industry applications, depositional model of Campos Basin, SE Brazil, 311–13
Irminger Basin, Greenland
    bathymetric maps, 177
    location map, 174
    oceanographic currents, 175–8

Japan Arc, Okushiri Ridge, hemipelagites, 326
Japan Sea, hemipelagites, 326

King Arthur Canyon, 340
    sedimentation rates, 343

Labrador Sea, Eirik Ridge, correlation with Greenland SE glaciated margin, 200–1
Labrador Sea Water (LSW) flow, 236, 273
Late Glacial period, subdivision, 281–4
Levantine Sea, Herodotus Basin, 19–41
Levantine Sea–Nile Cone, hemipelagites, 326
levee systems, Amazon Fan, 133–44
Lewis, West Lewis Basin, seismic characterization of Palaeogene sequences, 217–28
Libyan–Egyptian Shelf, megaturbidite, 19

Makran continental margin
    bathymetry and setting, 318
    composition of hemipelagites, 320–1
    sediment facies and structures, 318–20
mass-wasting events, 1–4
    Var river, 148–9
MAST II PALAEOFLUX program, turbidite fluxes in Mediterranean Sea, 19–41
mechanical stability of levee systems, 133–44
Mediterranean Deep Water (MDW) flow, 236
Mediterranean Ridge, 24, 38
Mediterranean Sea
    Herodotus Basin, Levantine Sea, 19–41
    Var Submarine Sedimentary System, 3, 145–66
    Western, hemipelagites, 325

megaturbidite, Libyan–Egyptian Shelf, 19
modelling
    age, Ocean Margin Exchange (OMEX) project, 340–2
    deposition, Campos Basin Brazil, 287–316
    deposition of hemipelagites in deep-water environments, 335
    soil mechanics: gravitational compaction, 135–7
Mozambique Basin, hemipelagites, 326

nanofossils
    dinoflagellates, 279–80
    Rockall Trough, 278–9
    see also foraminifera
Neoglobigerina pachyderma, carbon dating, 258, 263, 340
Neogloboquadrina pachyderma, 281
Nice
    Baie des Anges, mass-wasting events, 154, 157
    see also Var Submarine Sedimentary System
Nice Airport, Var River mass-wasting events, 148–57
    anthropogenic action, 155
    location map, 148–57
Nile Cone
    isopach map, 38
    turbidites, 19
Nonion zaandamae, 263, 264
North Atlantic Deep Water (NADW) flow, 235-6, 273
Northwest European Margin see European Northwest Margin
Norway, Storegga Slide region, 13–16
Norwegian Basin, 170
Norwegian Sea
    stadial/interstadial cycling, 263
    turbidites from Faeroes continental margin, 13
Norwegian Sea Deep Water (NSDW) flow, 5, 236, 273
Nova Scotia, hemipelagites, 325

Ocean Margin Exchange (OMEX) project, 339–50
    age model, 340–2
    Kasten core sampling data, 341
    methods and sampling, 340
    results and discussion, 340–8
    sedimentation rates, 343
Oman continental margin
    bathymetry and setting, 321–3
    composition of hemipelagites, 324
    microfabric, SEM, 324
    sediment facies and structures, 323
oxygen isotopes, Dansgaard–Oeschger cycles, 6, 257–63

Pacific Plate
    translation during Plio–Pleistocene, 43–66
    see also Alaska, Gulf of
Palaeogene seismic characterization, Rockall Trough, 217–28
Peach Slide, Hebrides Slope, 70–8, 85, 102
pelagites, diagnostic criteria for recognition and differentiation, 330–2
Peru margin, hemipelagites, 327
Phycosiphon, 329
piedmont glacier sediments, 64

Planolites, 304, 329
Pleistocene–Holocene alongslope and downslope
  currents, SE Brazil, Campos Basin, 4, 287–316
Porcupine Seabight, 343
Proudman camera, 94

Queen Charlotte Terrace margin, 46

rivers
  hydrology data and interpretation, 149–57
  hyperpycnal events, 149–50
  mass-wasting events, 148–9
Rockall Trough
  Barra Fan, debrites, 67–80
  bathymetric setting, 270
  biostratigraphy, 278–81
  Cenozoic sediments, 229–54
    Eastern, 240–3
    North central, 238–40
    Northwest, 238
    Western, Feni Ridge, 233–4, 236–8
  Feni Ridge, 233–4, 236–8
  and Hatton–Rockall Basin, 243–7
    regional setting, 233–6
  ice-rafted debris (IRD), 278
  Late Quaternary stratigraphy and change, 269–86
  nanofossils, 278–9
  northern, stratigraphy and pricipal sedimentary
    units, 282
  oceanographic setting, 273–4
  palaeoceanographic maps, last deglaciation, 283
  Palaeogene seismic characterization, 217–28
    Middle–Upper Palaeocene, 219–22
    Lower–Middle Eocene, 222–3
    Middle–Upper Eocene, 223
    Upper Eocene–Lower Oligocene, 224–5
    Middle–Upper Oligocene, 225
  present-day deep water circulation, 274
  regional setting, 271–3
  sediment-drift development, 229–54
  sedimentology, 274–7
  stratigraphical-range chart, 231
  Sula Sgeir Fan, Hebrides Slope, 240–3
Rockall–Faeroes Plateau, 5–17, 9–15
Rona Wedge *see* Faeroe–Shetland channel
Rosemary Bank, 270

sand–silt ripples, 96
sands, industry applications, depositional model of
  Campos Basin, SE Brazil, 311–13
São Paulo Plateau, 288
São Tomé eddy, Campos Basin, SE Brazil, 293, 295
São Tomé submarine canyon, 289
scour crescents, current-induced features, 95–6
sediment transport
  depositional model of Campos Basin, SE Brazil,
    311–13
  plate motion, 61–2
  rapid transfer events, 157–64
  rise and slope, 60–1
  sediment-drift development
    regional seismic expression, 236
    Rockall Trough, 229–54
  Var system, 145–66

sediments
  bulk density, formula, 135
  Cenozoic, preservation, 229
  compaction curve, theoretical, 137
  compressibility, 136
  compression curves, SCL and ICL, 126–9
  cyclic sedimentation, Faeroe Drift, 255–67
  industry applications, depositional model of
    Campos Basin, SE Brazil, 311–13
  Late Glacial to Recent fluxes, 339–50
  mass spectroscopy (AMS) age dating, 16, 340–2
  mechanical stability of levee systems, 133–44
  overpressured, Darcy's Law, 136
  porosity and burial depth, 137
  sedimentation, 1–4
  soil mechanics: gravitational compaction, 135–7
  terrigenous, glaciers, 59–64
  void index, 126–9
  void ratio, 136
seismic recording
  Baie des Anges, Nice, and mass-wasting events,
    157
  Faeroes
    continental margin, 167–70
    European North Atlantic Margin project, 5–17
  Greenland SE
    continental rise, 185–8
    continental shelf, 178–85
  Herodotus Basin, Levantine Sea, 22–3
  profiles, 8–9
  Rockall Trough, 217–28
  TOBI, 7–9
seismic stratigraphy, Faeroes continental margin,
  167–70
Shetland, West Shetland Slope, 117
shingling cycles, 64
Sierra Leone Rise, hemipelagites, 326
Sigmoilopsis schlumbergeri, 263
slide processes, 76–8
slide transfer pathways, timing and volumes, 73–6
soil mechanics: gravitational compaction, 135–7
South Atlantic Central Water (SACW), 291
South Shetland Isles, location map, 206
South Shetland Trench, 205
stability, 1–4
Storegga Slide region, w Norway, 13–16
stripe–slip faulting, Gulf of Alaska, 43
strontium dating, 68–9
submarine canyons, sediment delivery systems, 43–66
submarine landslide, defined, 67
Sula Sgeir Fan, Hebrides Slope, 105–16
  geotechnical profile, 120–2
  location map, 106
  Rockall Trough, 240–3

Terzaghi, principle of effective stress, 136
Thalassinoides, 304
thermal subsidence basins, 170
TOBI deep-tow side-scan sonar mosaic, 5–17, 89
  blow-up of debris flow, 12
TOPAS system, 205
tsunamis, Alika landslide, 154
turbidites
  chemical analysis, 19–21, 22

Cyprus–Eratosthenes seamount, 19
diagnostic criteria for recognition and
    differentiation, 330–2
Herodotus Basin, Levantine Sea, types, 24–31
Libyan–Egyptian Shelf, 19
Nile Cone, 19

unifites, diagnostic criteria for recognition and
    differentiation, 330–2

Var Submarine Sedimentary System, 3, 145–66
    geological and geographical setting, 146–9
    hydrology, 149–57
    mass-wasting events, 148–64
        features summarized, 162
        interpretation, 151–2, 154

    sedimentary impact, 152–7
    sedimentation processes (seaward) during flood
        periods, 160
Vedde Ash, 281, 342
void index, 126–9
void ratio, 136

Weddell Sea, abyssal plain, hemipelagites, 327
Wyville–Thompson Ridge, 119, 217–27, 270
    airgun profiles, 241
    Upper Eocene–Lower Oligocene fan geometry,
        226–7

Yakutat Terane margin, 57–9

Zoophycos, 329